Condensed MATTER THEORIES

VOLUME 5

Condensed MATTER THEORIES

VOLUME 5

Edited by
V. C. Aguilera-Navarro
Instituto de Física Teórica—UNESP
São Paulo, Brazil

Plenum Press • New York and London

ISBN 978-1-4612-7888-7

LC 87-656591

Proceedings of the 13th International Workshop on
Condensed Matter Theories, held August 6–12, 1989,
in Campos do Jordão, São Paulo, Brazil

ISBN-13: 978-1-4612-7888-7 e-ISBN-13: 978-1-4613-0605-4
DOI: 10.1007/978-1-4613-0605-4

Softcover reprint of the hardcover 1st edition 1990

A Division of Plenum Publishing Corporation
233 Spring Street, New York, N.Y. 10013

International Advisory Committee:

V. C. Aguilera-Navarro (Brazil)
J. S. Arponen (Finland)
E. P. Bashkin (U.S.S.R.)
R. F. Bishop (U.K.)
C. E. Campbell (U.S.A.)
J. W. Clark (U.S.A.)
J. Dabrowski (Poland)
M. de Llano (U.S.A.)
R. Guardiola (Spain)
J. Keller (Mexico)
E. V. Ludeña (Venezuela)
F. B. Malik (U.S.A.)
A. Plastino (Argentina)
S. Rosati (Italy)
C. W. Woo (Hong Kong)
A. Zuker (France)

Organizing Committee for the Thirteenth Meeting

V. C. Aguilera-Navarro, Chairman (São Paulo, Brazil)
D. Galetti (São Paulo, Brazil)
S. S. Mizrahi (São Carlos, Brazil)
A. C. Nunes (Brasília, Brazil)
S. M. Rezende (Recife, Brazil)
F. Souza Barros (Rio de Janeiro, Brazil)
R. F. Bishop (Manchester, U.K.)
J. W. Clark (St. Louis, U.S.A.)
M. de Llano (Fargo, U.S.A.)
F. B. Malik (Carbondale, U.S.A.)

Editing of this volume was sponsored by:

Fundunesp (São Paulo, Brazil)
CLAF (Rio de Janeiro, Brazil)
U.S. Army Research Office (Durham, North Carolina, U.S.A.)
CNPq (Brasília, Brazil)
FAPESP (São Paulo, Brazil)

PREFACE

This volume gathers the invited talks of the XIII International Workshop on Condensed Matter Theories which took place in Campos do Jordão near São Paulo, Brazil, August 6-12, 1989. It contains contributions in a wide variety of fields including neutral quantum and classical fluids, electronic systems, composite materials, plasmas, atoms, molecules and nuclei, and as this year's workshop reflected the natural preoccupation in materials science with its spectacular prospect for mankind, room temperature super-conductivity. All topics are treated from a common viewpoint: that of many-body physics, whether theoretical or simulational.

Since the very first workshop, held at the prestigious Instituto de Física Teórica in São Paulo, and organized by the same organizer of the 1989 workshop, Professor Valdir Casaca Aguilera-Navarro, the meeting has taken place annually six times in Latin America, four in Europe and three in the United States. Its principal objective has been to innitiate and nurture collaborative research networks of scientists interested in the multidisciplinary aspects of many-body theory applied to problems in condensed-matter physics.

Financial as well as moral support is gratefully appreciated by all: of the CLAF in Rio, the CNPq in Brasília, the FAPESP and the FUNDUNESP in São Paulo, and the U.S. Army Research Office in Durham, NC, USA.

Manuel de Llano
Fargo, ND, USA

ACKNOWLEDGEMENTS

The help of several people in the organization of the workshop and this book is gratefully acknowledged; John W. Clark, Carminda C. Landim, Paulo Milton B. Landim, M. de Llano, F.B. Malik, and the conference secretary Maria Inez G. Macieira. The financial support from the Centro Latinoamericano de Física (Rio), Fundação para o Desenvolvimento da Unesp (São Paulo), Fundação de Amparo a Pesquisa do Estado de São Paulo (São Paulo), Conselho Nacional de Desenvolvimento Científico e Tecnológico (Brasília), U.S. Army Research Office (Durham, NC), and Banco Itaú S.A. (São Paulo) is also acknowledged.

V.C.A.N.

CONTENTS

QUANTUM AND CLASSICAL FLUIDS

---

*Asterisk next to name identifies the speaker

FORMAL METHODS

THOMAS-FERMI EQUATION OF STATE–THE HOT CURVE

George A. Baker, Jr. and J. D. Johnson

Theoretical Division, Los Alamos National Laboratory
University of California, Los Alamos, N. M. 87545, USA

ABSTRACT

We derive the high-temperature limit of the equation of state based on the Thomas-Fermi statistical theory of the atom. The resulting "hot curve" is in fact the ideal Fermi gas. We expand the thermodynamic properties of this gas in powers of the fugacity and use this expansion to construct a representation of the pressure, accurate to about 0.1 %. This representation is compared with the actual theory for aluminum and the "hot curve" is found to represent it well over a large region of interest in applications.

1. INTRODUCTION AND SUMMARY

The Thomas-Fermi (T-F) statistical theory of the atom[1] as well as the modifications due to Dirac[2] have long been used as a basic starting point for the computation of approximations to the equations of state.[3,4] In order to make use of this procedure, computer programs have been written to compute the numerical content of the theory. They consume a sufficient amount of computer time, even today, so that it is impractical to use them to compute, *ab initio*, the value of the pressure, internal energy, *etc.*, every time that a new value is required inside an application computer program. Besides, as these efforts represent only approximate equations of state, some adjustment is necessary to bring them into accord with physical reality. Consequentially, to date largely empirical fits have been used to represent the equations of state for the purposes of applications.

In this work, we are concerned with beginning an analysis of the physical structure of the equations of state of real matter. As a start, we will study the Thomas-Fermi model equation of state which represents a fair amount of the physics, at least in some regions. One method which is normally fruitful, is to consider various limits. There are currently two which are known. The first is the low-density limit. Here there is complete ionization when the system is in equilibrium and the pressure for an element of nuclear charge Z is

$$P\Omega/N = (Z+1)kT, \tag{1.1}$$

the ideal gas equation of state. Here P is the pressure, Ω is the volume of the system, N is the number of atoms, k is Boltzmann's constant and T is the absolute temperature. The second limit[3] is the low-temperature limit, or the "cold curve." Here the pressure is of the form,

$$P\Omega/N = Z^{\frac{7}{3}}\phi(Z\Omega/N), \tag{1.2}$$

where $\phi(x)$ is a well defined function. If we think of the temperature-density, quarter-plane, these results give the limiting behavior of the T-F model along the zero-temperature and the zero-density edges. There remain the high-density and the high-temperature regions to examine for physical structure.

One might think that in the high-temperature limit it would be appropriate to describe the system in purely classical terms. Indeed if such were the case, Baker[5] has proven that the pressure would be of the form,

$$P\Omega/N = kT\,f(\Omega T^3/N, Z). \tag{1.3}$$

The Debye-Hückel correction[6] is of just this form. Also Baker has shown for this case that the internal energy has the particularly simple form,

$$u = 3P\Omega - \frac{3}{2}(Z+1)NkT. \tag{1.4}$$

The statistical mechanics of Coulombic systems have been much studied.[7] It is now well known that there does not exist a classical (*i.e.* Planck's constant $h = 0$) gas because atoms with a Coulomb interaction collapse to $E = -\infty$. Thus if we are to ever introduce a Coulomb attraction between the atomic nucleus and the electrons, we must necessarily include some account of the quantum effects that are needed to stabilize the system. As is also well known there are two important physical lengths to be considered. The first is the de Broglie length which is proportional to $h/\sqrt{mkT}$, where m is the electron mass, and which measures in a noninteracting gas the importance of quantum effects. The Coulomb interaction does not by itself provide the second length and the difficulty of its long range can not be circumvented by studying dilute systems because it contains no parameter with the dimensions of a length. The second length is the Debye screening length which is proportional to e^2/kT. This length is however a statistical effect and should follow from the theory, but unfortunately is not there *ab initio.* Thus when we look to the high-temperature and high-density regions, if we consider the cases where $\Omega/N >> (e^2/kT)^3$, then we can hope to start with a noninteracting electron gas (with a background gas of atomic nuclei) as the basic system.

In the second section, we derive the limit of Thomas-Fermi theory when the Debye screening length is negligible compared to the interparticle distance, and the de Broglie length remains arbitrary. We find that it correctly reduces to the ideal Fermi gas. We call this limit the "hot curve," because it is reached if one either fixes the density and lets the temperature go to infinity, or much less restrictively, it is also reached if one fixes the de Broglie length and then lets the temperature go to infinity. In the third section we review the theory of the ideal Fermi gas and describe how to calculate its properties in a practical manner. We derive lengthy fugacity series and find that the pressure function can be approximated to within, say 0.1%, by a low-order, two-point Padé approximant. In the final section we compare the ideal gas approximation to results for aluminum and map out its region of validity to various degrees of accuracy.

2. HIGH TEMPERATURE LIMIT OF THOMAS–FERMI THEORY

Thomas-Fermi theory has been applied to compute equations of state at finite temperature by Feynman *et al.*[3] They begin with an application of the statistical analysis of Fermi and Dirac which leads to the equation

$$\rho = \int_0^\infty \frac{2 \cdot 4\pi p^2 dp/h^3}{\exp[(p^2/2m - eV)/kT + \eta] + 1}, \tag{2.1}$$

where $-eV$ is the potential energy. We follow them in defining for convenience the auxiliary functions

$$I_n(\eta) = \int_0^\infty \frac{y^n dy}{\exp(y - \eta) + 1}. \tag{2.2}$$

Then one uses Poisson's equation to determine V(r). It yields

$$\frac{1}{r}\frac{d^2}{dr^2}(rV(r)) = \frac{16\pi^2}{h^3} e(2mkT)^{\frac{3}{2}} I_{\frac{1}{2}}\left(\frac{eV(r)}{kT} - \eta\right). \tag{2.3}$$

Note that in the case of no interaction that the right-hand side of (2.3) vanishes (e=0) and so the equation implies that $V = a + b/r$ where a and b are constants. In order to simplify the above equation, Feynman *et al.*[3] introduce dimensionless variables. First they define a length scale,

$$c = \left(\frac{h^3}{32\pi^2 e^2 m(2mkT)^{\frac{1}{2}}}\right)^{\frac{1}{2}} \propto T^{-\frac{1}{4}}, \tag{2.4}$$

where $s = r/c$. Then since η is independent of r, (2.3) becomes

$$\frac{d^2\beta}{ds^2} = sI_{\frac{1}{2}}(\beta/s), \tag{2.5}$$

where

$$\beta/s = (eV(r)/kT) - \eta. \tag{2.6}$$

The boundary conditions of (2.5) become, as at the origin $V(r)$ must behave as Ze/r,

$$\beta(0) = \alpha = Ze^2/kTc \propto T^{-\frac{3}{4}}. \tag{2.7}$$

The scheme employed is to suppose that each atom is confined to a sphere of volume equal to the volume per particle. This is clearly an approximation. The other boundary condition is to require that the number of electrons in the sphere is exactly equal to the nuclear charge. A little manipulation serves to show that the condition,

$$\frac{d\beta}{ds} = \beta/s \text{ at } s = b, \tag{2.8}$$

imposes this normalization in the sphere of radius $r = cb$. Feynman *et al.*[3] derive, among other things, the formula for the pressure as

$$P\Omega/N = \frac{2}{9}(ZkT) \cdot \frac{b^3}{\alpha} I_{\frac{3}{2}}\left(\frac{\beta_b}{b}\right), \tag{2.9}$$

where β_b is the value of β on the boundary $s = b$.

In a parallel way we may set out the corresponding formulae for the ideal Fermi gas. In this case the electron density is simply given by (2.1) with $e = 0$. As η is independent of r, one sees immediately by (2.6) that the equation for the density

(2.5) is simply satisfied. Since by (2.4) and (2.7) both the length and magnitude scales depend on the electronic charge $e = 0$, the normalization equation (2.8), in leading order, is automatically satisfied, and so does not determine the number of electrons in this limit. Returning to (2.1), we may impose the normalization condition by integrating the density over a sphere of radius r. It gives

$$Z = \frac{16\pi^2}{3} I_{\frac{1}{2}}(-\eta) \left[\frac{r\sqrt{2mkT}}{h}\right]^3 , \tag{2.10}$$

which implies η. In this limit, the pressure equation (2.9), becomes,

$$P\Omega/N = \frac{2}{9}(ZkT)\left(\frac{r^3}{c^3\alpha}\right) I_{\frac{3}{2}}(-\eta), \tag{2.11}$$

a parametric expression for the pressure in terms of the η of (2.10). Note is made that $c^3\alpha$ is independent of the electronic charge $e = 0$, so this form is valid in this noninteracting limit. Comparison with the results of Huang[8] for the ideal Fermi gas, reveal complete agreement, when it is remembered that for our case the spin, $s = \frac{1}{2}$.

Now we are ready to consider the "hot curve" limit of the Thomas-Fermi theory. In the basic equations of the theory, (2.5, 7-8), we make the following change of variables,

$$\sigma = s/\alpha^{\frac{1}{3}}, \ \gamma = \beta/\alpha^{\frac{1}{3}}. \tag{2.12}$$

We thus obtain

$$\frac{d^2\gamma}{d\sigma^2} = \alpha^{\frac{2}{3}}\sigma I_{\frac{1}{2}}\left(\frac{\gamma}{\sigma}\right), \tag{2.13}$$

$$\gamma(0) = \alpha^{\frac{2}{3}}, \tag{2.14}$$

$$\frac{d\gamma}{d\sigma} = \frac{\gamma}{\sigma}, \text{ at the boundary.} \tag{2.15}$$

In the limit $\alpha \to 0$ (by (2.7) this limit is equivalent to $T \to \infty$), we obtain the result that $\gamma = A\sigma$ solves (2.13-15). Again, as at (2.10) above, we have an undetermined normalization constant to be determined because in our high-temperature limit (2.15) is satisfied automatically. Again referring to (2.1) we obtain the normalization condition,

$$Z = \frac{16\pi^2}{3h^3}[r\sqrt{2mkT}]^3 I_{\frac{1}{2}}\left(\frac{\gamma}{\sigma}\right), \tag{2.16}$$

which determines A and thus the solution of (2.13-15). When we note the comparison $A = -\eta$, we find that this limiting solution is the same as the one we obtained for the ideal (noninteracting) Fermi gas. This result completes our demonstration of the proposition that the "hot curve" for Thomas-Fermi theory is the ideal Fermi gas!

3. PROPERTIES OF THE IDEAL FERMI GAS

The basic theory of the ideal Fermi gas is described by Huang.[8] To establish a correspondence between the results of the previous section and more standard notation, we note that in (2.16) $\gamma/\sigma = A$; therefore we introduce the notation $z = e^{-A}$. We can then rewrite (2.16) and (2.11) as

$$\frac{ZN}{\Omega} = \frac{3Z}{4\pi r^3} = 2\left(\frac{2\pi mkT}{h^2}\right)^{\frac{3}{2}} \frac{2}{\sqrt{\pi}} \int_0^\infty \frac{zy^{\frac{1}{2}}e^{-y}dy}{1+ze^{-y}}, \tag{3.1}$$

$$\frac{P}{kT} = 2\left(\frac{2\pi mkT}{h^2}\right)^{\frac{3}{2}} \frac{4}{3\sqrt{\pi}} \int_0^\infty \frac{zy^{\frac{3}{2}}e^{-y}dy}{1+ze^{-y}}, \tag{3.2}$$

where P is the pressure due to the electrons only and does not take account of the effect of the motion of the center of mass of the atom. If we introduce the further notation,

$$\lambda = \left(\frac{h^2}{2\pi mkT}\right)^{\frac{1}{2}}, \tag{3.3}$$

$$f_{\frac{3}{2}}(z) = \frac{2}{\sqrt{\pi}} \int_0^\infty \frac{zy^{\frac{1}{2}}e^{-y}dy}{1+ze^{-y}} = \sum_{l=1}^{\infty} \frac{(-1)^{l+1}z^l}{l^{\frac{3}{2}}}, \tag{3.4}$$

$$f_{\frac{5}{2}}(z) = \frac{4}{3\sqrt{\pi}} \int_0^\infty \frac{zy^{\frac{3}{2}}e^{-y}dy}{1+ze^{-y}} = \sum_{l=1}^{\infty} \frac{(-1)^{l+1}z^l}{l^{\frac{5}{2}}}, \tag{3.5}$$

where the series expansions are convergent for $|z| \leq 1$. We may now rewrite (3.1-2) as

$$\zeta = \frac{ZN\lambda^3}{2\Omega} = f_{\frac{3}{2}}(z), \tag{3.6}$$

and

$$\frac{P\Omega}{ZNkT} = \frac{f_{\frac{5}{2}}(z)}{f_{\frac{3}{2}}(z)}, \tag{3.7}$$

where ζ is the de Broglie density. The procedure to calculate the pressure of the ideal Fermi gas is now, in principle, quite straightforward. Eq. (3.6) is solved for z and then that value is substituted into (3.7).

To evaluate these expressions numerically we choose the following method. First we revert the series expansion (3.6) to give $z(\zeta)$ as a series in ζ. Then we substitute it into (3.7) to obtain

$$\frac{P\Omega}{ZNkT} = g(\zeta). \tag{3.8}$$

We have calculated the leading 36 terms of the series expansion. The method used is the classical Lagrange formula for the reversion of series.[9] The only point of difficulty is that a large number of decimal places are lost in the computation in this case. We have therefore taken the precaution of using at least 58 decimal places to carry out these computations. The results are listed in Table 1.

The above series expansion was derived for $|z| \leq 1$, but the above series plainly corresponds to a larger range. In the limit as $z \to \infty$ Huang shows that

$$f_{\frac{3}{2}}(z) \asymp \frac{4}{3\sqrt{\pi}}(\log z)^{\frac{3}{2}}\left[1 + \frac{\pi^2}{8(\log z)^2} + \cdots\right] + O(z^{-1}). \tag{3.9}$$

From the identity,[8] $z\frac{d}{dz}f_{\frac{5}{2}}(z) = f_{\frac{3}{2}}(z)$ one can easily also derive the asymptotic behavior of $f_{\frac{5}{2}}(z)$, and thus from (3.7) the asymptotic behavior of $g(\zeta)$. We obtain,

$$g(\zeta) \asymp \frac{2}{5}\left(\frac{3\sqrt{\pi}}{4}\right)^{\frac{2}{3}} \zeta^{\frac{2}{3}} \text{ as } \zeta \to \infty. \tag{3.10}$$

TABLE 1. $(P\Omega/ZNkT)$ as a series in the de Broglie density

0	1.0000000000	0000000000	0000000000	0000000000	0000000000000000E+000
1	1.7677669529	6636881100	2110905262	1225982120	8984422118509147E-001
2	-3.3000598199	1683655758	8617889323	8790328003	89171139305782E-003
3	1.1128932846	6542504524	9253533917	1305775999	1875768224181E-004
4	-3.5405040951	9736538278	3050093233	4626176046	46439677965E-006
5	8.3863470395	6925729619	7125848681	6218474298	427436245E-008
6	-3.6620617873	4852703663	1688233937	9045907824	8643167E-010
7	-1.0280607154	3957929799	3273512206	9735581999	5254513E-010
8	7.0550978435	7263454626	0275709452	8261969773	09158E-012
9	-2.6859639507	9285424406	0526716388	7926863588	4377E-013
10	4.0571834908	0612166197	1056127182	3091151601	35E-015
11	2.7970439770	9162019148	3071234746	1358106846	6E-016
12	-2.8379673439	5952590529	6631787032	9726025304	E-017
13	1.3992940717	5922219970	7552151122	203412696E	-018
14	-3.6303052861	0821033013	0082398676	2418074E-0	20
15	-6.0257400821	7251347692	8112664253	67093E-022	
16	1.2989538153	2549763684	7035089386	73544E-022	
17	-8.1719971340	6344259697	7319803759	795E-024	
18	2.9413082494	4946667164	3606073469	73E-025	
19	-2.0285711098	2088612486	4658243931	E-027	
20	-5.7410636166	7615749309	984730023E	-028	
21	4.8461575378	3763503589	33968480E-	029	
22	-2.2369786852	5871386652	1846940E-0	30	
23	4.7888680538	7474310454	78772E-032		
24	2.0304880286	8391265410	8553E-033		
25	-2.7811009124	7360566430	414E-034		
26	1.6149810555	1163427972	12E-035		
27	-5.2554355032	5730228297	E-037		
28	-1.3309033541	33284697E-	039		
29	1.4721238409	86015824E-	039		
30	-1.1062516681	9956070E-0	40		
31	4.7267873838	86169E-042			
32	-7.6386716803	536E-044			
33	-6.5324794996	62E-045			
34	7.1193401844	5E-046			
35	-3.8268661579	E-047			
36	1.097950074E	-048			

With this information and the series of Table 1, we may construct a two point Padé approximant[10] to $[g(\zeta)]^3$ of the form $[N+2/N]$ which is exact through order ζ^{2N+1} at the origin, and is also asymptotically correct as $\zeta \to \infty$. We find excellent convergence for this method and that for $0 \leq \zeta < \infty$ we get an accuracy of about 0.1 percent for $g(\zeta)$ from the approximation,

$$g(\zeta) \approx \left[\frac{1 + 0.61094880\zeta + 0.12660436\zeta^2 + 0.0091177644\zeta^3}{1 + 0.080618739\zeta}\right]^{\frac{1}{3}}. \tag{3.11}$$

Thus the total pressure would be (including the center of mass motion)

$$P = \frac{NkT}{\Omega}\{1 + Zg(\zeta)\}. \tag{3.12}$$

In the case where the temperature is fixed and $\Omega \to \infty$, the low-density limit, not only does the Debye density go to zero, as required to obtain the ideal Fermi gas limit of Thomas-Fermi theory, but also $\zeta \to 0$. In this case, as $g(0) = 1$, (3.12) reduces to (1.1) and thereby supplies an alternate derivation of the low-density limit of Thomas-Fermi theory.

As Huang[8] points out, the internal energy, U, for this case follows simply from (3.12) as,

$$U = \frac{3}{2}P\Omega. \tag{3.13}$$

Epstein[11] shows from the thermodynamic relation $dS = (dU + PdV)/T$, the above results, and Nernst's heat postulate that the entropy of the ideal Fermi gas is simply given by

$$S_e = ZNk\left(\frac{5}{2}g(\zeta) - \log z(\zeta)\right), \tag{3.14}$$

where the limit as $T \to 0$ is the limit $\zeta \to \infty$ by (3.6) and as Epstein further points out $S_e \to 0$ in this limit. If we add the contribution of the motion of the center of mass to the entropy, we get

$$S = Nk\left[-(Z+1)\log\zeta + \frac{5}{2} + Z\left(\frac{5}{2}g(\zeta) - \log[z(\zeta)/\zeta]\right)\right] + \text{constant}, \tag{3.15}$$

The Helmholtz free energy is now given directly by $A = U - TS$. The Gibbs thermodynamic potential is also directly given and is $G = U - TS + P\Omega$.

It now remains to give a representation of $\log z(\zeta) = \log\zeta + \log[z(\zeta)/\zeta]$ to complete the representation of the thermodynamic quantities for the ideal Fermi gas. Since $\log z \asymp \zeta^{\frac{2}{3}}$, the problem of deriving a representation for $\log[z(\zeta)/\zeta]$ should be similar to that of the representation (3.11). We give in Table 2 the necessary series coefficients in ζ for $z(\zeta)$ to work on this representation, but we will leave it for the future. Thermodynamic consistency depends on the equation between the two representations

$$g(\zeta) + \zeta g'(\zeta) = \zeta\frac{d\log z(\zeta)}{d\zeta}. \tag{3.16}$$

TABLE 2. The fugacity z as a series in the de Broglie density

1	1.0000000000	0000000000	0000000000	0000000000	00000000000000000E+000
2	3.5355339059	3273762200	4221810524	2451964241	7968844237018294E-001
3	5.7549910270	1247451636	1707316601	4181450799	416243291041327E-002
4	5.7639604009	1025440341	8852781947	0758923518	58214221729707E-003
5	4.0194941515	2300959555	6172119656	7773364832	0998466829345E-004
6	2.0981898872	2604799054	4860297423	5099614729	957102872728E-005
7	8.6021310842	6030566004	3913343164	3181688359	0277772573E-007
8	2.8647148623	7664872936	8242210245	0573640824	266032220E-008
9	7.9528314678	5241689019	4817612245	1032872937	5035650E-010
10	1.8774425910	0567756220	4988130993	7541387605	437996E-011
11	3.8247968264	1809029592	4653344686	7070280382	2264E-013
12	6.8432943010	1907998578	8027623030	3596055059	29E-015
13	1.0762104093	0537917245	5417733813	6774703889	3E-016
14	1.5124110216	1988369105	9052478125	978137640E	-018
15	2.0715738792	9770436279	3713783632	7032961E-0	20
16	1.3846671521	9900108771	8574969994	14568E-022	
17	5.3288541784	7605238410	1301497951	755E-024	
18	3.5079551301	2368023505	6432045696	E-027	
19	-5.9656175104	9257472195	3065263300	E-027	
20	5.2969138512	2627670501	874181389E	-028	
21	-2.5226985875	2718441504	10473445E-	029	
22	6.0209616883	8744484633	512535E-03	1	
23	1.8543035351	4383646428	76522E-032		
24	-3.0176817670	7158240262	6353E-033		
25	1.8757233170	6238133052	809E-034		
26	-6.7714760730	2256395698	9E-036		
27	3.7182598930	255841378E	-038		
28	1.4954203444	742341364E	-038		
29	-1.2728642729	99664053E-	039		
30	6.0377265821	589225E-04	1		
31	-1.3644496192	99721E-042			
32	-5.3539191733	757E-044			
33	7.8650740191	78E-045			
34	-4.7690907071	0E-046			
35	1.6535692458	E-047			
36	-2.3890246E-0	50			
37	-4.2646358E-0	50			

An alternate procedure would be to determine $z(\zeta)$ directly from this equation subject to the boundary condition $\lim_{\zeta \to 0} z(\zeta)/\zeta = 1$. This equation is an identity in the exact theory and not an extra condition.

From the theoretical point of view the most satisfactory proceedure would be to construct a sufficiently accurate representation of, say, the Helmholtz free energy A that would provide adequately accurate derivatives $(\frac{\partial A}{\partial V})_T = -P$, and $(\frac{\partial A}{\partial T})_V = -S$. Using (3.13), (3.15) (ignoring the constant), and integrating (3.16) we have for the Helmholtz free energy,

$$\begin{aligned} A = & - P\Omega + (Z+1)NkT \log \zeta + ZNkT \log[z(\zeta)/\zeta] \\ = & NkT \left[(Z+1)(\log \zeta - 1) + Z \int_0^{\zeta} [g(\eta) - 1] \frac{d\eta}{\eta} \right], \end{aligned} \tag{3.17}$$

for which the series expansion in ζ can be easily derived from Table 1. The inability to assign an absolute entropy for the ordinary ideal gas, leaves A uncertain by a linear term in T. It remains to be seen which of the procedures outlined above are computationally most efficient.

4. COMPARISON OF IDEAL FERMI GAS TO THOMAS-FERMI THEORY

We now show the extent of agreement for aluminum between the ideal Fermi gas and the Thomas-Fermi theory. We use the computer program of D. A. Liberman[12] to compute the T-F numbers. We present the results in the figures as contours of percentage differences (electron properties only).

For the pressure, Figure 1 shows in temperature-density parameter space the 1%, 10%, and 30% contours, as one goes from the top curve of the figure to the bottom, respectively. The expected feature is that for high-temperature and/or low density the ideal gas is accurate. The 10% contour, for example, will serve as our "hot envelope," that is to say, the limit of the validity of the "hot curve" approximation. For low-temperature and high-density the ideal Fermi gas is again a good representation of the T-F theory because the electrons are being forced to the pressure-ionized, degenerate, free electron gas. Since as the density increases the kinetic energy per atom is forced by the Pauli principle to increase proportional to the density to the two-thirds power (relativistic corrections are ignored here) and the potential energy is expected to increase only as the one-third power of density, the free-electron-gas energy becomes dominate. This effect is begining to be evident in the behavior of the 30% contour. The ranges of temperature and density shown are those of interest for a great many applications. Thus the ideal Fermi gas well reproduces the T-F pressure over a substantial region.

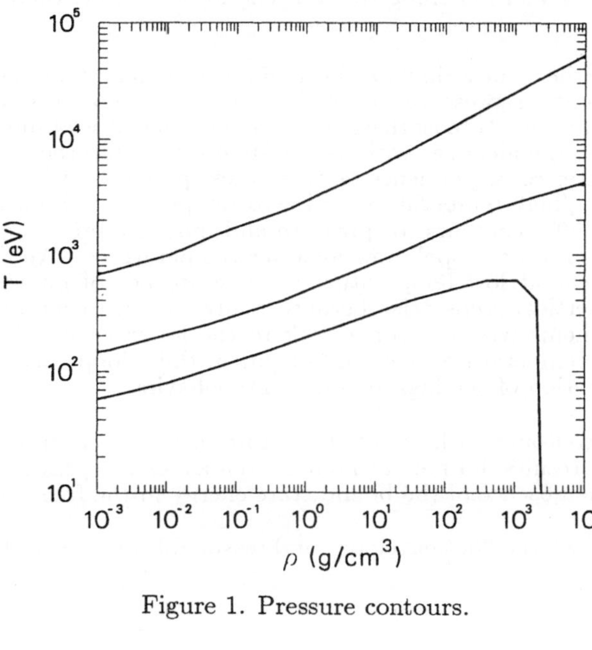

Figure 1. Pressure contours.

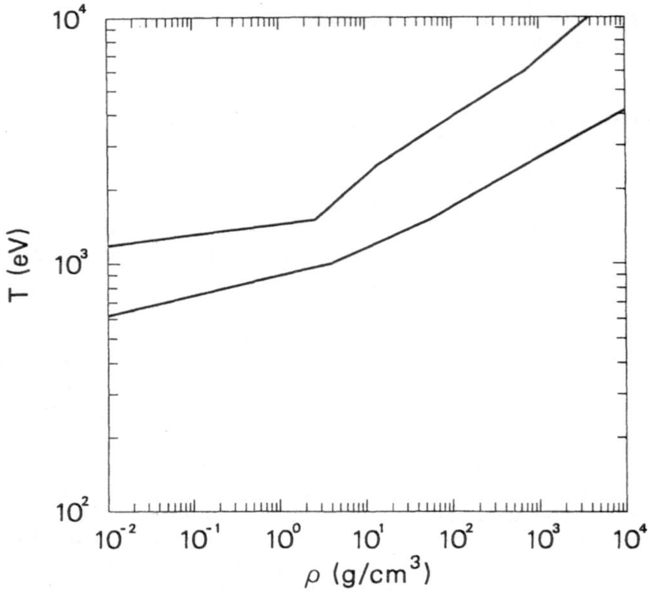

Figure 2. Energy contours.

Figure 2 shows the results for the internal energy. Here we see only the 10% and 30% contours because the ideal Fermi gas does not represent the T-F energy as well as it does the pressure. This result is at least partly due to what, in effect, is an extra term present in the T-F energy and not in the T-F pressure. The bound electrons do not contribute to the pressure but do have a large effect on the energy, for the temperature and density both small. Since the free gas has no bound electrons, there is more difficulty in matching the T-F energy. However, there is again a "hot envelope."

We did one other study that was beyond our original intent. Our goal is really not to find an analytic representation of the T-F theory, but to obtain a fit to the T-F with the zero-temperature isotherm subtracted. Thus it is of interest to compare just such a result to the ideal gas with its zero-temperature isotherm subtracted. We expect an even better correspondence between these pressures, with exact agreement both at low-density/high-temperature and zero temperature. Figure 3 shows again the 1%, 10%, and 30% contours for pressure and indeed there is improvement over Figure 1 with the "hot envelope" now at lower temperatures. We do not show the contours that appear at low temperature as they are not of interest to us in this study. The odd vertical steps arise because really the two contours at that point loop back under themselves and come back to the lower curves due to the forced agreement at zero temperature. We did not put in these loops because we felt that was a misrepresentation of the high-temperature behavior.

The energy contours with zero-temperature isotherm subtracted are not presented because the results did not turn out as well as for the pressure. This result is again caused by the absence of the bound state energy in the free Fermi gas.

In general we see the "hot envelope" and reasonable agreement between the free

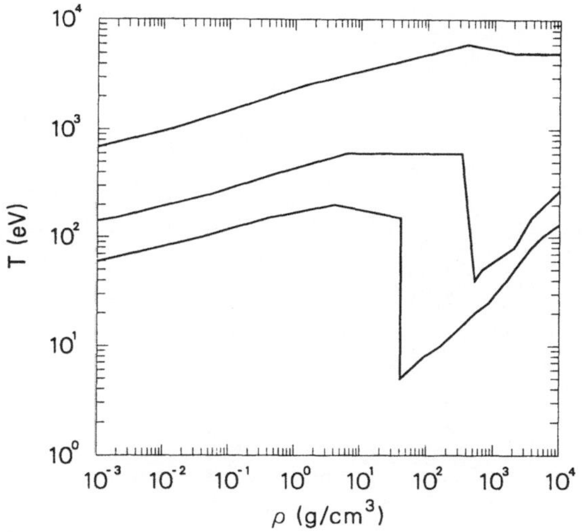

Figure 3. Pressure contours for the zero temperature isotherm subtracted.

Fermi gas and T-F theory for a large region of pressure. We understand the difference between the pressure and internal energy.

This work was performed under the auspices of the U. S. Department of Energy. In addition, one of the authors (G.B.) is happy to acknowledge partial travel support from the U.S. Army Research Office and from Fundunesp to permit his attendence at the *XIII International Workshop on Condensed Matter Theories*.

REFERENCES

1. L. H. Thomas, Proc. Cambridge Phil. Soc. **23**, 542 (1927); E. Fermi, Z. Physik **48**, 73 (1928).
2. P. A. M. Dirac, Proc. Cambridge Phil. Soc. **26**,376 (1930).
3. R. P. Feynman, N. Metropolis, and E. Teller Phys. Rev. **73**, 1561 (1949).
4. R. D. Cowan and J. Ashkin, Phys. Rev. **105**, 144 (1957).
5. G. A. Baker, Jr., Am. J. Phys. **27**, 29 (1959).
6. A. S. Eddington, "The Internal Constitution of Stars" (Dover, 1959, New York).
7. N. G. Van Kampen, in "Fundamental Problems in Statistical Mechanics" edited by E. G. D. Cohen (Wiley, 1968, New York) pg. 306.
8. K. Huang, "Statistical Mechanics" (Wiley, 1963, New York).
9. E. T. Copson, "An Introduction to the Theory of Functions of a Complex Variable" (Oxford Univ. Press, 1948, London).
10. G.A. Baker, Jr, "Essentials of Padé Approximants" (Academic, 1975, New York); G. A. Baker, Jr. and P. R. Graves-Morris, "Padé Approximants, Part I: Basic Theory and Part II: Extensions and Applications" part of the "Encyclopedia of Mathematics and its Applications, Vols. 13 & 14" (Cambridge Univ. Press, 1981, London).
11. P. S. Epstein, "Textbook of Thermodynamics" (Wiley, 1937, New York).
12. D. A. Liberman, private communication. This computer code is a straightforward programing of the T-F theory as presented in reference 3.

NEW MECHANISM OF TRANSPORT PHENOMENA IN SPIN-POLARIZED QUANTUM SYSTEMS

Eugene P. Bashkin

Academy of Sciences
Institute for Physical Problems
117334 Moscow, USSR

The presence of spin polarization in a quantum many-body system significantly influences its macroscopic behavior and especially its transport properties.[1] The influence of polarization on transport may be traced to the substantial contribution of the exchange interaction to the scattering of identical particles with spins. The magnitude of the scattering amplitude for two colliding particles depends on their total spin. Magnetic polarization produces a change in the occupation numbers of the different spin states, in particular, a change in the ratio of the number pairs of scattering particles with even total spin to the number of pairs with odd spin.

Fluctuations of transverse magnetism in spin-polarized quantum systems play an extremely important role in transport. As a rule, spin modes are strongly damped in the case of unpolarized systems. Weakly-damped oscillations of transverse magnetization practically always exist in the presence of spin polarization when the symmetry of a system changes – at least for uniform Larmor precession in a not-too-small magnetic field. Accordingly, when describing transport phenomena in spin-polarized systems, one must take into account not only the interparticle scattering, but also the interaction of paramagnetic atoms with collective spin modes. Hitherto, the latter interaction has been described in terms of a relativistic Zeeman interaction $H = -\beta\sigma\cdot\mathbf{B}(\mathbf{r},t)$, where β is the magnetic moment of the particle, σ the Pauli spin operator (restricting consideration to particles of spin 1/2), and $\mathbf{B}(\mathbf{r},t)$ is the macroscopic magnetic field induced by the fluctuating magnetization in a spin mode. Under the conditions for which giant opalescence occurs (sufficiently high degree of polarization, with weak enough external magnetic field), even this relatively weak relativistic coupling leads to significant and interesting phenomena.[2]

On the other hand, there exist much stronger exchange interactions between particles with spin and the magnetic-fluctuation field.[3] To illustrate the situation, we may consider a rarefied gas at low temperature, in the sense $T \ll \hbar^2/ma^2$, where m is the mass of a gas particle and a the scattering length (whose absolute value is assumed to be of the order of the particle size). (In the case of a degenerate Fermi gas, the degeneracy temperature should be substituted for T in specifying the appropriate conditions.) As is well known, under the stated condition the Fermi pseudopotential may be constructed using perturbation theory, and all final results may be expressed in terms of the real scattering amplitude a. First-order corrections (in the small parameter $ma^2T/\hbar^2 \ll 1$) to the self-energy of a particle of the medium are determined by the simplest diagrams (Fig. 1(a)). These

Condensed Matter Theories, Volume 5
Edited by V.C. Aguilera-Navarro
Plenum Press, New York, 1990

diagrams make the analytic contribution

$$\delta\Sigma = gN(\mathbf{r},t) - \beta^{-1} g\sigma\cdot\mathbf{M}(\mathbf{r},t) \quad , \tag{1}$$

where $g \equiv 2\pi\hbar^2 a/m$, N is the particle density, and M is the macroscopic magnetic moment per unit volume. The imaginary part of $\delta\Sigma$, describing the damping of single-particle excitations, appears only in second order in $ma^2T/\hbar^2$. The first term in Eq. (1) determines the interaction of a gas particle with the field of density oscillations (the phonons). The second term is the Hamiltonian of interest, describing as it does the interaction of a paramagnetic particle with the fluctuation field of macroscopic magnetization. Fluctuations of transverse magnetization (with respect to the spin-polarization vector) are not coupled to the density oscillations. Hence we can restrict attention to the second term of (1). Since the exchange approximation conserves the total magnetic moment, the Hamiltonian in question admits only the two types of inelastic processes illustrated in Fig. 2(b),

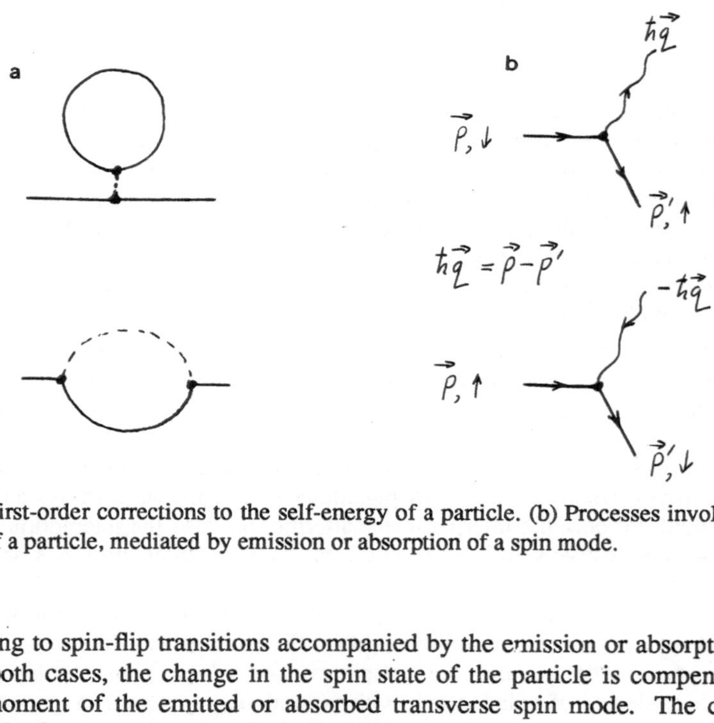

Fig. 1. (a) First-order corrections to the self-energy of a particle. (b) Processes involving spin-flip transitions of a particle, mediated by emission or absorption of a spin mode.

corresponding to spin-flip transitions accompanied by the emission or absorption of a spin mode. In both cases, the change in the spin state of the particle is compensated by the magnetic moment of the emitted or absorbed transverse spin mode. The cross section derived for the first process, using the indicated interaction, is

$$d\sigma(\downarrow\rightarrow\uparrow) \equiv d\sigma_1 = \frac{2m}{\rho N}\left(\frac{g}{\beta\hbar}\right)^2 [S_{xx}(\mathbf{q},\omega) - iS_{xy}(\mathbf{q},\omega)]\frac{d^3p'}{(2\pi\hbar)^3} \quad , \tag{2}$$

$$\hbar\omega = [p^2 - (p')^2]/2m + \hbar\Omega_H \quad , \tag{3}$$

where $\Omega_H = 2\beta H/\hbar$ is the Larmor frequency of the uniform precession in the external magnetic field H and $S_{ij}(\mathbf{q},\omega)$ constitutes the dynamic magnetic structure factor of the gas. The cross section $d\sigma(\uparrow\rightarrow\downarrow) \equiv d\sigma_2$ can be obtained from Eq. (2) by changing the sign of the second term in brackets and replacing Ω_H by $-\Omega_H$ in Eq. (3). Normally, all q-

dependent terms, including both real and imaginary corrections in the spectrum of a spin fluctuation, can be neglected in comparison to Ω_H. Here we also neglect the attenuation associated with the relativistic dipole-dipole interaction, i.e., we suppose that $\Omega_H \tau_s \gg 1$, where τ_s is of the order of the longitudinal relaxation time. Then, after some simple algebra, we obtain the following expressions for the total cross sections:

$$\sigma_1 = 16\pi a^2 \alpha (1 - e^{-\hbar\Omega_H/T})^{-1} \quad , \qquad \sigma_2 = 16\pi a^2 \alpha (e^{\hbar\Omega_H/T} - 1)^{-1} \quad , \tag{4}$$

where $\alpha \equiv (N_\uparrow - N_\downarrow)/N$ is the degree of polarization. One can readily check that the Einstein relations for radiation are satisfied by (4).

In thermodynamic equilibrium under polarization of the system by an external magnetic field, the cross sections have the limiting behavior

$$\sigma_1 = \sigma_2 = 8\pi a^2 \quad , \qquad \beta H \ll T \quad , \tag{5}$$

which is just the usual result for elastic scattering of a pair of identical fermions. Thus, even in a common situation, the contribution of the effect in question cannot be neglected.

The phenomenon becomes much more significant in the case of giant opalescence, for which

$$\sigma_1 = \sigma_2 = 8\pi a^2 \alpha T/\beta H \quad . \tag{6}$$

The cross sections σ_1 and σ_2 can then greatly exceed the elastic gas-kinetic results. This means that in contrast to previous theories,[4,5] our treatment predicts that the transport coefficients decrease, rather than increase, upon polarizing the system (at least when α is not very large). Hence, the curve describing the dependence on polarization of one or another of the familiar transport properties should show a minimum at some specific α. The heat conductivity κ and viscosity η of a gas under the given conditions can be calculated exactly,[6] with the results

$$\kappa = \left(\frac{2T}{\pi m}\right)^{1/2} \frac{1}{24\pi a^2 \alpha} \frac{2\beta H}{T} \quad ,$$

$$\eta = \left(\frac{2T}{\pi m}\right)^{1/2} \frac{m}{30\pi a^2 \alpha} \frac{2\beta H}{T} \quad . \tag{7}$$

The mechanism described above is applicable to the case of a slow neutron beam propagating within a target of polarized nuclei. The corresponding cross sections are

$$\sigma_1 = \pi (f_+ - f_-)^2 \left(1 + \frac{m_n}{M}\right) \alpha (1 - e^{-\hbar\Omega_H/T})^{-1} \quad ,$$

$$\sigma_2 = \pi (f_+ - f_-)^2 \left(1 + \frac{m_n}{M}\right) \alpha (e^{\hbar\Omega_H/T} - 1)^{-1} \quad , \tag{8}$$

where m_n is the neutron mass, M is the nuclear mass, and f_+ and f_- are the respective values of the scattering amplitude for triplet and singlet two-neutron states. To illustrate the magnitude of the effect, we may consider some typical experimental conditions. For instance, in a magnetically dilute silicon system containing just a few ^{29}Si nuclei,[7] the cross sections for the inelastic neutron scattering at $T = 4.2$ K and $H \cong 10$ G turn out to be enormous: $\sigma_1 \cong \sigma_2 \sim 10^5 - 10^6$ barns. For a target of gaseous ^{3}He↑ under normal experimental conditions[8] ($T \sim 1$ K, $\alpha \cong 0.5$, $H \cong 10$ G), the magnitudes of σ_1 and σ_2 corresponding to neutron scattering are also of the order of 10^5 to 10^6 barns, and thus even larger than the cross section for absorption of neutrons by ^{3}He nuclei.

REFERENCES

1. S. Stringari, ed., *Spin-Polarized Quantum Systems* (World Scientific, Singapore, 1989).
2. E. P. Bashkin, Pis'ma Zh. Eksp. Teor. Fiz. **44**, 322 (1986) [JETP lett. **44**, 414 (1986)].
3. E. P. Bashkin, Pis'ma Zh. Eksp. Teor. Fiz. **49**, 320 (1989).
4. E. P. Bashkin and A. E. Meyerovich, Adv. Phys. **30**, 1 (1981).
5. C. Lhuillier and F. Laloë, J. Phys. (Paris) **43**, 197, 225, 833 (1982).
6. E. P. Bashkin, Zh. Eksp. Teor. Fiz. **96** (1989), in press.
7. L. S. Vlasenko, N. V. Zavaritskii, S. V. Sorokin, and V. G. Fleisher, Zh. Eksp. Teor. Fiz. **91**, 1496 (1986) [Sov. Phys. JETP **64**, 881 (1986)].
8. P. J. Nacher, M. Leduc, G. Trenec, and F. Laloë, J. Phys. Lett. (Paris) **43**, L-525 (1982).

CORRELATED WAVE FUNCTIONS THEORY OF THE SPECTRAL FUNCTION

Omar Benhar

INFN, Sezione Sanitá, Physics Laboratory,
Istituto Superiore di Sanitá. I-00161 Roma, Italy

Adelchi Fabrocini

Dept. of Physics, University of Pisa and
INFN, Sezione di Pisa,I-56100 Pisa, Italy

Stefano Fantoni

Dept. of Physics, University of Lecce and
INFN, Sezione di Lecce, I-73100 Lecce, Italy

Abstract. *A microscopic theory based on orthogonal correlated basis functions is presented for the single particle spectral function of an infinite Fermi system. The method is used to calculate the nucleon spectral function $P(\mathbf{k}, E)$ for a realistic model of nuclear matter in which spin-isospin and tensor correlations are fully taken into account. $P(\mathbf{k}, E)$ is analyzed in terms of a single-particle strength, completely determined by two-body breakup processes, and a background, mainly provided by three-body breakup processes. The strength of single-particle states close to the Fermi surface can be measured by $(e, e'p)$ reactions in kinematical conditions corresponding to low missing energy E, whereas the background requires a wide range of E values, extended up to several hundreds of MeV. The relations between $P(\mathbf{k}, E)$, the momentum distribution $n(\mathbf{k})$ and the response function $S(\mathbf{q}, \omega)$ at high momentum transfers are discussed .*

1. Introduction

Scattering experiments are a valuable tool to study many-body systems: neutron

Condensed Matter Theories, Volume 5
Edited by V.C. Aguilera-Navarro
Plenum Press, New York, 1990

scattering in liquid Helium and electron scattering in nuclei have provided a large amount of information about the structure of the wave functions. The main ingredient in the analysis of the experimental data is the response function and its study in the high momentum trasfer region, in nuclear matter, is the subject of this contribution.

When the electron momentum transfer q is large ($q \sim 1GeV$) enough, a widely used approximation is the *Plane Wave Impulse Approximation* (PWIA), that corresponds to assuming that the outgoing nucleon can be represented by a plane wave. In PWIA, the cross section for the $(e, e\prime N)$ experiments is given by [1]

$$\frac{d^4\sigma}{d\epsilon_2 d\epsilon_N d\Omega_2 d\Omega_N} = \left(\frac{d\sigma}{d\Omega}\right)_{eN} (\epsilon_N + m) p P(\mathbf{k}, E), \tag{1.1}$$

where ϵ_N and $\mathbf{p}$ are the energy and the momentum of the knocked-out nucleon, ϵ_i and $\mathbf{k}_i$, $(i = 1, 2)$ those of the incident and scattered electrons and $(\frac{d\sigma}{d\Omega})_{eN}$ is the off-shell electron-nucleon cross section. The nucleon spectral function $P(\mathbf{k}, E)$ is defined as the probability of removing a nucleon with momentum $\mathbf{k} = \mathbf{k}_1 - \mathbf{k}_2 - \mathbf{p}$ from the target nucleus leaving the final system with excitation energy $E = \epsilon_1 - \epsilon_2 - \epsilon_N - E_R$, with E_R being the recoil energy of the residual system. $P(\mathbf{k}, E)$ is given by

$$P(\mathbf{k}, E) = \frac{\sum |< \bar{0}|a^{\dagger}_{\mathbf{k}}|\bar{n}(A-1) >|^2}{< \bar{0}|\bar{0} >< \bar{n}(A-1)|\bar{n}(A-1) >} \delta(E_n(A-1) - E_0(A) - E), \tag{1.2}$$

where $a^{\dagger}_{\mathbf{k}}$ is the creation operator of a nucleon with momentum $\mathbf{k}$, $|\bar{0} >$ is the ground state of the A-nucleon system with energy $E_0(A)$ and $|\bar{n}(A-1) >$ is the $n-th$ intermediate state of a (A-1)-nucleon system with energy $E_n(A-1)$.

The momentum distribution $n(\mathbf{k})$, defined as

$$n(\mathbf{k}) = \frac{< \bar{0}|a^{\dagger}_{\mathbf{k}} a_{\mathbf{k}}|\bar{0} >}{< \bar{0}|\bar{0} >}, \tag{1.3}$$

is related to $P(\mathbf{k}, E)$ via the sum rule

$$n(\mathbf{k}) = \int_{E_{min}}^{\infty} dE P(\mathbf{k}, E), \tag{1.4}$$

where $E_{min} = E_n(A-1) - E_0(A)$. Microscopic calculations of the spectral function have been performed for ^3He [2,3] and, more recently, for nuclear matter [4], which is a suitable system to study correlation and final state effects in the perspective of addressing more fundamental problems, like the modification of the nucleon form factors due to the presence of the nuclear medium.

These calculations, based on a non relativistic model of nucleons interacting via a realistic hamiltonian of the type

$$H = -\frac{\hbar^2}{2m} \sum_{i=1,A} \nabla_i^2 + \sum_{j>i=1,A} v_{ij} + \sum_{k>j>i=1,A} v_{ijk}, \tag{1.5}$$

predict large N-N correlation effects on $n(\mathbf{k})$ of nuclear matter [5], complex [6-8] and light [9,10] nuclei. As a result, the occupation probability of single particle states inside the Fermi sea results to be quenched with respect to the mean field theory estimates. Pandharipande et al. [11] have shown that such a depletion is in fair agreement with recent elastic and inelastic electron-nucleus scattering experiments [12-15].

$P(\mathbf{k}, E)$ is most conveniently separated into two parts, one corresponding to the one nucleon emission processes and the other, due to multiparticle emission processes, giving a *background* contribution. The nuclear matter analysis have shown that the one nucleon emission part, which contributes for $k < k_F$ only, is intimately related [16] with the hole-state strengths, whereas the background is very much spread out in energy implying that high values of the removal energy components may be important to estimate $n(\mathbf{k})$ [4].

Inclusive $(e, e\prime)$ experiments performed in the region of negative values of the y-scaling variable [17], give important information on the spectral function and, consequently on the momentum distribution [18,19]. Realistic hamiltonians of the type given in eq.(1.5) must include a three-nucleon interaction. It is known that nuclear hamiltonians containing two-nucleon interactions only underbind the $A = 3,4$ nuclei and give too large an equilibrium density for nuclear matter. The results presented in this contribution have been obtained by using the Urbana [20] two-body interaction and the Urbana TNI model of ref.[21] for the three-nucleon interaction.

Progresses in the solution of the Faddeev equations [22], in the Green Function Monte Carlo method [23] and in the variational theory [9,10] have allowed for accurate calculations of the ground state wave functions in ligth nuclei. In the variational method, the important scalar as well spin-isospin and tensor correlations induced by the strongly repulsive and state-dependent N-N interaction must be included in order to get a bound nucleus. A realistic wave function is provided by

$$|0) = \frac{G|0]}{[0|G^{\dagger}G|0]^{1/2}}, \tag{1.6}$$

where $|0]$ is the uncorrelated ground state and G is a correlation operator of the form

$$G = S \prod_{j>i=1,A} F(i,j), \tag{1.7}$$

$$F(i,j) = \sum_{n} f^{n}(r_{ij}) O^{n}(i,j), \tag{1.8}$$

where S is the symmetrizer of $\prod_{i<j} F(i,j)$ and the operators $O^n(i,j)$ include the four central components $(1, \sigma_i \cdot \sigma_j, \tau_i \cdot \tau_j, \sigma_i \cdot \sigma_j \quad \tau_i \cdot \tau_j)$ for $n = 1,4$ and both the isoscalar and the isovector tensor components for $n = 5,6$. Detailed calculations with such correlated wave functions in complex nuclei are not yet possible, althought they have ben carried out in some reasonable approximation [8] and new methods to treat ^{16}O are being developed [24,25]. However it is possible to perform variational calculations in nuclear matter by using hypernetted and operator chain summation techniques [26-28].

Correlated basis theories [29-32] provide for a consistent and unified treatment of the ground and the excited states of complex nuclei and nuclear matter. They are based upon the following set of correlated states

$$|n) = \frac{G|n]}{[n|G^\dagger G|n]^{1/2}}, \tag{1.9}$$

where $|n]$ is the generic state of an uncorrelated base; in the case of nuclear matter $|n]$ is an eigenstate of a Fermi gas hamiltonian at a given density ρ. The correlation functions $f^n(r_{ij})$ are determined variationally [21,33], by minimizing the energy expectation value of the hamiltonian (1.5) on $|0)$. The CB states (1.9) are not orthogonal to each other. They can be orthogonalized in such a way [32] that the diagonal matrix elements of the hamiltonian on $|n)$ (the variational estimates) are preserved. The resulting set of orthonormal states $|n>$, which are denoted as OCB states, can be used in standard perturbation theories.

In the following, nuclear matter results, obtained in the framework of correlated basis function theory, are presented and the relations with the momentum distribution and the response function are discussed, the aim beeing the understanding of the N-N correlations in the hadronic matter. In section 2 the microscopic calculations of $P(\mathbf{k}, E)$ are briefly reviewed, whereas section 3 is devoted to a presentation of the results obtained for the momentum distribution, the hole-state strengths and the response function.

2. Spectral function

The nucleon spectral function (1.2) can be written in the following form

$$P(\mathbf{k}, E) = \frac{1}{\pi}\Im\frac{<\bar{0}|a_{\mathbf{k}}^\dagger[H - E_0 - E - i\eta]^{-1}a_{\mathbf{k}}|\bar{0}>}{<\bar{0}|\bar{0}>}, \tag{2.1}$$

A convenient perturbative scheme to calculate $P(\mathbf{k}, E)$ is obtained by splitting the hamiltonian into an unperturbed part H_0 ,diagonal in the OCB states, and the remainder, namely

$$H = H_0 + H_I, \tag{2.2}$$

$$<i|H_0|j>= \delta_{ij} <i|H|i>= \delta_{ij}H_{ii}, \tag{2.3}$$

$$<i|H_I|j>= (1 - \delta_{ij}) <i|H|j>= \bar{H}_{ij}, \tag{2.4}$$

where $|\quad>$ are OCB states. The quantity $H - E_0$ in eq.(2.1) is split into $(H_0 - E_0^v)+(H_I - \Delta E_0)$, with $E_0^v = H_{00}$ being the variational estimate of the ground state energy and $\Delta E_0 = E_0 - E_0^v$ the perturbative correction to it, and then expanded in the interaction operator term $H_I - \Delta E_0$ with the result:

$$(H - E_0 - E + i\eta)^{-1} = (H_0 - E_0^v - E + i\eta)^{-1}\sum_n(-)^n[(H_I - \Delta E_0)(H_0 - E_0^v - E + i\eta)^{-1}]^n. \tag{2.5}$$

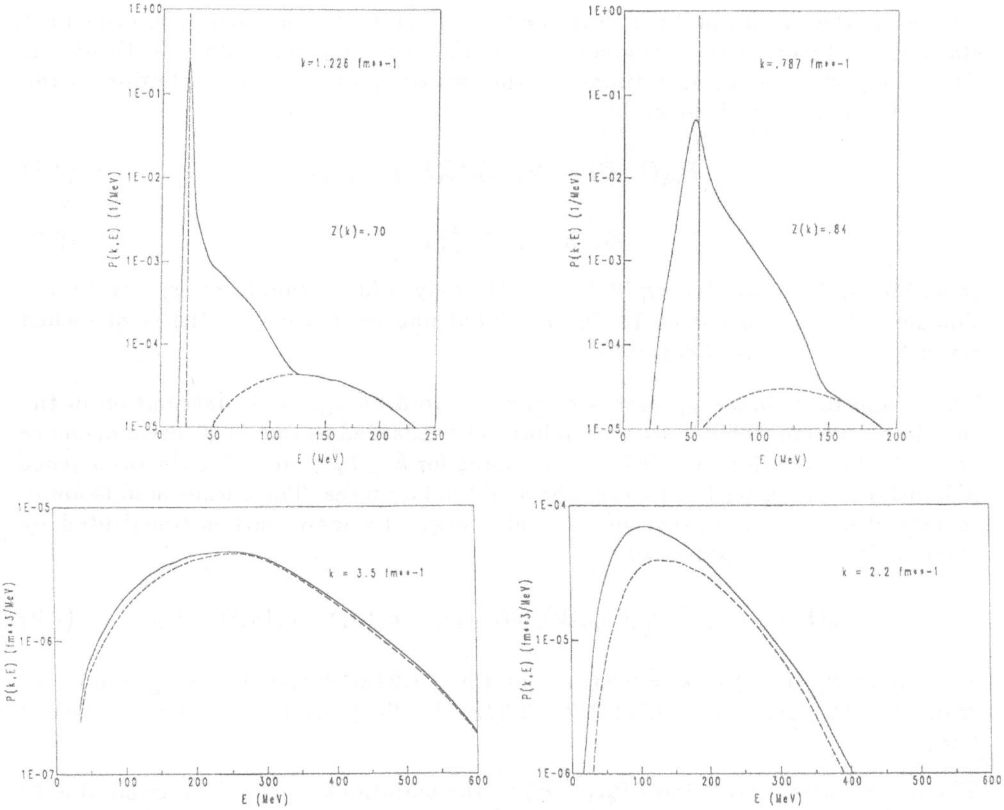

Fig. 1-*Spectral function of nuclear matter at $k_F = 1.33 fm^{-1}$ for several k-values. The dashed lines refer to the variational estimates and the solid lines to the full calculations including the perturbative corrections. The strenghts $Z(k \leq k_F)$ of the quasi-hole are also reported.*

Similarly, the ground state $|\bar{0}>$ is expressed in terms of the OCB states as:

$$|\bar{0}>= \sum_n (-)^n [(H_0 - E_0^v)^{-1}(H_I - \Delta E_0)]^n |0> . \tag{2.6}$$

The spectral function is then formally obtained by inserting $\sum_n |n><n| = 1$ between each pair of contiguous operators.

Several terms appearing in both the numerator and the denominator of $P(\mathbf{k}, E)$ are higly divergent in the thermodynamic limit. The divergencies arise from the factors ΔE_0 which are of order A and from the *unlinked* parts of the non diagonal matrix elements $< 0|a_{\mathbf{k}}^\dagger|n >$ or $\bar{H}_{mn}$. It has been proved that all these divergent terms cancel out exactly and that the remaining connected terms can be summed up by means of hypernetted and operator chain summations [4].

At the lowest order of the perturbative expansion we obtain the *variational* estimate of the spectral function $P^v(\mathbf{k}, E)$ by using OCB states in eq.(1.2).$P^v(\mathbf{k}, E)$ splits into two parts: one is obtained when the $1h$ OCB state $|\mathbf{k}>$ is the only intermediate state $|n>$;the other term is given by the $nh-(n-1)p$ OCB states with $n>1$. The first part provides an estimate of the two-body break-up contribution to the spectral function and is given by

$$P^{v}_{1h}(\mathbf{k},E)=|\Phi_{\mathbf{k}}(\mathbf{k})|^{2}\delta(E+e_{v}(k)), \tag{2.7}$$

$$\Phi_{\mathbf{k}}(n)=<0|a^{\dagger}_{\mathbf{k}}|n>. \tag{2.8}$$

In an uncorrelated matter, eq.(2.7) gives the only contribution to the spectral function and $\Phi_{\mathbf{k}}(\mathbf{k})$ is equal to 1. In correlated nuclear matter, $\Phi_{\mathbf{k}}(\mathbf{k})$ is quenched respect to 1 due to N-N correlations.

Three- and more-break up processes give rise to a background contribution in the correlated system. This contribution has to be regarded as *the most direct evidence of the N-N correlations* and it is nonvanishing for $k\geq k_F$, indicating the occurrence of knock-out processes from states above of the Fermi sea .The background is found to extend over a wide range of removal energy. Its main part is constituted by three-body break-up processes :

$$P^{v}_{2h-1p}(\mathbf{k},E)=\sum|\Phi_{\mathbf{h}_i\mathbf{h}_{i'}\mathbf{p}_i}(\mathbf{k})|^{2}\delta(e_{v}(p_i)-e_{v}(h_i)-e_{v}(h_{i'})-E), \tag{2.9}$$

Fig.1 gives $P^{v}_{1h}(\mathbf{k},E)$ at $k=0.8fm^{-1}$ and $k=1.2fm^{-1}$ and the background contribution $P^{v}_{2h-1p}(\mathbf{k},E)$ at $0.8,1.2,2.2,3.5fm^{-1}$. They are denoted by the dashed lines.

The perturbative correction $\delta P_{gr}(\mathbf{k},E)$ to the variational spectral function, due to $2h2p$ OCB admixtures in the ground state, is discussed in refs. [4,16] and it satisfies the sum rule

$$\int_{-e_v(k_F)}^{\infty}dE\delta P_{gr}(\mathbf{k},E)=\delta n_{gr}(k), \tag{2.10}$$

where $\delta n_{gr}(\mathbf{k})$ is the analogous correction to the variational momentum distribution $n_v(\mathbf{k})$, that has been found to be relevant, mainly around the Fermi surface [5].

Perturbative corrections $\delta P_{int}(\mathbf{k},E)$ due to $2h1p$ OCB states admixtures in $|\mathbf{k}>$ have a vanishing energy integral , so they contribute to the shape of the spectral function, giving a width to the single particle peaks, but do not affect the momentum distribution. In Fig. 1 the results for $P(\mathbf{k},E)$ with the perturbative corrections are shown at several values of k.

3. Momentum distribution, hole-state strength and response

The nucleon momentum distribution $n(\mathbf{k})$ can be calculated by the spectral function, via the sum rule (1.4), or by the direct use of eq.(1.3). The variational estimate $n_v(\mathbf{k})$ is obtained when only OCB states are inserted into eq.(1.3) and its expression is:

$$n_v(\mathbf{k})=\frac{<0|a^{\dagger}_{\mathbf{k}}a_{\mathbf{k}}|0>}{<0||0>}=\eta(N_c(k)+\Theta(k_F-k)N_d(k))+\Delta N_{comm}(k), \tag{3.1}$$

$\Delta N_{comm}(k)$ coming from the non commutativity of the operators $F(i,j)$ [5].

It has been proved [4] that $\Phi_{\mathbf{k}}(\mathbf{k})$ coincides with the discontinuous part of $n_v(\mathbf{k})$, $\eta N_d(k)$. It follows that it has to be regarded as the variational estimate of the quasi-particle strength $Z(k)$. In fact, $Z(k_F)$ is given by the discontuinity of the momentum distribution at the Fermi surface [34]. A substantial evidence of a sizeable quenching of the single particle states for nuclei in the lead region has been provided for by both elastic and inelastic electron scattering experiments [12,35].

In sect.2 has been shown that perturbative corrections to $n_v(\mathbf{k})$ come only from $\delta P_{gr}(\mathbf{k},E)$. $\delta P_{int}(\mathbf{k},E)$ gives a totally vanishing contribution, but part of it corresponds to the depletion of the quasi-particle strength due to the $2h1p$ admixtures in $|\mathbf{k}>$. The integral over the energy of this part gives an extra correction $\delta Z_{int}(k)$ to $Z(k)$, which then results to be

$$Z(k \leq k_F) \approx |\Phi_{\mathbf{k}}(\mathbf{k})|^2 + \delta n_{gr}(k) + \delta Z_{int}(k). \tag{3.2}$$

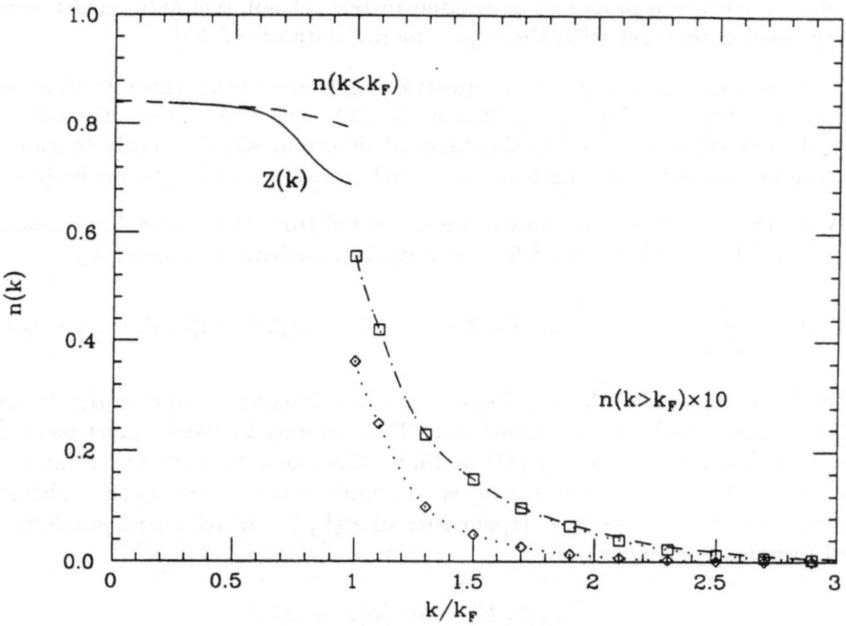

Fig. 2-*Removal energy integral of the nuclear spectral function* $n_{\overline{E}}(k) = \int_{-e_v(k_F)}^{\overline{E}} dEP(k,E)$ *in nuclear matter. The dotted lines with open diamonds and squares refer to* $\overline{E} = 100, 300MeV$ *respectively. The solid line gives the quasi-hole strength* $Z(k)$, *whereas the dashed line is the full* $n(k)$.

It turns out that $\delta n_{int}(k_F^-) = \delta n_{gr}(k_F^+)$ with the consequence that the total (variational + perturbative) background contribution is a continuous function of k. The difference between $n(k)$ and $Z(k)$ is ~ 0.1 at $k \sim k_F$ and slightly decreases for $k \to 0$. If $Z(k)$ is also corrected for the coupling of the single particle states with

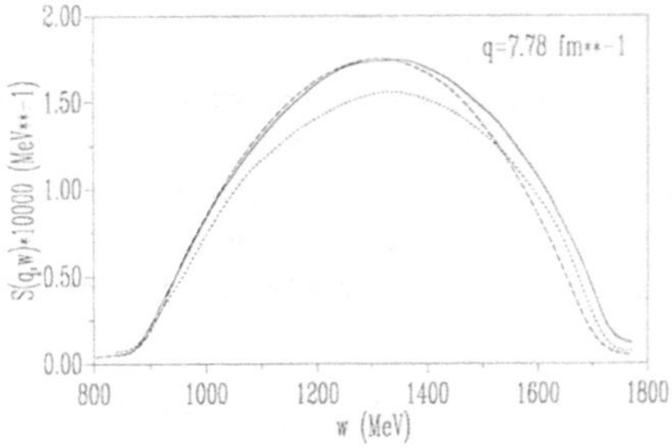

Fig. 3-*$S_P(\mathbf{q},\omega)$ (solid line) compared with $S(\mathbf{q},\omega)$ (dashed line) and with $S_{IA}(\mathbf{q},\omega)$ (dotted line) for nuclear matter at the equilibrium density.*

the surface vibration modes as estimated in refs.[11,36], one gets results which are in very good agreement with the experimental data of ref.[35].

The integral over the energy of the spectral function and the quasi particle strength $Z(k)$ are displayed in Fig. 2, together with $n(k)$. The figure shows the relevance of the high removal energy tail of the spectral function, which is entirely due to N-N correlations and is a common feature of nuclear matter and light nuclei [8-10].

In PWIA the response function can be evaluated from the spectral function and, if the classical kinematics is used for the outgoing nucleon, it is given by

$$S_P(\mathbf{q},\omega) = \int_{-e_v(k_F)}^{\infty} dE \int d\mathbf{k} P(\mathbf{k},E)\delta(\omega - E - e_v(|\mathbf{k}+\mathbf{q}|))\Theta(|\mathbf{k}+\mathbf{q}| - k_F). \quad (3.3)$$

In the above expression the knocked-out nucleon is considered completely uncorrelated from the remaining $A-1$ nucleons. The response in PWIA must be compared with the full response $S(\mathbf{q},\omega)$ [37] at high values of q to have an estimate of the final states effects. Also interesting is to compare with the response obtained by neglecting the missing energy dependence of $P(\mathbf{k},E)$, which corresponds to the so called *impulse approximation*:

$$P_{IA}(\mathbf{k},E) = n(\mathbf{k})\delta(E + e(k)), \quad (3.4)$$

$$S_{IA}(\mathbf{q},\omega) = \int d\mathbf{k} n(\mathbf{k})\delta(\omega - E - e_v(|\mathbf{k}+\mathbf{q}|))\Theta(|\mathbf{k}+\mathbf{q}| - k_F). \quad (3.5)$$

In Fig.3 $S_P(\mathbf{q},\omega)$, $S(\mathbf{q},\omega)$ and $S_{IA}(\mathbf{q},\omega)$ are compared at $q = 7.78 fm^{-1}$. At this momentum transfer, the PWIA response is in good agreement with the full calculation, while $S_{IA}(\mathbf{q},\omega)$ underestimates the response, mainly in the quasi elastic peak.

References

[1] T.deForest, Jr., J.D.Walecka, Advances in Phys. **15** (1966) 1;
A.E.L.Dieperink and T.de Forest Jr., Ann. Rev. Nucl. Science **25** (1975) 1;
S.Frullani and J.Mougey, Adv. Nucl. Phys. **14** (1984) 1;

[2] C.Ciofi degli Atti, E.Pace and G.Salme', Phys. Rev. **C21** (1980) 805;

[3] A.E.L. Dieperink, T. de Forest Jr., I.Sick and R.A.Brandeburg, Phys. Lett. **B63** (1976) 261;

H.Maier-Hajduk, Ch.Hajduk, P.U.Sauer and W.Theis, Nucl. Phys. **A395** (1983) 332;

[4] O.Benhar, A.Fabrocini and S.Fantoni, Nucl. Phys. **A**(1989) in press; *Electron-Nucleus Scattering*, A. Fabrocini *et al.* eds. ,World Scientific, Singapore,1989, 330;

[5] S.Fantoni and V.R.Pandharipande, Nucl. Phys. **A427** (1984) 473;

[6] J.G.Zabolitsky and W.Ey, Phys. Lett. **B76** (1978) 527;

[7] J.W.Van Orden, W.Truex and M.K.Banerjee, Phys. Rev. **C21** (1980) 2628;

[8] O.Benhar, C.Ciofi degli Atti, S.Liuti and G.Salme', Phys. Lett. **B177** (1986) 135;

[9] R.Schiavilla, V.R.Pandharipande and R.B.Wiringa, Nucl. Phys. **A449** (1986) 219;

[10] C.Ciofi degli Atti, E.Pace and G.Salme', Phys. Lett. **141B** (1984) 14;

[11] V.R.Pandharipande, C.N.Papanicolas, J.Wambach, Phys. Rev. Lett. **53** (1984) 1133;

[12] B.Frois and C.N.Papanicolas, Ann. Rev. Nucl. Part. Sci. **37** (1987) 4133 and references therein;

[13] J.M.Cavedon *et al.* Phys. Rev. Lett. **49** (1982) 978;

B.Frois *et al.*, Nucl. Phys. **A396** (1983) 409;

[14] C.N.Papanicolas *et al.*, Phys. Rev. Lett. **58** (1987) 2296. ;

[15] J.Lichtenstadt *et al.* Phys. Rev. **C20** (1979) 497;

[16] O.Benhar, A.Fabrocini and S.Fantoni, preprint (1989)INFN-ISS89/2;

[17] I.Sick, in *Momentum Distribution*, R.N.Silver and P.E.Sokol ,Plenum Press, NY, 1988, in press;

[18] I.Sick, D.Day and J.S.Mc Carthy, Phys. Rev. Lett. **45** (1980) 871;

[19] E.Pace and G.Salme', Phys. Lett. **B110** (1982) 411;

[20] I.E.Lagaris and V.R.Pandharipande, Nucl. Phys. **A359** (1981) 331;

[21] I.E.Lagaris and V.R.Pandharipande, Nucl. Phys. **A359** (1981) 349;

[22] C.R.Chen *et al.* Phys. Rev. **C33** (1986) 1740;

[23] J.Carlson, Phys. Rev. **C36** (1987) 2026; Phys. Rev. (1988) in press;

[24] V.R.Pandharipande, private communication;

MOMENTUM DISTRIBUTIONS IN ^{3}He–^{4}He MIXTURES

J. Boronat and A. Polls

Departament d'Estructura i Constituents de la Matèria
Universitat de Barcelona, E-8028 Barcelona, Spain

A. Fabrocini

Dipartimento di Fisica and Istituto Nazionale di Fisica Nucleare
Universita di Pisa, I-56100 Pisa, Italy

Abstract: We report variational calculations, in the framework of the HNC/FHNC theory, of the one-body density matrices and one-particle momentum distributions for ^{3}He–^{4}He mixtures at the zero temperature limit. The variational wave function used to describe the ground state of the mixture is a simple generalization of the trial wave functions for pure phases. We study the dependence on the ^{3}He concentration (x_3), of the condensate fraction of the ^{4}He ($n_0^{(4)}$) and the ^{3}He pole strength (Z_F) in the mixture along the isobar $P = 0$ atm. The differences between the results based on the Lennard-Jones and the Aziz potential are also discussed. It is found that the values of the condensate fraction do not depend strongly on the treatment of the ^{3}He statistics. On the other hand, the Fermi character of the ^{3}He is essential for the existence of the discontinuity in the ^{3}He momentum distribution in the mixtures, the correlations between ^{4}He and ^{3}He being the main source of the small values of Z_F. We predict a small increase of the ^{4}He condensate fraction when x_3 increases. This is a consequence of the fact that the total density of the mixture slightly decreases when the ^{3}He concentration increases.

1. INTRODUCTION

Determination of the momentum distribution of atoms in quantum liquids is a challenging problem of fundamental interest.[1] In the past several years, different experimental techniques, notably neutron scattering, have been used to obtain information on the momentum distribution in Helium liquids, ^{4}He and ^{3}He (see ref. 2 and references cited therein). The momentum distributions of the ^{4}He (^{3}He) are strongly affected by the Bose (Fermi) statistical behaviour of the atoms. The macroscopic occupation of the zero momentum state as measured by the condensate fraction $n_0^{(4)}$ characterizes the momentum distribution of liquid ^{4}He and is of fundamental importance in understanding the superfluid behaviour of liquid ^{4}He. On the other hand, the discontinuity Z_F at the Fermi momentum k_F is a characteristic of the ^{3}He system when is considered as a normal Fermi liquid.

In this paper we shall consider the very interesting case of ^3He–^4He mixtures where, due to the fermion-boson nature of the mixture, both of the quantities Z_F and $n_0^{(4)}$ are present at the same time. The projected neutron scattering experiments on ^3He–^4He mixtures have motivated us to perform a microscopic theoretical study of n_0 and Z_F as functions of the ^3He concentration (x_3) along the P = 0 atm isobar at zero temperature.

First we present the HNC/FHNC equations for calculating the momentum distributions or equivalently the one-body density matrices of the two components of the mixture.[3] Next, and in order to evaluate the elementary diagrams we generalize the scaling approximation used in pure phases.[4–7] In the second part we analyze the preliminary results obtained with wave functions containing only two-body correlations.

2. HNC/FHNC/S FOR JASTROW WAVE FUNCTIONS

We consider a homogeneous, isotopic mixture of N_3 ^3He atoms and N_4 ^4He atoms, confined in a box of volume Ω with densities $\rho_3 = N_3/\Omega$, $\rho_4 = N_4/\Omega$ and $\rho = \rho_3 + \rho_4$ and concentrations $x_3 = N_3/(N_3 + N_4)$ and $x_4 = N_4/(N_3 + N_4)$. In the end we let N_3, N_4 and Ω go to ∞, keeping the densities constant. The Hamiltonian of such a system is

$$H = \sum_{\alpha=3,4} -\frac{\hbar^2}{2m_\alpha} \sum_{i_\alpha=1,N_\alpha} \nabla^2_{i_\alpha} + \sum_{\alpha,\beta=3,4} \frac{1}{2} \sum_{\substack{i_\alpha=1,N_\alpha \\ j_\beta=1,N_\beta}} V^{(\alpha,\beta)}\,(i_\alpha,j_\beta) \tag{2.1}$$

where $V^{(\alpha,\beta)}$ are the interaction potentials between the particles of the α and β types and i_γ ($\gamma = \alpha,\beta$) runs over the coordinates of the γ–type particles.

To describe the ground state of the mixture we adopt a correlated wave function which is a simple generalization of the Jastrow wave function used to describe the pure phases:

$$\Psi(1...N_4,N_4 + 1,...N_4 + N_3) = \prod_{\substack{\alpha,\beta=3,4 \\ \alpha\leq\beta}} \prod_{i<j} f_{ij}^{(\alpha,\beta)}\, \phi(1...N_3) \quad . \tag{2.2}$$

In this expression, $\phi(1...N_3)$ is the Slater determinant of plane waves defined for the Fermi component of the mixture and $f^{(\alpha,\beta)}(r_{ij})$ are correlation functions, induced by the interactions, between the i–particle of α type and the j–particle of β type (α,β = 3,4 and $i_\alpha < j_\beta$).

It is well known that variational wave functions with only two-body correlations underbind helium liquids and that an improvement in the binding energy as well as in the location of the saturation density is achieved by incorporating three-body correlations into the wave function. However, in this preliminary study of the x_3 dependence of n_0 and Z_F we have restricted ourselves to two-body correlations. In fact, at a given density, the introduction of three-body correlations modifies the momentum distribution only slightly while lowering the kinetic energy by approximately 0.5 K. Therefore, a calculation including only two-body correlations should give the main features of the x_3 dependence of $n_0^{(4)}$ and Z_F, provided the calculation is performed at the experimental densities.

With the wave function (2) we calculate the one-body density matrices $\rho_\alpha(\mathbf{r}_1,\mathbf{r}'_1)$. The explicit definition of this quantity for the ^4He component (α = 4) is

$$\rho_4(\mathbf{r}_1,\mathbf{r}'_1) = N_4 \frac{\int \Psi^*(1...,N_4 + N_3)\Psi(1'...,N_4 + N_3)d\mathbf{r}_2...d\mathbf{r}_{N_4} \cdots d\mathbf{r}_{N_4+N_3}}{\int |\Psi(1...,N_4 + N_3)|^2\, d\mathbf{r}_1 \cdots d\mathbf{r}_{N_4+N_3}} \quad . \tag{2.3}$$

In homogeneus mixtures, with particle densities ρ_α, we have $\rho_\alpha(1,1') = \rho_\alpha(r)$ with $r = |\mathbf{r}_1 - \mathbf{r}'_1|$ and the normalization condition $\rho_\alpha(0) = \rho_\alpha$.

The momentum distribution of component α, or rather the occupation probability for single-particle states with momentum $\mathbf{k}$ and given spin projection (if relevant), can be obtained as the Fourier transform of the corresponding density matrix,

$$n_\alpha(k) = \delta_{\alpha 4} N_4 n_0^{(4)} \delta_{k0} + \frac{1}{\nu_\alpha} \int d\mathbf{r} \exp(i\mathbf{k}\mathbf{r})[\rho_\alpha(r) - \delta_{\alpha 4} n_0^{(4)} \rho_4] \tag{2.4}$$

where $n_0^{(4)} = \rho_4(\infty)/\rho_4$ is the condensate fraction of the ^{4}He in the mixture. Here ν_α stands for the average over the spin components (*i.e.*, $\nu_4=1, \nu_3=2$). Henceforth we shall omit subindices when they are not necessary.

A cluster analysis of $\rho_\alpha(r)$ in powers of $\omega^{(\alpha,\beta)} = f^{(\alpha,\beta)}-1$ and $h^{(\alpha,\beta)} = [f^{(\alpha,\beta)}]^2-1$, similar to that carried out for the pure phases,[8,9] gives the following structural decomposition for $\rho_\alpha(r)$:

$$\rho_\alpha(r) = \rho_\alpha n_0^{(\alpha)} N^{(\alpha)}(r) \tag{2.5}$$

where massive resummations of the diagrams defined in ref. 8 are implied. These resummations may be accomplished in practice by using HNC/FHNC techniques.[3,9] The strength factor $n_0^{(\alpha)}$ is given by

$$n_0^{(\alpha)} = \exp[2\Gamma_\omega^{(\alpha)} - \Gamma_d^{(\alpha)}] \quad , \tag{2.6}$$

while

$$N^{(\alpha)}(r) = [\delta_{\alpha 4}+\delta_{\alpha 3}(l(k_F r) - \nu (N_{\omega_c\omega_c}(r) + E_{\omega_c\omega_c}(r)))] \exp[N_{\omega\omega}^{(\alpha)}(r) + E_{\omega\omega}^{(\alpha)}(r)] \ . \tag{2.7}$$

In Eq. (2.7), $l(x) = 3j_1(x)/x$ and k_F is the Fermi momentum defined by $k_F = (6\pi^2\rho/\nu_3)^{1/3}$.

We discuss first the calculation of $N^{(\alpha)}(r)$. The nodal contributions $N_{\omega_c\omega_c}$ and $N_{\omega\omega}^{(\alpha)}$ can be obtained from the convolution integrals

$$N_{\omega\omega}^{(\alpha)} = \sum_{\lambda=3,4} \rho_\lambda \sum_{x,y} (g_{\omega x}^{(\alpha,\lambda)} - N_{\omega x}^{(\alpha,\lambda)} - \delta_{xd} | g_{y\omega}^{(\lambda,\alpha)} - \delta_{yd}) \tag{2.8a}$$

and

$$N_{\omega_c\omega_c}^{(3)}(r) = \rho_3(g_{\omega_c c} + l(k_F r_{12})/\nu - N_{\omega_c c} | g_{c\omega_c} + l(k_F r_{21'})/\nu)$$
$$+ \rho_3(-l/\nu \,|\, 2(g_{c\omega_c}+l/\nu - N_{c\omega_c}) - (g_{cc}+l/\nu - N_{cc})) \ . \tag{2.8b}$$

The notation $(A(ij)|B(jk))$ stands for the convolution product,

$$(A(ij)|B(jk)) = \int d\mathbf{r}_j \, A(r_{ij}) \, B(r_{jk}) \quad . \tag{2.9}$$

The summations over x and y extend to all possible connections allowed by the diagrammatic rules of the HNC/FHNC theory.[10,11]

The relevant spatial distribution functions are given by

$$g_{\omega d}^{(\alpha,\beta)}(r) = f^{(\alpha,\beta)}(r) \exp[N_{\omega d}^{(\alpha,\beta)}(r) + E_{\omega d}^{(\alpha,\beta)}(r)] \quad , \tag{2.10a}$$

$$g_{\omega e}^{(\alpha,3)}(r) = g_{\omega d}^{(\alpha,3)}(N_{\omega e}^{(\alpha,3)} + E_{\omega e}^{(\alpha,3)}) \quad , \tag{2.10b}$$

$$g_{\omega_c c}^{(3,3)} = g_{\omega d}^{(3,3)}(r)(-l(k_F r)/\nu \, N_{\omega_c c}^{(3,3)}(r) + E_{\omega_c c}^{(3,3)}(r)) \quad . \tag{2.10c}$$

and the nodal functions $N_{\omega x}^{(\alpha,\beta)}$'s are the solutions of the following integral equations:

$$N_{\omega x}^{(\alpha,\beta)} = \sum_{\lambda=3,4} \rho_\lambda \sum_{z,y} (g_{\omega z}^{(\alpha,\lambda)} - N_{\omega z}^{(\alpha,\lambda)} - \delta_{zd} | g_{yx}^{(\lambda,\beta)} - \delta_{yd}) \quad , \tag{2.11a}$$

$$N^{(3,3)}_{\omega_c c} = (g_{\omega_c c} + 1/\nu \, | g_{cc} - N_{cc}) + (N_{\omega_c c} | 1/\nu) \quad . \tag{2.11b}$$

The functions $g^{(\alpha,\beta)}_{dd}$, $g^{(\alpha,3)}_{de}$, $g^{(3,3)}_{ee}$ and $g^{(3,3)}_{cc}$ are defined in ref. 3. The quantities $E^{(\alpha,\beta)}_{\omega d}$, $E^{(\alpha,3)}_{\omega e}$ and $E^{(3,3)}_{\omega_c c}$ are the contributions of the elementary diagrams. The set of Eqs. (2.10) and (2.11) are the HNC/FHNC equations for the one body-density matrices of the mixture. The equations can be solved once a given prescription for the contribution of the elementary diagrams has been chosen. Neglect of the elementary diagrams defines the HNC/FHNC/0 approximation. The next logical step is the HNC/FHNC/4 approximation, which sums only elementary diagrams having four points. Figure 1 shows different types of four-point elementary diagrams. A circle (triangle) represents the position of a ^{4}He (^{3}He) atom, being solid [empty] when they are internal [external]. Each internal point implies an integration over its coordinates and a density factor ρ_α. External points, such as 1 and 1′ in $\rho_\alpha(1,1')$, are not integrated over. A solid line represents a $g^{(\alpha,\beta)}_{\omega d}-1$ link, depending on the type (α,β) of the particles at the end points. A dashed line is a $g^{(\alpha,\beta)}_{dd}-1$ link, a directed solid line is a $g^{(3,3)}_{\omega_c c}$ or $g^{(3,3)}_{cc}$ link according to whether the exchange loop involves or does not involve the points 1 or 1′, and a closed loop of directed solid lines connecting only two points is a $g^{(3,3)}_{ee}$ link. For example, the contribution of the second diagram of the fourth row of Fig. 1 is

$$E^{(4,4)}_{\omega\omega,4} = \rho_3^2/2 \int dr_2 \int dr_3 (g^{(4,3)}_{\omega d}(r_{12})-1)(g^{(4,3)}_{\omega d}(r_{13})-1) g^{(3,3)}_{ee}(r_{23}) (g^{(4,3)}_{\omega d}(r_{1'2})-1)(g^{(4,3)}_{\omega d}(r_{1'3})-1) \; , \tag{2.12}$$

the factor 1/2 is the symmetry factor which must be associated with that diagram to avoid double counting.

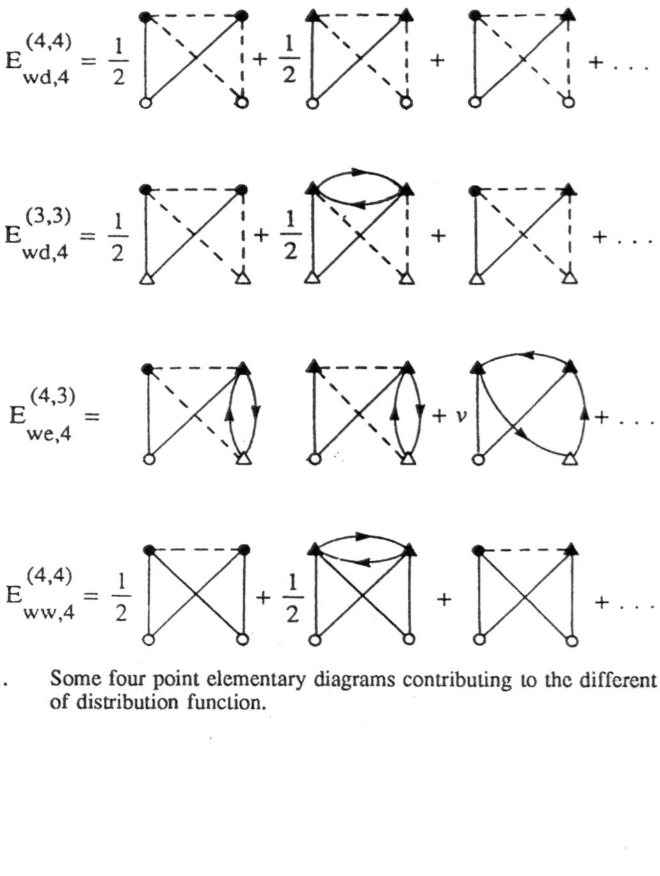

Fig. 1. Some four point elementary diagrams contributing to the different types of distribution function.

To calculate the strength factors $n_0^{(\alpha)}$, we need the quantities Γ_ω and Γ_d which are given by

$$\Gamma_x^{(\alpha)} = \sum_{\lambda=3,4} \rho_\lambda \int dr (X_{xd}^{(\alpha,\lambda)}(r) + \delta_{\lambda 3} X_{xe}^{(\alpha,3)}(r))$$
$$- \sum_{\lambda=3,4} \rho_\lambda \int dr (g_{xd}^{(\alpha,\lambda)}(r) + \delta_{\lambda 3} g_{xe}^{(\alpha,3)}(r)) E_{xd}^{(\alpha,\lambda)}(r)) - \rho_3 \int dr\, g_{xd}^{(\alpha,3)}(r) E_{xd}^{(\alpha,3)}(r)$$
$$- \frac{1}{2} \sum_{\lambda=3,4} \rho_\lambda \int dr (g_{xd}^{(\alpha,\lambda)}(r) - 1 + \delta_{\lambda 3}\, g_{xe}^{(\alpha\lambda)}(r)) N_{xd}^{(\alpha,\lambda)}(r)$$
$$- \frac{1}{2} \rho_3 \int dr (g_{xd}^{(\alpha,3)}(r) - 1) N_{xe}^{(\alpha,3)}(r) + E_\omega^{(\alpha)} \quad , \tag{2.13}$$

where $X_{xy}^{(\alpha,\lambda)} = g_{xy}^{(\alpha,\lambda)} - N_{xy}^{(\alpha,\lambda)} - \delta_{yd}$ $(x = \omega, d\ ;\ y = d, e)$ and $E_\omega^{(\alpha)}$ is the sum of the one-point elementary diagrams.[7]

Once the density matrices are known, the momentum distributions are computed by means of Eq. (2.4). We thus arrive at the results

$$n_{(4)}(k) = N_4 n_0^{(4)} \delta_{k0} + \rho_4 n_0^{(4)} \int dr \exp[ikr](\exp[N_{\omega\omega}^{(4)}(r) + E_{\omega\omega}^{(4)}(r)] - 1) \tag{2.14}$$

and

$$n_{(3)}(k) = n_0^{(3)} (\Theta(k_F - k)\, n_{disc}(k) + n_{cont}(k)) \tag{2.15}$$

with

$$n_{disc}(k) = 1 + 2\tilde{X}_{\omega_{cc}} + \frac{\tilde{X}_{\omega_c c}^2}{1 - \tilde{X}_{cc}} - \tilde{X}_{cc} \tag{2.16}$$

and

$$n_{con}(k) = - \frac{\tilde{X}_{\omega_{cc}}^2}{1 - \tilde{X}_{cc}} - \rho_3 \int dr \exp[ikr]((\exp[N_{\omega\omega}^{(3)}(r) + E_{\omega\omega}^{(3)}(r)] - 1)$$
$$\times (-l(k_F r)/\nu + N_{\omega_c\omega_c}^{(3)}(r) + E_{\omega_c\omega_c}^{(3)}(r)) + E_{\omega_c\omega_c}^{(3)}(r)) \quad , \tag{2.17}$$

where $X_{yc} = g_{yc} - N_{yc} + l/\nu$ for $y = \omega_c, c$ and $\tilde{y}^{(\alpha,\alpha)}(k)$ stands for the Fourier transform of $y^{(\alpha,\alpha)}(r)$ times ρ_α.

The momentum distributions must satisfy the normalization conditions

$$\frac{\nu_\alpha}{(2\pi)^3 \rho_\alpha} \int dk\, n(k) = 1 \tag{2.18}$$

or equivalently

$$\frac{\rho_\alpha(0)}{\rho} = 1 \quad . \tag{2.19}$$

As in the case of the pure phases,[6,12] we have the more stringent conditions

$$n_0^\alpha \exp[N_{\omega\omega}^{(\alpha)}(0) + E_{\omega\omega}^{(\alpha)}(0)] = 1 \quad , \tag{2.20a}$$
$$N_{\omega_c\omega_c}(0) + E_{\omega_c\omega_c}(0) = 0 \quad . \tag{2.20b}$$

Together these relations ensure the fulfillment of the normalization condition (2.19).

In the HNC/FHNC/0 approximation, the normalization conditions are seriously violated; the effect of the elementary diagrams must be taken into account if Eqs. (2.20) are to be satisfied. One way of simulating the contribution of the elementary diagrams, which has proven useful in treating the pure phases,[4–7] is the scaling approximation. An extension of the scaling approximation to ^{3}He–^{4}He mixtures at low x_3 concentrations is defined as follows:

$$E_{xd}^{(\alpha,\beta)}(r) = (1 + s_{xd})\, E_{xd,4}(r) \qquad (x = d,\omega) \quad ,$$
$$E_{yz}^{(\alpha,\beta)} = 0 \qquad (yz = de,\, ee,\, cc,\, \omega e,\, \omega_c c) \quad . \tag{2.21}$$

The four-point elementary diagrams $E_{dd,4}(r)$ and $E_{\omega d,4}(r)$ on the right hand side of the first member of (2.21) are calculated using an average distribution function specified by:

$$\bar{g}_y(r) = x_4^2\, g_{yd}^{(4,4)} + 2x_3x_4(g_{yd}^{(4,3)} + g_{ye}^{(4,3)}) + x_3^2(g_{yd}^{(3,3)} + 2g_{ye}^{(3,3)} + g_{ye}^{(3,3)}) \quad , \tag{2.22}$$

with $y = d,\omega$. Scaling of the other types of elementary diagrams has been carried out based on the relations

$$E_{\omega\omega}^{(\alpha)}(r) = (1 + s_{\omega\omega}^{(\alpha)})\, E_{\omega\omega,4}(r) \quad ,$$
$$E_{\omega_c\omega_c} = (1 + s_{cc})\, E_{\omega_c\omega_c,4}(r) \quad .$$

For the one point elementary diagrams, we adopt the same prescription as used in refs. 6 and 7, namely

$$E_y^{(\alpha)} = \left[1 + \frac{3}{2}\, s_{yd}\right] E_{y,4} \qquad (y = \omega,d) \quad ,$$

where the $E_{y,4}$ has been calculated with the average bond (2.20).

The scaling factor s_{dd} is fixed in the same way as in refs. 4 and 5 by requiring that two alternative forms of the kinetic energy of the mixture (the Jackson-Feenberg and the Pandharipande-Bethe expressions) agree with each other:

$$T_{JF}(s_{dd}) = T_{PB}(s_{dd}) \quad .$$

The scaling factors $s_{\omega\omega}^{(\alpha)}$ and s_{cc} are determined from the normalization conditions (2.20), while $s_{\omega d}$ is fixed by requiring equality between the total kinetic energy calculated by integrating the momentum distribution (T_{MD}) and the JF kinetic energy, where

$$T_{MD} = \frac{\hbar^2}{2m_4}\, \frac{1}{(2\pi)^3\, \rho_4} \int dk\, k^2 n_4(k) + \frac{\hbar^2}{2m_3}\, \frac{\nu}{(2\pi)^3\, \rho_3} \int dk\, k^2 n_3(k) \quad .$$

Afterwards, the comparison of the partial kinetic energies $T_{MD}^{(\alpha)}$ and $T_{JF}^{(\alpha)}$ is used to check the accuracy of the approximation.

3. RESULTS AND CONCLUSIONS

We present results for the two most commonly used interactions *i.e.*, the Lennard-Jones (LJ) and the Aziz potentials. In ^{3}He–^{4}He mixtures, the interaction is the same between any pair of particles. Based on this fact, we adopt the average correlation approximation (ACA). This approximation consists in taking the same correlation function for all pairs of particles. The ACA has been carefully analyzed for the impurity problem[13] and should be a reasonable assumption for small x_3 values.

We consider two types of correlation functions, namely (a) the McMillan form (f_{SRM}, short-range McMillan)

$$f(r) = \exp\left[-\frac{1}{2}\left(\frac{b}{r}\right)^5\right] \tag{3.1}$$

and (b) a correlation function (f_{LRA} long-range analytical)[15]

$$f(r) = \exp\left[-\frac{1}{2}\left(\frac{b}{r}\right)^5\right]\left[A + B\exp\left[-\alpha\frac{(r-D)^2}{r^4}\right]\right] . \tag{3.2}$$

The latter function has a long-range behaviour that can be adjusted to reproduce the experimental structure function at very low momenta.

All calculations have been performed at the experimental saturation densities of the mixture. Along the isobar P=0 atm, the density of the mixture, measured in σ units ($\sigma = 2.556$Å), decreases from $\rho = 0.3648\ \sigma^{-3}$ (*i.e.*, the saturation density of pure ^{4}He) at $x_3 = 0$ to $\rho = 0.3582\ \sigma^{-3}$ at $x_3 = 0.066$ (maximum solubility). The density of ^{3}He component changes from zero to $\rho_3 = 0.0236\ \sigma^{-3}$.

Table I gives the results for $n_0^{(4)}$ and Z_F for the LJ and the Aziz potentials at several ^{3}He concentrations. The corresponding saturation densities are $\rho = 0.3648$, 0.3629, 0.3609, 0.3582 all are in units of σ^{-3}. The calculations for the LJ potential have been performed with the f_{SRM} while for the Aziz potential we have used f_{LRA}. The condensate fraction $n_0^{(4)}$ shows an small increment when x_3 increases. This is a consequence of the fact that the total density of the mixture slightly decreases when x_3 increases as a consequence of the larger zero point motion of the ^{3}He. Most probably this variation is not large enough to be experimentally detected, being inside of the present experimental error bars of $n_0^{(4)}$ for the pure ^{4}He. We also give the results of $n_0^{(4)}$ in the boson-boson approximation; this approximation is equivalent to consider the underlying boson-boson mixture and consists in taking a wave function (Eq. (2.1)) without the Slater determinant. The effect of the Fermi statistics on $n_0^{(4)}$ is seen to be almost negligible. On the other hand, the Fermi statistics is crucial for the stability of the mixture.[16] The condensate fraction values produced by the wave functions determined for the LJ potential are systematically higher than the ones for the Aziz potential. This is in agreement with the fact that the LJ potential gives smaller kinetic energies than the Aziz potential.

Table I

x_3 dependence of $n_0^{(4)}$ and Z_F along the isobar P = 0 atom. (BB) and (FB) mean boson-boson and Fermi-boson approximation that are defined in the text.

	LJ			Aziz		
x_3	$n_0^{(4)}$(BB)	$n_0^{(4)}$(FB)	Z_F	$n_0^{(4)}$(BB)	$n_0^{(4)}$(FB)	Z_F
0	10.23	-	-	7.86	-	-
0.02	10.37	10.39	0.104	7.98	8.00	0.080
0.04	10.51	10.52	0.106	8.11	8.13	0.082
0.066	10.70	10.69	0.108	8.28	8.31	0.085

The values of Z_F, the discontinuity of the ^3He momentum distribution at the Fermi surface, are very small. This indicates a large depletion of the Fermi sea. Since the density of the ^3He component is very small, the correlations with the ^4He atoms must be the responsible for the small values of Z_F. In fact, Z_F is considerably smaller than in pure liquid ^3He (where is is roughly 0.25), giving rise to an enormous energy-dependent effective mass of the ^3He quasiparticles ($m_E^* \sim 12\ m_3$, for $x_3 = 0.04$). In spite of the fact that ρ_3 is an increasing function of x_3, Z_F increases with x_3 because the decrement of the total density dominates over the increment of the partial density of the ^3He component.

The scaling approximation for calculating the momentum distribution has been tested in the case of ^4He by comparing with Monte Carlo and molecular dynamics calculations using an f_{SRM} Satisfactory agreement is obtained. Nevertheless, recent Monte Carlo calculations[17] seem to indicate a larger discrepancy between the n_0 evaluations in spite of good agreement for the energies. Further investigations are needed to clarify the origin of the differences and the drawbacks of the different methods for calculating n_0.

Figure 2 shows the distribution functions $g^{(4,4)}$ and $g^{(3,3)}$ at $x_3 = 0.02$. The exchange effects are the responsible for the small values of $g^{(3,3)}$. It is interesting to study how the peak of $g^{(3,3)}$ grows when x_3 increases.

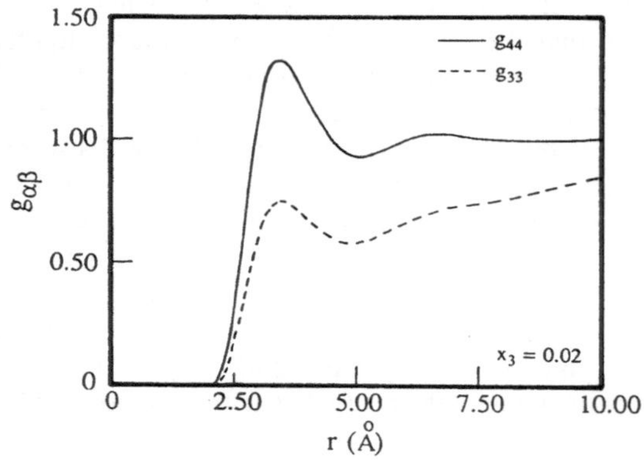

Fig. 2. Radial distribution functions $g^{(4,4)}$ and $g^{(3,3)}$. For $\rho = 0.3629\sigma^3$ and $x_3 = 0.02$

Figures 3 and 4 show the density matrices for two concentrations. The x_3 dependence of $\rho_4(r)/\rho$ is very small. The effects of the Fermi statistics makes the x_3 dependence of $\rho_3(r)$ more sizeable. Due to the small values of ρ_3 it is necessary to go out to large of r before $\rho_3(r)$ starts to oscillate around zero. The ^3He momentum distributions at the same two concentrations are shown in Fig. 5. The corresponding Fermi momenta at these concentrations are $0.235Å^{-1}$ and $0.347Å^{-1}$, which are to be compared with the value $k_F = 0.79Å^{-1}$ for pure ^3He at saturation density. When x_3 increases, the Fermi momentum increases, the depletion decreases and the discontinuity increases.

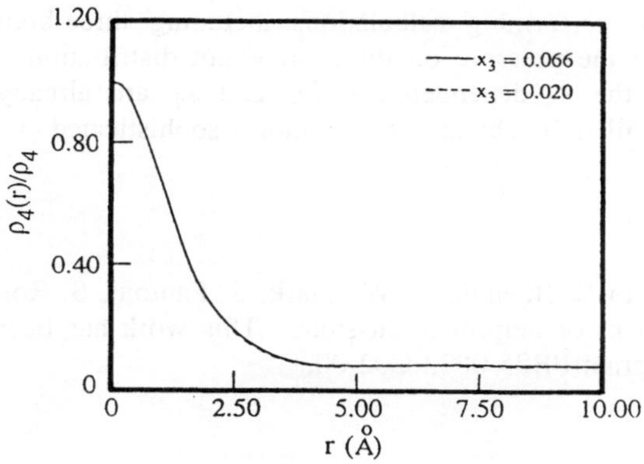

Fig. 3. $\rho_4(r)/\rho$ in HNC/FHNC/S approximations.

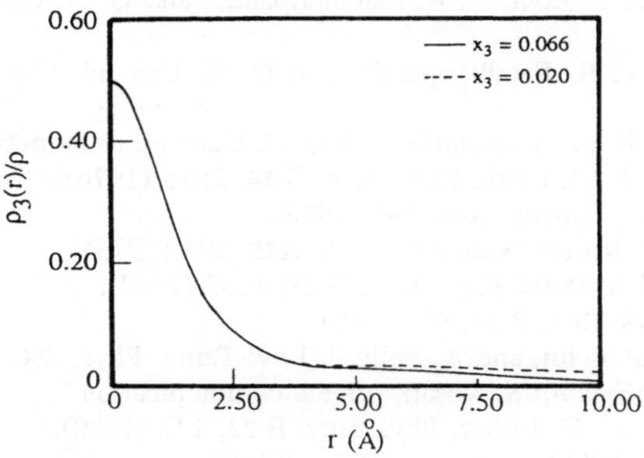

Fig. 4. $\rho_3(r)/\rho$ in HNC/FHNC/S approximation at $x_3 = 0.066$ and 0.020. Notice that the plotted numbers have been divided by the degeneracy $\nu_3 = 2$

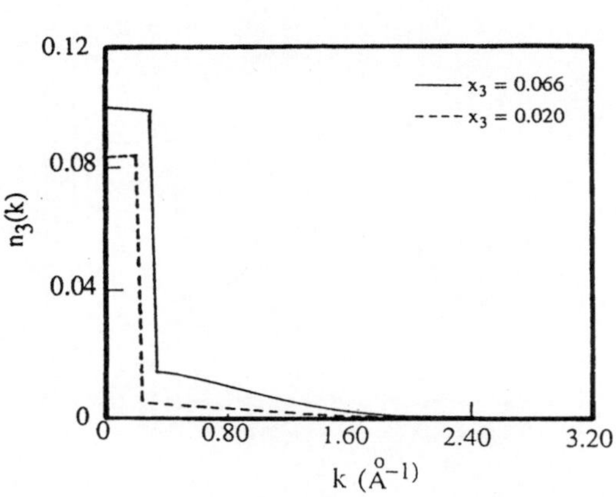

Fig. 5. ^{3}He momentum distributions at $x_3 = 0.066$ and 0.020. The vaues of k_F are 0.347Å^{-1} and 0.235Å^{-1} respectively.

Currently we are performing calculations including three-body correlations and studying the effects of the pressure on the momentum distributions. Nevertheless, the main conclusions on the x_3 dependence of n_0 and Z_F are already contained in the present study and will not change when more sophisticated wave functions are employed.

Acknowledgments

We are indebted to E. Buendia, J. W. Clark, S. Fantoni, S. Rosati, S. Vitiello, M. Viviani and P. Whitlock for helpful discussions. This work has been supported in part by CAYCIT (Spain), grant PB85-0072-C02-00.

References

1. *Momentum Distributions*, edited by R. N. Silver (Plenum, New York, in press).
2. P. E. Sokol, Can. J. Phys. **65**, 1393 (1987).
3. A. Fabrocini and A. Polls, Phys. Rev. B **26**, 1438 (1982).
4. Q. N. Usmani, S. Fantoni, and V. R. Pandharipande, Phys. Rev. B **26**, 6123 (1982).
5. E. Manousakis, S. Fantoni, V. R. Pandharipande, and Q. N. Usmani, Phys. Rev. B **28**, 3770 (1983).
6. E. Manousakis, V. R. Pandharipande, and Q. N. Usmani, Phys. Rev. B **31**, 7022 (1985).
7. A. Fabrocini, V. R. pandharipande, and Q. N. Usmani, to be published.
8. M. L. Ristig and J. W. Clark, Phys. Rev. B **14**, 2875 (1976).
9. S. Fantoni, Nuovo Cimento **A44**, 191 (1978).
10. S. Fantoni and S. Rosati, Nuovo Cimento **A25**, 593 (1975).
11. A. Fabrocini and A. Polls, Phys. Rev. B **25**, 4533 (1982).
12. M. F. Flynn, Phys. Rev. B **33**, 91 (1986).
13. J. Boronat, A. Fabrocini, and A. Polls, J. Low Temp. Phys. **74**, 347 (1989).
15. E. Buendia, M. Viviani, S. Rosati, private communication.
16. R. D. Guyer and M. D. Miller, Phys. Rev. B **22**, 142 (1980).
17. S. Vitiello and P. Whitlock, private communication.

FINITE TEMPERATURE PROPERTIES FOR THE ELECTRON GAS WITH LOCALIZATION UP TO 3 DIMENSIONS

A. Calles and A. Cabrera

Depto. de Física, Facultad de Ciencias, UNAM
Apdo. Post. 70-646, 04510 México D.F.

We performed finite temperature calculations for the electron gas in the density range $1 \le r_s \le 100$ up to 60K in temperature. A kind of Hartree-Fock equations were solved using the screened coulombic interaction in the deformed jellium model scheme. Properties such as total ground state energies, specific heats , γ values, temperature and density transitions are discussed. No transition to the localized state were found after T = 60K. Our results are compared with experimental values and other results reported in the literature for the case of zero temperature.

INTRODUCTION

The idea that electronic energy of a metal could be stabler by a charge density wave (CDW) was first put forward some 30 years ago by Peierls[1] and Frölich[2]. They considered the case of a one dimensional metal, in which the electrons are restricted to move only in one dimension (1D) by confining them to a linear chain of atoms. Since then, 1D calculations in systems candidates to present CDW have received a lot of attention. Besides, in recent years a new mechanism of conductivity, CDW conduction[3], has been discovered and extensively investigated.

Also a variety of quasi-one-dimensional materials exhibit a growth of electronic correlations as the temperature is lowered. Scattering experiments, as well as measurements of thermodynamic properties such as the specific heat, provide detailed information on the manner the electronic coupling behaves as the temperature is lowered. While a variety of theoretical approaches have been intensively used to explore the properties of the purely one- dimensional problem, much less work has been developed about the two and three-dimensional behaviour of the electronic coupling problem.

Condensed Matter Theories, Volume 5
Edited by V.C. Aguilera-Navarro
Plenum Press, New York, 1990

Therefore, it is valuable both from theoretical and experimental standpoints to predict the properties and phases of an electron gas for a complete range of densities and temperatures for the two and three-dimensional cases.

The relation of these kind of calculations with Wigner's crystalization[4] has been discussed frequently in the literature. There is a closer comparison between the charge spheres used by Wigner with a model that localize charge in three dimensions.

In this work we present the calculation of localized solutions in the form of CDW[5] for the electron gas using a Hartree-Fock (HF) like wave functions in the jellium model scheme.

In previous works[6,7], we proposed a general form for the wave function which contains as particular case the Overhauser's functions. The proposed form was a modified Bloch function which gave the Plane Wave (PW) solution or the CDW as a direct result from a Hartree-Fock calculation. As it is concerned for the localization we did the calculation for the one-dimensional case. Low dimensional calculations were sufficient because real systems where the first CDW or SDW were detected, were precisely low dimensional systems.

There is also a discussion about whether or not the localization means a phase transition and in such a case which order is implied. Then, the study of certain physical properties could help to the discussion of this problem. For instance, temperature dependent properties such as the specific heat can do it well. In the reference [7] we did such a discussion for the 1D case.

It is well known that Hartree-Fock calculations at finite temperature does not exist for the electron gas, so it is necessary going to a superior approximation (i.e. the RPA) or cutting the long range of the coulombic potential which is the responsible for the divergences in some energy integrals. We prefer to screen the coulombic potential in the scheme proposed within the Thomas-Fermi theory.

As it is concerned with the jellium model, used by several authors in order to calculate CDW, we see how such a model can arise from a variational principle which can be applied at zero and at finite temperatures.

THE MODEL

The jellium consists of particles repelling one another with a Coulomb force and immersed in a field of an external charge distribution call the background. The background charge density is not concentrated at a point, but rather spread with density $\eta(R)$ through the volume V. It can be regarded as a model of highly condensed matter with the background charge coming from an assemblage of ions and fast moving electrons.

The non-relativistic hamiltonian for the electron gas is:

$$\hat{H} = \hat{H}_{ee} + \hat{H}_{bb} + \hat{V}_{eb} \qquad (1)$$

where the first term represents the electronic Hamiltonian, the second one is the background Hamiltonian and the third is the interaction between electrons and the background, that is:

$$\hat{H}_{ee} = \sum_{k} T_{kk}\, a_k^+ a_k^+ + \frac{1}{2} \sum_{kk'} V_{kk'kk'}\, a_k^+ a_{k'}^+ a_{k'} a_k \qquad (2a)$$

$$\hat{H}_{bb} = \frac{1}{2} \int \eta(\underline{R})\, V(\underline{R}-\underline{R}')\, \eta(\underline{R}')\, d\underline{R}\, d\underline{R}' \qquad (2b)$$

$$\hat{V}_{eb} = - \sum_{k} \int V_{kk}(\underline{R})\, \eta(\underline{R})\, d\underline{R}\, a_k^+ a_k \qquad (2c)$$

a_k^+ and a_k are the creation and anihilation operators respectively, T_{kk} and $V_{kk}(\underline{R})$ are the matrix elements of kinetic energy and one particle operators in the basis orbitals $\varphi_k(\underline{r})$ and finally $V_{kk'kk'}$ are the matrix elements for the two electron operator.

The total energy of the system as function of the temperature can be written as:

$$E = \sum_{k} T_{kk} \langle \hat{n}_k \rangle_T - \sum_{k} \int \eta(\underline{R})\, V_{kk}(\underline{R})\, d\underline{R}\, \langle \hat{n}_k \rangle_T +$$

$$+ \frac{1}{2} \sum_{kk'} (V_{kk'kk'} - V_{kk'k'k}) \langle \hat{n}_k \rangle_T \langle \hat{n}_{k'} \rangle_T +$$

$$+ \frac{1}{2} \int \eta(\underline{R})\, V(\underline{R}-\underline{R}')\, \eta(\underline{R}')\, d\underline{R}\, d\underline{R}' \qquad (3)$$

where $\langle \hat{n}_k \rangle_T$ is the Fermi distribution function which reduces to the step function when T= 0.

The Hartree-Fock equations, which can be obtained from a variation of the energy respect to the basis orbitals, are:

$$T \varphi_k(\underline{r}) + V(\underline{r})\, \varphi_k(\underline{r}) + \sum_{k'} V_{kk'}(\underline{r}) \langle \hat{n}_k \rangle_T\, \varphi_k(\underline{r}) -$$

$$- \sum_{k'} V_{kk'}(\underline{r}) \langle \hat{n}_k \rangle_T\, \varphi_{k'}(\underline{r}) = \varepsilon_k\, \varphi_k(\underline{r}) \qquad (4)$$

On the other hand, the best choice[8] for the background density such that the energy, equation (3), is an extremum (in fact a minimum) is:

$$\int \eta(\underline{R}')\, V(\underline{X}-\underline{R}')\, d\underline{R}' = \sum_{k} V_{kk}(\underline{X}) \langle \hat{n}_k \rangle_T \qquad (5)$$

when substituted in equation (3), the energy reduces to:

$$E = \sum_{k} T_{kk} \langle \hat{n}_k \rangle - \frac{1}{2} \sum_{kk'} V_{kk'k'k} \langle \hat{n}_k \rangle_T \langle \hat{n}_{k'} \rangle_T \qquad (6)$$

where the only contribution for the electron-electron interaction energy comes from the exchange term. This means that the direct term plus the electron-background interaction plus the background-background energy adds to zero. This precisely define the so call deformable jellium model used by Overhauser[9-12] in his CDW and SDW calculations.

We have a reduced energy equation with only two terms. We just required the orbitals to satisfy the theorem in reference [8]. In order to determine the orbitals we must solve the HF equation that comes from the variation of the reduced energy, equation (6), respect to the orbitals with the usual orthonormalization condition. After doing the variation, the reduced Hartree-Fock equations are:

$$T\, \varphi_k(\underline{r}) - \sum_{k'} V_{kk'}(\underline{r})\, \langle \hat{n}_k \rangle_T\, \varphi_k(\underline{r}) = \varepsilon_k\, \varphi_k(\underline{r}) \qquad (7)$$

The potential we use is known as the screened coulombic interaction with the following form:

$$V(r) = \frac{e^{-\mu_o r}}{r} \qquad (8)$$

where μ_o is the screening parameter. Within the Thomas-Fermi[13] model we take as the screening parameter $\mu_o = 0.815/\sqrt{r_s}$.

We propose to use a kind of Bloch function as solution to the equation (7) in the form:

$$\varphi_k(\underline{r}) = \frac{1}{\sqrt{V}}\, e^{-\underline{k}.\underline{r}} \sum_{n=N_1}^{N_2} C_n\, e^{iq_o \underline{n}.\underline{r}} \qquad (9)$$

where $\underline{n} = n_x\,\hat{i} + n_y\,\hat{j} + n_z\,\hat{k}$, with n_x, n_y and n_z integers and $q_o \geq 2k_F$, k_F being the Fermi radius defined in the momentum space. A particular case to this equation is the Plane Wave function (PW) known as the trivial solution. Other particular cases to this equation are those in the form[14,15]:

$$C_1 \exp(i\, \underline{k}.\underline{r})\, [1 + \alpha\, (\exp(-i\, \underline{q}.\underline{r})] \qquad \underline{k}.\underline{q} > 0 \qquad (10a)$$

$$C_2 \exp(i\, \underline{k}.\underline{r})\, [1 + \alpha\, (\exp(-i\, \underline{q}.\underline{r})]^2 \qquad \underline{k}.\underline{q} > 0 \qquad (10b)$$

$$C_3 \exp(i\, \underline{k}.\underline{r})\, [1 + \alpha \cos(\underline{q}.\underline{r})] \qquad (10c)$$

These kind of functions are CDW solutions to the electron gas problem. Depending on the density and temperature of the system the function, equation (9), can be stabler than PW solution. In general it is convenient, but not necessary, to use symmetrical values for the integers involved in equation (9) in order to have the momentum conservation which is not satisfied by some of the CDW solutions, equations (10).

The coefficients C_n in equation (9) are obtained through the solution of the algebraic equation that comes from the substitution of equation (9) in the reduced HF equations. The eigenvalue matrix equation for the coefficients has to be solved self-consistently. The solutions to the algebraic equation for the coulombic and screened interactions, in the one dimensional case, were reported in reference [6] at zero temperature and in reference [7] at finite temperature.

RESULTS AND DISCUSSION

In this section we present the results for different properties when the localization is considered in 1, 2 and 3 dimensions.

The HF solution for the the coefficients can be such that $C_0= 1$ and $C_n= 0$ ($\underline{n}\neq 0$) then we have the trivial solution, otherwise if $C_n\neq 0$ ($\underline{n}\neq 0$) the equation (9) is a generalized Overhauser's type function.

In order to have an idea of how important is the strength of the localized wave, we define a kind of order parameter that is related with the electron charge density at the center of the wave in the following way:

$$\delta= 1 - \rho(0) \tag{11}$$

where ρ is the the electron charge density defined as:

$$\rho(\underline{r})= e^2 \sum_k |\varphi_k(\underline{r})|^2 \langle\hat{n}_k\rangle_T =$$

$$= e^2 \sum_{nn'} C_n C_{n'} \cos[q_o(\underline{n}-\underline{n}').\underline{r}] \tag{12}$$

note that there is no explicit temperature dependence in $\rho(\underline{r})$ due to the form of the function we chose. However there is an implicit dependence through the C_n coefficients because they depend parametrically on the temperature.

We also performed the calculation for the specific heat of the system:

$$C_v= \left(\frac{\partial E}{\partial T}\right) \tag{13}$$

The plots in figures 1a, 1b and 1c shown the results for the energy difference between the PW and the HF solutions, the order parameter δ and specific heat C_v respectively, for the 1D case as function of the distance r_s and temperature. At T= 0 the transition to the CDW occurs at $r_s\simeq$ 32. The energy difference at r_s= 100 is about 1 meV. The plots indicates a weak transition from the normal state to the CDW.

Similarly the plots in figures 2a, 2b and 2c show the results for the energy difference between the PW and the HF solutions, the order parameter δ and specific heat C_v respectively, for the 2D case as function of the distance r_s and temperature. The transition to the CDW is stronger than the 1D case. The energy difference in the plot 2a indicates approximately 2 meV, doubling the 1D case. The transition to the CDW at zero temperature occurs at $r_s\simeq$ 37. The anomalies in the specific heat, plot 2c, are not as weak as in the previous case.

Finally the plots in figures 3a, 3b and 3c show the results for the 3D case in the same region of distances and temperatures as before. The figure 3a shows that the CDW state is twice as much as the 2D case because at r_s= 100 the value is 4.1 meV. The transition at zero temperature occurs at $r_s\simeq$ 38, roughly speaking at the same density as the 2D case. The specific heat plot indicates a sharp transition in the region near to $r_s\simeq$ 40.

In Table 1 are shown the γ values calculated with our model, with the free electron model and the experimental one. As it can the seen, the theoretical calculations with our model are close to experimental values for metals like: Mg, Ca, Sr and Ba from the IIA group, Hg from the IIb group, Al and In from the IIIB group and Sn and Pb from the IVB group.

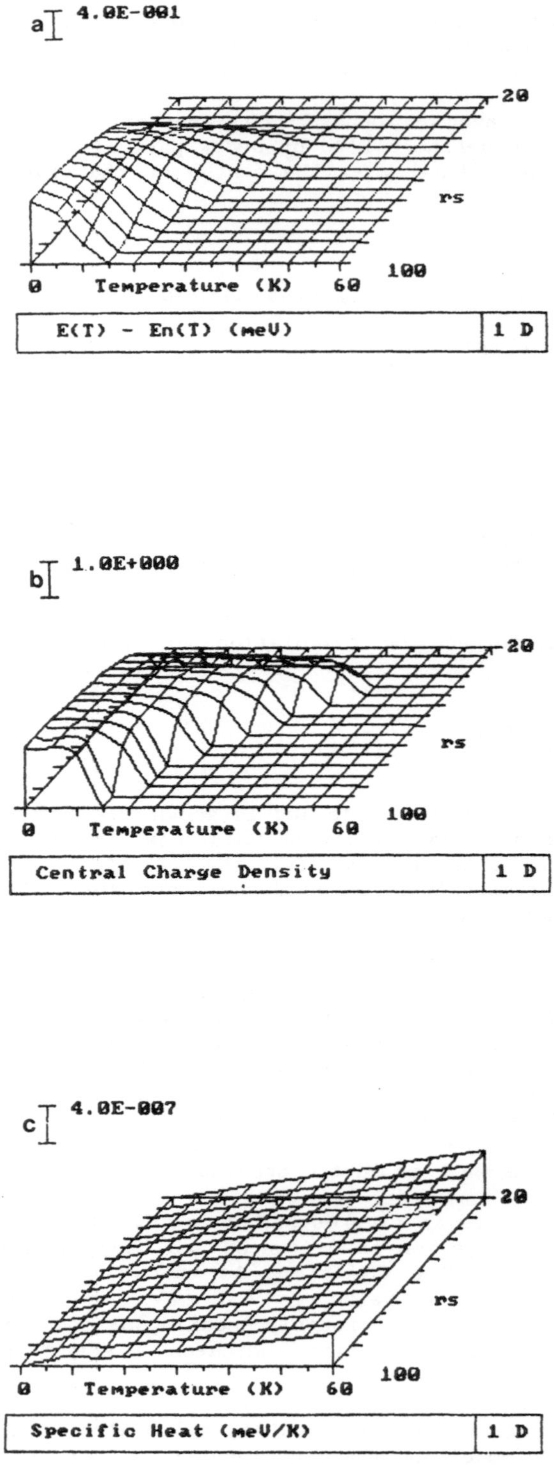

Figure 1. 1D properties in the range $20 \leq r_s \leq 100$ and $0 \leq T \leq 60$.
(a) Energy difference between the PW and the HF energies;
(b) Order parameter δ (Central Charge Density);
(c) Specific Heat.

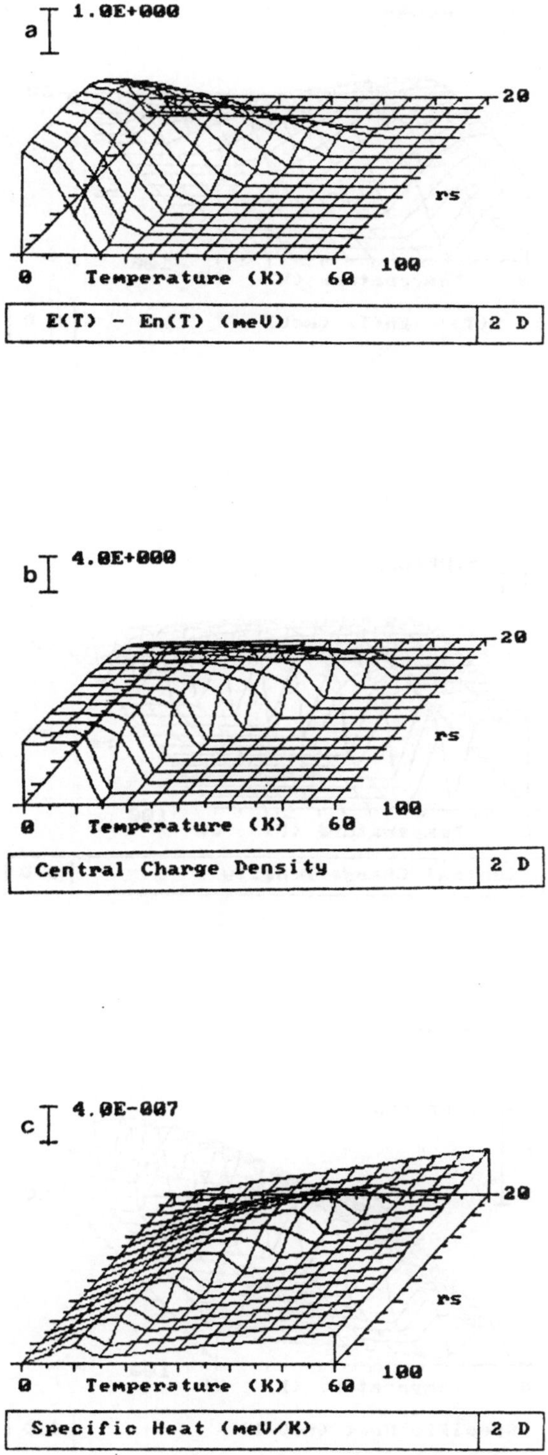

Figure 2. 2D properties in the range $20 \le r_s \le 100$ and $0 \le T \le 60$.
(a) Energy difference between the PW and the HF energies;
(b) Order parameter δ (Central Charge Density);
(c) Specific Heat.

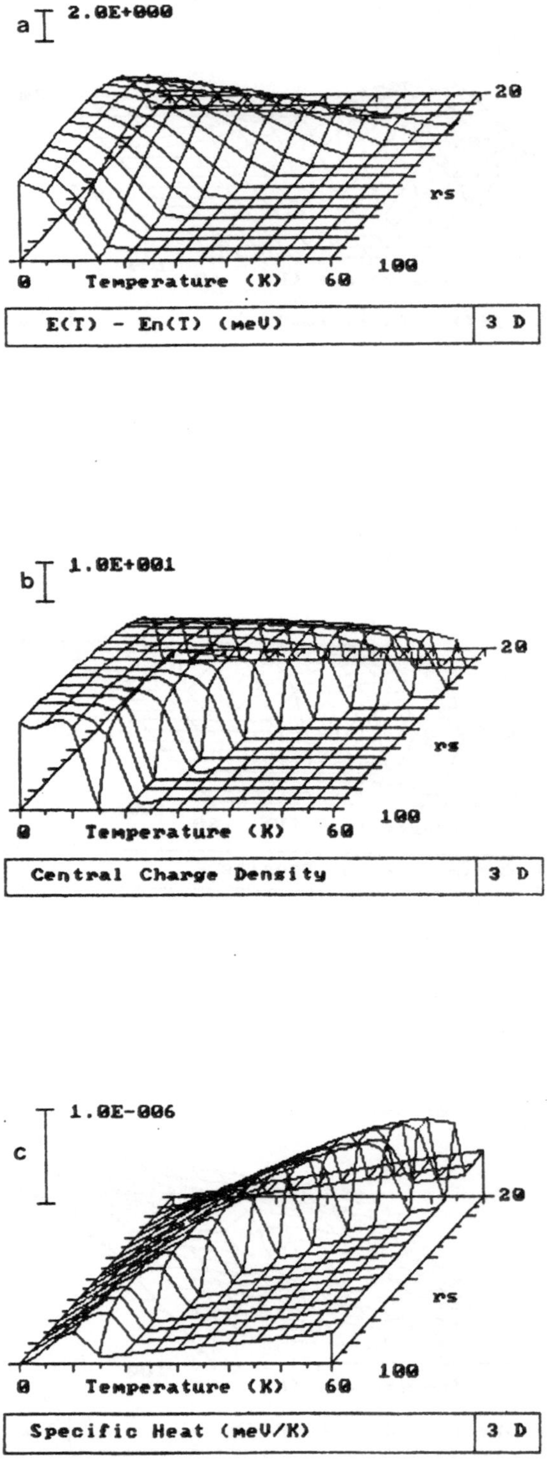

Figure 3. 3D properties in the range $20 \leq r_s \leq 100$ and $0 \leq T \leq 60$.
(a) Energy difference between the PW and the HF energies;
(b) Order parameter δ (Central Charge Density);
(c) Specific Heat.

Table 1. γ coefficients for all metals according to their group. In the last three columns are the free electron, the present calculation and the experimental values in 10^{-4} cal/(mol K^2). The third column Z are the valences of the metals.

Group	Metal	Z	r_s	Free[a] electron	Ours	Exp.[a]
IA	Li	1	3.25	1.8	2.8	4.2
IA	Na	1	3.93	2.6	4.4	3.5
IA	K	1	4.86	4.0	7.5	4.7
IA	Rb	1	5.20	4.6	8.8	5.8
IA	Cs	1	5.62	5.3	10.7	7.7
IB	Cu	1	2.67	1.2	1.8	1.6
IB	Ag	1	3.02	1.5	2.4	1.6
IB	Au	1	3.01	1.5	2.3	1.6
VA	Nb	1	3.07	1.6	2.5	20.0
IIA	Be	2	1.87	1.2	1.6	0.5
IIA	Mg	2	2.66	2.4	3.5	3.2
IIA	Ca	2	3.27	3.6	5.7	6.5
IIA	Sr	2	3.57	4.3	7.0	8.7
IIA	Ba	2	3.71	4.7	7.7	6.5
VIIIA	Fe	2	2.12	1.5	2.1	12.0
VIIB	Mn	2	2.14	1.5	2.1	40.0
IIB	Zn	2	2.30	1.8	2.5	1.4
IIB	Cd	2	2.59	2.3	3.3	1.7
IIB	Hg	2	2.65	2.4	3.5	5.0
VA	Al	3	2.07	2.2	2.9	3.0
VA	Ga	3	2.19	2.4	3.3	1.5
VA	In	3	2.41	2.9	4.2	4.3
VA	Tl	3	2.48	3.1	4.5	3.5
IVA	Sn	4	2.22	3.3	4.6	4.4
IVA	Pb	4	2.30	3.6	5.0	7.0
VA	Bi	5	2.25	4.3	5.9	0.2
VA	Sb	5	2.14	3.9	5.3	1.5

[a]Taken from reference [16].

CONCLUSIONS

The solutions we obtained have a parametric temperature dependence through the coefficients C_n.

The CDW solutions are stabler as the dimensionality increases. In particular the r_s's transition values at zero temperature are 32, 37 and 38 for 1, 2 and 3D respectively. The figure for the 2D case coincides with the report of Tanar and Ceperly[17] who performed Green's-function Monte Carlo calculations for the two-dimensional electron gas at zero temperature.

The transition densities r_s are temperature dependent such that the larger the temperature the smaller the r_s transition value. No transition at all was found for temperatures larger than 60K and the largest transition temperature was found at r_s= 40. It is worthwhile to note that the 3 cases have the same cualitative transition behaviour. They differ in the strength of the transition such that it is stronger as the dimensionality increases.

The plots for the specific heats indicate a second order transition in the temperature. Above the transition temperatures the specific heat has the typical linear metalic behaviour.

Finally, our calculations for the γ values, for most of the metals reported in Table 1, are much better than the free electron calculation and we have good figures for some of them as compared with the experiment.

One of us (A. Calles) thanks to Prof. V. Aguilera for suggesting the 3D calculations and to Prof. M. de Llano for encouraging comments related to this work. Partial support is due the Programa Universitario de Superconductividad de Alta Temperatura de Transición at UNAM, México.

REFERENCES

1. R.E. Peierls, Quantum Theory of Solids, Oxford Press, (1955).
2. H. Frölich, Proc. Roy. Soc., A223: 296 (1954).
3. R.M. Fleming and C.C. Grimes, Phys. Rev. Lett., 42, 1423 (1979).
4. E. Wigner, Trans. Faraday Soc., 34: 678 (1938).
5. A.W. Overhauser, Phys. Rev., 167,3: 691 (1968).
6. A. Cabrera, A. Calles, R.M. Méndez-Moreno and M.A. Ortíz, Tech. Rep., 1-85, Fac. de Ciencias, UNAM (1985).
7. A. Cabrera y A. Calles, Rev. Mex. Fís., 32,4: 573 (1986).
8. A. Cabrera y A. Calles, Rev. Mex. Fís., 33,2: 194 (1987).
9. A.W. Overhauser, Phys. Rev. B, 3,10 : 3173 (1971).
10. A.W Overhauser, Phys. Rev., 128,3: 1437 (1962).
11. V.C. Aguilera-Navarro, M. de Llano, S. Peltier and A. Plastino, Phys. Rev. A, 15,3: 1256 (1977).
12. R. Barrera, M. Grether and M. de Llano, J. Phys. C., 12: L7 15 (1979).
13. P. Debye and E. Hückel, Phys. Z., 24,185: 305 (1923).
14. L. Döhnert, M. de Llano and A. Plastino, Phys. Rev. A., 17, 2: 767 (1978).
15. R. Barrera, M. de Llano, S. Peltier and A. Plastino, Phys. Rev. B., 18, 6: 2931 (1978).
16. Neil W. Ashcroft and N. David Marmin, "Solid State Physics", Holt, Rinehart and Winston, New York (1976).
17. B. Tanar and D.M. Ceperly, Phys. Rev. B, 39,8: 5005 (1989)., also is interesting to see: D. Ceperly and B.J. Alder, Phys. Rev. Lett., 45: 566 (1980).

GENERALIZED MOMENTUM DISTRIBUTIONS OF QUANTUM FLUIDS

John W. Clark

McDonnell Center for the Space Sciences
and Department of Physics
Washington University, St. Louis, MO 63130 USA

Manfred L. Ristig

Institut für Theoretische Physik
Universität zu Köln, D-5000 Köln 41 BRD

INTRODUCTION

The two-body density matrix $\rho_2(\mathbf{r}_1, \mathbf{r}_2, \mathbf{r}'_1, \mathbf{r}'_2)$ of the ground state of a quantum fluid is a rich repository of information about its correlation structure. For example, the restricted version $\rho_1(\mathbf{r}_1, \mathbf{r}_2, \mathbf{r}'_1) = \rho_2(\mathbf{r}_1, \mathbf{r}_2, \mathbf{r}'_1, \mathbf{r}_2)$ plays a crucial role in the description of final-state interactions in a novel theory of the deep-inelastic neutron scattering from the helium liquids,[1] capturing the correlation effects which give rise to deviations from the impulse approximation out to very high momentum and energy transfers. This contribution will focus on the microscopic evaluation of $\rho_2(\mathbf{r}_1, \mathbf{r}_2, \mathbf{r}'_1)$ for uniform Fermi and Bose fluids. For simplicity, we assume that the ground state of either system is adequately described by a Jastrow trial function of the appropriate symmetry.

The formal structural analysis of the two-body density-matrix elements is most efficiently carried out in the configuration-space representation, and will be pursued using well-established cluster-expansion procedures,[2,3] followed by hypernetted-chain resummation of cluster diagrams.[4,5] On the other hand, the physical interpretation of the results is more vividly expressed in the momentum representation, i.e., in terms of the *generalized momentum distribution*

$$n(\mathbf{p}, \mathbf{q}) = \sum_{\hat{k}} < \Psi | a^\dagger_{\hat{k}+\mathbf{q}} a^\dagger_{\hat{p}-\mathbf{q}} a_{\hat{p}} a_{\hat{k}} | \Psi > \quad , \tag{1}$$

which is related to $\rho_2(\mathbf{r}_1, \mathbf{r}_2, \mathbf{r}'_1)$ by Fourier transformation. (Here, $\hat{k}$ denotes the single-particle orbital with quantum numbers $\mathbf{k}, \sigma$, where σ is the spin projection, while $\hat{k} - \mathbf{q} = (\mathbf{k} - \mathbf{q}, \sigma)$.) Using the formal results obtained for the structure of $\rho_2(\mathbf{r}_1, \mathbf{r}_2, \mathbf{r}'_1)$ – involving its closed expression in certain sums of irreducible cluster diagrams – we have been able to achieve a clean decomposition of $n(\mathbf{p}, \mathbf{q})$ into contributions from various scattering processes occurring in the medium. In comparing Bose and Fermi cases, certain features arising from exchange (Pauli kinematic effects)

Condensed Matter Theories, Volume 5
Edited by V.C. Aguilera-Navarro
Plenum Press, New York, 1990

will also be evident. Quantitatively, the contributions from these assorted physical effects are determined by a set of form factors, which are susceptible to evaluation by hypernetted-chain techniques. We shall present numerical results for the form factors depending on a single momentum variable, and test the quality of an estimate for $n(\mathbf{p}, \mathbf{q})$ used by Silver[1] as input for his theory of final-state effects in deep-inelastic neutron scattering.

The microscopic treatment of $n(\mathbf{p}, \mathbf{q})$ will be developed for a Fermi system of arbitrary single-particle level degeneracy ν. The corresponding results for the Bose case may then be obtained by taking the limits $\nu \to \infty$ and $k_F \to 0^+$ while keeping the density $\rho = \nu k_F^3/6\pi^2$ constant.

The role of $n(\mathbf{p}, \mathbf{q})$ in deep-inelastic neutron scattering becomes more tangible when we write this quantity as

$$n(\mathbf{p}, \mathbf{q}) = < \Psi | \rho_{\mathbf{q}} a^{\dagger}_{\hat{p}-\mathbf{q}} a_{\hat{p}} | \Psi > - n(p) \quad , \tag{2}$$

where $\rho_{\mathbf{q}}$ is the density fluctuation operator (with $\mathbf{q} \neq 0$) and $n(p)$ is the single-particle momentum distribution function. The expectation value (2) may be interpreted as a transition matrix element for scattering of a particle out of orbital $\hat{p} = (\mathbf{p}, \sigma')$ into another orbital $\hat{p} - \mathbf{q} = (\mathbf{p} - \mathbf{q}, \sigma')$, the process being mediated by a density fluctuation of wave vector $\mathbf{q}$. Thus, an evaluation of $n(\mathbf{p}, \mathbf{q})$ amounts to a calculation of the rightmost single-atom vertex in the Fig. 1.

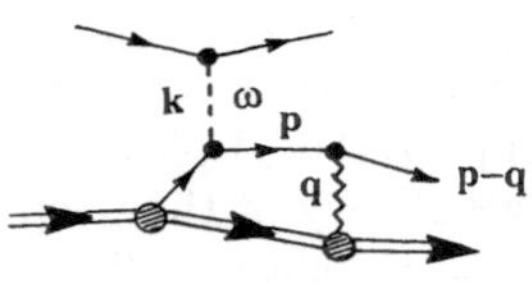

Fig. 1. Deep-inelastic scattering of neutrons from a (normal) helium liquid, involving final-state interactions mediated by a density fluctuation of the target system (a phonon of wave vector $\mathbf{q}$).

STRUCTURAL RESULTS FOR $\rho_2(\mathbf{r}_1, \mathbf{r}_2, \mathbf{r}_1')$

The ground-state wave function of the N-particle Fermi system, supposed to constitute a uniform fluid, is taken in the Jastrow-Slater form $\Psi = C^{-1}\Pi_{i<j} f(r_{ij})\Phi$, where Φ is the usual Slater determinant of plane-wave orbitals, filling a Fermi sea characterized by Fermi wave number k_F and level degeneracy ν, while $f(r_{ij})$ is a Jastrow pair correlation function to be determined by minimization of the energy expectation value. The constant C is introduced to normalize Ψ to unity. The results to be described below can be generalized in a straightforward manner to the case that the Jastrow correlating factor is replaced by a Feenberg function.[6]

The first step of the formal development is to recognize that the generalized momentum distribution function (1) may be expressed as follows

$$n(\mathbf{p}, \mathbf{q}) = \delta_{\mathbf{q}0}(N-1)n(p) + (1 - \delta_{\mathbf{q}0}) < \Psi | N(\hat{p}, \mathbf{q}) | \Psi > \tag{3}$$

in terms of the one-body momentum distribution $n(p)$ and the expectation value of a symmetric sum of two-body operators,

$$N(\hat{p}, \mathbf{q}) \;=\; \sum_{i<j}^{N} [e^{i\mathbf{q}\cdot\mathbf{r}_i} o_{\hat{p}-\mathbf{q},\hat{p}}(j) + e^{i\mathbf{q}\cdot\mathbf{r}_j} o_{\hat{p}-\mathbf{q},\hat{p}}(i)] \quad . \tag{4}$$

The action of the one-body operator $o_{\hat{p}-\mathbf{q},\hat{p}}(i)$ is specified by

$$o_{\hat{p}-\mathbf{q},\hat{p}}|\hat{k}>=\delta_{\hat{p}\hat{k}}|\hat{p}-\mathbf{q}> \quad , \qquad <\hat{k}|o_{\hat{p}-\mathbf{q},\hat{p}}=\delta_{\hat{k},\hat{p}-\mathbf{q}}<\hat{p}| \quad . \tag{5}$$

Standard procedures are available for cluster-expanding the expectation value of a sum of two-body operators, the most familiar case being the potential energy corresponding to a sum of pair potentials.[7] These procedures have been applied to yield explicit results for the cluster contributions to $n(\mathbf{p},\mathbf{q})$ through three-body order, plus selected four-body contributions. The generalized Ursell-Mayer diagrammatic representation introduced in Ref. 2 facilitates the bookkeeping. At two-body order there are 7 cluster diagrams, at three-body there are 138, etc., the number growing very rapidly. The analysis becomes more transparent when we go over to the position representation, inverting the Fourier relation

$$n(\mathbf{p},\mathbf{q})=\frac{1}{\nu}\frac{\rho}{N}\int \rho_2(\mathbf{r}_1\mathbf{r}_2\mathbf{r}_1')\mathrm{e}^{-i\mathbf{p}\cdot(\mathbf{r}_1-\mathbf{r}_1')}\mathrm{e}^{-i\mathbf{q}\cdot(\mathbf{r}_1-\mathbf{r}_2)}d\mathbf{r}_1d\mathbf{r}_2d\mathbf{r}_1' \tag{6}$$

to obtain the corresponding cluster expansion for the two-body density-matrix elements $\rho_2(\mathbf{r}_1\mathbf{r}_2\mathbf{r}_1')$. There is a very simple recipe for accomplishing this digrammatically. One simply removes the arrows representing specific plane-wave orbitals and changes the field points which they originally intersected into root points. It then becomes evident that most of the diagrams are reducible (factorizable), i.e., they consist of products of simpler graphs.

In the context of extensive technical experience with cluster expansions, the diagrams which have been explicitly generated are sufficient to reveal the structure of $\rho_2(\mathbf{r}_1\mathbf{r}_2\mathbf{r}_1')$ out to infinite cluster order. An exact representation of this quantity is provided by the expression

$$\rho_2(\mathbf{r}_1\mathbf{r}_2\mathbf{r}_1')=\rho_{2D}(\mathbf{r}_1\mathbf{r}_2\mathbf{r}_1')[L(\mathbf{r}_1\mathbf{r}_1')+L(\mathbf{r}_1\mathbf{r}_2\mathbf{r}_1')] \quad . \tag{7}$$

The first factor collects the direct-direct portions of the full set of diagrams contributing to $\rho_2(\mathbf{r}_1\mathbf{r}_2\mathbf{r}_1')$. By definition,[7] direct-direct diagrams do not have exchange lines attached to any of the root (or reference) points $\mathbf{r}_1$, $\mathbf{r}_2$, and $\mathbf{r}_1'$. The complementary set of graphs contains only diagrams with exchange lines beginning and/or ending at two or three reference points. Of these graphs, the ones with exchange lines at two reference points combine to form the two-point exchange factor $L(\mathbf{r}_1\mathbf{r}_1')$, while the graphs with exchange lines at all three reference points compose the three-point function $L(\mathbf{r}_1\mathbf{r}_2\mathbf{r}_1')$. Either of these exchange functions vanishes if any one of the coordinates in its argument recedes to an infinite distance from the others. In the Bose limit ($\nu\to\infty$, $k_F\to 0^+$, ρ constant), the function $L(\mathbf{r}_1\mathbf{r}_1')$ approaches unity and the three-point function $L(\mathbf{r}_1\mathbf{r}_2\mathbf{r}_1')$ goes to zero. As expected, only the direct-direct contribution ρ_{2D} survives. Returning to (7) at finite ν, this component may be compared, at a diagrammatic level, with the structural result which was derived for the Bose-fluid $\rho_2(\mathbf{r}_1\mathbf{r}_2\mathbf{r}_1')$ in Ref. 8. We arrive thereby at the representation

$$\rho_{2D}(\mathbf{r}_1\mathbf{r}_2\mathbf{r}_1')=\rho\rho_{1D}(\mathbf{r}_1\mathbf{r}_1')f(|\mathbf{r}_1-\mathbf{r}_2|)f(|\mathbf{r}_1'-\mathbf{r}_2|)\exp[-P(\mathbf{r}_1\mathbf{r}_2)-P(\mathbf{r}_1'\mathbf{r}_2)-P(\mathbf{r}_1\mathbf{r}_2\mathbf{r}_1')] \quad . \tag{8}$$

The generating functions $P(\mathbf{r}_1\mathbf{r}_2)=P(|\mathbf{r}_1-\mathbf{r}_2|)\equiv P(r)$ and $P(\mathbf{r}_1\mathbf{r}_2\mathbf{r}_1')$ are irreducible quantities – sums of irreducible diagrams – just as in the Bose case. To give at least a hint of their character, Fig. 2 shows the leading cluster contributions to these functions. We may remind the reader of the relevant diagrammatic conventions.[2,8] Open circles represent root points (lower left circle: $\mathbf{r}_1$, lower right circle: $\mathbf{r}_1'$, upper circle: $\mathbf{r}_2$). Solid circles denote field points and imply an integration $\rho \int$. A wavy [dashed] line stands for the correlation bond $f(r_{ab}) - 1$ [respectively, $f^2(r_{ab}) - 1$], where $r_{ab} = |\mathbf{r}_a - \mathbf{r}_b|$, and $\mathbf{r}_a$ and $\mathbf{r}_b$ are the coordinate points (solid or open circles) connected by the bond. The diagrams shown explicitly in Fig. 2 are in fact the same as the leading diagrams which occur for the Bose fluid. However, as in the familiar examples of the momentum distribution function $n(p)$ and the one-body density matrix $\rho_1(\mathbf{r}_1\mathbf{r}_1')$, additional, non-Bose diagrams will arise from the systematic introduction of exchange insertions at the field points of the Bose diagrams of higher orders.[7,2] These insertions are characterized by the presence of exchange lines at one or both end points. An exchange line, drawn solid with an arrow attached, represents the Slater statistical bond $l(k_F r_{ab})$, $l(x) = 3x^{-3}\sin x - x\cos x$.

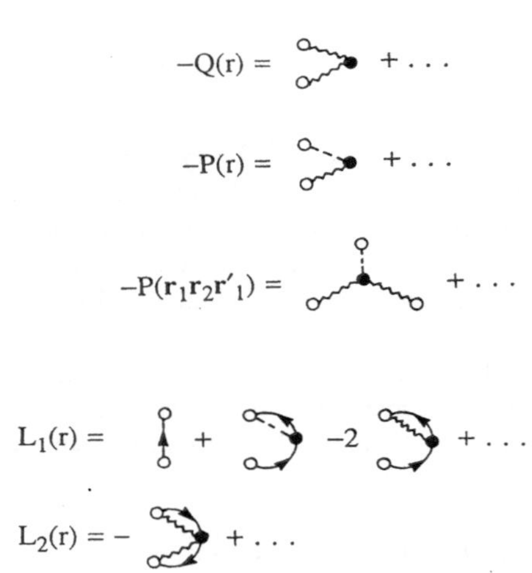

Fig. 2. Leading cluster diagrams of the irreducible direct functions $Q(r)$, $P(r)$, and $P(\mathbf{r}_1\mathbf{r}_2\mathbf{r}_1')$ and of the exchange functions $L_1(r)$ and $L_2(r)$.

The function $\rho_{1D}(\mathbf{r}_1\mathbf{r}_1')$ appearing as a factor in (8) is just the direct-direct component of the full Fermi one-body density matrix $\rho_1(\mathbf{r}_1\mathbf{r}_1')$. The structure of ρ_{1D} is well known from previous work.[2–4] This quantity is generated by the irreducible phase-phase correlation function $Q(\mathbf{r}_1\mathbf{r}_1')$ (see Fig. 2, and Figs. 8 and 10 of Ref. 2), according to

$$\rho_{1D}(\mathbf{r}_1\mathbf{r}_1') = \rho n_o \exp[-Q(\mathbf{r}_1\mathbf{r}_1')] \quad , \tag{9}$$

where $n_o = \exp Q(\mathbf{r}_1\mathbf{r}_1)$ is an overall strength factor.

At this point we have identified three sums of irreducible diagrams, namely $Q(r)$, $P(r)$, and $P(\mathbf{r}_1\mathbf{r}_2\mathbf{r}_1')$ (where, and throughout, $r \equiv |\mathbf{r}_1 - \mathbf{r}_2|$). As in the Bose case analyzed earlier[8] (and in the more familiar example of the generating function $U(r)$ of the hypernetted-chain representation $g(r) = f^2(r)\mathrm{e}^{-U(r)}$ of the Bose radial distribution function[7]), these functions may be decomposed into nodal and elementary components. Fig. 2 shows in fact the leading diagrams of nodal type; elementary diagrams first appear in the next cluster order. In practice, the diagram sums $Q(r)$, $P(r)$, and $P(\mathbf{r}_1\mathbf{r}_2\mathbf{r}_1')$ may be evaluated approximately, to all orders, by solving sets of Fermi-hypernetted-chain (FHNC) equations,[7,4,5] incorporating, stepwise, larger and larger classes of elementary diagrams. The simplest approximation, denoted FHNC/0, involves the neglect of all elementary components.

Fermi exchange effects arise implicitly from insertions at the field points of the

Bose diagrams contributing to $Q(r)$, $P(r)$, and $P(\mathbf{r}_1\mathbf{r}_2\mathbf{r}_1')$. Exchange manifests itself more explicitly in the second factor of the structural relation (7). We now examine the functions $L(\mathbf{r}_1\mathbf{r}_1')$ and $L(\mathbf{r}_1\mathbf{r}_2\mathbf{r}_1')$ in more detail.

Consider that, due to the disappearance of correlations at large distances, the quantity $\rho_2(\mathbf{r}_1\mathbf{r}_2\mathbf{r}_1')$ must reduce simply to $\rho\rho_1(\mathbf{r}_1\mathbf{r}_1')$ when $\mathbf{r}_2 \to \infty$. On the other hand, consider that the one-body density matrix $\rho_1(\mathbf{r}_1, \mathbf{r}_1')$ is known to have the structure[2,4,5]

$$\rho_1(\mathbf{r}_1\mathbf{r}_1') = \rho_{1D}(\mathbf{r}_1\mathbf{r}_1')[L_1(\mathbf{r}_1\mathbf{r}_1') + L_2(\mathbf{r}_1\mathbf{r}_1')] \tag{10}$$

in terms of the direct-direct component ρ_{1D} of (9) and the exchange functions L_1 and L_2 (denoted N_1 and N_2 in Refs. 2,4). The exchange functions L_1 and L_2 are in turn known, both formally and numerically, within FHNC theory.[4,5] The diagrams contributing to these functions through three-body cluster order are indicated in Fig. 2. The foregoing considerations, together with the vanishing of the function $L(\mathbf{r}_1\mathbf{r}_2\mathbf{r}_1')$ as $\mathbf{r}_2 \to \infty$, lead to the identification

$$L(\mathbf{r}_1\mathbf{r}_1') = L_1(\mathbf{r}_1\mathbf{r}_1') + L_2(\mathbf{r}_1\mathbf{r}_1') \quad . \tag{11}$$

The structure of the three-point exchange function $L(\mathbf{r}_1\mathbf{r}_2\mathbf{r}_1')$ cannot be determined from asymptotic properties; it is instead inferred from the raw results for the cluster expansion of $\rho_2(\mathbf{r}_1\mathbf{r}_2\mathbf{r}_1')$, along with the resolution (7). Many diagrams are seen to factorize, consistent with the following expression in terms of irreducible diagram sums $P_{\alpha\beta}$, $P_{\alpha\beta\gamma}$:

$$\begin{aligned} L(\mathbf{r}_1\mathbf{r}_2\mathbf{r}_1') = &-\nu^{-1}l(r)l(r') \\ &+ l(r)[P_{cc}(r') + P_{dcc}(\mathbf{r}_1'\mathbf{r}_2\mathbf{r}_1)] + l(r')[P_{cc}(r) + P_{dcc}(\mathbf{r}_1\mathbf{r}_2\mathbf{r}_1')] \\ &+ l(\mathbf{r}_1\mathbf{r}_1')[P_{de}(r) + P_{de}(r') + P_{ded}(\mathbf{r}_1\mathbf{r}_2\mathbf{r}_1')] \\ &+ P_{cdc}(\mathbf{r}_1\mathbf{r}_2\mathbf{r}_1') + P_{cdc}^2(\mathbf{r}_1\mathbf{r}_2\mathbf{r}_1') + P_{cec}(\mathbf{r}_1\mathbf{r}_2\mathbf{r}_1') \\ &- \nu[P_{cc}(r) + P_{dcc}(\mathbf{r}_1\mathbf{r}_2\mathbf{r}_1')][P_{cc}(r') + P_{dcc}(\mathbf{r}_1'\mathbf{r}_2\mathbf{r}_1)] \quad . \end{aligned} \tag{12}$$

The two- and three-point irreducible exchange functions $P_{\alpha\beta}$ and $P_{\alpha\beta\gamma}$ are classified according to the presence or absence of exchange lines at the root points. The category to which a given function belongs is indicated by its subscripts $\alpha\beta$ or $\alpha\beta\gamma$, according to the conventional scheme[7]: *cc* (circular), *de* (direct-exchange), *dcc* (direct-circular), *ded* (direct-exchange-direct), *cdc* (circular-direct-circular), and *cec* (circular-exchange-circular). The direct-direct functions $P(r)$ and $P(\mathbf{r}_1\mathbf{r}_2\mathbf{r}_1')$ fit in the same scheme, with d subscripts (omitted for simplicity). All of these functions may be separated into their nodal and elementary parts, and the techniques of Fermi-hypernetted-chain theory invoked for their numerical evaluation. It is to be noted that the two-point functions $P_{\alpha\beta}$ arise already in the theory of the Fermi one-body density matrix and have been thoroughly studied within the FHNC framework.[4,5]

The requirement that the fully-diagonal portion $\rho_2(\mathbf{r}_1\mathbf{r}_2\mathbf{r}_1)$ coincide with $\rho^2 g(r_{12})$, where $g(r_{12})$ is the radial distribution function for the assumed wave function, imposes certain relations on the quantities $P(r)$, $P(\mathbf{r}_1\mathbf{r}_2\mathbf{r}_1')$, $P_{\alpha\beta}$, and $P_{\alpha\beta\gamma}$, namely

$$\begin{aligned} -2P(r) - P(\mathbf{r}_1\mathbf{r}_2\mathbf{r}_1) &= N_{dd}(r) + E_{dd}(r) \quad , \\ P_{cc}(r) + P_{dcc}(\mathbf{r}_1\mathbf{r}_2\mathbf{r}_1) &= N_{cc}(r) + E_{cc}(r) \quad , \\ 2P_{de}(r) + P_{ded}(\mathbf{r}_1\mathbf{r}_2\mathbf{r}_1) &= N_{de}(r) + E_{de}(r) \quad , \\ P_{cdc}(\mathbf{r}_1\mathbf{r}_2\mathbf{r}_1) &= N_{de}(r) + E_{de}(r) \quad , \\ P_{cec}(\mathbf{r}_1\mathbf{r}_2\mathbf{r}_1) &= N_{ee}(r) + E_{ee}(r) \quad . \end{aligned} \tag{13}$$

The nodal (N) and elementary (E) diagram sums on the right-hand side are just those (with the corresponding subscripts) which arise in the FHNC analysis of $g(r)$ (cf. Ref. 7).

FORM FACTORS

We now determine the structure of the generalized momentum distribution function $n(\mathbf{p},\mathbf{q})$ of (1) by exploiting the structural results (7) (8), (9), (11), and (12) for the two-body density-matrix elements $\rho_2(\mathbf{r}_1\mathbf{r}_2\mathbf{r}_1')$ appearing in Fourier integral (6). To arrive at a decomposition of $n(\mathbf{p},\mathbf{q})$ which accomplishes a clean separation of contributions from differing physical processes, the function $\rho_2(\mathbf{r}_1\mathbf{r}_2\mathbf{r}_1')$ is first decomposed into a part containing all terms generated purely by two-point functions, and a remainder in which the terms also depend on the irreducible three point functions,

$$\rho_2(\mathbf{r}_1\mathbf{r}_2\mathbf{r}_1') = \rho_2^{(2)}(\mathbf{r}_1\mathbf{r}_2\mathbf{r}_1') + \rho_2^{(3)}(\mathbf{r}_1\mathbf{r}_2\mathbf{r}_1') \quad , \tag{14}$$

the notation being transparent. The last term vanishes if the various three-point functions $P(\mathbf{r}_1\mathbf{r}_2\mathbf{r}_1')$ and $P_{\alpha\beta\gamma}(\mathbf{r}_1\mathbf{r}_2\mathbf{r}_1')$ are set equal to zero.

Next, we appeal to the Fermi-hypernet equations resulting from the FHNC analysis of the one-body density matrix.[4,5] These equations relate the bare correlation function $f(r)$ to the spatial distribution functions defined by the direct and nodal diagram sums $X_{Q\alpha\beta}(r)$ and $N_{Q\alpha\beta}(r)$ introduced in Refs. 4,5, $N_{Qdd}(r)$, in particular, being identified with the nodal part of $-P(r)$:

$$\begin{aligned} g_{Qdd}(r) &= 1 + F_{dd}(r) = 1 + X_{Qdd}(r) + N_{Qdd}(r) \quad , \\ g_{Qcc}(r) &= 1 + F_{cc}(r) = 1 + X_{Qcc}(r) + N_{Qcc}(r) \quad , \\ g_{Qde}(r) &= 1 + F_{de}(r) = 1 + X_{Qde}(r) + N_{Qde}(r) \quad . \end{aligned} \tag{15}$$

The hypernet equations read

$$\begin{aligned} f(r)\mathrm{e}^{-P(r)} &= g_{Qdd}(r) \quad , \\ f(r)\mathrm{e}^{-P(r)}P_{cc}(r) &= \nu^{-1}l(r)F_{dd}(r) + F_{cc}(r) \quad , \\ f(r)\mathrm{e}^{-P(r)}P_{de}(r) &= F_{de}(r) \quad . \end{aligned} \tag{16}$$

Eqs. (16) permit us to eliminate the bare correlation factor $f(r)$ from each term of the explicit expression for the decomposition (14), with the results

$$\begin{aligned} \rho_2^{(2)}(\mathbf{r}_1\mathbf{r}_2\mathbf{r}_1') = {} & \rho\rho_1(\mathbf{r}_1\mathbf{r}_1')g_{Qdd}(r)g_{Qdd}(r') \\ & + \rho\rho_{1D}(\mathbf{r}_1\mathbf{r}_1')l(\mathbf{r}_1\mathbf{r}_1')[g_{Qdd}(r)F_{de}(r') + g_{Qdd}(r')F_{de}(r)] \\ & - \nu\rho\rho_{1D}(\mathbf{r}_1\mathbf{r}_1')[\nu^{-1}l(r) - F_{cc}(r)][\nu^{-1}l(r') - F_{cc}(r')] \end{aligned} \tag{17}$$

and

$$\begin{aligned} \rho_2^{(3)}(\mathbf{r}_1\mathbf{r}_2\mathbf{r}_1') = {} & \rho_2^{(2)}(\mathbf{r}_1\mathbf{r}_2\mathbf{r}_1')\{\exp[-P(\mathbf{r}_1\mathbf{r}_2\mathbf{r}_1')] - 1\} \\ & + \rho\rho_{1D}(\mathbf{r}_1\mathbf{r}_1')g_{Qdd}(r)g_{Qdd}(r')\exp[-P(\mathbf{r}_1\mathbf{r}_2\mathbf{r}_1')] \\ & \times [l(r)P_{dcc}(\mathbf{r}_1'\mathbf{r}_2\mathbf{r}_1) + l(r')P_{dcc}(\mathbf{r}_1\mathbf{r}_2\mathbf{r}_1') \\ & \quad + l(\mathbf{r}_1\mathbf{r}_1')P_{ded}(\mathbf{r}_1\mathbf{r}_2\mathbf{r}_1') + P_{cdc}(\mathbf{r}_1\mathbf{r}_2\mathbf{r}_1') + P_{cdc}^2(\mathbf{r}_1\mathbf{r}_2\mathbf{r}_1') \\ & \quad + P_{cec}(\mathbf{r}_1\mathbf{r}_2\mathbf{r}_1') - \nu P_{dcc}(\mathbf{r}_1\mathbf{r}_2\mathbf{r}_1')P_{dcc}(\mathbf{r}_1'\mathbf{r}_2\mathbf{r}_1) \\ & \quad - \nu P_{cc}(r)P_{dcc}(\mathbf{r}_1'\mathbf{r}_2\mathbf{r}_1) - \nu P_{cc}(r')P_{dcc}(\mathbf{r}_1\mathbf{r}_2\mathbf{r}_1')] \quad . \end{aligned} \tag{18}$$

(Here $r = |\mathbf{r}_1 - \mathbf{r}_2|$ (as usual) and $r' = |\mathbf{r}_1' - \mathbf{r}_2|$.)

Inserting (14) into the integral (6), and using the results (17) and (18), we obtain the decomposition

$$\begin{aligned} n(\mathbf{p},\mathbf{q}) = N\delta_{\mathbf{q}0}n(p) & \\ & + F_{dd}(q)[n(p) + n(|\mathbf{p}-\mathbf{q}|)] \\ & + F_{de}(q)[n_{Dl}(p) + n_{Dl}(|\mathbf{p}-\mathbf{q}|)] \\ & - n_o[\theta(k_F - p) - F_{cc}(p)][\theta(k_F - |\mathbf{p}-\mathbf{q}|) - F_{cc}(|\mathbf{p}-\mathbf{q}|)] \\ & + n^{(2)'}(\mathbf{p},\mathbf{q}) + n^{(3)'}(\mathbf{p},\mathbf{q}) \quad . \end{aligned} \tag{19}$$

Thus, the component $n(\mathbf{p},\mathbf{q}) - N\delta_{\mathbf{q}0}n(p)$ of the generalized momentum distribution depending nontrivially on two momenta is written as a sum of (i) separable contributions involving form factors

$$F_{\alpha\beta}(q) = \rho \int F_{\alpha\beta}(r)e^{i\mathbf{q}\cdot\mathbf{r}}d\mathbf{r} \tag{20}$$

and either the one-body momentum distribution

$$n(p) = \nu^{-1} \int \rho_1(r)e^{i\mathbf{p}\cdot\mathbf{r}}d\mathbf{r} \quad , \tag{21}$$

a modified momentum distribution

$$n_{Dl}(p) = \nu^{-1} \int \rho_{1D}(r)l(r)e^{i\mathbf{p}\cdot\mathbf{r}}d\mathbf{r} \quad , \tag{22}$$

or the strength factor n_o [second, third, and fourth terms of (19), respectively], (ii) a non-separable integral $n^{(2)'}(\mathbf{p},\mathbf{q})$ involving only two-point quantities, and (iii) another three-point integral $n^{(3)'}(\mathbf{p},\mathbf{q})$ generated from the component (18) of $\rho_2(\mathbf{r}_1\mathbf{r}_2\mathbf{r}_1')$. Explicitly, the fifth term is

$$n^{(2)'}(\mathbf{p},\mathbf{q}) = \frac{1}{\nu}\frac{\rho}{N} \int K(\mathbf{r}_1\mathbf{r}_2\mathbf{r}_1')e^{-i\mathbf{p}\cdot(\mathbf{r}_1-\mathbf{r}_1')}e^{-i\mathbf{q}\cdot(\mathbf{r}_1-\mathbf{r}_2)}d\mathbf{r}_1 d\mathbf{r}_2 d\mathbf{r}_1' \quad , \tag{23}$$

where

$$\begin{aligned} K(\mathbf{r}_1\mathbf{r}_2\mathbf{r}_1') = \rho\rho_1(\mathbf{r}_1\mathbf{r}_1')F_{Qdd}(r)F_{Qdd}(r') & \\ & + \rho\rho_{1D}(\mathbf{r}_1\mathbf{r}_1')l(\mathbf{r}_1\mathbf{r}_1')[F_{Qdd}(r)F_{de}(r') + F_{Qdd}(r')F_{de}(r)] \\ & - \nu\rho[\rho_{1D}(\mathbf{r}_1\mathbf{r}_1') - \rho n_o][\nu^{-1}l(r) - F_{cc}(r)][\nu^{-1}l(r') - F_{cc}(r')] \, . \end{aligned} \tag{24}$$

The contribution (23) may be reduced to a three-dimensional integral in momentum variables.

In momentum space, the sequential relation

$$\int \rho_2(\mathbf{r}_1\mathbf{r}_2\mathbf{r}_1')d\mathbf{r}_2 = (N-1)\rho_1(\mathbf{r}_1\mathbf{r}_1') \tag{25}$$

becomes

$$n(\mathbf{p},\mathbf{q}=0) = (N-1)n(p) \quad . \tag{26}$$

Specializing (19) to $\mathbf{q} = 0$, and substituting into (26), there results a condition on the form factors. If we employ this condition in (19) itself, we may recast the decomposition of $n(\mathbf{p}, \mathbf{q})$ in the form

$$\begin{aligned} n(\mathbf{p},\mathbf{q}) = (N-1)\delta_{\mathbf{q}0}n(p) \\ + (1-\delta_{\mathbf{q}0})F_{dd}(q)[n(p) + n(|\mathbf{p}-\mathbf{q}|)] \\ + (1-\delta_{\mathbf{q}0})F_{de}(q)[n_{Dl}(p) + n_{Dl}(|\mathbf{p}-\mathbf{q}|)] \\ - n_o(1-\delta_{\mathbf{q}0})[\theta(k_F - p) - F_{cc}(p)][\theta(k_F - |\mathbf{p}-\mathbf{q}|) - F_{cc}(|\mathbf{p}-\mathbf{q}|)] \\ + (1-\delta_{\mathbf{q}0})n^{(2)'}(\mathbf{p},\mathbf{q}) + (1-\delta_{\mathbf{q}0})n^{(3)'}(\mathbf{p},\mathbf{q}) \quad . \end{aligned} \tag{27}$$

This expression achieves the desired separation of contributions from the various scattering processes underlying the generalized momentum distribution function (cf. Ref. 8). The first term reproduces the trivial result for dynamically and statistically uncorrelated particles (except that the momentum distribution function $n(p)$ appearing in (27) is that for the fully correlated Fermi system). The correlations prevailing in the interacting fluid permit the scattering of a fermion from orbital $\hat{p}$ to another orbital $\hat{p} - \mathbf{q}$, with the intervention of a phonon to conserve momentum. The effect of this process and the corresponding time-reversed mechanism are described by the second term in (27). The associated exchange scattering effects are embodied in the third term, which is proportional to the exchange form factor $F_{de}(q)$. To interpret the fourth line of (27), we note that if the particles are noninteracting, but the statistical correlations are turned on, the totally uncorrelated result $(N-1)\delta_{\mathbf{q}0}n(p)$ must be corrected by a Pauli kinematic term $-(1-\delta_{\mathbf{q}0})\theta(k_F - p)\theta(k_F - |\mathbf{p} - \mathbf{q}|)$. For the present case of interacting fermions, the dynamical correlations, manifested in virtual single-particle excitations out of the Fermi sea, lead to tails on the step distributions (the F_{cc} terms). The dynamical correlations also produce an overall quenching of the effect, through the strength factor n_o $(0 \leq n_o \leq 1)$. The last line of (27) contains terms of "higher order" which act to correct the various processes just considered.

Upon taking the Bose limit of (27), we recover the corresponding decomposition of the generalized momentum distribution of a Bose fluid as derived in Ref. 8, namely

$$\begin{aligned} n(\mathbf{p},\mathbf{q}) = \delta_{\mathbf{q}0}(N-1)n(p) \\ + (1-\delta_{\mathbf{q}0})(\delta_{\mathbf{p}0} + \delta_{\mathbf{p}\mathbf{q}})Nn_oF_1(q) \\ + (1-\delta_{\mathbf{q}0})(1-\delta_{\mathbf{p}0})(1-\delta_{\mathbf{p}\mathbf{q}})n'(\mathbf{p},\mathbf{q}) \quad , \end{aligned} \tag{28}$$

the non-condensate portion (with $n'(p) = n(p) - Nn_o\delta_{\mathbf{p}0}$) being given by

$$\begin{aligned} n'(\mathbf{p},\mathbf{q}) = F_1(q)[n'(p) + n'(|\mathbf{p}-\mathbf{q}|)] \\ + \frac{\rho^2}{N}\int \rho_1(\mathbf{r}_1\mathbf{r}_1')F_1(\mathbf{r}_1\mathbf{r}_2\mathbf{r}_1')e^{-i\mathbf{p}\cdot(\mathbf{r}_1-\mathbf{r}_1')}e^{-i\mathbf{q}\cdot(\mathbf{r}_1-\mathbf{r}_2)}d\mathbf{r}_1 d\mathbf{r}_2 d\mathbf{r}_1' \quad . \end{aligned} \tag{29}$$

Only the direct contributions to (27) survive. Thus the third and fourth lines are to be omitted, and the form factor $F_{dd}(q)$ is to be identified with the function $F_1(q)$ studied in the earlier paper. The function $F_1(\mathbf{r}_1\mathbf{r}_2\mathbf{r}_1')$ entering (29) is specifically

$$F_1(\mathbf{r}_1\mathbf{r}_2\mathbf{r}_1') = F_1(r)F_1(r') + [1 + F_1(r)][1 + F_1(r')]\{\exp[-P(\mathbf{r}_1\mathbf{r}_2\mathbf{r}_1')] - 1\}. \tag{30}$$

The three addends of (28) have the following interpretations as contributions from distinct scattering processes taking place in the many-body medium (see Fig. 3): The trivial first term corresponds to the null scattering process 3(a). The third term accounts for 3(b), in which a boson is scattered from orbital $\mathbf{p}$ outside the condensate to another non-condensate orbital $\mathbf{p}-\mathbf{q}$ by transferring momentum $\hbar\mathbf{q}$ to a density fluctuation. The other two processes shown, 3(c) and 3(d), are described by the second line of (28) and involve the zero-momentum condensate. They correspond, respectively, to creation of a particle of momentum $-\hbar\mathbf{q}$ by the condensate and to absorption of a particle of momentum $\hbar\mathbf{p}$ into it, a phonon being created in each case to conserve momentum.

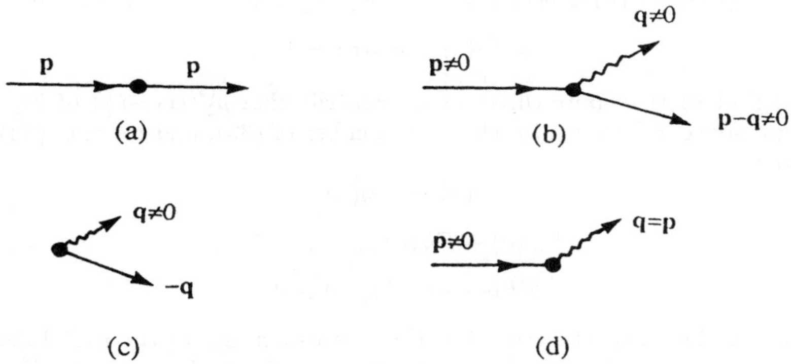

Fig. 3. Scattering processes contributing to the generalized momentum distribution function $n(\mathbf{p},\mathbf{q})$ of a Bose fluid with a macroscopic zero-momentum condensate (cf. Eq. (28) and following text).

Finally, we should address the issue of the practical validity of the sequential relation (25) or (26). In the Bose limit, we may split this relation into two conditions, one applying at $p=0$ (the condensate condition) and the other applying at finite p (and involving terms smaller by a factor $1/N$). These conditions read $1+2F_1(q=0)=0$ and $n^{(2)'}(\mathbf{p},0)+n^{(3)'}(\mathbf{p},0)=0$, respectively. As discussed in Ref. 8, they constrain the choice of correlation factor in the Bose case. In particular, it may be shown that the condensate condition is satisfied for *optimal* Jastrow correlations, even at the HNC/0 level of approximation (i.e., neglecting elementary contributions), but fails for a generic $f(r)$. In the Fermi case, the situation is rather different: the sequential relation (26) is fulfilled identically for *any* Jastrow trial function, and indeed, even if this choice is extended to include multi-body correlations of Feenberg type. The universal satisfaction of (26) is a consequence of the presence of Pauli exchange correlations, implied by the Slater determinant Φ. Technically, this property may be attributed to a Fermi cancellation phenomenon of the type encountered in the FHNC analysis of the static structure function $S(q)$ (cf. Ref. 7). While formally exact, the required cancellation, and hence the sequential relation, is violated to some degree in any finite-order approximation within the HNC/n sequence.

NUMERICAL RESULTS FOR SINGLE-MOMENTUM FORM FACTORS

In correcting the impulse approximation to deep-inelastic scattering from the helium liquids for final-state interactions, Silver[1] has used the simple approximation

$$\rho_2(\mathbf{r}_1\mathbf{r}_2\mathbf{r}_1') \simeq \rho\rho_1(\mathbf{r}_1\mathbf{r}_1')g(|\mathbf{r}_1-\mathbf{r}_2|) \tag{31}$$

for the required two-body density-matrix elements. This approximation may be tested microscopically within variational theory, by hypernetted-chain evaluation of the functions $\rho_2(\mathbf{r}_1\mathbf{r}_2\mathbf{r}_1')$, $\rho_1(\mathbf{r}_1\mathbf{r}_1')$, and $g(r)$.

In the framework of the Jastrow variational theory developed here for Fermi systems of arbitrary level degeneracy, Silver's approximation (31) corresponds to certain rough first estimates of various quantities appearing in the contributions (17) and (18) to (14), en route to the construction of $n(\mathbf{p},\mathbf{q})$ via (27). In more detail, it is equivalent to the replacements

$$\rho_1(\mathbf{r}_1\mathbf{r}_1') \simeq \rho_{1D}(\mathbf{r}_1\mathbf{r}_1')l(\mathbf{r}_1\mathbf{r}_1') \quad , \tag{32}$$

$$F_{dd}(r) + F_{de}(r) \simeq g(r) - 1 \quad , \qquad F_{dd}(r') \simeq 0 \quad , \qquad F_{de}(r') \simeq 0 \tag{33}$$

$$\nu^{-1}l(r) - F_{cc}(r) \simeq 0 \quad , \tag{34}$$

and to neglect of the "pure-three-point" contribution $\rho_2^{(3)}(\mathbf{r}_1\mathbf{r}_2\mathbf{r}_1')$ of Eq. (18). In momentum space, relation (32), the first member of (33), and relation (34) become, respectively,

$$n(p) \simeq n_{Dl}(p) \quad , \tag{35}$$

$$F_{dd}(q) + F_{de}(q) \simeq S(q) - 1 \quad , \tag{36}$$

$$\theta(k_F - p) - F_{cc}(p) \simeq 0 \quad , \tag{37}$$

where $S(q)$ is the static structure function corresponding to the radial distribution function $g(r)$. As may be seen by taking the Bose limit, the corresponding replacements implied in the Bose case are

$$F_1(r) \simeq g(r) - 1 \quad , \qquad F_1(r') \simeq 0 \quad , \tag{33'}$$

$$F_1(q) \simeq S(q) - 1 \quad , \tag{36'}$$

and neglect of the three-point function $F_1(\mathbf{r}_1\mathbf{r}_2\mathbf{r}_1')$.

The most striking feature of Silver's approximation is, of course, its violation of time-reversal invariance, evident in the asymmetric treatment of at least one of the pairs $F_{dd}(r)$, $F_{dd}(r')$ and $F_{de}(r)$, $F_{de}(r')$ by (33), and of the pair $F_1(r)$, $F_1(r')$ by (33').

To make a quantitative judgment of the efficacy of approximations (32)-(37), (33'), and (36'), we have calculated the distribution functions and form factors entering these relations, for liquid ^{4}He and liquid ^{3}He at their respective equilibrium densities ($\rho = 0.0218$ A^{-3} and $\rho = 0.01658$ A^{-3}). For the Bose liquid we have used a Jastrow correlation factor $f(r)$ optimized by a paired-phonon analysis[9]; a Schiff-Verlet form[10] $f(r) = \exp[-(b/r)^5/2]$ was chosen in the Fermi case, with $b = 2.9547$ A (cf. Ref. 11). The numerical evaluations were carried out in the Bose or Fermi hypernetted-chain approximation (HNC/0 or FHNC/0) in which the elementary-diagram contributions to the various quantities are set zero. This approximation should be adequate for the immediate task of testing the ansatz (31).

Our results are summarized in Figs. 4-6.

Figure 4 compares the form factor $F_1(q)$ (solid curve) with the overshoot $S(q)-1$ of the static structure function (dashed curve), testing (36') [or the first member of (33')] in liquid ^{4}He. The two functions have roughly similar shapes, but depart substantially at small q. At $q = 0$, the numerical result for $F_1(q)$ goes to the correct asymptotic result $-1/2$ (as it should, since the 'condensate' condition derived from the sequential relation must be met in HNC/0 when $f(r)$ is optimized). On the other hand, the estimate (36') deviates from the correct limit by a wide margin, since $S(0)$ must vanish for the optimized $f(r)$.

Figures 5 and 6 address the situation in liquid ^{3}He. The shortcomings of the estimate (36) [or, equivalently, the first member of (33)] are revealed in Fig. 5, which shows $S(q) - 1$ and the sum of form factors $F_{dd}(q) + F_{de}(q)$. Within the FHNC/0 approximation, the departure from assumption (36) is clearly exposed and again is particularly apparent at small momenta. The exclusion principle alters the $q = 0$ limits seen in the Bose case (Fig. 4), by virtue of the Fermi cancellation effect mentioned in the preceding section. For a Fermi system described by a Slater-Jastrow

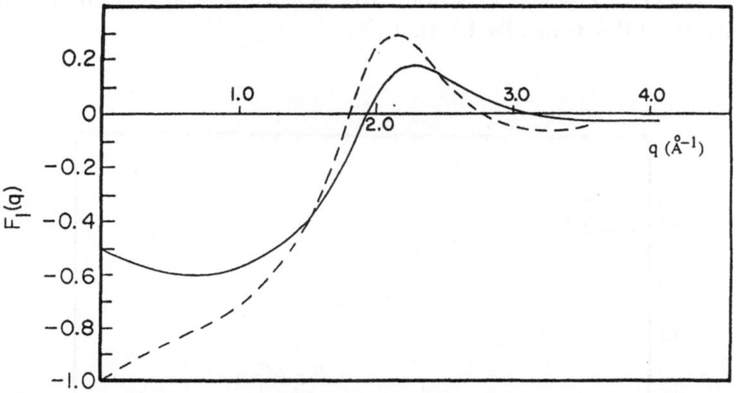

Fig. 4. Test of (36'). Form factor $F_1(q)$ for creation of a particle out of the condensate, in HNC/0 (solid curve) and Silver's (dashed curve) approximations, for liquid ^{4}He at equilibrium density, described by an optimized Jastrow wave function. Static structure function of (36') was evaluated in HNC/0 approximation.

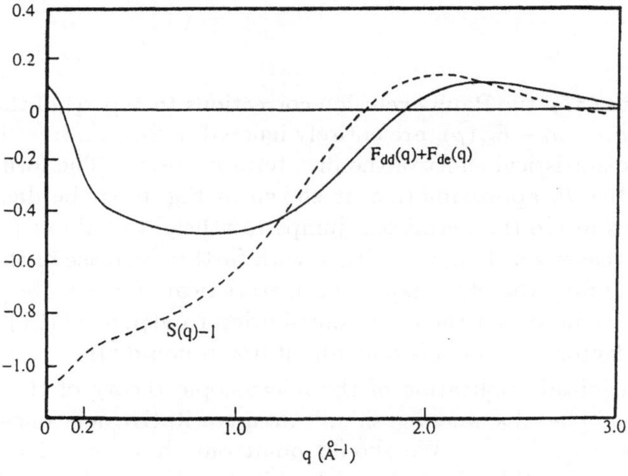

Fig. 5. Test of (36). Sum of form factors $F_{dd}(q) + F_{de}(q)$ (solid curve) compared with overshoot of static structure function $S(q)$ (dashed curve), for liquid ^{3}He at equilibrium density, described within a Jastrow-FHNC/0 approximation.

wave function, this cancellation phenomenon guarantees, at $q = 0$, the properties $X_{de}(q) = X_{ee}(q) = -1$ and, consequently, $S(q) = 0$ and $F_{dd}(q) + F_{de}(q) = 0$. The latter properties are (approximately) reflected in our numerical results for $S(q)$ and $F_{dd}(q) + F_{de}(q)$. However, one does see, in Fig. 5, slight deviations from the correct limiting value of zero, which result from use of the FHNC/0 approximation. The standard FHNC approximants (/0, /4, etc.) are known to disobey the Fermi cancellation rules as a result of the neglect or inconsistent treatment of elementary diagrams.[7]

Fig. 6 shows the momentum distributions $n(p)$ and $n_{Dl}(p)$ and tests the replacement (35) [equivalent to (32)]. The strength factor associated with both of distributions is $n_o = 0.2212$. The two functions are seen to have very similar behavior, but their magnitudes differ typically by 10-15%.

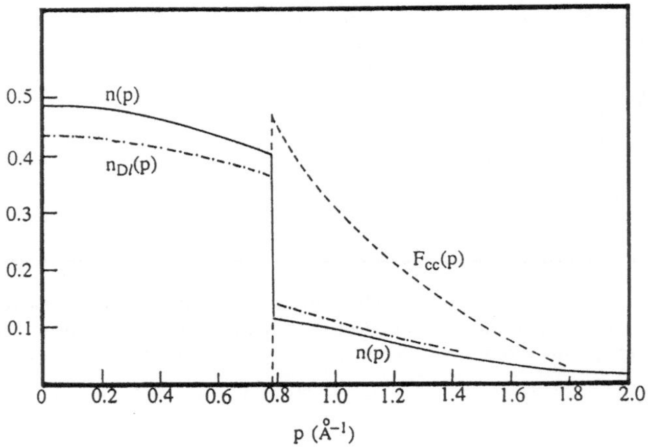

Fig. 6. Test of (35) and (37). Momentum distribution functions $n(p)$ (solid curve) and $n_{Dl}(p)$ (dot-dashed curve) of liquid ^{3}He at equilibrium density, described within a Jastrow-FHNC/0 approximation. Dashed curve shows circular-exchange function $F_{cc}(p)$.

According to (37), the Pauli exclusion corrections to $n(\mathbf{p}, \mathbf{q})$ of the circular type, involving $\theta(k_F - p) - F_{cc}(p)$, are entirely ignored in Silver's treatment – even the trivial kinematic statistical effect of the first term is absent. The form factor $F_{cc}(p)$, evaluated in FHNC/0 approximation, is shown in Fig. 6 as the dashed line. This function vanishes inside the Fermi sea, jumps to a height of about 0.5 at the Fermi surface, and decreases slowly in magnitude with further increase of the wave number p. In general one may therefore expect such statistical effects to be important. On the other hand, we note that their net contribution to $n(\mathbf{p}, \mathbf{q})$ of (27) is proportional to the strength factor n_o, which is only about 0.2 in liquid ^{3}He.

In brief, an initial application of the microscopic theory of the density-matrix elements $\rho_2(\mathbf{r}_1\mathbf{r}_2\mathbf{r}'_1)$ has documented significant quantitative deficiencies of the simple estimate proposed by Silver. We should point out, however, that the final-state corrections evaluated in Silver's theory of deep-inelastic scattering at high momentum and energy transfers may be insensitive to the errors we have noted, and, in particular, to the behavior of the sum of form factors $F_{dd}(q) + F_{de}(q)$ at small q. This possibility is currently under investigation.[12]

FURTHER WORK

We have not reported numerical data on the non-separable terms $n^{(2)'}(\mathbf{p},\mathbf{q})$ and $n^{(3)'}(\mathbf{p},\mathbf{q})$ of (27), nor on the integral term in (29) involving the three-point function $F_1(\mathbf{r}_1\mathbf{r}_2\mathbf{r}_1')$. While these more complicated objects may all be calculated from quantities generated in the FHNC/0 or HNC/0 treatment, their detailed evaluation will be deferred until a scaling or interpolation procedure[13,14] has been implemented for the incorporation of elementary-diagram corrections. At the same time, the Jastrow ansatz will be supplemented by triplet correlations. Work in these directions is currently in progress.[15]

It is worth mentioning that some HNC/0 results for the quantity $F_1(\mathbf{r}_1\mathbf{r}_2\mathbf{r}_1')$ were presented in Ref. 8., results which demonstrate the failure of the HNC/0 treatment to fulfill the condition $\rho_2(\mathbf{r}_1\mathbf{r}_2\mathbf{r}_1) = \rho^2 g(r_{12})$ and thereby provide another reminder of the necessity of including effects of elementary diagrams.

While our derivation of the decomposition (27) for the Fermi generalized momentum distribution was predicated on the Slater-Jastrow choice for the wave function Ψ, the corresponding decomposition (28) for the Bose system (derived here as a limiting case of the Fermi result) was originally obtained for very general Ψ, including the exact ground-state wave function.[8] Indeed, Ref. 8 contains a general asymptotic analysis of the full Bose two-body density matrix in configuration space, as well as the restricted version $\rho_2(\mathbf{r}_1\mathbf{r}_2\mathbf{r}_1')$ considered here.

A more detailed presentation of some of the results of this contribution may be found in a longer article.[16] Further studies along the same lines will be concerned with the full Fermi two-body density matrix.

ACKNOWLEDGEMENTS

This research was supported in part by the Spanish CICyT under its sabbatical program, by the Condensed Matter Theory Program of the Division of Materials Research of the U. S. National Science Foundation under Grant No. DMR-8519077, and by the Theoretical Physics Institute of the University of Minnesota. We thank Richard Silver for many informative discussions and Fernando Arias de Saavedra and Enrique Buendia for furnishing the numerical data for our analysis. JWC is grateful to the Army Research Office, Durham, for providing travel funds.

REFERENCES

1. R. N. Silver, in *Condensed Matter Theories*, Vol. 3, ed. J. S. Arponen, R. F. Bishop, and M. Manninen (Plenum, New York, 1988), p. 131; R. N. Silver, Phys. Rev. B **37**, 3794 (1988); R. N. Silver, Phys. Rev. B **38**, 2283 (1988).
2. M. L. Ristig and J. W. Clark, Phys. Rev. B **14**, 2875 (1976).
3. M. L. Ristig, Nucl. Phys. **A317**, 163 (1979).
4. M. L. Ristig, in *From Nuclei to Particles*, Proceedings of the International School of Physics "Enrico Fermi", Course LVII, Varenna 1981, ed. A. Molinari (North Holland, Amsterdam, 1982), p. 340.
5. S. Fantoni, Nuovo Cimento **A44**, 191 (1978).
6. J. W. Clark, Nucl. Phys. **A328**, 587 (1979).

7. J. W. Clark, in *Progress in Particle and Nuclear Physics*, Vol. 2, ed. D. H. Wilkinson (Pergamon, Oxford, 1979), p. 89.
8. M. L. Ristig and J. W. Clark, Phys. Rev. B, in press.
9. E. Feenberg, *Theory of Quantum Fluids* (Academic, New York, 1969).
10. D. Schiff and L. Verlet, Phys. Rev. **160**, 208 (1967).
11. K. E. Kürten and J. W. Clark, Phys. Rev. B **30**, 1342 (1984).
12. R. N. Silver, private communication.
13. Q. N. Usmani, B. Friedman, and V. R. Pandharipande, Phys. Rev. B **25**, 4502 (1982); Q. N. Usmani, S. Fantoni, and V. R. Pandharipande, Phys. Rev. B **26**, 6123 (1982); M. Pouskari and A. Kalos, Phys. Rev. B **30**, 152 (1984); E. Manousakis, V. R. Pandharipande, and Q. N. Usmani, Phys. Rev. B **31**, 7022 (1985); M. F. Flynn, Phys. Rev. B **33**, 91 (1986).
14. A. Fabrocini and S. Rosati, Nuovo Cimento **D1**, 567 (1982); **D1**, 615 (1982); M. Viviani, E. Buendia, A. Fabrocini, and S. Rosati, Nuovo Cimento **D8**, 561 (1986); M. Viviani, E. Buendia, S. Fantoni, and S. Rosati, Phys. Rev. B **38**, 4523 (1988); S. Rosati, M. Viviani, and E. Buendia, contribution to this volume.
15. F. Arias, E. Buendia, and M. Viviani, private communication.
16. M. L. Ristig and J. W. Clark, to be published.

GROUND STATE ENERGY AND LANDAU PARAMETERS OF SPIN-POLARIZED DEUTERIUM USING GREEN'S FUNCTION METHODS

C. W. Greeff, B. E. Clements, E. F. Talbot and H. R. Glyde

University of Delaware
Department of Physics and Astronomy
Newark, Delaware 19716

ABSTRACT

The Galitskii-Feynman-Hartree-Fock (GFHF) theory is outlined and applied to a system of fully polarized deuterium atoms. The effective interaction in this theory is the GFHF T-matrix. We give a calculation of the single particle energies, $\epsilon(\vec{k})$, and the ground state energy as a function of density. The calculation of $\epsilon(\vec{k})$ uses a new evaluation of the self energy which includes the effects of two hole intermediate states in the effective interaction. We also calculate the Landau parameters both from the T-matrix and from an induced interaction which uses the T-matrix as input. The good agreement of our ground state with the Monte Carlo calculations of Panoff and Clark and the internal consistency of our Landau parameters indicate that the T-matrix gives a good account of the effective interaction in this system.

1. INTRODUCTION

Galitskii-Feynman-Hartree-Fock (GFHF) theory is a formalism, based on the Green's function formulation of many-body theory, which is suitable for the description of particles which interact through potentials with strongly repulsive cores, such as ^{3}He and electron spin polarized Deuterium ($D\downarrow$). Here we present an outline of the theory together with an application to a system of completely spin-polarized deuterium atoms ($D_1^\downarrow$). This includes a calculation of the ground state energy, the single particle energies, and the Landau parameters. We give an improved calculation of the self-energy which includes the effects of two-hole propagation in the intermediate states. Our previous calculations had neglected this contribution to the real part of the self-energy. The calculation of the Landau parameters uses first the T-matrix and then the T-matrix plus an induced interaction. The induced interaction is introduced by following a formalism due to Kadanoff and Baym.[1] We also study the effect of using complex single particle energies in the iterations.

Section 2 of the paper is a brief introduction to GFHF theory including the T-matrix, self-energy and ground state energy. In particular we emphasize the use of the analytic properties of the T-matrix in including hole-hole propagation

Condensed Matter Theories, Volume 5
Edited by V.C. Aguilera-Navarro
Plenum Press, New York, 1990

in the evaluation of the self-energy. In section 3 we describe our model of fully spin-polarized deuterium and present results of the application of GFHF theory to this system. Section 4 gives a short description of our formulation of the induced interaction and presents our results for the Landau parameters. Finally, we give our conclusions as well as some directions for further study.

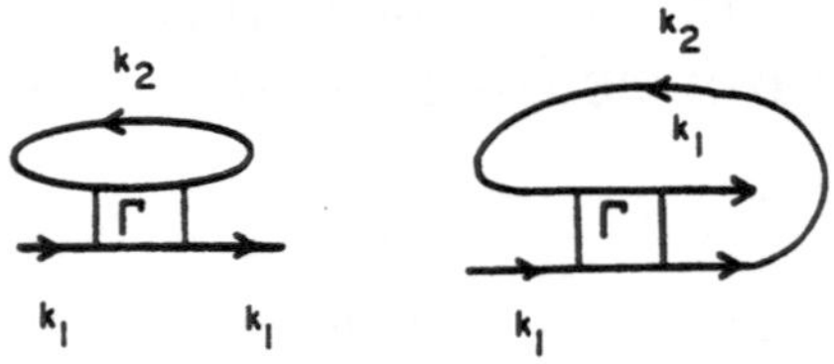

Fig. 1. GFHF Self-Energy.

2. GFHF THEORY

The GFHF self-energy (fig. 1) is obtained from the usual Hartree-Fock self-energy by replacing the bare interaction by the sum of the ladder diagrams, which we refer to as the GFHF T-matrix. This is motivated by analogy with the scattering of two particles in empty space. In that case these diagrams represent the Born series summed to all orders, which allows for the modification of the wavefunction due to the interparticle potential. In the same way the T-matrix is expected to give a good representation of the residual interaction between particles in the fluid since the short-range correlations due to the strong repulsive cores are taken into account.

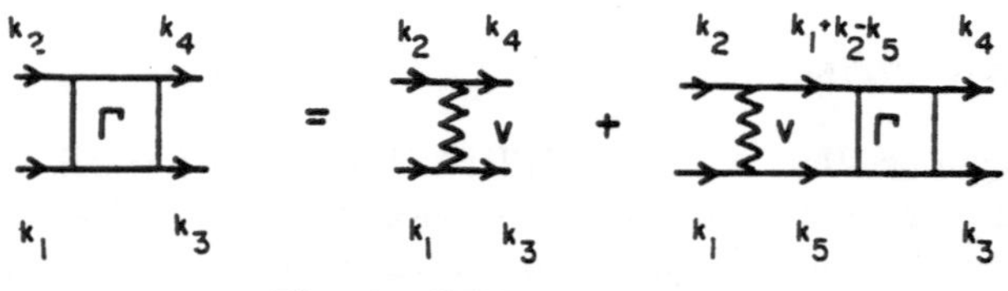

Fig. 2. GFHF T-Matrix.

The GFHF T-matrix is obtained by summing the ladder diagrams as in figure 2. Thus we have the equation,

$$\Gamma(k_1, k_2; k_3, k_4) = V(k_1 - k_3) + i \int \frac{d^4 k_5}{(2\pi)^4} V(k_1 - k_5) \times G(k_5) G(k_1 + k_2 - k_5) \Gamma(k_5, k_1 + k_2 - k_5; k_3, k_4). \quad (1)$$

where $V(k)$ is the Fourier transform of the interatomic potential. Noting that Γ depends on the frequency only through $E = \omega_1 + \omega_2$ and using Hartee-Fock form for the Green's functions,

$$G(\vec{k},\omega) = \frac{1-n(\vec{k})}{\omega - \epsilon(\vec{k}) + i\eta} + \frac{n(\vec{k})}{\omega - \epsilon(\vec{k}) - i\eta}, \tag{2}$$

the frequency integral in (1) can be done explicitly, giving

$$\begin{aligned}\Gamma(\vec{k_1},\vec{k_2};\vec{k_3},\vec{k_4},E) &= V(\vec{k_1}-\vec{k_3}) \\ &+ \int \frac{d^3k_5}{(2\pi)^3} V(\vec{k_1}-\vec{k_5}) \left[\frac{(1-n_5)(1-n_{1+2-5})}{E-\epsilon_5-\epsilon_{1+2-5}+i\eta} - \frac{n_5 n_{1+2-5}}{E-\epsilon_5-\epsilon_{1+2-5}-i\eta}\right] \\ &\times \Gamma(\vec{k_5},\vec{k_1}+\vec{k_2}-\vec{k_5};\vec{k_3},\vec{k_4},E)\end{aligned} \tag{3}$$

where ϵ_1 and n_1 are shorthand for $\epsilon(\vec{k_1})$ and $n(\vec{k_1})$, the single particle energy and occupation number respectively. Equation (3) must be solved numerically as detailed in ref. (2). The above equations define the direct T-matrix, which is diagonal in the spin indices,

$$\Gamma_{\sigma_1\sigma_2\sigma_3\sigma_4} = \Gamma \delta_{\sigma_1\sigma_3}\delta_{\sigma_2\sigma_4}. \tag{4}$$

One also defines an exchange-symmetrized interaction,

$$\begin{aligned}\Gamma^{SY}_{\sigma_1\sigma_2\sigma_3\sigma_4}(k_1,k_2;k_3,k_4) &= \Gamma(k_1,k_2;k_3,k_4)\delta_{\sigma_1\sigma_3}\delta_{\sigma_2\sigma_4} \\ &\quad -\Gamma(k_1,k_2;k_4,k_3)\delta_{\sigma_1\sigma_4}\delta_{\sigma_2\sigma_3}.\end{aligned} \tag{5}$$

Or, more compactly,

$$\Gamma^{SY}_{\sigma_1\sigma_2\sigma_3\sigma_4} = \Gamma^D \delta_{\sigma_1\sigma_3}\delta_{\sigma_2\sigma_4} - \Gamma^E \delta_{\sigma_1\sigma_4}\delta_{\sigma_2\sigma_3}. \tag{6}$$

From fig.(1) the self-energy is given by

$$\Sigma(\vec{k_1},\omega_1) = -i\sum_{\sigma_2}\int \frac{d^4k_2}{(2\pi)^4}\Gamma^{SY}_{\sigma_1\sigma_2\sigma_1\sigma_2}(k_1,k_2;k_1,k_2)G(k_2) \tag{7}$$

The spin sum replaces Γ^{SY} with $g\Gamma^s = g\Gamma^D - \Gamma^E$ where Γ^s is the spin symmetric interaction (in the sense of the Landau parameters) and g is the number of spin states. In evaluating the frequency integral one must use the analytic properties of Γ which follow from,

$$\Gamma(E) = \int_0^\infty \frac{dx}{2\pi}\left[\frac{\Gamma_1(x)}{E-2\mu-x+i\eta} - \frac{\Gamma_2(x)}{E-2\mu+x-i\eta}\right] \tag{8}$$

where Γ_1 and Γ_2 are spectral weights for the propagation of two-particle and two-hole states respectively. Here μ is the chemical potential. Thus we can write (7) symbolically as

$$\begin{aligned}\Sigma &= -i\int \frac{d^4k_2}{(2\pi)^4}(\Gamma_{pp}+\Gamma_{hh})(G_p+G_h) && (9)\\ &= \Sigma_{pph} + \Sigma_{hhp} && (10)\end{aligned}$$

where

$$\Sigma_{pph} = -i\int \frac{d^4k_2}{(2\pi)^4}\Gamma_{pp}G_h \tag{11}$$

$$\Sigma_{hhp} = -i\int \frac{d^4k_2}{(2\pi)^4}\Gamma_{hh}G_p. \tag{12}$$

The terms involving $\Gamma_{pp}G_p$ and $\Gamma_{hh}G_h$ vanish because their poles are entirely in the lower and upper half-planes respectively. Carrying out the integration explicitly we have,

$$\begin{aligned} Re\Sigma(\vec{k_1},\omega_1) &= g\int \frac{d^3k_2}{(2\pi)^3}\Big[Re\Gamma^s(\vec{k_1},\vec{k_2};\vec{k_1},\vec{k_2},\omega_1+\epsilon_2)n(k_2) \\ &\quad -\mathcal{P}\int_{-\infty}^{2\mu}\frac{dE}{\pi}\frac{Im\Gamma^s(\vec{k_1},\vec{k_2};\vec{k_1},\vec{k_2},E)}{\omega_1+\epsilon_2-E}\Big] \end{aligned} \tag{13}$$

$$\begin{aligned} Im\Sigma(\vec{k_1},\omega_1) &= g\int \frac{d^3k_2}{(2\pi)^3}Im\Gamma^s(\vec{k_1},\vec{k_2};\vec{k_1},\vec{k_2},\omega_1+\epsilon_2) \\ &\quad \times\left[\Theta(\mu-\epsilon_2)-\Theta(2\mu-\omega_1-\epsilon_2)\right]. \end{aligned} \tag{14}$$

The second term in (13) corresponds to $Re\Sigma_{hhp}$. In previous works on this theory we had neglected this term on the basis of a phase space argument.[3] We refer to the case without $Re\Sigma_{hhp}$ as the Brueckner-Hartree-Fock (BHF) case because the Brueckner G-matrix has no hole-hole spectral weight. This stems from the fact that in the equation for the Brueckner G-Matrix the second term in square brackets in our eqn. (3) is absent.[4] Here we present calculations of both the Brueckner-Hartree-Fock and full GFHF self energies.

One would like to use the resulting self energy as input to the next iteration of a self-consistent calculation. However the use of the full freqency dependent self-energy would make the resulting frequency integrals intractable. We therefore define single particle energies by,

$$\epsilon(\vec{k}) = \epsilon^0(\vec{k}) + \Sigma(\vec{k},\epsilon(\vec{k})) \tag{15}$$

or,

$$\epsilon(\vec{k}) = \epsilon^0(\vec{k}) + Re\Sigma(\vec{k},\epsilon(\vec{k})) \tag{16}$$

which are put back into the Green's functions, (2). The first of these, (15), gives a complex $\epsilon(\vec{k})$ and thus results in a spectral function for G which is Lorentzian, centered at the quasi-particle energy. The second, (16), gives a delta function for the spectral weight, corresponding to quasi-particles with infinite lifetimes.

Our program, then, consists of iterating equations (3), (13), (14) and (15) or (16) until the resulting single particle energies are self-consistant. In performing these iterations one has four choices corresponding to the Brueckner or full self-energies and to the real or complex single particle energies. Actually we do not consider the case of the full self-energy using complex $\epsilon(\vec{k})$. This is because, in this case, the two-hole contribution to $Im(\Gamma)$ does not cut off sharply at $E = 2\mu$ and the expression (13) for the self energy is not valid.

From the general relation for the ground state energy and the form (2) for the Green's functions one obtains the following relation for the energy per particle,

$$E = \int \frac{d^3k}{(2\pi)^3}\left[\epsilon^0(\vec{k}) + \frac{1}{2}Re\Sigma(\vec{k},\epsilon(\vec{k}))\right]n(\vec{k}). \tag{17}$$

Corresponding to the three options mentioned above for the single particle energies, there are three possibilities for the ground state energy. We denote these by E, E_1 and E_{1c} corresponding to the GFHF, BHF and BHF with complex $\epsilon(\vec{k})$ cases respectively.

3. ELECTRON SPIN-POLARIZED DEUTERIUM

We consider a model of doubly spin-polarized deuterium. In this model the electrons of the deuterium atoms are considered to be completely polarized by an external magnetic field, thus preventing combination into D_2 molecules. For the interatomic potential between two spin down deuterium atoms, we use the Silvera fit as given by Friend and Etters.[5]

The nuclei have a net spin of one so in general there are 3 spin states. Here we consider only the case where just one nuclear spin state is occupied (nuclear polarization). This may best represent the experimental situation since otherwise hyperfine interactions will unpolarize the electrons resulting in combination into D_2.

In fig. 3 we show the full GFHF ground state energy $E = E_{pph} + E_{hhp}$, the energy $E_1 = E_{pph}$ in which only the Brueckner term is retained and the energy E_{1c} in which only the Brueckner term is retained but complex single particle energies are used in the iterations. The difference between E and E_1 displays the contribution from Σ_{hhp}, the two hole, one particle term. The contribution from Σ_{hhp} is small at low density and increases as n increases. On the basis of a hole-line expansion we expect $\Sigma_{hhp} << \Sigma_{pph}$ at low density. The Σ_{hhp}, however, is clearly significant at high density, $n \gtrsim 4 \times 10^{-3}$ Å^{-3}, in $D_1^{\downarrow}$. From the difference between E_1 and E_{1c} we see that including the imaginary part of $\epsilon(\vec{p})$ in the iterations does not make much difference to E_1. We could not use complex $\epsilon(\vec{p})$ in the full E since the $Re\Sigma_{hhp}$ term (second term in (13)) is valid only for a real single particle spectrum. Otherwise the energy integration does not cut off at 2μ and one cannot separate $Re\Gamma_{hh}$. We attempted to use a complex $\epsilon(\vec{p})$ in the full E and obtained an $E(n)$ which rose rapidly as n increased.

In fig. 4 we compare the full GFHF E with the variational Monte Carlo (MC) values obtained by Panoff an Clark.[6] The agreement is good, especially given the extremely large cancellation between the kinetic and potential energies. We found similar results for $D_3^{\downarrow}$ but in that case the GFHF energy lies somewhat below the MC value.

The GFHF energy could be well fitted by the function

$$E(n) = E_0 + Ax^2 + Bx^3 + Cx^4 \tag{18}$$

where $x = (n - n_0)/n_0$, $n_0 = 3.26 \times 10^{-3}$ Å^{-3} is the saturation density, $E_0 = .414$, $A = 1.54$, $B = 1.16$ and $C = -.88$ in K. The single particle energy curve, $\epsilon(\vec{p})$, yields a total effective mass $m^*(\vec{p}) \approx 0.9$ independent of $\vec{p}$ with no peaking or exceptional behavior at the Fermi surface.

4. INDUCED INTERACTION AND LANDAU PARAMETERS

In this section we report on recent progress in calculating the induced interaction for spin polarized deuterium. We begin with a brief discussion of the manner in

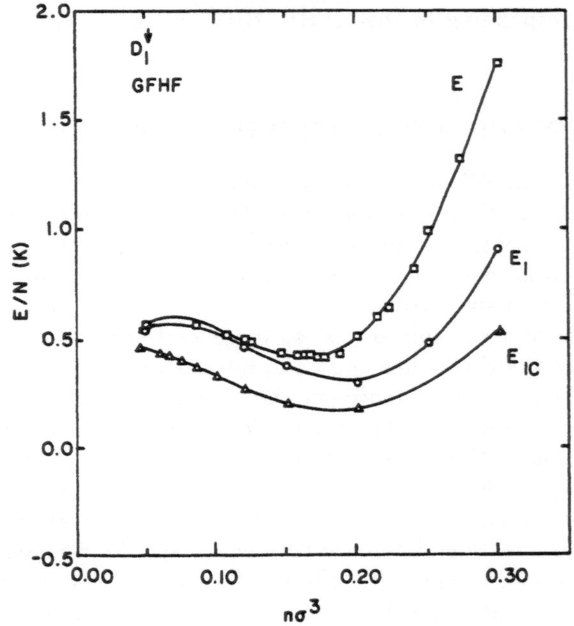

Fig. 3. Ground state energy of Spin Polarized Deuterium ($\sigma = 3.689$ Å).

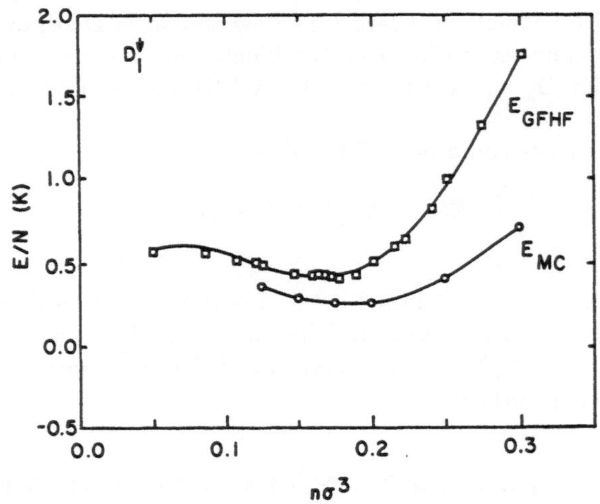

Fig. 4. Ground state energy of Spin Polarized Deuterium ($\sigma = 3.689$ Å).

which we introduce the induced interaction into the GFHF formalism and close with a comparison of the Landau parameters calculated directly from the T-matrix, from the T-matrix plus induced interaction, and indirectly by first calculating the compressibility.

Baym and Kadanoff[1] have stressed the importance of using approximations in which number, energy, momentum and angular momentum conservation laws are built in. They showed that for the dynamic susceptibility, $\chi(\vec{Q},\omega)$, which describes the system's response to an external perturbation in which momentum $\vec{Q}$ and energy ω are transferred to the system, to be conserving it must be derivable from an integral equation which we depict diagramatically in fig. 5. The $\chi(\vec{Q},\omega)$ follows upon integrating $\bar{\chi}$ over the 4-momenta K and K' and $I(k_1, k_2; k_3, k_4)$ is related to the self energy through,

$$I(k_1, k_2; k_1, k_2) = i\frac{\delta\Sigma(k_1)}{\delta G(k_2)} \tag{19}$$

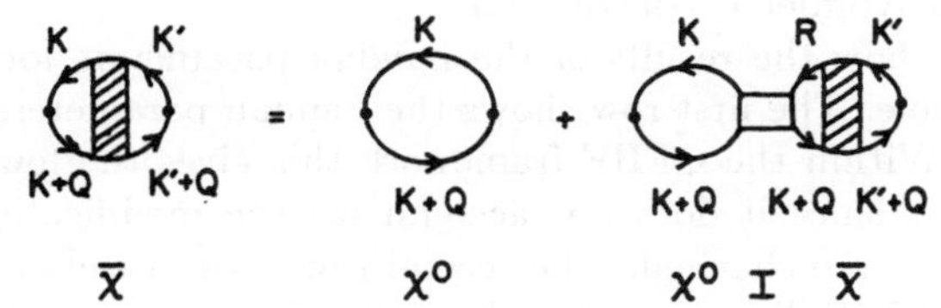

Fig. 5. Diagrammatic Structure of $\chi(\vec{Q},\omega)$

(Since we are considering the case of $D_1^{\downarrow}$ with only one spin state, we leave off spin indices.) Using the GFHF approximation to the self-energy (fig. 1) one finds:

$$\begin{aligned} I(K+Q,R;K,R+Q) &= \Gamma(K+Q,R;K,R+Q) \\ &+ i\int\frac{d^4P}{(2\pi)^4}\Gamma(K+Q,P+R;R+Q,P+K) \\ &\times G(P+K)G(P+R)\Gamma(P+K,R;P+R,K). \end{aligned} \tag{20}$$

Invoking eqns. (2) and (8) allows the frequency integrations to be immediately carried out. From simple phase space arguments we assert that the largest term resulting from the frequency integration can be expressed in the form of an induced interaction as first introduced by Babu and Brown.[7] In the present work we shall focus on the Landau limit where $\vec{Q} \to 0$, $\omega \to 0$ and $|\vec{K}| = |\vec{R}| = k_f$, the Fermi wave vector. In that case I depends only on the Landau angle, Θ_L, between the vectors $\vec{K}$ and $\vec{R}$:

$$
\begin{aligned}
I(\hat{K},\hat{R};\hat{K},\hat{R}) &= \Gamma(\hat{K},\hat{R};\hat{K},\hat{R}) - \int \frac{d^3P}{(2\pi)^3}\Gamma^2(\vec{K},\vec{P}-\vec{q}/2;\vec{K}-\vec{q},\vec{p}+\vec{q}/2,2\epsilon_f) \\
&\quad \times \left[\frac{n(\vec{P}+\vec{q}/2)-n(\vec{P}-\vec{q}/2)}{\epsilon(\vec{p}+\vec{q}/2)-\epsilon(\vec{P}-\vec{q}/2)}\right]
\end{aligned}
\tag{21}
$$

where $\vec{q} = \vec{K} - \vec{R}$. Finally, we Legendre expand $I(\hat{K},\hat{R};\hat{K},\hat{R})$ and express the result in terms of the well known Landau parameters,

$$
F_l^s = \frac{dn}{d\epsilon}\frac{2l+1}{2}\int_{-1}^{1} d(cos\Theta_L)P_l(cos\Theta_L)I^s(cos\Theta_L)
\tag{22}
$$

Here $(dn/d\epsilon) = n(\frac{3}{2\epsilon_f})$ is the density of states per unit volume at ϵ_f. The superscript s denotes the spin symmetric contribution.

In table 1 we show the results of the Landau parameters for $D_1^{\downarrow}$ at saturation volume; 185 cc/mole. The first row shows the Landau parameters using the GFHF T-matrix alone. Within the GFHF framework this gives the lowest order approximation to the F's since it does not account for the modification of Γ when the occupation numbers are changed. The second row is the Landau parameters corresponding to the GFHF T-matrix plus the induced interaction obtained from (22). The final row gives the Landau parameter F_0^s obtained from

$$
(n\kappa)^{-1} = n(\frac{dn}{d\epsilon})^{-1}(1+F_0).
\tag{23}
$$

Here $(n\kappa)^{-1} = V\partial^2E/\partial V^2$ is the compressibility ($(n\kappa)^{-1} = n^2\partial^2E/\partial n^2 = 2A$ at saturation from (18)). The calculation of F_0^s from $E(n)$ and from (22) is of higher order.

We find that the induced interaction in (21) is generally quite small for the spin-polarized case and we attribute this to a freezing out of the spin fluctuations. Including induced terms F_0^s is less negative in agreement with the compressibility result.

In table 1, we also show the effective mass calculated within the various approximations. The effective masses displayed in rows 1 and 2 are determined from $m^* = (1 + F_1^s/3)$ and row 3 follows from the total derivative of the single particle spectrum, $\epsilon(\vec{k})$. These results are in reasonable agreement with the value of 0.87 determined by variational Monte Carlo studies of Dave et al.[8]

We close this section by noting that although much semi-phenomenological work has been devoted to calculating the induced interaction,[9] few first principle calcu-

Table 1. Landau Parameters and Effective Masses

	F_0^s	F_1^s	F_2^s	F_3^s	F_4^s	F_5^s	m^*
T-matrix	-1.36	-0.25	-0.52	0.94	0.13	0.32	0.92
T-matrix plus Induced Intersection	-0.94	-0.53	-0.49	0.52			0.82
Compressibility and $\epsilon(k)$ Calculations	0.1 ± 0.3						0.9

lations exist. Perhaps the most notable exceptions are the calculations of Polls et al.[10], using a Brueckner G-matrix and the correlated basis function method used by Krotscheck[11] , both for spin-polarized ^{3}He. In ref. 10, the induced interaction is obtained by solving the crossed channel Bethe-Salpeter equation. In that formalism the Baym-Kadanoff result is just the first iteration of the Bethe-Salpeter equation. Thus the present work is more of an attempt to use strict conserving approximations whereas the work of ref. 10 is more consistent with a parquet approach to diagram resummation.

5. CONCLUSION

Given the good values for the ground state energy and the Landau parameters, we expect that the GFHF T-matrix is a good starting point for the description of interactions in $D_1^\downarrow$. In particular, the small difference between F_0^s as calculated from the T-matrix and from the compressibility indicates that collective effects are small. In part this is because spin fluctuations are frozen out in the fully polarized fluid. We expect, then, that our induced interaction, when extended to finite Q and ω, should give a good description of the dynamics in this system. This in turn can be used to give an estimate of the influence of collective behavior on single particle properties such as $\epsilon(\vec{k})$ and m^*. These are lines we intend to pursue.

Regarding the use of a complex $\epsilon(\vec{k})$, we note that, from fig. 3, the use of complex single particle energies does not make a large difference to the ground state energy. However, our inability to use complex energies in the calculation of the full GFHF self-energy is indicative of fundamental difficulties. For example, the Green's function (2) with complex $\epsilon(\vec{k})$ does not obey the Lehmann representaion, which requires that $ImG(\vec{k},\omega)$ change sign when ω crosses μ. Thus we conclude that there is no consistent way to keep $Im\Sigma(\vec{k},\omega)$ without maintaining its full frequency dependence. We noted earlier that this would give rise to substantial computational difficulties in an iterative procedure. However there is no essential difficulty if, having determined $\epsilon(\vec{k})$ self-consistently as we have here, one then uses this as input to a calculation of the full frequency dependent self-energy. This would allow a determination of such properties as spectral functions, momentum distribution and the residues of the quasi-partricle poles at this level of approximation.

ACKNOWLEDGEMENTS

The authors wish to thank Dr. Pitor Findeisen for his computational assistance. Support from the U. S. Department of Energy, Office of Basic Energy Sciences, under contract No. DE-FG02-84ER45082 is also gratefully acknowledged.

REFERENCES

1. G. Baym and L. P. Kadanoff, Phys. Rev. 124, 287 (1961).

2. H. R. Glyde and S. I. Hernadi, Phys. Rev. B28, 141 (1983).

3. H. R. Glyde and S. I. Hernadi, Phys. Rev. B29, 3873 (1984) and in Condensed Matter Theories, Vol. 1, ed. F. B. Malik (Plenum, N.Y. (1986).

4. J. Hüfner and C. Mahaux, Ann. Phys. N.Y. 73, 525-577 (1972).

5. D. G. Friend and R. E. Etters, J. Low Temp. Phys. 39, 409 (1980).

6. R. M. Panoff and J. W. Clark, Phys. Rev. 36, 5527 (1987).

7. S. Babu and G. E. Brown, Ann. Phys. N.Y. 78, 1-38 (1973).

8. R. D. Dave, J. W. Clark and R. M. Panoff, Preprint.

9. C. Sanchez-Castro and K. S. Bedell, J. Low Temp. Phys. 75, 95 (1989) and references therein.

10. W. H. Dickhoff, H. Müther and A. Polls, Phys. Rev. B36, 5138 (1987).

11. E. Krotscheck, Phys. Rev. A26, 3536 (1982).

QUANTUM MOLECULAR DYNAMICS SIMULATION OF ELECTRON BUBBLES IN A DENSE HELIUM GAS

Rajiv K. Kalia and P. Vashishta

Materials Science Division, Argonne National Laboratory, Argonne, IL 60439, USA

S. W. de Leeuw

Universiteit van Amsterdam, Amsterdam, The Netherlands

John Harris

Institut für Festkörperforschung der Kernforschungsanlage, D-5170 Jülich, West Germany

1. INTRODUCTION

In recent years, mixed quantum-classical systems consisting of excess electrons interacting with classical many-body systems at finite temperatures have been studied extensively with computer-simulation techniques[1]. The simulation methods for these systems include the path integral Monte Carlo[1] or molecular dynamics[2] and dynamical simulated annealing[3]. The latter can only provide the ground-state static properties of the quantum particles. The path integral approach has been used successfully to calculate the equilibrium properties, but the study of time correlation functions[4] is not reliable at long times. However, the recently developed quantum molecular dynamics method, which deals directly with the time-dependent Schrödinger equation, contains all the dynamical information for quantum particles.

2. QUANTUM MOLECULAR DYNAMICS METHOD

Quantum molecular dynamics method[5] provides the real-time dynamics of quantum and classical particles in mixed systems at finite temperatures through the numerical solutions of the time-dependent Schrödinger equation for quantum particles and Newton's equations of motion for classical particles. To understand this technique, consider a quantum particle of mass m described by the wave function, $\psi(\mathbf{r},t)$, interacting with a classical system of N particles with masses M_i and positions $\{\mathbf{R}_i(t)\}$. For the quantum particle we have

Condensed Matter Theories, Volume 5
Edited by V.C. Aguilera-Navarro
Plenum Press, New York, 1990

$$i\frac{\partial\psi(\bar{r},t)}{\partial t} = \left(-\frac{\nabla^2}{2m} + V(\bar{r})\right)\psi(\bar{r},t)\,; \quad V(\bar{r}) = \sum_{i=1}^{N} v(\bar{r} - \bar{R}_i) \tag{1}$$

where v is the interaction potential between the quantum and classical particles and we use atomic units, $\hbar = e = 1$. In the Born-Oppenheimer approximation, the positions of the classical particles evolve from Newton's equations of motion:

$$M_i \ddot{\bar{R}}_i = -\nabla_i U(\{\bar{R}_i\}) - \nabla_i \int d\bar{r}\, |\psi(\bar{r},t)|^2\, v(\bar{r} - \bar{R}_i) \tag{2}$$

where U is the potential energy for the classical particles. Choosing a small time step Δt, the solution of Eq. (1) can be written as[6]

$$\psi(\bar{r},t+\Delta t) = e^{i\Delta t\nabla^2/4m}\, e^{-i\Delta t V}\, e^{i\Delta t\nabla^2/4m}\, \psi(\bar{r},t) + O[(\Delta t)^3] \tag{3}$$

Equation (3) is solved with fast Fourier transform (FFT) techniques[6]. First, note

$$e^{i\Delta t\nabla^2/4m}\psi(\bar{r},t) = \sum_{\bar{k}} \psi(\bar{k},t)\, e^{-i\Delta t k^2/4m} e^{i\bar{k}.\bar{r}} \tag{4}$$

which means multiplying the Fourier transform of $\psi(\mathbf{r},t)$, i. e., $\psi(\mathbf{k},t)$ with exp(-iΔt $\mathbf{k}^2$/4m), followed by an inverse FFT. Next, the outcome of Eq. (4) is multiplied by exp(- iΔt V). Finally, the FFT of the resultant of the last step is multiplied with exp(-iΔt $\mathbf{k}^2$/4m) and then the inverse FFT is taken. These three steps are repeated to obtain the time evolution of the wave function. The classical equations of motion can be integrated numerically with one of several available algorithms[7].

For systems with broken symmetry due to the presence of surfaces or an external electric field, the use of periodic boundary conditions in the broken-symmetry direction is inappropriate and so the first and third steps in the time-stepping operation in Eq. (3) cannot be executed with the FFT method. For these physical situations a new QMD algorithm[8] has been developed.

3. AN EXCESS ELECTRON IN A DENSE HELIUM GAS

An injected electron in a sufficiently dense helium gas tends to localize in the form of a bubble as a result of the strongly repulsive electron-helium interaction[9] at short and intermediate distances. Recently we investigated the electron bubble formation using the QMD scheme[10]. For the electron in helium problem, the helium particles interact with each other via a two-body Lennard-Jones potential with parameters ε = 10.22K and σ = 2.576 Å, and with the electron via a pseudopotential[9].

The QMD simulations[10] were performed at 77K for reduced helium densities $n = \rho\sigma^3$ = 0.1, 0.17, and 0.25 which correspond to ρ = 0.61, 1.0, and 1.46 x 10^{22} cm^{-3}. Systems with n = 0.1 and 0.17 contained 512 helium atoms while the simulations at the highest helium density were carried out with 64 and 140 particles. These gave similar results though the smaller system showed evidence of finite-size effects. The electron wave packet was propagated with fast Fourier transforms on 32^3 grid points and with time step, Δt, in the range of 0.2 - 0.5 a.u. The total energy was conserved to better than 0.1% over 10^6 time steps. Some simulations with 64^3 grid points gave identical results. Classical molecular dynamics for helium atoms was performed in the canonical ensemble[11].

4. RESULTS

Figure 1 shows the electron-helium radial distribution function, G(r), measured relative to the center-of-mass of the electron wave packet for n = 0.25. At this high helium density, there are no helium atoms up to r ~ 12 a.u. and this excluded-volume effect is an indication of an electron bubble. The excluded volume is also present at n = 0.17. The shape parameter[10] for the electron bubble indicates that the bubble is spherical. From the excluded-volume region in G(r) we estimate the size of the bubble to be ~ 12 a.u. At the lowest density, n = 0.1, G(r) does not display the excluded-volume behavior as the wave packet de-localizes.

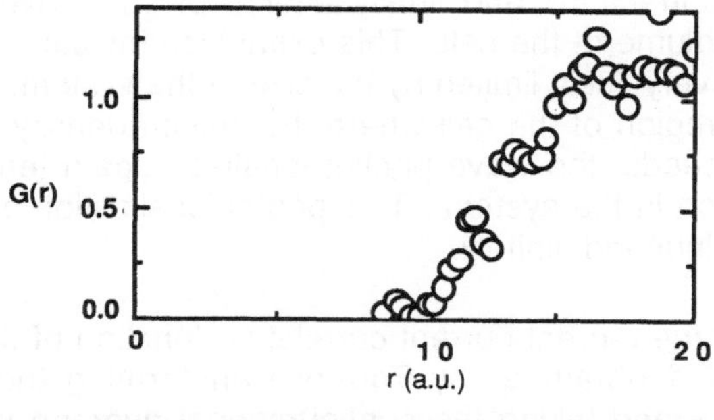

Figure 1. Electron-helium radial distribution function, G(r), for n = 0.25.

Next, let us examine the time variation of the extent of the self-trapped electron. At n = 0.17, the participation ratio remains small and almost constant over the entire simulation, indicating that the electron is localized. The behavior is quite similar at n = 0.25, but very different at n = 0.10. Figure 2 shows the participation ratio, $p(t) = (\Omega \int d\mathbf{r}\ |\psi(\mathbf{r},t)|^4)^{-1}$, normalized to the volume of the MD cell for n = 0.1. This simulation was started by expanding the length scale of the final configuration of the n = 0.17 run to a value corresponding to n = 0.1 and projecting out the new electron ground state, which corresponds to an almost spherical bubble. Therefore the participatin ratio starts

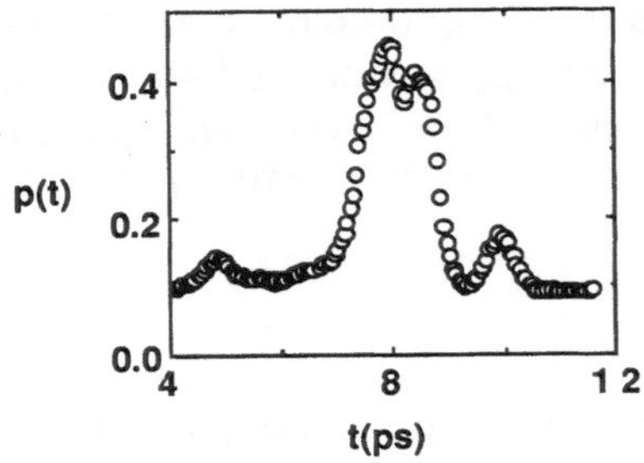

Figure 2. Participation ratio for n = 0.1 as a function of time.

out at a value determined primarily by the bubble size at n = 0.17 scaled by the increase in the cell size in changing the system density from n = 0.17 to 0.1. It is observed that the bubble quickly expands to a volume that is a sizeable fraction of the cell volume. After 4 ps the electron wave packet undergoes large expansions and contractions and finally after 8 ps the wave packet occupies almost the entire volume of the cell. This expansion indicates de-localization of the electron that is very likely limited by the size of the system. The wave packet attempts to find a region of the cell where the helium density is less. Thus, as the simulation proceeds, the wave packet localizes again temporarily because of thermal fluctuation in the system. This partial localization and de-localization is expected to continue indefinitely.

Figure 3 shows the current-current correlation function of the electron, $\chi''(\omega)$, for n = 0.25 and 0.1 obtained by Fourier transforming the time correlation function $<\nabla\psi_t.\nabla\psi_0>$ and taking the configurational average within the Franck-Condon[6] approximation. For n = 0.25, $\chi''(\omega)$, which is simply related to the optical absorption spectrum, displays a significant structure which shifts to a lower energy as the helium density is reduced to 0.17. These peaks are present throughout the simulations. At n = 0.1, when the wave packet is de-localized, the structure disappears and only a background is observed.

At T = 77K we have also calculated[8] the electron mobility at a helium density of 1.25×10^{22} cm^{-3}. Applying an electric field of 2.6×10^5 volt/cm in the x direction, it is observed that the electron drifts with a velocity of $(2.2 \pm 0.5) \times 10^4$ cm/s. The applied field is small since it does not produce any noticeable changes in G(r). From this simulation we obtain the electron mobility to be (0.08 ± 0.02) cm^2/volt-s which is in good agreement with the extrapolated experimental value[12] (0.1 cm^2/volt-s).

5. CONCLUSION

In this paper we have presented the QMD simulation technique for studying the dynamics of quantum particles in mixed systems at finite temperatures. This technique is applied to simulate the behavior of an excess electron in helium gas at 77K. It is found that the electron localizes in the form of a nearly spherical bubble of approximately 7 Å in radius above a critical density of 0.6 x 10^{22} cm^{-3}. The bubble possesses quasi bound excited states and intra-bubble dipole transitions between these states give rise to a pronounced structure in the optical absorption spectrum. Below the critical density the electron percolates through the helium gas and displays a featureless excitation spectrum.

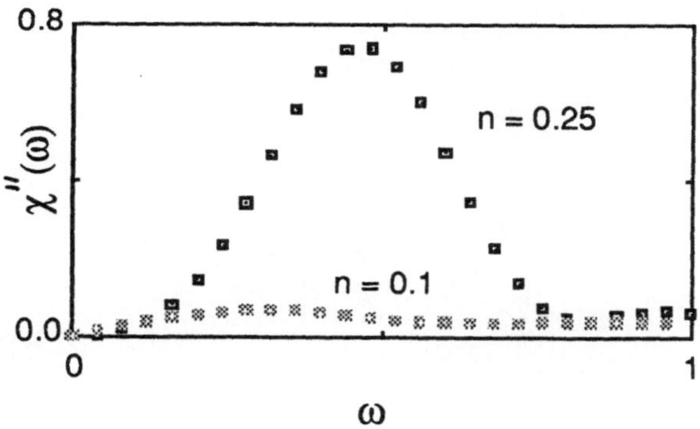

Figure 3. Imaginary part of the current-current correlation function as a function of energy in electron volts for helium densities n = 0.25 and 0.1. The peak at n = 0.25 is due to intra-bubble transitions.

ACKNOWLEDGEMENTS

We would like to thank L. H. Yang, M. H. Degani, and J. P. Rino for useful discussions. This work was supported by the U. S. DOE, BES-Materials Sciences Contract No. W-31-109-ENG-38. One of us (R.K.K) would also like to acknowledge grants of CPU time on the MFE Cray 2 at Livermore through a Grand Challenge proposal. R. K. K. and P. V. would like to acknowledge partial travel support from the U. S. Army Research Office.

REFERENCES

1) B. J. Berne and D. Thirumalai in Annual Reveiew of Physical Chemistry, vol 37, eds. H. L. Strauss, G. T. Babcock, and C. Bradley Moore, (Annual Reviews Inc. Palo Alto, 1986) pp. 401-424.
2) M. Parrinello and A. Rahman, J. Chem. Phys. **80**, 860 (1984).
3) R. Car and M. Parrinello, Phys. Rev. Lett. **55**, 2471 (1985).
4) J. D. Doll and D. L. Freeman, J. Phys. Chem. **92**, 3278 (1988) .
5) A. Selloni, P. Carnevali, R. Car, and M. Parrinello, Phys. Rev. Lett. **59**, 823 (1987).
6) M. D. Feit, J. A. Fleck, and A. Steiger, J. Comp. Phys. **47**, 412 (1982) .
7) A. Rahman, Correlation Functions and Quasiparticle Interactions in Condensed Matter, ed. J. Woods Halley, (Plenum, N.Y., 1978) pp. 417-433.
8) R. K. Kalia, P. Vashishta, and S. W. de Leeuw, J. Chem. Phys. **90**, 6802 (1989).
9) N. R. Kestner, J. Jortner, M. H. Cohen, and S. A. Rice, Phys. Rev. **140,** A56 (1965).
10) R. K. Kalia and J. Harris, to be published.
11) S. Nosé, Mol. Phys. **52**, 255 (1984) .
12) A. Bartels, Appl. Phys. **8**, 59 (1975) .

QUANTUM LIQUID FILMS: A GENERIC MANY-BODY PROBLEM

E. Krotscheck†, J. L. Epstein†, and M. Saarela‡
†Center for Theoretical Physics, Department of Physics
Texas A&M University, College Station, Texas 77843
‡Department of Theoretical Physics, University of Oulu
Linnanmaa, SF-90570 Oulu 57, Finland.

This contribution reports on recent progress in the investigation of the structure and excitations of quantum liquids adsorbed to a solid surface, the states of impurity atoms in such films, and the interaction between them. The subtitle of my talk is *"A Generic Many-Body Problem"*, it indicates that I will, to some extent, emphasize the methodological aspect of the problem.

We start the microscopic description of a many-body system of N particles in a given volume Ω in an external potential with an empirical Hamiltonian

$$H = \sum_{i=1}^{N} \left\{ -\frac{\hbar^2}{2m}\nabla_i^2 + U_{sub}(\mathbf{r}_i) \right\} + \frac{1}{2} \sum_{1 \le i < j \le N} v(|\mathbf{r}_i - \mathbf{r}_j|). \tag{1}$$

To be specific, we consider a system of ^{4}He atoms at zero temperature. Ideally, one would like to solve the Schrödinger equation for this Hamiltonian. But, even with substantial computational resources, this is only possible only for a few systems with simple geometries and simple interactions. Therefore, we must resort to approximations. However, obvious physical consequences of the exact problem are not necessarily satisfied by an approximate theory.

A microscopic theory must be able to deal with the short-ranged repulsion between particles and include the long-ranged correlations correctly. Considering N particles in a given volume Ω, the theory must also be able to *decide* whether particles fill the given volume uniformly, or only a part of it. There may be a regime of average densities N/Ω where *both* states are possible. This is where the *pressure* of the uniform phase is negative, but the *compressibility* is positive. However, if the uniform phase is diluted to the density point where the compressibility becomes negative, the theory for the uniform phase should cease to have solutions. In other words, the theory must describe the response of the system to long wavelength excitations correctly.

The correct treatment of the bulk phase is prerequisite to determining whether a system of particles will or will not be adsorbed to a substrate. The requirement of the correct inclusion of short- and

Condensed Matter Theories, Volume 5
Edited by V.C. Aguilera-Navarro
Plenum Press, New York, 1990

long-ranged correlations is needed when we consider the structure of the adsorbed atoms. Heuristically, one would think of the ^{4}He atoms as hard spheres. These spheres would first form a layer closest to the surface, then, as their number is increased, a second layer, and so on. Eventually, the zero-point motion of the atoms will win over the attraction to the substrate, and far from the substrate the ^{4}He particles will behave as if they are in the bulk. Next, we add some attraction to the "hard spheres", which is strong enough to form a many-body bound state. The particles will wish to stick together even when the attractive substrate is not there. In the presence of the attractive substrate, it is not clear whether the plane geometry (i.e. a geometry that is translationally invariant in the directions parallel to the surface) is energetically favourable. The question is to what extent the liquid will "wet" the surface. Formally, we again encounter the problem of stability against long-wavelength excitations.

Summarizing the above qualtitative arguments, we find that, in order to adequately describe the physics discussed above, the many-body theory to be used must at least qualitatively implement the following features:

- The theory must describe short-ranged correlations in order to account for a layer-structure of the adsorbed liquid.

- The theory must include long-ranged correlations in order to give us the right geometry.

- Finally, our theoretical description of the system should be flexible enough to take any symmetry breaking fully into account.

These physical requirements translate into the simple *dictum* that a theory should be internally consistent such that it has only *stable* solutions. Phrased in the formal language of many-body theory, our theory must contain a self-consistent description of short- and long ranged correlations, which is accomplished by a self-consistent summation of ring- and ladder diagrams. For simpler systems like the bulk liquid one may get away with simpler theories for a while, but usually the "crimes" will catch up and lead to unphysical predictions like a negative compressibility. Even if numerical comparison with experiments look good, it is clear that something is *fundamentally wrong* with such a theory.

Of course, everything we do should eventually be compared with experiments. Unfortunately, quantities that can be calculated easily are often hard to measure and vice versa. From the way the experiments are done, ground-state energy measurements are difficult. Neutron scattering experiments probing the liquid surface are difficult since the penetration depth of the neuton is about 100 Å, whereas the surface width of ^{4}He is of the order of 10 Å. If one is interested in exploring the layer structure, one has to look for experiments that are sensitive to the surface structure. Therefore, it may also be necessary to extend the theory in order to address the quantities that can be measured more easily.

Going back to the formal problem of a self-consistent summation of ring- and ladder diagrams, it has been known for some time that this goal is accomplished by both the optimized hypernetted-chain[1] (HNC/EL) and the parquet-diagram[2] theory, which ultimately lead to the same equations to be solved[3]. Here we adopt the HNC/EL version of the theory since it is more widely developed and can more easily tested by variational Monte-Carlo calculations. One starts with a variational *ansatz* for the ground-state wave function of the N particle system of ^{4}He atoms with coordinates $\mathbf{r}_1, \ldots, \mathbf{r}_N$:

$$\Psi_0(\mathbf{r}_1,\ldots,\mathbf{r}_N) = \exp\frac{1}{2}\left\{\sum_{1\le i\le N} u_1(\mathbf{r}_i) + \sum_{1\le i<j\le N} u_2(\mathbf{r}_i,\mathbf{r}_j)\right\}. \tag{2}$$

The one-body function $u_1(\mathbf{r})$ describes the spatial structure of the system, and the two-body function $u_2(\mathbf{r}_i,\mathbf{r}_j)$ describes the short- and long-ranged correlations between pairs of particles. These functions are determined by minimization of the ground-state energy-expectation value E_0.

$$\frac{\delta E_0}{\delta u_1(\mathbf{r})} = 0, \qquad \frac{\delta E_0}{\delta u_2(\mathbf{r}_i,\mathbf{r}_j)} = 0. \tag{3}$$

In many cases the calculation of the variational energy expectation value E_0 cannot be performed exactly. Using *approximate* energy functionals may, of course, render the Euler equations (3) meaningless. It is therefore important to use an energy functional that has meaningful stationary points under the variational problems (3). It has been known for some time that the HNC hierarchy of approximations has meaningful variational minima. To show that the HNC/EL equations also provide a correct description of the short- and the long ranged correlations, we give the HNC/EL equations for the homogeneous Bose liquid[4]. These equations are most conveniently formulated in terms of the pair correlation function $g(r)$, and the static structure function $S(k) = 1 + \rho\int d^3r[g(r)-1]e^{i\mathbf{k}\cdot\mathbf{r}}$:

$$S(k) = \left\{\sqrt{1+\frac{4m\rho}{\hbar^2k^2}\tilde{V}_{ph}(k)}\right\}^{-\frac{1}{2}} \tag{4}$$

with

$$V_{ph}(r) = g(r)v(r) + \frac{\hbar^2}{m}|\nabla\sqrt{g(r)}|^2 + [g(r)-1]w_I(r). \tag{5}$$

The "induced interaction[5]" $w_I(r)$ is most conveniently represented in Fourier-space:

$$\tilde{w}_I(k) = -\frac{\hbar^2k^2}{4m\rho}[2S(k)+1][1-S^{-1}(k)]^2 \tag{6}$$

(The three-dimensional Fourier-transform is denoted by a tilde). It is worth noting here that

$$V_{ph}(|\mathbf{r}-\mathbf{r}'|) = \frac{\delta^2E_0}{\delta\rho_1(\mathbf{r})\delta\rho_1(\mathbf{r}')}, \tag{7}$$

where the variational derivative is taken for constant $u_2(\mathbf{r},\mathbf{r}')$. In Eq. (4) we recover the RPA expression for the static form factor. The HNC/EL theory supplements the RPA with a microscopic theory of the particle- hole interaction $V_{ph}(r)$.

An alternative way to formulate the Euler-Lagrange equation is[5]

$$\left[-\frac{\hbar^2}{m}\nabla^2 + v(r) + w_I(r)\right]\sqrt{g(r)} = 0. \tag{8}$$

Eq. (8) is the Boson-version of the Bethe-Goldstone equation, in which the bare interaction has been supplemented by the "induced interaction" $w_I(r)$.

Eqs. (4) and (8) are algebraically equivalent, they merely suggest different iteration paths for the determination of the optimal $g(r)$. The self-consistent solution of Eqs. (4-6) takes three Fourier transforms per iteration, compared with two Fourier transforms per iteration for the solution of the

HNC equations with a parametrized Jastrow function $u_2(r)$. Thus, the efficiency of the optimization algorithm makes the use of a parametrized trial function completely pointless. The same is true for three-body correlations and for the Fermion version of the HNC/EL approach.

Returning to the physics, we see that the HNC/EL theory includes both the ring and the ladder diagrams exactly, and the mixed diagrams approximately[2]. In particular, Eq. (4) shows that the theory has, as required, no *uniform* solution with negative compressibility, c.f. Eq. (7). The appropriate symmetry breaking (a droplet or a plane surface geometry, for example) is put into the wave function (1) by including a one-body function $u_1(\mathbf{r})$ and breaking the translational invariance of $u_2(\mathbf{r}_i, \mathbf{r}_j)$. If system is, within the assumed geometry, stable against small perturbations, then the theory has solutions. We conclude that the HNC/EL theory satisfies all the minimum requirements that are needed to deal with the problem of a quantum liquid surface.

The simplest symmetry breaking that can be treated with reasonable computational effort is the plane surface or the spherical geometry. The HNC/EL equations for the inhomogeneous system are straightforward generalizations of Eqs. (4-6), the equations and the algorithm for their iterative numerical solution has been given in Ref. 6.

The physical system considered here is a number of helium atoms interacting via the Aziz potential[7]. The atoms are adsorbed to a substrate which is described by an external field $U_{sub}(z)$. A simple form for $U_{sub}(z)$ is the potential obtained by averaging Lennard-Jones interactions between helium and substrate atoms over a half space[8]. One obtains

$$U_{sub}(z) = e\left[\frac{1}{15}\left(\frac{s}{z}\right)^9 - \left(\frac{s}{z}\right)^3\right]. \tag{9}$$

Given the substrate potential, the two-body interaction, and the surface coverage

$$n = \int dz \rho_1(z), \tag{10}$$

the physical problem is completely defined. In our geometry, all two-body quantities depend on the distances z_i of both particles from the substrate, and their separation $r_{\parallel}$ parallel to the surface. The Euler equations for the ^{4}He background were solved numerically in the HNC approximation, specifying only the surface coverage n. Some typical examples of density profiles are shown in Fig. 1. Most remarkable are the stong density-oscillations of the ^{4}He-background, which are due to the geometric core-exclusion between the ^{4}He particles.

It was already pointed out that a direct measurement of the surface profile by neutron scattering experiments is very difficult. It also turns out that the energy per particle does not depend strongly on the layer structure of the system[9], and the chemical potential has only a weak modulation, c.f. Fig. 2.

One might expect that the sound velocity exhibits a stronger dependence on the layer structure of the surface, but the analysis of our theoretical calculations and the experiments shows that the third sound velocity c_3 (or the derived quantity mc_3^2) depends strongly on the underlaying substrate potential (Fig. 3) which makes conclusions on the structure of the film very difficult.

So far, we have seen that it is difficult to probe the layer structure of an adsorbed ^{4}He film directly. A moderately easy probe of the surface structure is a ^{3}He impurity. Experimental interest focusses presently on the following areas:

Fig. 1. The density $\rho_1(z)$ of the background film of ^{4}He atoms (dotted lines) and of the ^{3}He impurity, $\rho_1^I(z)$ (solid lines) are shown for surface coverages of $n = 0.15, 0.20, 0.25, 0.30$, and 0.35 ^{4}He atoms/Å^2. The ^{4}He densities of all films shown are virtually identical within the first layer, whose density maximum is about 0.08 atoms/Å^3. The ^{3}He impurity density is normalized such that $\int dz \rho_1^I(z) = 1$.

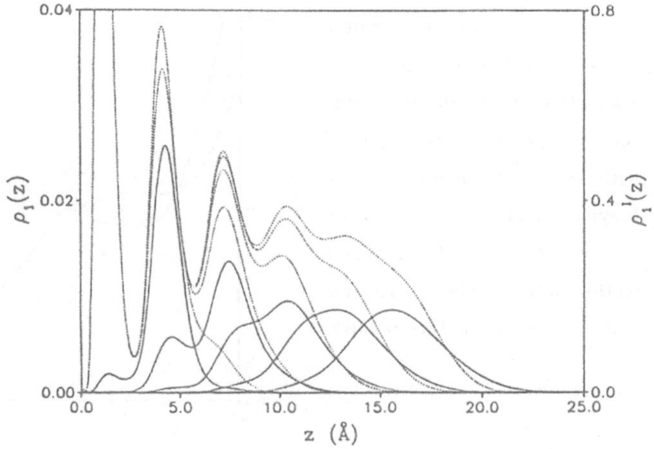

Fig. 2. The chemical potential μ of ^{4}He atoms is shown, as a function of the surface coverage n, for three different substrate potentials of different strength. The upper curve corresponds to a weakly attractive glass substrate, the lowest one to graphite, and the middle curve to a potential of average strength.

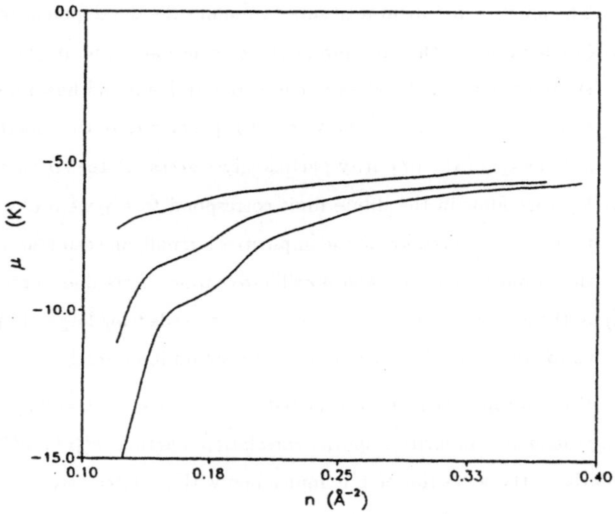

(a) Measurements of the binding energy and the specific heat of ^{3}He impurities[11], and

(b) Measurement of magnetic properties of ^{3}He impurities adsorbed to ^{4}He surfaces[12].

Qualitatively, one should expect the following effects:

- For a *thin* film of ^{4}He atoms (one or two layers thick), the substrate potential dominates, and the ^{3}He impurity atom will be in the outermost layer.
- For a *thick* film of ^{4}He atoms (five or more layers), the larger zero-point motion of the ^{3}He impurity will dominate, and the impurity atom will appear to "swim" on the ^{4}He background.
- If the outermost layer of the ^{4}He film is full, it will be hard for the ^{3}He impurity to move. The "backflow" effect and hence the effective mass of the impurity atom will be large.
- If the outermost layer of the ^{4}He film is incomplete, the backflow, and hence the effective mass, will be smaller.

Figure 3. mc_3^2 for the medium (weakly oscillating curve) and the strong (strongly oscillating curve) substrate potential are compared with the experimental data of Maynard and Chan (Ref. 10, crosses). See Ref. 9 for details on the analysis of the experiments and the calculational procedures.

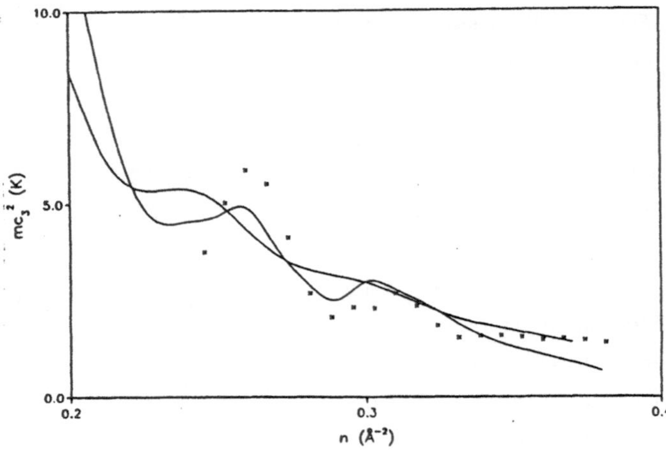

The last two points also depend on how deep the impurity atom penetrates into the ^{4}He surface.

The above experimental situations require the development of three levels of theoretical tools:

(a) Single-impurity theories aim at the calculation of impurity binding-energy and effective mass. The static correlations between particles determine if and how many bound states the impurity particle can have at the surface, and whether it penetrates to the substrate. The calculation of the (complex) self-energy of the impurity particle gives access to the specific heat and the mobility.

(b) Mixture films in the dilute limit correspond to a system of *two* static impurities[13]. To the extent that the concentration of the impurities is small, one can ignore all higher-order correlations between the impurity particles and possible dynamic effects due to momentum-dependent correlations.

(c) A theory for two-component systems is needed for large "impurity" concentrations. This enables us to study the structure of quantum-liquid interfaces.

Formally, impurities may be added to the system by extending the wave function to include impurity-background and impurity-impurity correlation functions $u_1^I(\mathbf{r}^I)$, $u_2^{IB}(\mathbf{r}_i^I, \mathbf{r}_j)$, and $u_2^{II}(\mathbf{r}_i^I, \mathbf{r}_j^I)$. For Fermion impurities, the wave function is multiplied with a Slater determinant to ensure the antisymmetry with respect to exchange. To include a momentum dependence, one may either add a "backflow" function to the one-impurity correlations, or calculate the self-energy in CBF perturbation theory. The variational wave function of a system of N ^{4}He-background atoms and one ^{3}He impurity of momentum $\mathbf{q}$ is

$$\Psi_{\mathbf{q}}(\mathbf{r}^I, \mathbf{r}_1, \ldots \mathbf{r}_N) = \exp\left\{\frac{1}{2}u_1^I(\mathbf{r}^I) + i\mathbf{q}\cdot\mathbf{r}^I + \sum_{j=1}^{N}[\frac{1}{2}u_2^{IB}(\mathbf{r}^I, \mathbf{r}_i) + i\alpha_{\mathbf{q}}(\mathbf{r}^I, \mathbf{r}_i)]\right\}\Psi_0(\mathbf{r}_1, \ldots, \mathbf{r}_N). \quad (11)$$

$\alpha_{\mathbf{q}}(\mathbf{r}^I, \mathbf{r}_i)$ is the backflow correlation function[14] describing the current of ^{4}He atoms flowing around the moving impurity. The functions $u_1^I(\mathbf{r}^I)$, $u_2^{IB}(\mathbf{r}^I, \mathbf{r}_i)$, and $\alpha_{\mathbf{q}}(\mathbf{r}^I, \mathbf{r}_i)$ are determined by minimizing the energy $E_{\mathbf{q}}$ of the system consisting of the ^{4}He-background and an impurity particle with momentum $\mathbf{q}$:

$$\frac{\delta E_{\mathbf{q}}}{\delta u_1^I(\mathbf{r}^I)} = 0, \qquad \frac{\delta E_{\mathbf{q}}}{\delta u_2^{IB}(\mathbf{r}^I, \mathbf{r}_i)} = 0, \qquad \frac{\delta E_{\mathbf{q}}}{\delta \alpha_{\mathbf{q}}(\mathbf{r}^I, \mathbf{r}_i)} = 0. \quad (12)$$

The total energy of the whole system can then be written as

$$E_{\mathbf{q}} = E_0 + \epsilon_0 + \frac{\hbar^2 q^2}{2m_H}, \tag{13}$$

where E_0 is, as above, the ground state energy of the ^{4}He background system, ϵ_0 is the binding energy of an impurity with zero momentum, and m_H is the so-called hydrodynamic effective mass. It is the contribution to the effective mass arising from the interaction of the impurity particle with the background.

Similarly, the variational wave function for two impurity Fermions with coordinates $\mathbf{r}_1^I$, $\mathbf{r}_2^I$ is

$$\begin{aligned}\Psi_{N+2}^{II}(\mathbf{r}_1^I, \mathbf{r}_2^I; \mathbf{r}_1, \ldots \mathbf{r}_N) = \exp \frac{1}{2}\Bigg\{ & u_1^I(\mathbf{r}_1^I) + u_1^I(\mathbf{r}_2^I) + u_2^{II}(\mathbf{r}_1^I, \mathbf{r}_2^I) + \\ & + \sum_{1\leq i\leq N} [u_2^{IB}(\mathbf{r}_1^I, \mathbf{r}_i) + u_2^{IB}(\mathbf{r}_2^I, \mathbf{r}_i)] \Bigg\} \Psi_0(\mathbf{r}_1, \ldots, \mathbf{r}_N)\Phi(1,2),\end{aligned} \tag{14}$$

where $\Phi(1,2)$ is a 2-particle Slater determinant. We leave out the backflow correlations since we will consider only small momenta of the ^{3}He impurities. The only new unknown function is the impurity-impurity correlation function $u_2^{II}(\mathbf{r}_1^I, \mathbf{r}_2^I)$, which is again determined by minimization of the total energy.

Instead of describing the further manipulations in detail (see Refs. 15-17), let us turn to the results of our calculations. The solid lines in Fig. 1 show the impurity density in comparison with the background density. We see that the ^{3}He particle is, as predicted, inside the outermost ^{4}He layer for our calculation of a double-layer system. As the thickness of the background increases, the ^{3}He particle is pushed outward into the low-density regime of the film.

A more instructive picture is given by considering the distribution of ^{4}He particles in the vicinity of the impurity. Figs. 4 and 5 show the ^{4}He density assuming that the ^{3}He atom is located at $(x, y) = (0, 0)$, i.e. the quantity

$$\rho^B(z, r_\parallel) = \int dz^I \rho^{IB}(\mathbf{r}^I, \mathbf{r}), \tag{15}$$

where $\rho^{IB}(\mathbf{r}^I, \mathbf{r})$ is the impurity-background two-body density. For the very thin film we see that the ^{3}He atom is located within the outermost layer, whereas for the thick film, the ^{4}He background is only modestly deformed.

We have carried out extensive calculations for the static ground state of one ^{3}He impurity on a family of ^{4}He backgrounds ranging from a double layer ($n = 0.12\text{Å}^{-2}$) to a system of about five helium layers ($n = 0.35\text{Å}^{-2}$). The hydrodynamic effective mass has been calculated using the uniform limit[1] for the Euler-equation for the backflow function $\alpha_{\mathbf{q}}(\mathbf{r}^I, \mathbf{r}_i)$. In this limit, one obtains a closed-form expression for the effective mass which is identical to a second-order approximation for the self-energy in terms of the emission and re-adsorption of ripplons and phonons. Our theoretical description of the first layer should not be considered very realistic[5], but the uncertainty in the description of the first layer does not seriously affect our conclusions. This part of the system has a very high density and the impurity hardly penetrates (c.f. Fig. 1). Effectively, the impurity atom "sees" the first layer as if it were a solid.

Our results are compared in Fig. 6 with the experimental results of Refs. 11. We find that the hydrodynamic mass is somewhat lower than the experimental mass. We notice, however, that the

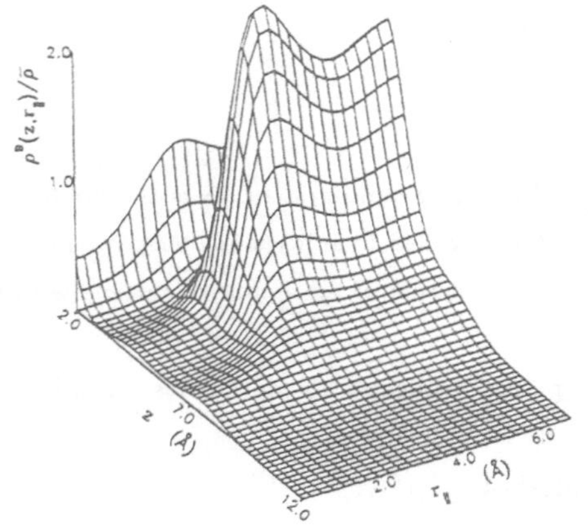

Fig. 4. The density of ^{4}He atoms in the vicinity of the ^{3}He impurity, $\rho^B(z,r_\parallel)$, is shown for the double layer background film with a surface coverage of $n = 0.15\text{Å}^{-2}$. The density is normalized to the *calculated* bulk equilibrium density, $\bar\rho^{HNC}$.

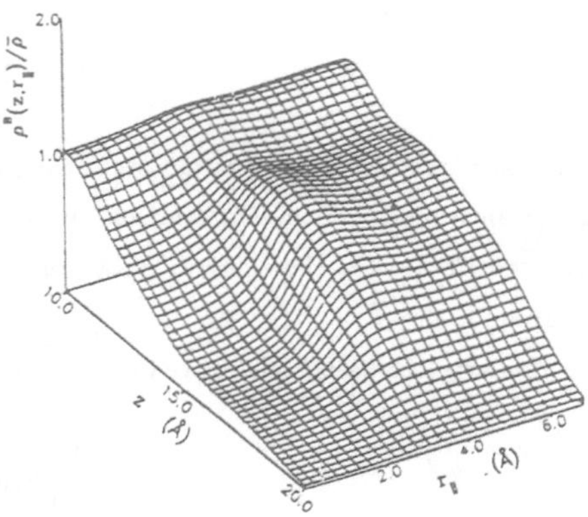

Fig. 5. Same as Fig. 4 for the background films with the largest surface coverage $n = 0.35\text{Å}^{-2}$.

experiment has not been done in the dilute limit, but for a ^{3}He density of 0.3 layers. Therefore we must include the effects of the quasiparticle interaction between the ^{3}He atoms.

The interaction between ^{3}He impurities is intelligently discussed in terms of a *local* effective interaction $V_{eff}(\mathbf{r},\mathbf{r}')$. This effective interaction includes both the direct Van der Waals force between the impurities and the exchange of ripplons and phonons through the ^{4}He background. It can be obtained by generalizing the theory of the impurity-impurity interaction[11] to the inhomogeneous case. In the case of a dilute system of impurities, we may take the impurity-impurity interaction as an effective interaction between any two ^{3}He atoms for a *finite* impurity population ("quasiparticle interaction"). The relation to Landau's Fermi-Liquid theory in two dimensions[18] is drawn in momentum space by identifying

$$\begin{aligned} f_c(q_\parallel) + f_\sigma(q_\parallel)\sigma_1\cdot\sigma_2 &\equiv V(0+) - \frac{1}{2}V(q_\parallel)(1+\sigma_1\cdot\sigma_2) \\ &= \int dz_1 dz_2 d^2r_\parallel \rho_1^I(z_1)\rho_1^I(z_2)V_{eff}(z_1,z_2,\mathbf{r}_\parallel)\left[1-\delta_{\sigma_1,\sigma_2}e^{i\mathbf{q}_\parallel\cdot\mathbf{r}_\parallel}\right], \end{aligned} \tag{16}$$

with the quasiparticle interaction. Given the interaction (16), we can calculate the Fermi-Liquid parameters of the two-dimensional Fermi liquid,[18]

$$\begin{aligned} N(0)f_c(|\mathbf{q}_1-\mathbf{q}_2|) &= \sum_{m=0}^{\infty} F_m^s \cos(m\phi) \\ N(0)f_\sigma(|\mathbf{q}_1-\mathbf{q}_2|) &= \sum_{m=0}^{\infty} F_m^a \cos(m\phi), \end{aligned} \tag{17}$$

where ϕ is the angle between $\mathbf{q}_1$ and $\mathbf{q}_2$, and $N(0) = m^*/\pi\hbar^2$ the density of states at the Fermi surface.

From these, we obtain the magnetic susceptibility $\chi(0)$, in units of the susceptibility of the free two-dimensional Fermi gas, χ_{30},

$$\chi(0)/\chi_{30} = (m_H/m_3)(1 + F_1^s/2)/(1 + F_0^a) \tag{18}$$

and the total effective mass

$$m^* = m_H(1 + F_1^s/2). \tag{19}$$

Figure 6 (solid line) shows the total effective mass m^* calculated from Eq. (19). For one active layer of ^{4}He the hydrodynamic mass is about 1.8 m_3, in reasonable agreement with both the data of Refs. 11 and the conclusion of Ref. 12 with their thinnest ^{4}He film. In this regime the correction due to the quasiparticle interaction becomes quite sizable with increasing ^{3}He concentration, but it is difficult to make a quantitative statement due to both the high density of the background and the rapid variation of the effective mass.

Fig. 6. The calculated effective mass m^* is shown in units of the bare ^{3}He mass m_3 as a function of the *inverse film thickness* D^{-1} (dashed line). The circles with error bars are the experimental data of Refs. 11. The upper scale shows the surface coverage corresponding to the film thickness D. The dashed line shows the hydrodynamic effective mass m_H for 0.3 layers of ^{3}He.

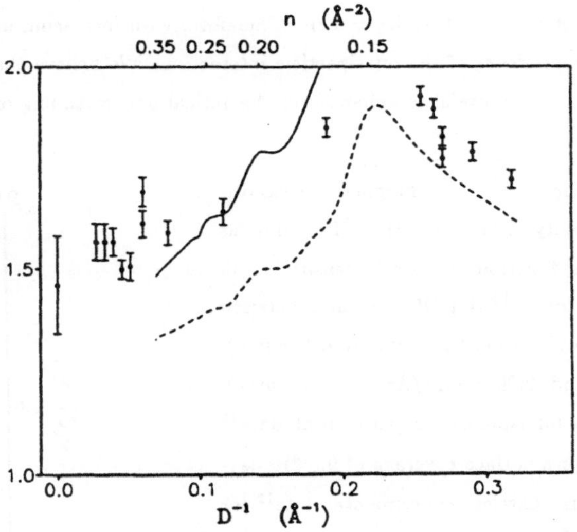

For all cases with thicker ^{4}He films, the accuracy of our theoretical prediction is quite satisfactory. The microscopic theory predicts a hydrodynamic mass that is consistent with the estimate $m_H/m_3 = 1.26 \pm 0.15$ given by Valles *et al.* (Ref. 12), but below the results of Refs. 11. The agreement with the latter data is improved when the corrections due to the quasiparticle interaction are included. The quasiparticle interaction between ^{3}He atoms in the surface gives a density-dependent correction to the effective mass. The contribution is about 10% for the case of 0.3 layers of ^{3}He.

There are slight oscillations of the ^{3}He effective mass as a function of the thickness of the underlying film. A weak plateau is seen around $D^{-1} \approx 0.1\text{Å}^{-1}$. This coincides with the regime where the third liquid layer is formed. Thus, we conclude that the impurity effective mass has weak layer structure; but that these oscillations dampen out rather fast. The effect of the migration of the impurity atom into the surface is much stronger than the modification of the backflow for different degrees of filling of the last layer so that it would be much harder to observe fluctuations of the ^{3}He effective mass for very thick films.

A comparison of the theoretical and the experimental magnetic susceptibility ratio $\chi(0)/\chi_{30}$ is shown in Fig. 7. Here, we have taken a hydrodynamic mass of $1.26 m_3$, which gives the best fit to experimental data[12]. This choice is consistent with our calculation of the effective mass since the experimental data refer to a much smaller ^{3}He coverage. We find a quite satisfactory agreement berween theory and experiment.

The agreement is not quite so good for smaller ^{4}He coverages. This is partly due to the fact that no attempt was made to re-adjust the effective mass. One must also be concerend about the accuracy of the theoretical description when the ^{3}He impurity penetrates into a regime of higher ^{4}He density, where HNC methods are intrinsically less accurate. However, HNC-type microscopic theories are, while using modest computational resources, quite capable of giving at least a semi-quantitative picture of the structure of complicated systems like the ones discussed here.

Finally, we stress that the good agreement between the experimental results and our calculation has been obtained with a *static, momentum-dependent* quasiparticle interaction. The calculation of the magnetic susceptibility involves phase-space integrals of the quasiparticle interaction over momenta between $q = 0$ and $q = 2k_F$. Therefore, with increasing impurity density one probes the *momentum-dependence* of the quasiparticle interaction. We believe that measurements of the type of Ref. 12 are extemely useful to enhance our theoretical understanding of the quasiparticle interaction in liquid ^{3}He.

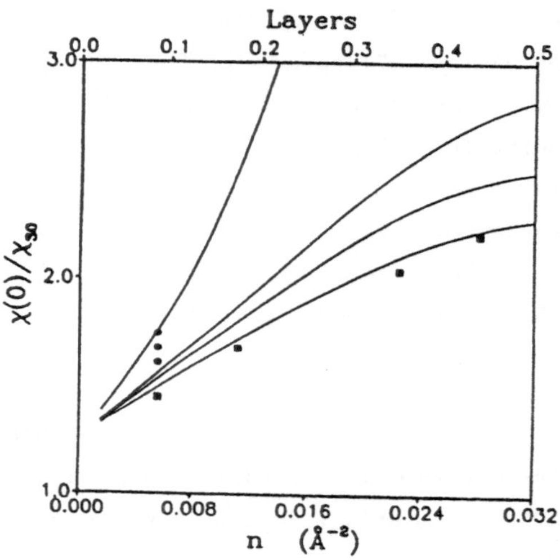

Fig. 7. Magnetic susceptibility χ/χ_{30} of the ^{3}He film as a function of areal density n in atoms/Å^2, for ^{4}He coverages between 0.15 atoms/Å^2 (uppermost curve), and 0.30 atoms/Å^2 (lowest curve). Solid squares: experimental data[12] for a surface coverage of 9.5 ^{4}He layers. Circles: experimental data[12] for a ^{3}He density of 0.088 layers, for ^{4}He coverages of 2.8, 3.4, 4.75, and 5.24 atoms/Å^2.

I hope that I have succeeded in this contribution to present the study of quantum liquid surfaces as an exciting field for both experimental theoretical research. Let me conclude by highlighting the most interesting physical and conceptual aspects of this field of research:

- Physically, one is able to construct very clean two-dimensional Fermi systems and investigate these systems over a wide regime of densities. This is in contrast to bulk ^{3}He, where experiments can be performed only in a narrow regime around the equilibrium density. By carrying out measurements on ^{3}He impurity films over a large density regime, one should be able to *measure the momentum dependence of the quasiparticle interaction.*
- Conceptually, quantum liquid films are many-body systems where state-of-the-art many-body theory is necessary for a thorough microscopic understanding. At this time I do not intend to suggest that

quantum liquid films be studied with other many-body methods. But looking back at the obvious requirements that must be satisfied by a theory in order to successfully deal with such systems, it may be worth asking

Can <u>your</u> favorite many-body theory deal (in principle and in practice) with these systems ?

Acknowledgements

This paper was written while one of us (EK) enjoyed the hospitatlity of the Theroretical Physics Institute at the University of Minnesota in Minneapolis. The work was supported, in part, by the National Science Foundation under Contract PHY-8806265 and the Robert A. Welch foundation under Grant A-1111. Participation in this workshop has been made possible by a travel grant from the U.S. Army research office. Stimulating discussions with C. E. Campbell are gratefully acknowledged.

References

1. E. Feenberg, *Theory of Quantum Fluids* (Academic, New York, 1969).
2. A. D. Jackson, A. Lande and R. A. Smith, Phys. Rep. **86**, 55 (1983); A. D. Jackson, A. Lande and R. A. Smith, Phys. Rev. Lett. **54**, 1469 (1985).
3. E. Krotscheck, A. D. Jackson, and R. A. Smith, Phys. Rev. **A33**, 3535 (1986).
4. C. E. Campbell and E. Feenberg, Phys. Rev. **188**, 396 (1969).
5. L. J. Lantto and P. J. Siemens, Phys. Lett. **68B**, 308 (1977).
6. E. Krotscheck, G.-X. Qian, and W. Kohn, Phys. Rev. **B31**, 4245 (1985).
7. R. A. Aziz, V. P. S. Nain, J. C. Carley, W. L. Taylor and G. T. McConville, J. Chem. Phys. **70**, 4330 (1979).
8. M. W. Cole, D. R. Frankl, and D. L. Goodstein, Rev. Mod. Phys **53**, 199 (1981).
9. J. L. Epstein and E. Krotscheck, Phys. Rev. **B37**, 1666 (1988).
10. J. D. Maynard and M. H. W. Chan, Physica **109-110B**, 2090 (1982)
11. X. Wang and F. M. Gasparini, Phys. Rev. **B38**, 11245 (1988); B. K. Bhattacharyya, M. J. DiPirro, and F. M. Gasparini, Phys. Rev. **B30**, 5029 (1984).
12. J. M. Valles, Jr., R. H. Higley, R. B. Johnson, and R. B. Hallock, Phys. Rev. Lett. **60**, 428 (1988).
13. J. C. Owen, Phys. Rev. Lett. **47**, 586 (1981).
14. J. C. Owen, Phys. Rev. **B23**, 2169 (1981).
15. E. Krotscheck, M. Saarela, and J. L. Epstein, Phys. Rev. **B38**, 111 (1988).
16. E. Krotscheck, M. Saarela, and J. L. Epstein, Phys. Rev. Lett. **61**, 1728 (1988).
17. J. L. Epstein, E. Krotscheck, and M. Saarela, preprint (1989).
18. S. M. Havens-Sacco and A. Widom, J. Low Temp. Phys. **40**, 357 (1980).

STRUCTURE AND DYNAMICS OF SUPERCOOLED FLUIDS

S.W. de Leeuw and M.J.D. Brakkee
Laboratory for Physical Chemistry, University of Amsterdam
Nieuwe Achtergracht 127, 1018 WS Amsterdam, (The Netherlands)

ABSTRACT

The results of molecular dynamics simulations of Lennard-Jones systems in the meta-stable supercooled state are presented. The variation of various time-correlation functions and transport coefficients with temperature is discussed and compared with recently developed mode-coupling theories. The temperature dependence of the viscosity and coefficient of self-diffusion is well described by a power law with an exponent $\mu = 1.8$, in close agreement with the prediction of mode coupling models.

INTRODUCTION

It is a well known experimental fact that most liquids can be cooled to temperatures well below their crystallization temperatures. The *meta*-stable supercooled liquid can be characterized by structural and mechanical relaxation times, which increase strongly with decreasing temperature. When some of these relaxation times become long relative to experimental time scales the liquid cannot respond to changes in temperature or pressure fast enough to achieve equilibrium. The liquid-glass transition is associated with a dramatic slowing down of structural and mechanical relaxation. This structural arrest manifests itself in the values of transport coefficients, such as the fluidity η^{-1} and the self-diffusion coefficient D, which become extremely small at the glass transition point. From this point of view the liquid-glass transition may be characterized as solidification without crystallization. For an overview of various phenomena associated with the glass transition we refer the reader to recent reviews by Jäckle [1] and Klinger [2].

Recently the application of self-consistent mode coupling theory has led to interesting developments in the dynamical theory of the liquid-glass transition. Interest in such approaches was stimulated by papers of Leutheusser [3] and Bengtzelius, Götze and Sjölander [4]. Within this theory a nonlinear feedback mechanism was uncovered which could lead to the ultimate structural arrest. The liquid-glass transition is then viewed as a transition from ergodic to non-ergodic behaviour. The theory also predicts scaling behaviour for the dynamical structure factor [5,6] near the glass transition. Recent neutron scattering studies [7] partly confirm these predictions. A similar feedback mechanism is present in a mode-coupling treatment of a set of nonlinear fluctuating hydrodynamic equations [8], so that these equations could exhibit a transition from ergodic to non-ergodic behaviour. This set of equations has the advantage of

Condensed Matter Theories, Volume 5
Edited by V.C. Aguilera-Navarro
Plenum Press, New York, 1990

being much simpler than the kinetic equations allowing a more detailed investigation of the liquid dynamics. In later studies however, a more careful analysis showed that the feedback mechanism, which causes the sharp singularity in the simpler mode coupling models, is still present, but eventually cut off by other nonlinearities, resulting in a rounded transition. The system remains ergodic [9].

The most extensive calculations within the frame-work of self-consistent mode coupling theory were carried out by Bengtzelius [10] for the glass transition in a Lennard-Jones (LJ) liquid. Experimental data for such a system can be obtained from molecular dynamics (MD) experiments. MD simulation is by now an established technique for studying in detail atomic motions in fluids. Previous simulations [11-16] have firmly established that the LJ system exhibits a glass transition when cooled at rates accessible in computer experiments. The high cooling rate however results in a rounded and somewhat more diffuse transition [11].

In this paper we want to discuss the results of MD experiments on LJ systems in the supercooled state. The main purpose of these calculations was an extensive comparison with the predictions of the self-consistent mode coupling theory. In addition one might hope that MD calculations yield detailed information about atomic motions in the supercooled state, which in turn may lead to a better understanding of the liquid-glass transition. In section 2 we discuss briefly the thermodynamic and structural properties of the LJ system in the meta-stable supercooled state. In section 3 the main results on the dynamical properties are presented. We close this paper with a brief discussion of the results in the light of the recently proposed mode coupling models.

2. MOLECULAR DYNAMICS EXPERIMENT

It has been argued convincingly by Fox and Andersen [11], that isobaric cooling simulations approach most closely the conditions in which most glass-forming experiments are carried out in the laboratory. We have therefore carried out a series of MD experiments in which the LJ liquid was isobarically cooled into the meta-stable regime. We used Nosé dynamics, in which the system is coupled to an external heat bath, to fix the temperature during the simulation. The technique was extended to allow for volume fluctuations so that pressure was also held constant during the simulations [17 -19]. In all our simulations the number N of particles was fixed at N = 958. Full details of these simulations are given in reference [17].

We shall employ the usual reduced units: energy is measured in units of the well depth ε of $\tau = (m\sigma^2/\varepsilon)^{\frac{1}{2}}$. The dimensionless density is $\rho^* = \sigma^3/\nu$, ν being the molar volume, temperature $T^* = k_B T/\varepsilon$ and pressure $p^* = p\sigma^3/\varepsilon$. In these units the time step of our simulations was $\Delta t = 0.004\tau$.

We carried out two sets of simulations. In the first set of simulations the temperature of the Nosé heath bath was lowered continuously at a rate proportional to the temperature. Thus at high temperatures, where the system can still follow changes in temperature, the cooling rate was most rapid. At a series of temperatures production runs of 10^4 time-steps were then carried out. In the second set we used stepwise cooling: the bath temperature was lowered in steps of 0.1 after which the system was equilibrated for several thousand time-steps, followed by a production run of 10^4 time-steps. For both sets of simulations

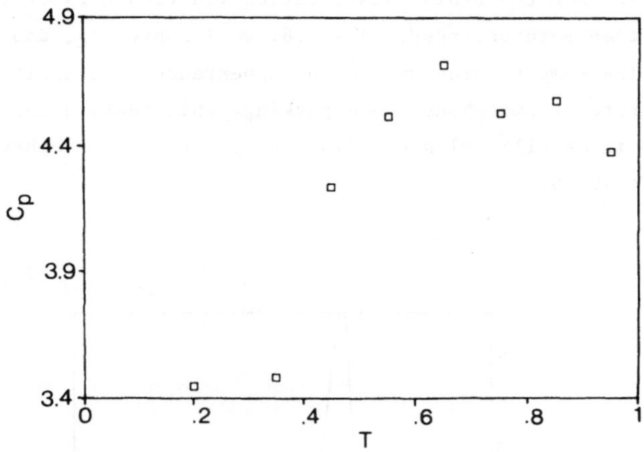

Fig. 1. Variation of specific heat C_p with temperature during a stepwise cooling quench.

several quenches were carried out and the final results for each set was obtained by averaging over different quenches. In all simulations the pressure was equal to p* = 3.2. The starting temperature for each quench was T* = 1.0, well in the liquid regime of the LJ-system.

During the simulations we monitored various thermodynamic quantities. In fig.(1) we show the variation of the specific heat C_p with temperature for the case of stepwise cooling. It shows the cross-over behaviour from typical liquid-like values at high temperatures to solid-like values at temperatures below T* ≅ 0.4. This behaviour is often taken as the signature of the glass transition, as the fluid cannot reach equilibrium within observational times. A similar behaviour is observed for the thermal expansion coefficient α_p, again a cross-over being observed near $T^*_c = 0.45$. Two points are worth mentioning here. Firstly, because of the high cooling rate (corresponding to $\cong 10^{12}$ K/sec in Argon) the temperature region over which the cross-over takes place, is rather broad: $\Delta T^* \cong 0.15$. This point has been discussed extensively by Fox and Andersen [11]. Secondly the cross-over temperature $T\eta$, defined as the temperature of which the viscosity of the fluid reaches a value of 10^{13}P. Because of the extremely high cooling rates one might expect the liquid to fall out of equilibrium at temperatures well above $T\eta$. Recent experiments [20], in which micro-size droplets of organic liquids were cooled at extremely high rates ($>10^5$ K/sec) seem to confirm this.

In fig.(2) we show the radial distribution function g(r) of the supercooled system at two temperatures, resp. $T^* = 0.8$, well above T^*_c, and $T^* = 0.2$, well below T^*_c. At low temperatures we see the appearance of a split second peak, a well-known feature of amorphous dense packing. This feature has been observed experimentally in metallic glasses [21]. The onset of this behaviour occurs at temperatures $T^* <0.55$.

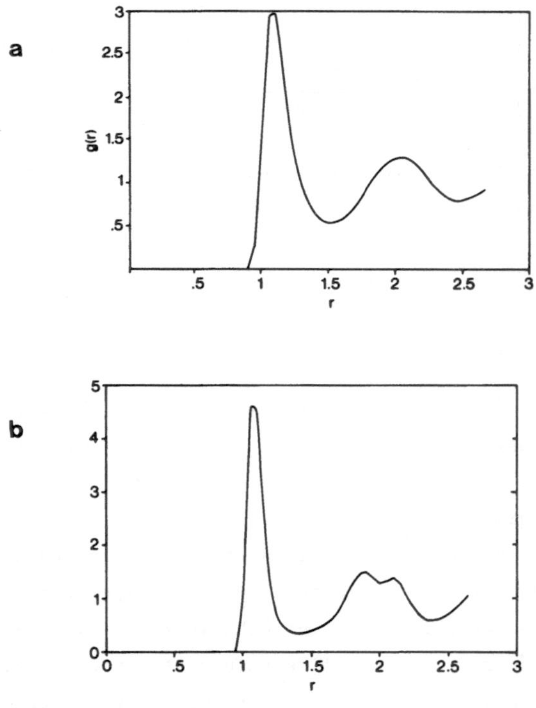

Fig. 2. Radial distribution function g(r) for a) $T^* = 0.8$ and b) $T^* = 0.2$

3. DYNAMIC PROPERTIES

It is evident from earlier discussion, that the glass transition as witnessed in the laboratory or in computer simulation experiments is primarily a dynamic phenomenon, characterized by the inability of the liquid to reach its (meta-stable) equilibrium state upon a change in temperature or pressure. It is therefore natural to focus on the dynamic properties of the supercooled liquid.

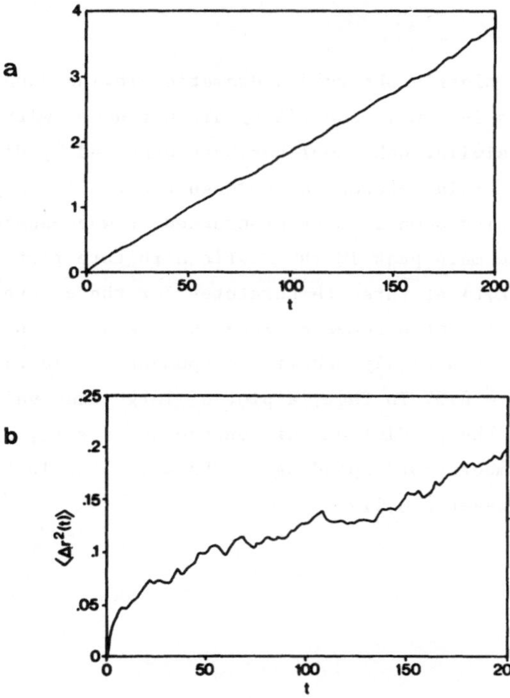

Fig. 3. Mean square displacement $\langle\Delta r^2(t)\rangle$ for two different temperatures: a) T* = 0.70; b) T* = 0.40.

We begin with the mean square displacement $\langle\Delta R^2(t)\rangle$ still displays a non-zero limiting slope, implying normal diffusive behaviour (albeit with a small coefficient of self-diffusion D). Mode coupling theory predicts a power law behaviour for the diffusion coefficient: $D \alpha (T-T_0)^{\mu}$. We find that our data fit this behaviour very well for temperatures T* >0.3, with $T^*_0 = 0.27$ and $\mu = 1.8$, a typical value of the theory [3,10,22]. Below this temperature deviations from this power law behaviour are observed and D is still non-zero at temperatures well below T^*_0.

The intermediate scattering function F(k,t), defined by:

$$F(k,t) = \frac{1}{N} < \sum_{l} \sum_{j} \exp (ik.(r_l(t)-r_j(0))>$$

plays a crucial role in mode coupling models of the glass transition. Within the simpler versions of the theory the structural arrest manifests itself in the form of a non-decaying part of F(k,t):

$$f_k = \lim_{t \to \infty} F(k,t)$$

In our simulations we clearly observed a dramatic slowing down in the decay of F(k,t) at temperatures $T^* < 0.40$ and F(k,t) did not decay within the time-scale of our simulation. A similar behaviour has been observed by Ullo and Yip [13] upon compression of a fluid interacting through the repulsive part of LJ-interactions only. The slowing down is most pronounced at wave-numbers corresponding to the position of the main peak in the static structure factor S(k).
In fig.(4) we plot F(k,t) at three temperatures for these wave-numbers. At $T^* = 0.6$ a slow component becomes noticeable. The decomposition of F(k,t) into a fast decaying and a slowly decaying component is clearly visible at the lowest temperature $T^* = 0.3$. In fact, a plot of large time values of $F(k,t=20\tau)$ follows qualitatively the predictions of Bengtzelius for f_k, a maximum being observed near wave-numbers corresponding to the main peak in S(k). Quantitative differences remain however [16,17]

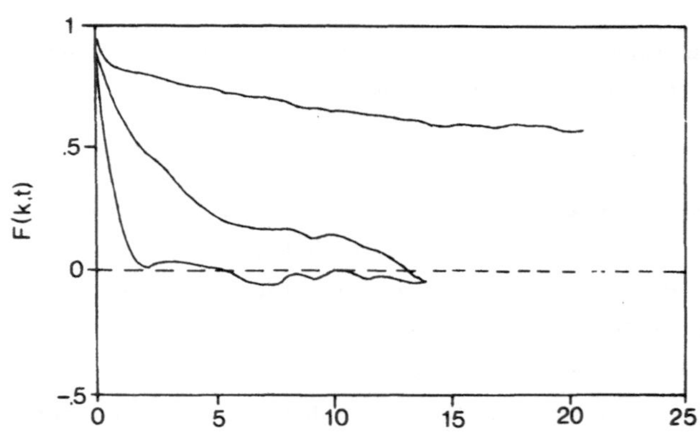

Fig. 4. Intermediate scattering function F(k,t) at three different temperatures: a) $T^* = 1.0$, $k\sigma = 6.8$; b) $T^* = 0.6$, $k\sigma = 7.0$; c) $T^* = 0.3$, $k\sigma = 7.2$.

The increasingly visco-elastic character of the liquid manifests itself strongly in the behaviour of the transverse current correlation function $C_t(k,t)$, defined by:

$$C_t(k,t) = \frac{1}{N} < \sum_l \sum_j (\hat{k}xv_l(t))(\hat{k}xv_j(0))\exp(ik.(r_l(t)-r_j(0)))>$$

The hats denote unit vectors. At high temperatures $C_t(k,t)$ decays monotonically to zero for the smallest wave-numbers accessible in our simulation, reflecting the fact that a liquid cannot sustain a shear deformation. As the temperature is lowered oscillatory behaviour sets in, indicating that the liquid can propagate shear waves at increasingly long wave-lengths. This behaviour of $C_t(k,t)$ is shown in fig.(6) for $T^* = 0.3$ at the smallest wave-number that we can reach in our simulations. The time-integral of $C_t(k,t)$ is proportional to the inverse shear viscosity η^{-1} in the hydrodynamic limit. An alternative, more reliable route to η is through the stress correlation function $C_\sigma(t)$. The viscosities, computed from the stress correlation function again follow a power law behaviour $\eta \propto (T-T_o)^{-\mu}$ with values of T_o and μ consistent our diffusion data and mode coupling models. We should note however, that at the lowest temperatures the stress correlation function decayed so slow, that extrapolation was necessary to compute the time-integral. The data are therefore not sufficiently accurate to discriminate between a power law behaviour and other (eg. Vogel-Fulcher) behaviour.

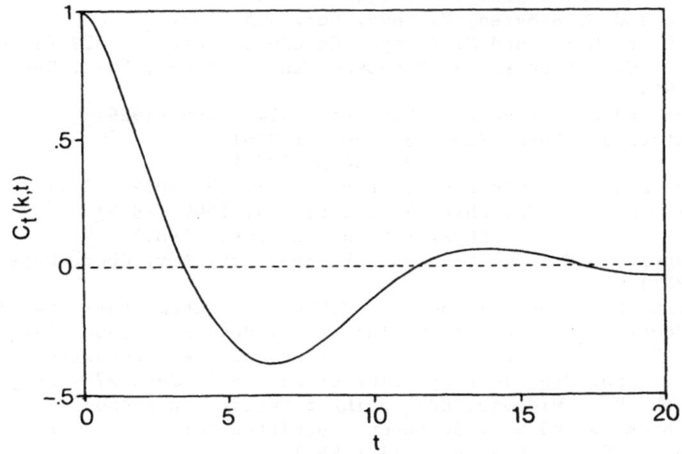

Fig. 5. Transverse current correlation function at the longest wave-numbers accessible in our simulations at $T^* = 0.3$, $k\sigma = 0.6$

4. DISCUSSION

In this paper we studied the variation of time correlation functions and transport properties of a meta-stable liquid with temperature. Our simulations confirm the power law behaviour for the coefficient of self diffusion and the viscosity predicted by self-consistent mode coupling theories, with values of the exponent $\mu \cong 1.8$ in close agreement with prediction. A similar result has been observed by Ullo and Yip [13] in their simulations of a fluid interacting through a truncated LJ-potential. Recently this power law behaviour has also been observed experimentally in a large class of glass-forming liquids [22]. However, unlike the experimental results we find T_o, the temperature at which the viscosity diverges, well below the temperature T_c at which the specific heat anomaly occurs. We attribute this to the extremely high cooling rate during our quenches, which shifts the specific heat anomaly to higher temperatures and broadens the temperature range over which it occurs.

5. ACKNOWLEDGEMENTS

This work is part of the research programme of the Foundation for Fundamental Research of Matter (FOM), supported by the Netherlands Foundation for Chemical Research (SON) and made possible by financial support from the Netherlands Organisation for Scientific Research (NWO). One of us, dr. S.W. de Leeuw, thanks Dr. S. Yip for useful discussions.

6. REFERENCES

1. J. Jäckle, Rep. Prog. Phys. 49, 171 (1986)
2. M. Klinger, Phys. Rep. 165, 275 (1988)
3. E. Leutheusser, Phys. Rev. A29, 2765 (1984)
4. U. Bengtzelius, W. Götze and A. Sjölander, J. Phys. C17, 5915 (1984)
5. W. Götze, Z. Phys. B60, 195 (1985)
6. W. Götze and L. Sjögren, Z. Phys. 865, 415 (1987)
7. W. Knaak, F. Mezei and B. Farago, Europhys. Lett. 7, 529 (1988)
8. S.P. Das, G.F. Mazenko, S. Ramaswamy and J. Toner, Phys. Rev. Lett. 54, 118 (1985)
9. S.P. Das and G.F. Mazenko, Phys. Rev. A34, 2265 (1986)
10. U. Bengtzelius, Phys. Rev. A34, 5059 (1986)
    A33, 3433 (1986)
11. J.R. Fox and H.C. Andersen, J. Phys. Chem. 88, 4019 (1984)
12. J.J. Ullo and S. Yip, Phys. Rev. Lett. 54, 1509 (1985)
    Phys. Rev. A (in press, 1989)
13. C.A. Angell, J.H.R. Clarke and L.V. Woodcock, Adv. Chem. Phys. 48, 397 (1981)
14. A. Rahman, M.J. Mandell and J.P. McTague, J. Chem. Phys. 64, 1564 (1976)
15. R. D. Mountain, and D. Thirumalai, Phys. Rev. A36, 3300 (1987)
16. M.J. D. Brakkee and S.W. de Leeuw, Proc. of the International Workshop on "Static and Dynamic Properties of Liquids", June 27 - July 4 1988, Dubrovnik. M. Davidovic, Ed; Springer Verlag (in press)
17. M.J.D. Brakkee and S.W. de Leeuw, submitted for publication.
18. S. Nosé, J. Chem. Phys. 81, 511 (1984)
19. H.C. Andersen, J. Chem. Phys. 72, 2384 (1980)
20. G.P. Johari, A. Hallbrucker and E. Mayer, J. Phys. Chem. 93, 2648 (1988)
21. R. Zallen, "Physics of Amorphous Solids", J. Whiley, New York, 1983, Chapter 2.

CORRELATED RPA CALCULATIONS FOR MODEL NUCLEAR MATTER

Eirene Mavrommatis

Physics Department, University of Athens
Panepistimioupoli, 15771 Athens, Greece

John W. Clark

McDonnell Center for the Space Sciences
and Department of Physics
Washington University, St. Louis, Missouri 63130

INTRODUCTION

One of the most exciting prospects on the current nuclear scene is the promise that precision high-energy electron scattering experiments will reveal new (and perhaps unforetold) aspects of nuclear structure and dynamics. The search is on for distinctive signatures of subnucleonic degrees of freedom, and especially for manifestations of the underlying quarkic substructure of nuclei. However, to reach any definite conclusions regarding such effects, it is necessary that we know, with precision, the values which are predicted for the measured quantities by the conventional picture of nuclei. In the conventional picture, a nucleus is composed of nucleons alone, moving nonrelativistically. The nucleonic constituents are considered to interact via bare potentials which reproduce the few-nucleon data while obeying certain constraints imposed by fundamental symmetries and by meson-exchange theory. Even at this rather superficial level, one is confronted with a very difficult many-body problem, essentially nonperturbative because of the strong short-range interactions among the nucleons. It should therefore be no surprise that mean-field theory (in old language, the shell model) fails in experimental settings where large momentum transfers take place and the high-momentum components of the nuclear wave function are being probed. A proper understanding of the quantitative implications of the conventional nuclear picture requires a careful and coherent treatment of the fine-scale spatial correlation structure and the collective properties which emerge from solution – or approximate solution – of the many-nucleon Schrödinger equation of the nucleus. Only when this refined microscopic description is achieved, and the conventional picture still found wanting, can a legitimate claim be made that the relevant experiments have penetrated to a more fundamental level of nuclear hadronic dynamics.

One focal point of the ongoing refinement of conventional nuclear theory is the longitudinal response function $R_L(q,\omega)$ of heavy nuclei in the quasielastic regime. In treating this property, mean-field theory has proven inadequate and it is necessary (but perhaps not sufficient) to incorporate the dynamical correlations among nucleons in a consistent and accurate manner. A microscopic understanding of the dynamic structure function $S(q,\omega)$ of the hypothetical problem of infinite, symmetrical nuclear matter should yield valuable

Condensed Matter Theories, Volume 5
Edited by V.C. Aguilera-Navarro
Plenum Press, New York, 1990

insights into the effects of correlations in the observed longitudinal response. In particular, Fantoni and Pandharipande[1] have argued for an approximate proportionality of $R_L(q,\omega)$ and $S(q,\omega)$ at large q, the proportionality factor being given by the absolute square of the proton form factor (see also Ref. 2).

The work of Fantoni and Pandharipande[1] and Fabrocini and Fantoni,[2] based on the method of correlated basis functions (CBF),[3,4] represents a substantial advance over all previous efforts on this problem, providing state-of-the-art evaluation of the longitudinal response function of nuclear matter (and more directly, its dynamic structure function $S(q,\omega)$), with a realistic two-nucleon interaction and simulated three-body potential as input. Within the CBF scheme, the calculation was performed, roughly speaking, at the Tamm-Dancoff level. The correlated random-phase approximation[5,6,4] (CRPA), which performs the ring sums within CBF perturbation theory, opens the way for considerable improvement of some aspects of this calculation, particularly at the lower momentum transfers where the effects of the particle-hole force are most evident. In this paper we present an initial application of CRPA theory to infinite nuclear matter. The goals of the calculation are modest: Preliminary to a full calculation with a realistic nucleon-nucleon interaction with elaborate state-dependence, we shall explore the predictions of CRPA for a simplified model of the nucleon-nucleon interaction which has frequently been used as a test case, namely the v_2 potential.[7] Moreover, we shall apply a simplified, local version of CRPA which has proven successful in applications to spin-polarized liquid ^{3}He and the electron gas. It is found that this approach already leads to results of qualitative or semi-quantitative value, which may be systematically improved as the techniques of CRPA, and more generally the CBF theory of dynamical response, are further developed.

In a broader context, microscopic evaluation of the density-density response function $\Pi(q,\omega)$ of nuclear matter (whose imaginary part gives $S(q,\omega)$), together with consistent evaluation of the self-energy $\Sigma(k,E)$, yield fundamental information about the elementary excitations of this hypothetical hadronic system. The properties of collective modes, typified by the zero-sound dispersion relation, may be extracted from $\Pi(q,\omega)$, while the nature of single-particle excitations is revealed by $\Sigma(k,E)$, from which one may derive an energy-dependent effective mass. These properties have obvious importance for a deeper understanding of nuclei. They are likewise basic to a description of the structure, dynamics, and thermal history of neutron stars – being essential to the evaluation of such quantities as the specific heat, viscosity, superfluid gap, etc. Since empirical constraints on the properties of neutron-star material are limited in the extreme, such astrophysical applications make it doubly important to hone and test our many-body calculational methods.

LOCAL CORRELATED RANDOM-PHASE APPROXIMATION

The correlated random-phase approximation (CRPA) extends ordinary RPA to strongly-interacting systems like liquid ^{3}He, nuclear matter, and nuclei. Since this approach has been developed in detail in other places,[5,6,4,8,9] we shall only describe its main features and display the working formulas of a local approximation proposed by Krotscheck.[6] Here, "local" implies that the particle-hole force is taken to depend only on the momentum transfer $\hbar q = |\mathbf{p}-\mathbf{h}|$ in the direct particle-hole channel (apart from momentum-conserving delta functions). We confine our attention to the uniform, infinite Fermi medium, for which the appropriate model states are Slater determinants of plane waves, and complete the specification of the underlying correlated basis by the adoption of a Jastrow correlating factor $F = \Pi_{i<j} f(r_{ij})$.

In terms of an irreducible particle-hole (*ph*) interaction U which CBF theory generates for a given bare (and possibly singular) two-body potential, correlated RPA for our problem looks exactly like the familiar RPA for a weak potential, aside from a minimal energy dependence of U which can usually be ignored. Implementation of the full CRPA approach including exchange nonlocalities is in principle straightforward, but in practice requires considerable numerical effort, entailing (for example) discretization of the matrix eigenvalue equations on a grid in momentum space.[8] To provide a simpler alternative, Krotscheck[6] has constructed a *local, energy-independent* approximation to U which is designed to preserve certain fundamental relations of the ingredients of U to the static structure function $S(q)$ of the Jastrow ground-state trial function and to its graphical derivative $S'(q)$. One of these relations is the optimization condition on the Jastrow pair correlation function $f(r)$, which reads

$$\frac{\delta <H>_o}{\delta \ln f^2(r)} \equiv \Delta(r) = 0 \quad . \tag{1}$$

In this expression, $<H>_o$ is the energy expectation value in the Jastrow trial ground state $\Pi f(r_{ij})\Phi_o$, where Φ_o represents the noninteracting Fermi sea. The proposed local approximation to the particle-hole interaction is simply

$$U(q) = \Delta(q) S^{-2}(q) + \frac{\hbar^2 q^2}{4m}[S^{-2}(q) - S_F^{-2}(q)] \quad , \tag{2}$$

where $S_F(q)$ is the static structure function of the noninteracting Fermi gas and

$$\Delta(q) = \rho \int d^3 r \exp(i\mathbf{q}\cdot\mathbf{r})\, \Delta(r) = \frac{\hbar^2 q^2}{4m}[S(q) - 1] + S'(q) \quad . \tag{3}$$

The Jastrow $S(q)$ entering (2)-(3) may be evaluated with good accuracy by solving the (nonlinear) FHNC/C equations, while its graphical derivative $S'(q)$, appearing in (3), may be obtained by solving the (linear) FHNC/C′ equations.[10] The vanishing of Δ is equivalent to the optimization condition. Hence for *optimal* Jastrow correlations the particle-hole force $U(q)$ depends only on $S(q)$ and properties of the noninteracting system. For optimal correlations, the choice (2) for $U(q)$ is just what is needed to regain $S(q)$ from the density-density response function through the fluctuation-dissipation theorem, if the collective approximation $\Pi_o^c = (\hbar^2 q^2/m)[\hbar^2\omega^2 - (\hbar^2 q^2/2mS_F(q))^2]^{-1}$ is used for the particle-hole propagator.[6,11]

Having adopted a local particle-hole force, one has quite standard algebraic RPA formulas which lead to the familiar RPA expressions for the density-density response function and self-energy[12,13]:

$$\Pi(q,\omega) = \frac{\Pi_o(q,\omega)}{1 - U(q)\Pi_o(q,\omega)} \quad , \tag{4}$$

$$\Sigma(k,\Omega) = U(0) + U_{\text{Fock}}(k) + \frac{i\hbar}{(2\pi)^4\rho}\int d^3q\, d\Omega\, G_o(\mathbf{k}-\mathbf{q}, \Omega-\Omega')U^2(q)\Pi(q,\Omega) \quad . \tag{5}$$

In (4), the response function $\Pi_o(q,\omega)$ is the particle-hole propagator of the free Fermi gas, i.e., the Lindhard function. In (5), G_o is the free single-particle Green's function, U_o is a constant (related to the chemical potential), and $U_{\text{Fock}}(k)$ is the Fock term of the particle-hole force $U(q)$.

The dynamic structure function and the properties of zero sound (if present) are derived from the relation (4) in the usual manner. Trivially,

$$S(q,\omega) = -\frac{1}{\pi}\,\text{Im}\,\Pi(q,\omega) \quad . \tag{6}$$

The zero-sound dispersion relation $\omega = \omega_{zs}(q)$ is determined by the roots of the denominator of (4), i.e. the roots of

$$1 - U(q)\,\text{Re}\,\Pi_o(q,\omega) = 0 \tag{7}$$

in the region where $\text{Im}\,\Pi_o(q, \omega) = 0$. The strength Z_{zs} of the zero-sound mode is given by

$$Z_{zs}^{-1}(q) = U^2(q) \left[\frac{d}{d\omega} \text{Re}\,\Pi_o(q, \omega) \right]_{\omega=\omega_{zs}} . \tag{8}$$

Note that we are considering only the density channel; more generally a spin-isospin decomposition can be made, and collective modes other than zero sound can be studied.

A version of the single-particle energy spectrum can be obtained from the self-energy (5) using the on-shell prescription

$$\varepsilon_k = t_k + \Sigma(k, \Omega) , \qquad \Omega = t_k = \hbar^2 k^2 / 2m . \tag{9}$$

The on-shell effective mass m^* is then given by

$$(m^*)^{-1} = k^{-1}\, d\varepsilon_k / dk \tag{10}$$

A qualitative shortcoming of the local correlated RPA (LCRPA), seen in the form (4) of the polarization propagator, is that the effects of dynamical correlations on the *continuous* portion of $S(q, \omega)$, in regions of the (q, ω) plane where $\Pi_o(q, \omega)$ vanishes, are not accessible. It is in fact just such effects which were examined many years ago by Czyż and Gottfried.[14] Thus any useful comparison with their work is precluded. Although the (q, ω) domain corresponding to individual $1p\,1h$ excitations is the same in LCRPA as in the free system, the RPA denominator in expression (4) introduces nontrivial correlation effects in that region; moreover, outside that region, for $\text{Im}\,\Pi_o(q,\omega) = 0$, zero sound may emerge as a distinct collective mode, corresponding to vanishing of the denominator.

Local CRPA will also suffer, at a quantitative level, from the static nature of the effective interaction $U(q)$ appearing in (4). By contrast, dynamic screening is known to be important in the electron gas at metallic densities.[15] Moreover, one does not expect the momentum dependence of the self-energy to be faithfully predicted within the LCRPA scheme, especially in the very delicate example of unpolarized liquid ^{3}He (Ref. 11). In spite of its shortcomings, LCRPA offers a simple and straightforward microscopic touchstone for phenomenological theories of comparable structure, such as the polarization-potential model.

In the polarization-potential approach of Aldrich and Pines,[16] which has been adapted to nuclear problems by Pines, Quader, and Wambach,[17] the density-density response function $\Pi(q, \omega)$ is expressed in a form similar to (4). However, the Lindhard function appearing in (4) is replaced by a more complicated propagator accounting both for a single-pair effective mass different from the bare value and for the presence of multipair excitations. Secondly, the static, local particle-hole interaction $f_s^i(q)$ of this approach is supplemented by a wave-vector and frequency-dependent contribution $(\omega^2/q^2) f_v^i(q)$ corresponding to backflow. Beyond these structural differences there is the important conceptual distinction that our particle-hole interaction is determined microscopically, whereas the polarization potentials of the Aldrich-Pines theory are determined by a combination of sum rules and phenomenology.

RESULTS FOR v_2 HOMEWORK MODEL

Based on the local correlated RPA scheme, we have carried out a numerical study of the dynamical response of a model of nuclear matter which is extremely simple, yet may capture important aspects of its correlation structure. The bare interaction between the nucleons is taken as the v_2 "homework-model" potential.[7] This two-nucleon potential consists of the central part of the $^3S_1 - ^3D_1$ component of the Reid soft-core interaction, assumed to act in *all* partial waves. It has seen wide use in tests of many-body methods (see, for example, Refs. 18,19). Specifically,

$$v_2(r) = [9924.3\exp(-4.2r) - 3187.8\exp(-2.8r) + 105.468\exp(-1.4r) - 10.463\exp(-0.7r)]/(0.7r) \quad . \tag{11}$$

Our calculations are based, primarily, on the parameterized Jastrow correlation factor

$$f(r) = \exp[-Ae^{-Br}(1-e^{-r/D})/r] \quad , \tag{12}$$

with parameters which minimize the Jastrow ground-state energy expectation value E_J at single-particle level degeneracy $\nu = 4$ and given density, E_J being computed by the Metropolis Monte Carlo algorithm.[20] Additionally, we have examined the particle-hole force generated by two versions of the correlation function

$$f(r) = (1-e^{-r^2/b^2})^n + gr^m e^{-r^2/\gamma^2} \tag{13}$$

studied by Benhar *et al.*[21] In the simpler version (B1), the parameter g, which measures the overshoot of $f(r)$ above unity, is set zero and the remaining parameters b and n fixed by minimization of the energy expectation value truncated at three-body cluster order. In the other version (B2), the three-body cluster approximation to the energy is minimized (approximately) with respect to the five parameters b, g, γ, m and n, subject to the sequential (or normalization) condition[22] as constraint.

The three correlation functions are labeled (in order) C, B1, and B2. None of these choices is optimal in the sense of being a solution of the Euler equation (1). The sequential condition imposed in the case of B2 is probably too stringent, while B1 is presumably too crude a choice, although both alternatives have the virtue of promoting good convergence of the cluster expansion of the Jastrow energy expectation value. The function C, being determined through a Monte Carlo evaluation, may provide a decent representation of the true optimal function, except in the large-r region which contributes little to the ground-state energy. The deviation at large r will, however, be reflected in the behavior of $U(q)$ at small q. In particular, it will produce a significant departure of $\Delta(q)$ from zero at low q values. In the present work, as in Ref. 9, we have chosen to set $\Delta(q)$ equal to zero, since the behavior of the $U(q)$ at small q is already suspect due to the local approximation itself. This conforms with a strategy suggested by Krotscheck,[6] who, in considering the Bijl-Feynman dispersion relation, found that it is a better approximation to assume that the correlation function has been optimized – even if the actual optimization has not been carried through – than to include the $\Delta(q)$ correction.

Numerical results with the C correlations have been obtained for symmetrical nuclear matter, i.e., level degeneracy $\nu = 4$, at $k_F = 1.39\ \text{fm}^{-1}$, corresponding to a density $\rho = 0.182\ \text{fm}^{-3}$ near nuclear saturation; and for pure neutron matter, i.e., $\nu = 2$, at $k_F = 1.75$, 2.25, 2.90 fm^{-1}, corresponding to three densities $\rho = 0.182$, 0.386, 0.822 fm^{-3} of relevance to the study of neutron stars. For both level degeneracies we have examined a range of wave-number transfers q from 0 to about 4 fm^{-1}. Results with B1 and B2 correlations are available only for symmetrical nuclear matter, at Fermi wave number $k_F = 1.4\ \text{fm}^{-1}$.

For reference, we have also generated nuclear-matter ($\nu = 4$) results for the case that the two-nucleon interaction is replaced by a pure hard-core potential of core radius $c = 0.5$ fm. As in an earlier calculation,[9] where $c = 0.4$ fm was studied, the simple form

$$\begin{aligned} f(r) &= 0 \ , & r &\le 0 \ , \\ &= 1 - \exp[-\mu(r-c)] \ , & r &> c \end{aligned} \tag{14}$$

is assumed for the Jastrow pair correlation function. An optimal determination of the parameter μ has been carried out by Flynn[23] within the FHNC/C approximation, for both core sizes ($c = 0.4$ and 0.5 fm).

We confine our discussion to a selection of results for symmetrical nuclear matter. (A more detailed report will be published elsewhere.) The CBF particle-hole interaction appropriate to the v_2 potential at $k_F = 1.39\ \text{fm}^{-1}$, for the C choice of $f(r)$, is shown in Fig. 1, along with the $c = 0.5$ fm hard-sphere result at the same density. Due to the difference between the respective correlation functions (the v_2 correlations being slightly stronger beyond 1 fm), the particle-hole force for v_2 is actually somewhat more repulsive at low-to-medium q's than that for the hard-core potential. Both potentials support a collective mode corresponding to zero sound. In both cases, this mode is found to emerge from the particle-hole continuum around 0.3-0.4 fm^{-1} and sink back into it at about 1.53 fm^{-1}. The zero-sound dispersion relation $\omega_{zs}(q)$ and the strength $Z_{zs}(q)$ of the zero-sound mode are plotted in Fig. 2, for the v_2 potential. The corresponding curves for the hard-core potential are nearly coincident with those for v_2.

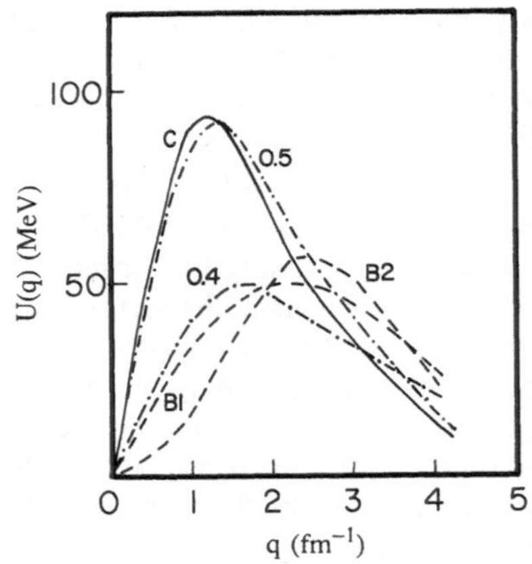

Fig. 1. Wave-number dependence of local particle-hole interaction $U(q)$ for v_2 model of nuclear matter, based on correlation functions C, B1, and B2 (as labeled); and for hard-sphere nucleons with $c = 0.4$ and 0.5 fm (as labeled), based on optimized correlation function (14). Density near 0.18 nucleons/fm^3.

Fig. 2. Zero sound dispersion relation $\omega_{zs}(q)$ (solid curve) and wave number dependence of zero sound strength $Z_{zs}(q)$ (dashed), for v_2 model of nuclear matter at $k_F = 1.39\ \text{fm}^{-1}$, based on C correlation function.

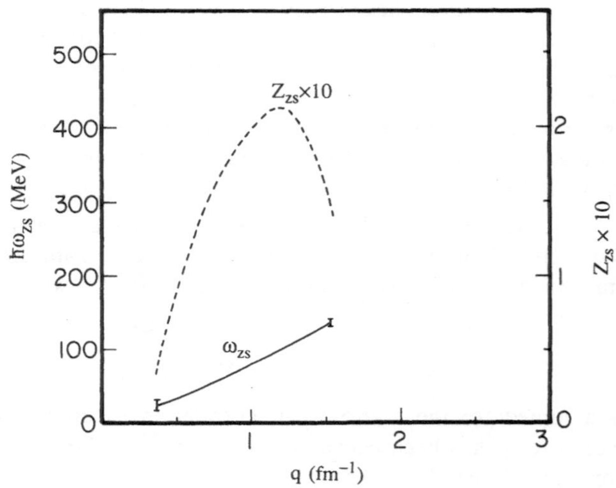

It may seem surprising that the effective interaction derived from the soft-core v_2 potential is so close to that for the hard-core potential. The concurrence implies that the two potentials are similarly effective in lifting particles from below to above the Fermi surface, at least when the momentum transferred to the particle-hole pair is not excessive and the density is near the saturation value for nuclear matter. In turn this observation suggests that a useful measure of the strength of the particle-hole force may be furnished by the wound parameter of Brueckner theory, which may be interpreted as the average depletion of the Fermi sea produced by the given bare interaction. Indeed, the wound parameters for the two potentials (v_2 and $c = 0.5$ fm hard core) are quite similar. As is well known,[18] the Jastrow variational analog of the Brueckner wound parameter is $\kappa = \rho \int d^3r\, [f(r) - 1]^2 [1 - \nu^{-1} l^2(k_F r)]$, where $l(x) = 3x^{-3}(\sin x - x\cos x)$. For the C choice of the correlation function, we have $\kappa = 0.232$, which is to be compared with the value $\kappa = 0.263$ for correlation function (14) with $\mu = \mu_{opt} = 2.5\ \text{fm}^{-1}$.

We may carry these considerations somewhat further by examining the curves in Fig. 1 labeled B1 and B2. The particle-hole force $U(q)$ for B1 or B2 is seen to differ substantially from that corresponding to correlation function C, being much weaker at smaller q values (below 2 fm^{-1}) and somewhat stronger at large q. The large differences cannot be due to the slight discrepancy in densities (0.185 fm^{-1} (B1 and B2) compared to 0.182 fm^{-1} (C)). Instead, it must be a reflection of the reduction in the flexibility of $f(r)$ entailed by a procedure employing a truncated cluster expansion, the range of variation being restricted to functions for which the cluster expansion of the energy is rapidly convergent. The substantial difference in predictions for $U(q)$ is (as expected) accompanied by substantial differences in the wound parameters for B1 and B2, from those for the correlation function C. The specific values are $\kappa = 0.146$ (B1) and $\kappa = 0.143$ (B2). Our thesis that the wound parameter is a major determinant of the overall behavior of $U(q)$, gains additional support from the observation that correlation functions B1 and B2, with approximately the same wound parameters, yield $U(q)$ results which are quantitatively as well as qualitatively similar. To complete the comparison, we have included in Fig. 1 a $U(q)$ curve for a $c = 0.4$ fm hard-sphere interaction at $k_F = 1.39\ \text{fm}^{-1}$, described by means of the correlation function (14) with $\mu = \mu_{opt} = 3\ \text{fm}^{-1}$. The corresponding wound parameter is $\kappa = 0.139$. Once again it is seen that correlation functions with similar wound parameters yield similar results for the particle-hole force $U(q)$.

The discussion now focuses on the results obtained for the C choice of correlations. The dynamic structure function $S(q,\omega)$ of the v_2 model of nuclear matter, derived from the LCRPA treatment, is plotted as a function of energy transfer $\hbar\omega$ in Figs. 3-6, at fixed q values of 1.47, 2.03, 2.76, and 3.87 fm^{-1}, respectively. These choices for q are close to values at which experimental data on the longitudinal response function $R_L(q,\omega)$ of medium- and large-A nuclei are available[24–26] and/or are (reasonably) close to q values examined in the microscopic calculation of Fantoni and Pandharipande.[1] In the figures we trace, for comparison, the Fermi-gas structure function $S_F(q,\omega)$ as well as the LCRPA result for the $c = 0.5$ hard-core. It has been argued by Fantoni and Pandharipande that the longitudinal dynamic structure function $S_L(q,\omega)$ of nuclear matter should not be very different from $S(q,\omega)$ over the range $q = 1.5$–$2.8\ \text{fm}^{-1}$, and, accordingly, that the theoretical nuclear-matter $S(q,\omega)$ multiplied by the square of a suitably chosen proton form factor, may be sensibly compared with the experimental response function $R_L(q,\omega)$ of heavy nuclei (cf. also Refs. 2,27). At a deeper level, such comparisons exploit the generally accepted view that meson-exchange currents, pion production, delta excitation, etc., should not be very important in the *longitudinal* response function, for the q values considered here.[28] In this view one must look elsewhere for a dramatic breakdown of the conventional picture of the nucleus as a nonrelativistic many-body problem with only nucleonic degrees of freedom.

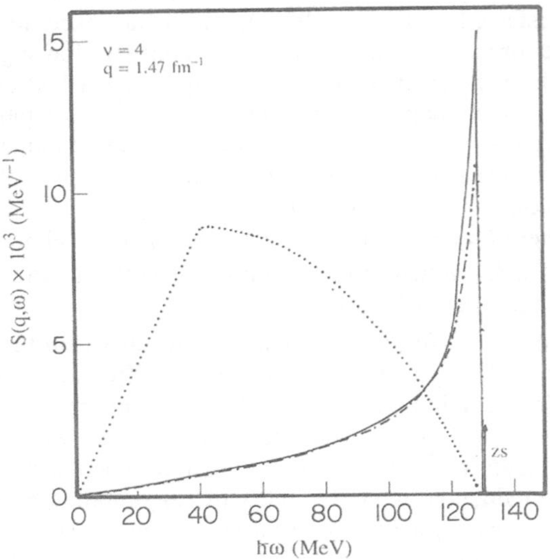

Fig. 3. Dynamic structure function $S(q,\omega)$ versus energy transfer $\hbar\omega$ at fixed wave number transfer $q = 1.47\ \text{fm}^{-1}$ and $k_F = 1.39\ \text{fm}^{-1}$. Solid curve: for v_2 model of nuclear matter based on C correlation function. Dot-dashed: for hard sphere nucleons with $c = 0.5$ fm. Dotted: for free nucleons. Zero sound contribution in v_2 case is indicated by vertical spike.

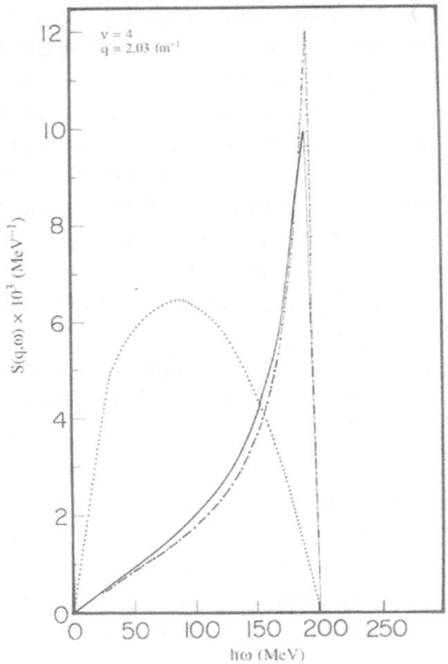

Fig. 4. Same as Fig. 3, except $q = 2.03\ \text{fm}^{-1}$ and there is no zero sound.

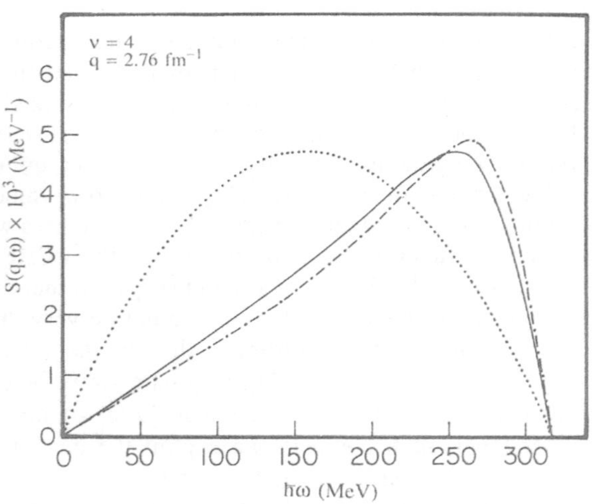

Fig. 5. Same as Fig. 3, except $q = 2.76\ \text{fm}^{-1}$ and there is no zero sound.

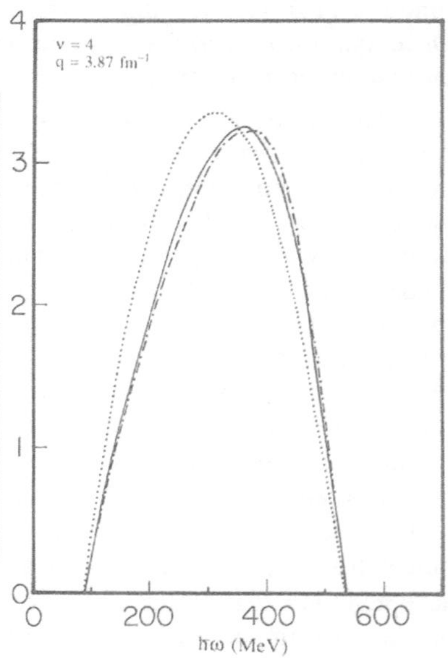

Fig. 6. Same as Fig. 3, except $q = 3.87\ \text{fm}^{-1}$ and there is no zero sound. Vertical scale as in Fig. 5.

As seen in Figs. 3-6, the LCRPA calculation shows the well-known quenching of the response at low energies, compared to the independent-particle-model result.[28–30] This effect is perhaps too severe in our results at the lower q values. The predictions for $S(q,\omega)$ are very similar for v_2 and hard-core potentials, suggesting that the response in the relevant q regime is insensitive to the choice of bare, state-independent central potential, provided the wound parameter is kept roughly the same. For a given q, the strength is shifted to values of $\hbar\omega$ higher than the experimental peak energy. This effect, as well as the excessive quenching at low ω, is presumably due to the overly repulsive character of the v_2 and hard-core potentials, which act equally in all partial waves, in contrast to the strong partial-wave dependence of realistic nucleon-nucleon interactions. Indeed, in the cases $q = 1.47\ \mathrm{fm}^{-1}$ and $q = 2.03\ \mathrm{fm}^{-1}$ (corresponding respectively to 290 MeV/c and 400 MeV/c), there is a very pronounced peak at the upper end of the allowed range of energy transfer, which is rather unphysical. As mentioned earlier, aside from zero sound, the LCRPA gives no response outside the (q,ω) region relevant for the noninteracting system, so the strength which would otherwise be distributed to higher energies seems to be "piling up" at the high-ω boundary. The similar trends seen in the results of Fantoni and Pandharipande and of Pines, Quader and Wambach are much milder. As we go to high q, the departures of the LCRPA result for $S(q,\omega)$ from the Fermi gas curve, and from the microscopic prediction of Fantoni and Pandharipande, become less noticeable. This is an obvious consequence of the decreasing importance of $U(q)$ at larger momentum transfers. Since, quite generally, we may expect the particle-hole force to become negligible in the large-q regime, predictions for $S(q,\omega)$ should be relatively insensitive to the choice of the bare interaction, or to the treatment within conventional nuclear many-body theory, at large enough q. Of course, excitation of subnucleonic degrees of freedom may become a significant factor in this domain.

For symmetrical nuclear matter at $k_F = 1.39\ \mathrm{fm}^{-1}$, a preliminary evaluation[31] of the LCRPA self-energy (5) using the techniques of Ref. 32 yields $m^* = 1.1m$ for the on-shell effective mass at the Fermi momentum and shows no significant enhancement of $m^*(k)$ in the vicinity of the Fermi surface (cf. Ref. 33). If the v_2 interaction is allowed to act only in S waves (more appropriate in the nuclear context), a value of $m^*(k_F)$ near $0.8m$ should be obtained.[34,19]

CONCLUSIONS

The density-density response function of symmetrical nuclear matter and pure neutron matter has been studied under simplifying conditions, leading to a microscopic theory – the local correlated random-phase approximation – requiring as input only the static structure function corresponding to a Jastrow description of the ground state. This theory has the virtues of easy application and straightforward interpretation of its predictions. Relative to an ordinary local RPA treatment, it takes account, approximately, of important dispersive, polarization, and geometrical effects arising from the strong interactions. In spite of its limitations, the theory has considerable value in establishing qualitative trends. For example, it is found that predictions for the dynamic structure factor do not depend very strongly on the choice of central, state-independent potential, so long as the wound parameter remains essentially unchanged.

One serious limitation of the theory as implemented here lies in the choice of the v_2 model interaction as the bare force. It is well known that state dependence of the nuclear interaction and especially the presence of a tensor component have important consequences for most nuclear properties. The dynamic response is no exception, particularly at low momentum transfer.[35] Nevertheless, before proceeding to a realistic interaction like Urbana or Argonne v_{14}, and to the introduction of three-nucleon interactions, the intrinsic limitations of the local correlated RPA approach should be quantified. As implemented here for parameterized correlation functions which do not strictly obey the Euler optimality condition, one such limitation is the neglect of the $\Delta(q)$ term in the formula (2) for the particle-hole force. We have argued that this is not likely to produce significant additional errors in the description of the particle-hole interaction, since the correlation function assumed in our primary calculations (of form (12)) is believed to be nearly optimal. Deviations from optimality will mainly affect $U(q)$ at small q – but in this region the local approximation must itself be considered questionable. These arguments notwithstanding, the importance of the $\Delta(q)$ term, for the correlations used here, is currently under numerical investigation, and optimal correlations will be implemented in future calculations.

The next step is to assess the effects of nonlocalities in the particle-hole force. Within the CBF description of elementary excitations, this will involve numerical solution of the full CRPA equations, with explicit inclusion of exchange.[8] Such a treatment is needed to remove the unphysical feature of local CRPA that (apart from possible collective modes) the response is confined to the (q, ω) domain in which the dynamic structure function of the noninteracting system is nonzero.

A truly quantitative description of the response and elementary excitations in nuclear systems will require a theory which goes beyond correlated RPA to the explicit inclusion of correlated multipair effects. Work in this direction has begun.

This research was supported in part by the Condensed Matter Theory Program of the Division of Materials Research, and by the Nuclear Theory Program of the Physics Division of the National Science Foundation, under Grant No. DMR-8519077. Support during the initial stages of the project was provided by the U. S. National Aeronautics and Space Administration under NASA Innovative Grant NAGW-122 to Washington University. J. W. C. thanks the Army Research Office for a travel grant.

REFERENCES

1. S. Fantoni and V. R. Pandharipande, Nucl. Phys. **A473**, 234 (1987).
2. A. Fabrocini and S. Fantoni, Nucl. Phys. A, in press.
3. E. Feenberg, *Theory of Quantum Fluids* (Academic, New York, 1969).
4. J. W. Clark and E. Krotscheck, Springer Lecture Notes in Physics **198**, 127 (1984).
5. J. M. C. Chen, J. W. Clark, and D. G. Sandler, Zeits. Physik **A305**, 223 (1982).
6. E. Krotscheck, Phys. Rev. A **26**, 3536 (1962).
7. V. R. Pandharipande, R. B. Wiringa, and B. D. Day, Phys. Lett. **57B**, 205 (1975).
8. N.-H. Kwong, Ph.D. thesis, California Institute of Technology (1982), unpublished.
9. E. Mavrommatis, R. Dave, and J. W. Clark, in *Condensed Matter Theories*, Vol. 2, ed. P. Vashishta, R. K. Kalia and R. F. Bishop (Plenum, New York, 1987), p. 249.
10. E. Krotscheck, R. A. Smith, J. W. Clark, and R. M. Panoff, Phys. Rev. B **24**, 6383 (1981).
11. E. Krotscheck, in *Quantum Fluids and Solids, Sanibel, Florida, 1983*, ed. E. D. Adams and G. G. Ihas (AIP, New York, 1983), p. 132.
12. G. E. Brown, *Many Body Problems* (North-Holland, Amsterdam, 1972).
13. W. M. Alberico, R. Cenni, and A. Molinari, Riv. del Nuovo Cim. **1**, 1 (1978).
14. W. Czyż and K. Gottfried, Ann. of Phys. **21**, 47 (1963).
15. F. Green, D. N. Lowy, and J. Szymanski, Phys. Rev. Lett. **48**, 638 (1982).
16. C. H. Aldrich III and D. Pines, J. Low Temp. Phys. **25**, 677 (1976); **32**, 689 (1978).
17. D. Pines, K. F. Quader, and J. Wambach, Nucl. Phys. **A477**, 365 (1988).
18. J. W. Clark, Prog. Part. Nucl. Phys. **2**, 89 (1979); and references cited therein.

19. A. Ramos, A. Polls, and W. H. Dickhoff, Nucl. Phys. A, in press.
20. D. Ceperley, G. V. Chester, and M. H. Kalos, Phys. Rev. B **16**, 3081 (1977).
21. O. Benhar, C. Ciofi degli Atti, A. Kallio, L. Lantto, and P. Toropainen, Phys. Lett. **60B**, 129 (1976); O. Benhar, C. Ciofi degli Atti, S. Fantoni, S. Rosati, A. Kallio, L. Lantto, and P. Toropainen, Phys. Lett. **64B**, 395 (1976).
22. J. W. Clark and M. L. Ristig, Phys. Rev. C **5**, 1553 (1972).
23. M. F. Flynn, private communication.
24. Z. E. Meziani *et al.*, Phys. Rev. Lett. **52**, 2130 (1984).
25. M. Deady *et al.*, Phys. Rev. C **33**, 1897 (1986).
26. C. C. Blatchley *et al.*, Phys. Rev. C **34**, 1243 (1986).
27. D. B. Day *et al.*, Phys. Rev. C **40**, 1011 (1989).
28. J. M. Laget, Springer Lecture Notes in Physics **137**, 148 (1981); Physics Reports **69**, 1 (1981).
29. J. W. van Orden, Ph.D. thesis, Stanford University (1978), unpublished.
30. R. Rosenfelder, Ann. of Phys. **128**, 188 (1980).
31. R. D. Dave, private communication.
32. J. P. Blaizot and B. L. Friman, Nucl. Phys. **A372**, 69 (1981).
33. E. Krotscheck, J. W. Clark, and A. D. Jackson, Phys. Rev. B **28**, 5088 (1983).
34. W. H. Dickhoff, private communication.
35. A. Dellafiore and F. Matera, Phys. Rev. C **40**, 960 (1989).

THEORY OF THE CRITICAL POINT OF ^{4}He

A. Meroni and L. Reatto

Dipartimento di Fisica
Università Degli Studi di Milano, 20133 Milano, Italy

and

K. J. Runge

Theory Center
Cornell University, New York, N. Y. 14853, USA

INTRODUCTION

We do not have yet a microscopic global theory of the thermodynamic properties of ^{4}He in the normal phase, inclusive of the region of the critical point. At a very simple level the effect of the attractive part of the interatomic forces has been considered [1] in a perturbative way starting from the properties of quantum hard spheres. This is instructive but critical fluctuations are very badly treated in this way and one just has a mean field theory.

With respect to the universal aspects of critical phenomena convincing arguments [2] say that quantum effects are irrelevant at the critical point so that the universality class is predicted, in agreement with experiment, to be that of the Ising model. Other quantities, for instance the critical temperature T_c and density ρ_c, are effected by quantum effects and in the case of ^{4}He these are very large as shown by the strong deviation from the law of corresponding states.

A complete theory of ^{4}He in the normal fluid phase is very difficult because one has to treat at the same time strong quantum effect and the regime of strong density fluctuations characteristic of the critical point. In the case of a classical fluid only recently [3] a liquid state theory has been developed which has the proper critical behavior, i.e. scaling laws and non classical critical exponents, starting from a realistic pair interaction. In this theory, the hierarchical reference theory (HRT), the interatomic potential $v(r)$ is written as the sum of a strong repulsive part $v_0(r)$ and of a weaker attractive part $w(r)$. The properties of a system with interaction $v_0(r)$, the reference system, are assumed to be known from some other theory and HRT gives the equations which determine how the free energy and the correlation functions are modified as $w(r)$ is selectively turned on in wavevector space.

Condensed Matter Theories, Volume 5
Edited by V.C. Aguilera-Navarro
Plenum Press, New York, 1990

We have extended HRT to treat the quantum case in an approximate way. The attractive part $w(r)$ of the interaction is a rather slowly varying function of r so that we expect that the quantum effects directly due to $w(r)$ are rather small. More precisely we neglect the commutator between w and the hamiltonian of the reference system and we end up with a problem which is formally equivalent to one in classical statistical mechanics. This equivalent classical system has the same attractive forces of the quantum system and the role of the "repulsive forces" is taken by the logarithm of the diagonal part of the density matrix of the reference system. These "repulsive forces" are not simply pair additive but the remarkable property of HRT is that it is not necessary to specify the interaction of the reference system but only its equation of state and correlation functions are needed. In this way we can apply the HRT approach to the quantum problem.

For classical fluids a very successful approximation [4] consists in truncation of the HRT equations at the first one, that of the free energy. In this case one needs to know just the chemical potential and the pair correlation function of the reference system which we have taken to be hard spheres. We obtain these quantities from a path integral computation and in this way we get a closed equation. In the next section we present our theory, then we discuss how to treat the reference system and finally we give our results.

THEORY

We consider a system of ^{4}He atoms interacting through a pair potential $v(r)$ which we write as the sum of a singular repulsive part $v_0(r)$ and of a regular attractive part $w(r)$. Now we can write the canonical partition function of the system

$$Q_N = Tr(e^{-\beta(T+V)}) = Tr(e^{-\beta(T+V_0)-\beta\Delta V}) \tag{1}$$

where T is the kinetic energy operator, $V = \sum_{i<j} v(r_{ij})$, $V_0 = \sum_{i<j} v_0(r_{ij})$, $\Delta V = \sum_{i<j} w(r_{ij})$ and β the inverse temperature, in the form

$$Q_N = Q_N^0 \langle Te^{-\int_0^\beta d\lambda \Delta V_\lambda} \rangle_0. \tag{2}$$

T denotes the "time" ordered product, the average with respect to the reference system is given by

$$\langle \ldots \rangle_0 = \frac{Tr(\ldots e^{-\beta(T+V_0)})}{Tr(e^{-\beta(T+V_0)})}$$

and we write

$$\Delta V_\lambda = e^{\lambda(T+V_0)} \Delta V e^{-\lambda(T+V_0)}. \tag{3}$$

ΔV is a rather slowly varying function of r so that one can use a semiclassical approximation even for ^{4}He. The zero order approximation for this problem is to neglect the non commutativity of ΔV with the kinetic part to get the result

$$Q_N = Q_N^0 \langle e^{-\beta \Delta V} \rangle_0 \tag{4}$$

If we introduce now the density matrix of the reference system $\rho_0(R, R', \beta)$ where R represents collectively the coordinates of the N–particle system this expression reduces to

$$Q_N = Q_N^0 \int dR \rho_0(R, R, \beta) e^{-\beta \Delta V(R)}. \tag{5}$$

Notice that for a Bose system ρ_0 is a non negative function. This expression is very significant because we have reduced the quantum problem to a computation which is formally equivalent to a classical problem of an assembly of N classical particles interacting through a potential which is the sum of $\beta^{-1} \ln \rho_0(R, R, \beta)$ and of an attractive pair potential which is the same of the quantum system. $\ln \rho_0$ can be cluster expanded so that this equivalent classical problem contains, in addition to a pairwise additive potential, also three body and higher order potentials and all these will be temperature dependent. These terms are not known but we will see that we do not need to know them explicitly.

In order to apply the HRT approach to this equivalent classical problem it is only necessary to assume that the cluster expansion of $\ln \rho_0$ truncates after a finite number of terms, i.e. many body terms of arbitrary large order as N increases should not be present. This is an unproved but very reasonable assumption. We take as reference system the system with interaction $\ln \rho_0$ and study how the properties change as ΔV is turned on. However ΔV is not introduced all in one step but gradually via the introduction of *Q-systems*. This is the key element which allows us to go beyond mean field.

The *Q-systems* interact through a potential which is the sum of the reference part $\ln \rho_0$ and of the attractive pair potential $w_Q(r)$ which is the Fourier transform of

$$\tilde{w}_Q(q) = \begin{cases} \tilde{w}(q) & \text{for } q > Q \\ 0 & \text{for } q < Q. \end{cases} \tag{6}$$

The *Q-systems* are an interpolation between the reference system ($Q = \infty$ and the fully interacting system ($Q = 0$), and the evolution of the excess Helmholz free energy as we change Q is given by the same equation [3] as in the case of the truly classical system, i.e.

$$-\frac{d\mathcal{A}_Q^{ex}}{dQ} = \frac{Q^2}{4\pi^2} \ln \left[1 + \frac{\rho\tilde{\phi}(Q)}{(1 - \rho\mathcal{C}_Q(Q))} \right] \tag{7}$$

where $\mathcal{A}^{ex}$ differs from A^{ex} by analytic terms, $\tilde{\phi}(k) = -\tilde{w}(k)/k_B T$ and $\mathcal{C}_Q$ is related to the direct correlation function c_Q of the *Q-system* by

$$\mathcal{C}_Q(k) \equiv c_Q(k) + \tilde{\phi}(k) - \tilde{\phi}_Q(k). \tag{8}$$

Eq. (7) is exact and the fact that in the present case the reference system has complicated many body forces and these interactions are temperature dependent is all buried in the boundary condition of eq. (7) at $Q = \infty$.

In the HRT the evolution with Q of the two body function $c_Q(k)$ is given by a differential equation involving the three and four body direct correlation functions; it has been shown [4] in the classical case that a very successful approximation is obtained if the equation (7) for the free energy is decoupled from the higher order equations with the position

$$\mathcal{C}_Q(r) = c_R(r) + \lambda_Q \phi(r). \tag{9}$$

If there is a hard core at d eq. (9) is taken only for $r > d$, while $\mathcal{C}_Q(r)$ for $r < d$ is determined by the condition of vanishing radial distribution function of the *Q-system*:

$$g_Q(r) \equiv 1 + \int \frac{d^3k}{(2\pi)^3} \frac{c_Q(k)}{1 - \rho c_Q(k)} e^{i\mathbf{k}\cdot\mathbf{r}} = 0 \quad \text{for} \qquad r < d. \tag{10}$$

$c_R(r)$ is the direct correlation function of the (quantum) reference system, $g_Q(r)$ is the radial distribution function of the *Q-system* and λ_Q is determined by the compressibility sum rule which reads

$$\mathcal{C}_Q(k=0) = \partial^2 \mathcal{A}_Q^{ex} / \partial \rho^2. \tag{11}$$

We assume that the ansatz (9) is appropriate also in the quantum case. Then we have the same problem of the classical case but the quantities pertaining to the reference system are given by the solution of the quantum problem characterized by the repulsive potential $v_0(r)$. There is evidence [5,6] also in quantum systems that the properties of such a soft repulsive potential can be parameterized into the properties of a hard sphere system of a properly chosen diameter d. The scattering length of $v_0(r)$ has been used as d however in the quantum case at $T > 0$ there is no stringent argument for this choice. Also in our case it is convenient to describe the reference system using the properties of a quantum hard sphere fluid so that we can use the extensive results obtained by Path Integral Monte Carlo (PIMC) described in the following section.

It has been shown [3] that the equation (7), with the condition (11) and the request of analyticity of $\mathcal{C}_Q(k)$ as a function of k as implied by (9) can be analyzed within the ϵ–expansion framework ($\epsilon = 4 - D$ where D is the dimensionality) and turns out to be correct to first order in ϵ, while a non–perturbative analysis in three dimensions gives a divergence of the adimensional isothermal compressibility $S(0)$ with an exponent $\gamma = 1.378$ which is 10% accurate (experimentally $\gamma \sim 1.24$). We recall that the HRT is the first truly liquid state theory which gives a non trivial critical behavior.

The numerical calculation is organized in the following manner: the partial differential equation (7) is integrated starting from $Q = \infty$ down to $Q = 0$ and on a density range 0–ρ_{max} where ρ_{max} is a density of order of the freezing density. For $r < d$ we expand $\mathcal{C}_Q(r)$ on a polynomial basis in the interval $(0, d)$, the coefficient are determined by the requirement that the equation (10) should be satisfied for each Q. This requirement gives a set of partial differential equations for the coefficient all coupled with the equation (7) due to the constraint (11). These equations for the coefficients of $\mathcal{C}_Q$ can be simplified and reduced to ordinary differential equations by neglecting the feedback of critical fluctuations on the short range behavior [4]. For this calculation a basis of five Legendre polynomial has been used. So we must solve the equation (7) coupled with these ordinary differential equations that guarantee the validity of the core condition. At low density the boundary condition for (7) is given by virial expansion while at ρ_{max} we put $\lambda_Q = 1$ which corresponds to an Optimized Random Phase (ORPA) closure. The initial condition at $Q = \infty$ requires the chemical potential μ_R, this because the actual equation which is solved is the density derivative of eq. (7), and $c_R(k)$ of the reference system at the temperatures of the computation and for all densities in the range 0–ρ_{max}. How we obtain these quantities is discussed in the next section.

REFERENCE SYSTEM

The method of Path Integral Monte Carlo (PIMC) was used to compute the required properties of the quantum hard-sphere reference system. Simply speaking, one uses the Monte Carlo method to sample paths $R(\tau)$ of the system in configuration space out of the probability distribution $e^{-S[R(\tau)]}$, where $S[R(\tau)]$ is Feynman's action

$$S[R(\tau)] = \int_0^\beta d\tau \quad \left[\frac{1}{2}\frac{m}{\hbar^2}\left(\frac{dR}{d\tau}\right)^2 + V_0(R(\tau))\right], \tag{12}$$

and m denotes the mass of the atom. Thermodynamical and structural properties of the liquid are then found as particular averages over the sampled paths. The full details of the simulations may be found in Runge and Chester.[5]

The reference system properties to be computed by PIMC are the chemical potential $\mu(\rho, T)$ and the static structure function $S(k; \rho, T)$. The chemical potential was obtained along a given isotherm by integrating the pressure with respect to density to obtain the helmholtz free energy per particle f:

$$f(\rho, T) = f_{\text{ideal}}(\rho, T) + \int_0^\rho k_B T \left(\frac{P(\rho', T)}{\rho' k_B T} - 1\right) \frac{d\rho'}{\rho'} \tag{13}$$

and then using the relation $\mu = f + P/\rho$. The integral was performed by least-squares fitting the PIMC pressure $P(\rho, T)$ to a polynomial in ρ and then doing the integral analytically.

The function $S(k)$ is the Fourier transform of the radial distribution function $g(r)$ in eq. (10). Thus, the relation to $c(k)$ is given by

$$S(k) = \frac{1}{1 - \rho c(k)} \tag{14}$$

The quantity $g(r)$ is the one most readily computed in the PIMC simulations. Unfortunately, since the simulations are performed on finite systems (108 hard-spheres in a periodic cube) $g(r)$ is known only in the range $0 < r < L/2$ where L is the side length of the box. Therefore, to Fourier transform $g(r)$ some sort of extension to $r > L/2$ must be done. We have used the method of Ceperley and Chester[7] where a sum of (usually two or three) decaying oscillating exponentials is fit to $g(r)$ for $r_c < r < L/2$. The value r_c is usually chosen to be somewhere between the first maximum and first minimum of $g(r)$. In performing the Fourier transform the PIMC data is used for $r < L/2$ and the fitting function for $r > L/2$. This procedure removes much of the 'ringing' that occurs in $S(k)$ for small values of k, however, for very small values of k our results occasionally have some spurious oscillations. We have regularized our Fourier transforms at small k by replacing $S(k)$ in the region $0 < k < k_c$ by the function

$$\tilde{S}(k) = a + bk^2 + ck^4. \tag{15}$$

The value a is fixed by the PIMC equation of state, since $S(0) = k_B T \left(\frac{\partial P}{\partial \rho}\right)_T^{-1}$. The parameters b and c are then selected so that $\tilde{S}(k)$ matches smoothly onto the Fourier transform $S(k)$ at k_c. We have taken $k_c d = 2.5$, where d is the hard-sphere diameter.

Finally, the PIMC simulations were performed on a 'grid' in the density-temperature plane consisting of six temperatures in the range $1.2 \leq T^* \leq 2.88$ and eight densities in the range $0 \leq \rho^* \leq 0.3$, where $T^* = k_B T/(\hbar^2/md^2)$ and $\rho^* = \rho d^3$. To obtain the chemical potential and $S(k)$ at an arbitrary point (ρ, T) in the above region, cubic splines in both the density and temperature variables were fit to the data on the grid points.

RESULTS

The equation (7) has been solved numerically with a Lax–Wendroff scheme [8] on a grid in density varying from 50 to 75 points for a range of reduced density $\rho^* = \rho d^3$ from $\rho^* = 0$ to $\rho^*_{max} = 0.3$ (the melting transition density for quantum hard spheres varies from $\rho^* = 0.27$ at $T = 0$ to $\rho^* \sim 0.35$ at $T = 4K$). At each step a stability check is performed according to Von–Neumann stability criterion and a reduction of the integration step in Q is performed when needed. The ordinary differential equations for the core condition is integrated with a simple discrete scheme.

The calculation has been performed both with standard Lennard–Jones (LJ) potential with the de Boer–Michels parameters [9]

$$\begin{aligned} v(r) &= 4\epsilon[(\sigma/r)^{12} - (\sigma/r)^6] \\ \epsilon &= 10.22K \\ \sigma &= 2.556\text{Å} \end{aligned} \tag{16}$$

and with the HFDHE2 potential of Aziz et al. [10]. The division of the potential in attractive and repulsive part has been done according to WCA criterion [11], i.e. if the potential has a minimum at $r = r_m$ of depth ϵ we write

$$v_0(r) = \begin{cases} v(r) + \epsilon & \text{for } r < r_m \\ 0 & \text{for } r > r_m \end{cases} \tag{17}$$

$$w(r) = \begin{cases} -\epsilon & \text{for } r < r_m \\ v(r) & \text{for } r > r_m \end{cases} \tag{18}$$

As reference system we substitute $v_0(r)$ with the hard sphere potential of a suitable diameter d. The critical point was located looking at the divergence of the adimensional compressibility $S(0)$. For example with the choice of d as the scattering length a of the corresponding repulsive part of the potential $v_0(r)$ (for LJ $a = 2.138\text{Å}$ while for HFDHE2 $a = 2.203\text{Å}$) we find

$$\begin{aligned} \rho_c &= 0.009\text{Å}^{-3} \\ T_c &= 4.1K \end{aligned} \tag{19}$$

for LJ and

$$\begin{aligned} \rho_c &= 0.007\text{Å}^{-3} \\ T_c &= 3.8K \end{aligned} \tag{20}$$

for HFDHE2.

A calculation based on a mean field theory has been performed recently by Runge et al. [1] with the same potentials and in that paper it was found that the critical point was located at $\rho_c = 0.0116\mathring{A}^{-3}$ for both the potential while the critical temperature was at $T_c = 6.4K$ for LJ and $T_c = 6.8K$ for HFDHE2. The corresponding experimental values are $\rho_c = 0.0105\mathring{A}^{-3}$ and $T_c = 5.2K$.

Both the critical density and the temperature given by our theory are low. It should be noticed, however, that the position of the critical point is strongly dependent on the choice of the effective diameter d. It turns out that it is possible to choose d to give roughly the correct location of the critical point both in density and in temperature. Using $d = 2.15\mathring{A}$ with HFDHE2 potential, the critical point is at $T_c = 5.48K$ and $\rho_c = 0.0102\mathring{A}^{-3}$.

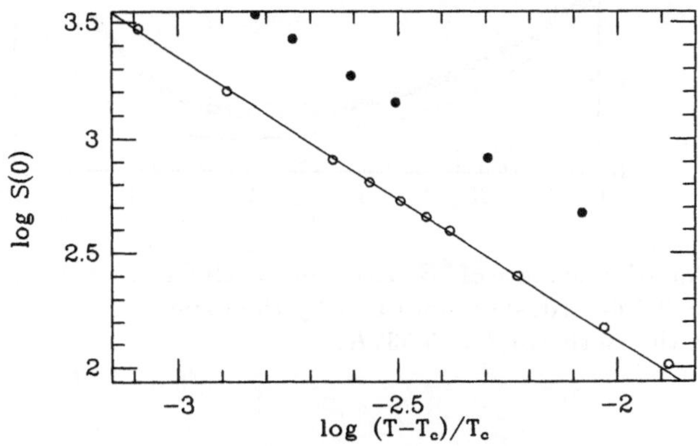

Fig. 1. Comparison between the adimensional isothermal compressibility factor $S(0)$ as given by the theory for the choice $d = 2.15\mathring{A}$ ($\circ$) and $d = 2.203\mathring{A}$ ($\bullet$) and by a fit to the experimental data [12] (continuous line) as a function of the reduced temperature $t = (T - T_c)/T_c$. The potential is HFDHE2.

This same choice gives also the best representation (in this theory) of the divergence of $S(0)$. The comparison of $S(0)$ with the experimental data [12] is presented in fig. 1. We remark that in the quantum case at $T > 0$ a well defined criterion for the choice of the hard sphere diameter is not known and this circumstance is particularly disappointing due to the strong dependence of the results on this parameter.

The choice of d as the scattering length of v_0 has been justified[6] at $T = 0K$ but there is no theoretical justification that this is the optimal choice also at T of the order of the critical temperature.

In the reduced temperature range shown in the figure one does not see the asymptotic theoretical value of the exponent γ (1.378) but an effective value which is in good agreement with experiment. This is similar to what was found in the classical case [4].

For the same choice of the diameters we compare in fig. 2 the equation of state for an isotherm slightly above T_c as given by theory and by experiment [13]. The agreement is not good and the disagreement is particularly strong at low and high density. This means that it is not so much the treatment of the critical region which is at fault but how we treat the interaction in a more global sense. One possible reason is that the closure (9) is less appropriate in the quantum than in the classical case. The other is that it is necessary to take into account in some way also the quantum effects due to the attractive forces.

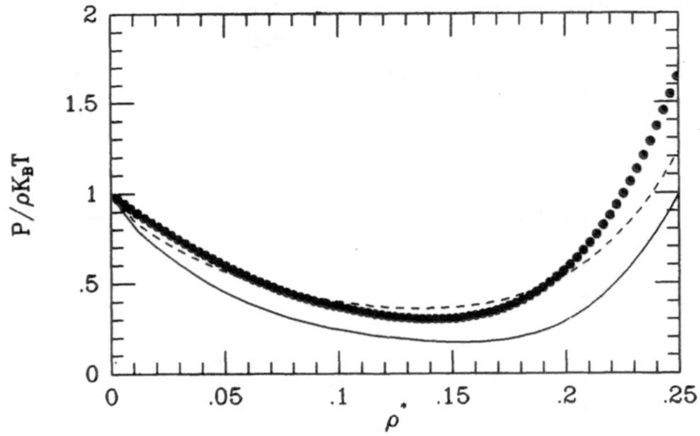

Fig. 2. Equation of state of ^{4}He as given by HRT for $d = 2.15\mathring{A}$ (continuous line) and $d = 2.203\mathring{A}$ (dashed line) and by the experiment [13] (•) as function of density on the isotherm $T = 5.537K$.

CONCLUSIONS

We have shown that the critical point of ^{4}He can be treated in the framework of HRT and for the first time we produce a microscopic theory which gives non classical critical phenomena for a quantum system. On the other hand this is only a first step toward a more complete approach. The question of the choice of the hard sphere diameter has to be clarified as well as the role of the closure (9) for the direct correlation function for a quantum system. Another step is to include the quantum effect due to the attractive part of the pair interaction, by taking into account the commutator between this attractive part and the reference system hamiltonian to first order.

Finally the equations have been studied only for temperature greater than the critical temperature T_c, because for $T < T_c$ the equation (7) becomes unstable in proximity of the coexistence region as in the classical case. Work is in progress to treat also $T < T_c$.

ACKNOWLEDGEMENTS

This work was partially supported by Consiglio Nazionale delle Ricerche under Piano Finalizzato "Sistemi Informatici e Calcolo Parallelo" and under Italia–U.S. Cooperative Science Program and by Ministero Pubblica Istruzione.

REFERENCES

1. K.J. Runge, G.V. Chester and M.H. Kalos, *Phys. Rev. B* **39**, 2707 (1989).
2. F.J. Wegner in C. Domb and M.S. Green *Phase Transitions and Critical Phenomena vol. 6* (Academic Press, 1976)
3. A. Parola and L. Reatto, *Phys. Rev. A* **31**, 2417 (1985).
4. A. Parola, A. Meroni and L. Reatto, *Phys. Rev. Lett.* **62**, 2981 (1989).
5. K.J. Runge and G.V. Chester, *Phys. Rev. B* **38**, 135 (1988).
6. M.H. Kalos, D. Levesque and L. Verlet, *Phys. Rev. A* **9**, 2178 (1974).
7. G.V. Chester and D.M. Ceperley, *Phys. Rev. A* **15**, 755 (1977).
8. W.H. Press, B.P. Flannery, S.A. Teukolsky and W.T. Vetterling, *Numerical Recipes* (Cambridge University Press, Cambridge, 1986).
9. J. de Boer and A. Michels, *Physica (Utrecht)* **5**, 945, 1983.
10. R.A. Aziz, V.P.S. Nain, J.S.Carley, W.L. Taylor and G.T. McConville, *J. Chem. Phys.* **70**, 4330 (1979).
11. J.T. Weeks D. Chandler and H.C. Andersen, *J. Chem. Phys.* **54**, 5237 (1971).
12. P. R. Roach, *Phys. Rev.* **170**, 213 (1968).
13. R.D. McCarty, *J. Phys. Chem. Ref. Data*, **2**, 923, 1973.

CORRELATIONS AND MOMENTUM DISTRIBUTION IN THE GROUND STATE OF LIQUID ^{3}He

Sergio Rosati and Michele Viviani

Department of Physics, University of Pisa and I.N.F.N.
Piazza Torricelli 2, 56100 Pisa, Italy

Enrique Buendia

Department of Physics, University of Granada
18003 Granada, Spain

Abstract. A flexible variational wave function describing the ground state of liquid ^{3}He is used to calculate the momentum distribution n(k) and the one-body density matrix $\rho_1(r)$ of the system. The trial wave function contains pair, triplet, backflow and spin-dependent correlations. The importance of the various correlations to determine n(k) has been explored and the results are compared to those available from variational and GFMC approaches and to the experimental data.

1. Microscopic calculation on liquid ^{3}He

At present the Green Function Monte Carlo method (GFMC) and the variational theory have been extensively applied to the study of the equation of state of bulk liquid helium. The aim is to understand if the correlations amongst helium atoms are well described and if the HFDHE2 potential as given by Aziz et al.[1] does accurately reproduce the helium interatomic pair potential. The confidence in the results so far obtained is satisfactory, but other checks on the evaluation of the elementary diagrams in the variational approach, and on the finite size effects in the GFMC calculations, could be very useful.

In this section the variational wave function adopted to describe the ground state of liquid ^{3}He at zero temperature is given, while the results obtained for the momentum distribution are presented in the next section. The wave function has the form

$$\Psi_0=F\Phi_0, \tag{1}$$

where the model function Φ_0 describes the non-interacting Fermi sea, and F is an appropriate correlation operator. The form chosen for F is the one determined in ref. 2

$$F=F_J F_T \mathcal{S} \{ \prod_{j>i=1}^{N} F_{SD}(i,j) \}, \tag{2}$$

where

$$F_J = \prod_{j>i=1}^{N} f_2(r_{ij}), \tag{3}$$

Condensed Matter Theories, Volume 5
Edited by V.C. Aguilera-Navarro
Plenum Press, New York, 1990

is the so-called Jastrow correlation factor,

$$F_T = \prod_{k>j>i=1}^{N} f_3(\mathbf{r}_{ij},\mathbf{r}_{ik}), \tag{4}$$

is a triplet correlation factor, and

$$S\{\prod_{j>i=1}^{N} F_{SD}(i,j)\}, \tag{5}$$

is a symmetrized product of state-dependent pair correlation operators. Such a correlation operator is chosen of the form

$$F_{SD}(i,j) = f_k(i,j)\, f_\sigma(i,j), \tag{6}$$

where $f_k(i,j)$ is the pair correlation operator generating the Feynman-Cohen backflow [3],

$$f_k(i,j) = \exp\{\, i\, \eta_B(r_{ij})\, \mathbf{r}_{ij} \cdot [\,\mathbf{K}(i) - \mathbf{K}(j)\,]\}, \tag{7}$$

with the operator $\mathbf{K}(i) = -\,i\,\nabla_i$ acting only on Φ_0. Finally, the spin-dependent correlation factor is given by

$$f_\sigma(i,j) = 1 + \eta_\sigma(r_{ij})\, \sigma_i \cdot \sigma_j\, . \tag{8}$$

The models listed as J, JT, JTB, JTS and JTBS correspond to the approximations where only f_2, or f_2, f_3, or f_2, f_3, f_k, or f_2, f_3, f_σ, or f_2, f_3, f_k, f_σ are taken different from unit. The dependence of the correlation factors on the interparticle distances has been determined in ref.2. The behavior of the pair correlation functions calculated at the equilibrium density $\rho_0=.277\sigma^{-3}$, σ=2.556 Å, for the JTBS model is shown in fig.1; in particular, the long-range tail $(\sim r^{-2})$ of the function $f_2(r)$ must be noticed. The energies per particle obtained for liquid

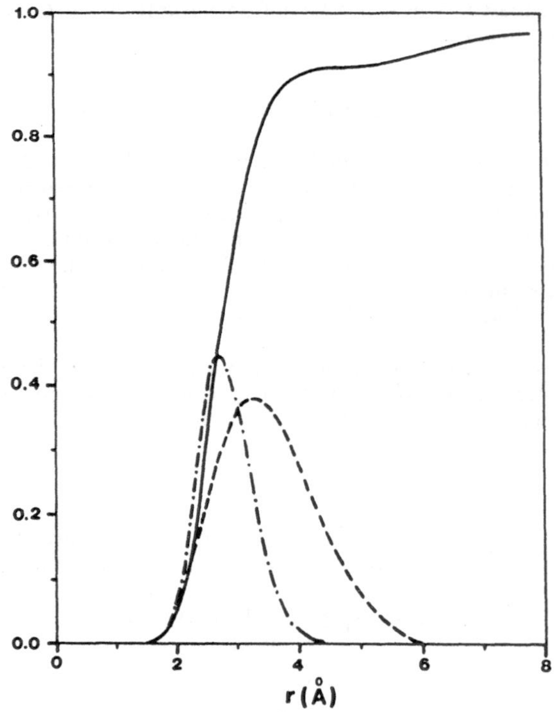

Fig1. Taken from ref.2. The correlation functions f_2 (solid line), $10 f_2 \eta_B$ (dot-dashed line) and $10 f_2 \eta_\sigma$ (dashed line).

^3He in ref.2, for different models and in correspondence to the Aziz potential, are reported in table I. As it can be seen by inspection of the table, the calculated JTBS energies compare favourably with the experimental values so that we can be confident that the wave function so determined be well suited for evaluating other quantities of interest as the momentum distribution of the system.

Table I. Extracted from ref.2. Energy per particle of liquid ^3He for three values of the density and different models (the JTBS values correspond to those listed as JTSB3 in ref.2). In the last column the experimental data are reported. All the energies are in K.

$\rho(\sigma^{-3})$	E(J)	E(JT)	E(JTB)	E(JTS)	E(JTSB)	E_{exp}
0.277	-1.31	-1.80	-2.31	-2.31	-2.46	-2.47
0.301	-1.00	-1.60	-2.17	-2.26	-2.38	-2.42
0.330	-0.38	-1.15	-1.73	-1.99	-2.15	-2.22

2. Momentum distribution in liquid ^3He at T=0

The one-body density matrix $\rho_1(\mathbf{r}_1, \mathbf{r}_1')$ for bulk liquid ^3He depends only on $r=|\mathbf{r}_1 - \mathbf{r}_1'|$ and is defined for the unit normalized state $|\Psi\rangle$ by

$$\rho_1(r) = N \sum_{spin} \int d^3r_2 \ldots d^3r_N \; \Psi^*(1,2, \ldots ,N)\, \Psi(1',2,\ldots ,N). \tag{9}$$

The momentum distribution and the one-body density matrix are related by Fourier transformation

$$n(k) = \frac{1}{\nu} \int d^3r \; \rho_1(r)\, e^{i\,\mathbf{k}\cdot\mathbf{r}}, \tag{10}$$

with $\nu=2$ being the spin multiplicity of unpolarized ^3He. The following properties are satisfied:

$$\lim_{r\to 0} \rho_1(r) = \frac{N}{\Omega} = \rho, \tag{11}$$

$$\int \frac{d^3k}{(2\pi)^3\rho}\, n(k) = 1, \tag{12}$$

$$-\frac{\hbar^2}{2m}\langle \nabla_1^2 \rangle = -\frac{\hbar^2}{2m}\frac{1}{\rho}\frac{d^2\rho_1(r)}{dr^2}\bigg|_{r=0} = \frac{\hbar^2}{2m}\int \frac{d^3k}{(2\pi)^3\rho}\, k^2\, n(k). \tag{13}$$

In the variational theory $\rho_1(r)$, and hence $n(k)$, can be calculated by means of the Monte Carlo method[4]. Alternatively, the Fermi Hypernetted Chain (FHNC) technique allows to sum up all the contributions from the cluster expansion of $\rho_1(r)$; this has been originally done by Fantoni[5] for wave functions containing only two-body correlations; the extension to the case where also triplet[6], backflow[7] and spin-dependent[8] correlations are present is rather straightforward. A crucial difficulty is found in numerically solving the FHNC equations due to the presence of the so-called elementary diagrams, namely highly connected diagrams which cannot be constructed by parallel and chain connection of the simplest bonds. One can disregard all the elementary contributions, or include only basic elementary diagrams involving up to n particles (FHNC/n approximation). However, numerical evaluations with n>4 are nearly prohibitive and for dense systems, like liquid helium, basic elementary diagrams with n>4 can produce appreciable corrections. Due to this motivation, various approximations have been devised to simulate the effect of elementary diagrams. Two approximations have been applied to the study of liquid helium with satisfactory results. The first one, called the scaling approximation[9,10] (FHNC/s), assumes that the sum of all the elementary diagrams is proportional to the contribution of the four-particle elementary diagrams. The second approximation, the interpolating integral equations method[11,2] (FHNC/α), interpolates between the FHNC/0 equations and the linearized version (Fermi Percus-Yevick equations), and it is the one adopted in this paper. Both the approximations contain adjustable parameters that are chosen so as to fulfill certain exact conditions.

When only pair correlations $f_2(r)$ are included in the variational wave function, from the definition (9) and after cluster expanding in powers of $w_2(r) = f_2(r) - 1$ and $h_2(r) = f_2(r)^2 - 1$, one gets the expression[5]

$$\rho_1(r) = \rho\, n_0\, N(r), \tag{14}$$

where the strength factor n_0 is given by

$$n_0 = \exp\,[2R_w - R_d]. \tag{15}$$

The quantities $N(r)$, R_w and R_d can be expressed in terms of the solutions(*) $g_{dd}(r)$, $g_{de}(r)$, $g_{ee}(r)$ and $g_{cc}(r)$ of the standard FHNC equations[12] and three more functions $g_{wd}(r)$, $g_{we}(r)$ and $g_{wcc}(r)$ which can be calculated by means of a different set of FHNC equations[5]. Together with the functions g_{xy} it is necessary to introduce nine functions $E^J(r)$ which are the sum of all the elementary contributions of a given type. The calculation of $E^J{}_{dd}(r)$, $E^J{}_{de}(r)$, $E^J{}_{ee}(r)$, $E^J{}_{cc}(r)$ has been discussed in ref.2, the remaining elementary functions $E^J{}_{ww}(r)$, $E^J{}_{wd}(r)$, $E^J{}_{we}(r)$, $E^J{}_{wcc}(r)$, $E^J{}_{wcwc}(r)$ have been calculated by an interpolating procedure; an unique parameter β has been used for $E^J{}_{wd}(r)$, $E^J{}_{we}(r)$, $E^J{}_{wcc}(r)$ whilst the parameters γ and δ have been introduced for $E^J{}_{ww}(r)$ and $E^J{}_{wcwc}(r)$, respectively. When also triplet correlations are considered, on the ground of the magnitude of the correction, only the elementary contributions involving four particles and triplet correlations have been taken into account, i.e.

$$E_{xy}(r) = E^J{}_{xy}(r) + E^T{}_{xy,4}(r). \tag{16}$$

The necessary modifications in the FHNC equations when backflow correlations are included in the wave function are given in ref.7 and those due to spin-dependent correlations in ref.8.

The normalization condition (12) can be rewritten as

$$n_0\, N(0) = \rho_1(0)/\rho = 1. \tag{17}$$

However, for Fermi liquids the following more stringent conditions hold[13]

$$n_0 \exp[G_{ww}(0) + E_{ww}(0)] = 1, \tag{18}$$

$$G_{wcwc}(0) + E_{wcwc}(0) = 0, \tag{19}$$

where the functions $G_{ww}(r)$ and $G_{wcwc}(r)$ represent the sum of all nodal contributions of the type ww and wcwc, respectively. The parameters β, γ and δ have been fixed so as to satisfy the conditions (18), (19) and (13), where in eq.(13) the left -hand side has been calculated by using the Jackson-Feenberg form for the kinetic energy.

The ingredients of the present calculation are similar to the ones used by Fabrocini et al.[7] but for the following main differences. The Euler equations satisfied by $f_2(r)$ in the two approaches are different, moreover in ref.7 the scaling approximation has been used and the scaling parameters for $E^J{}_{we}$ and $E^J{}_{wcc}$ have been set equal to zero; finally, in this paper the importance of spin-dependent correlations has been also explored.

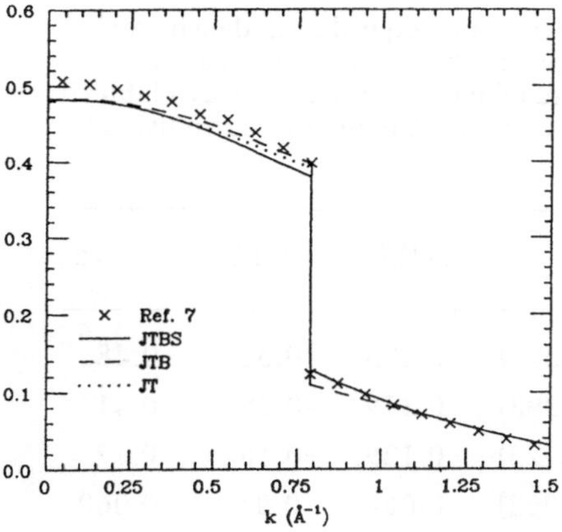

Fig.2. n(k) at the density ρ_0 given by the JT (dotted line), JTB (dashed line) and JTBS (continuous line) model. The crosses refer to the JTB results of ref.7.

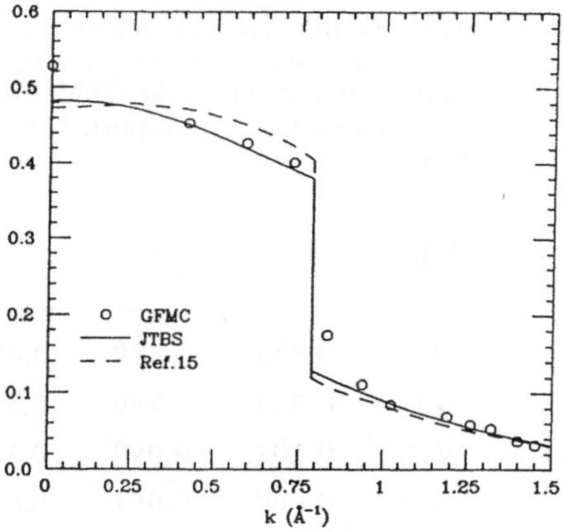

Fig.3. n(k) as given by the JTBS model (continuous line) is compared with the fixed-node GFMC results[14] (open circles) and the empirical data[15] (dashed line).

The results obtained for the momentum distribution with the JT, JTB and JTBS models are presented in fig.2 together with the corresponding JTB calculation or ref.7. In fig.3 the JTBS results are compared with the values calculated[14] by using the fixed-node Green's Function Monte Carlo method and with the n(k) obtained[15] by fitting the experimental[16] dynamical structure function S(k,E), at k=13.85 Å, with the impulse approximation. The JTBS one-body density matrix is shown in fig.4 together with the corresponding one of the GFMC approach[14].

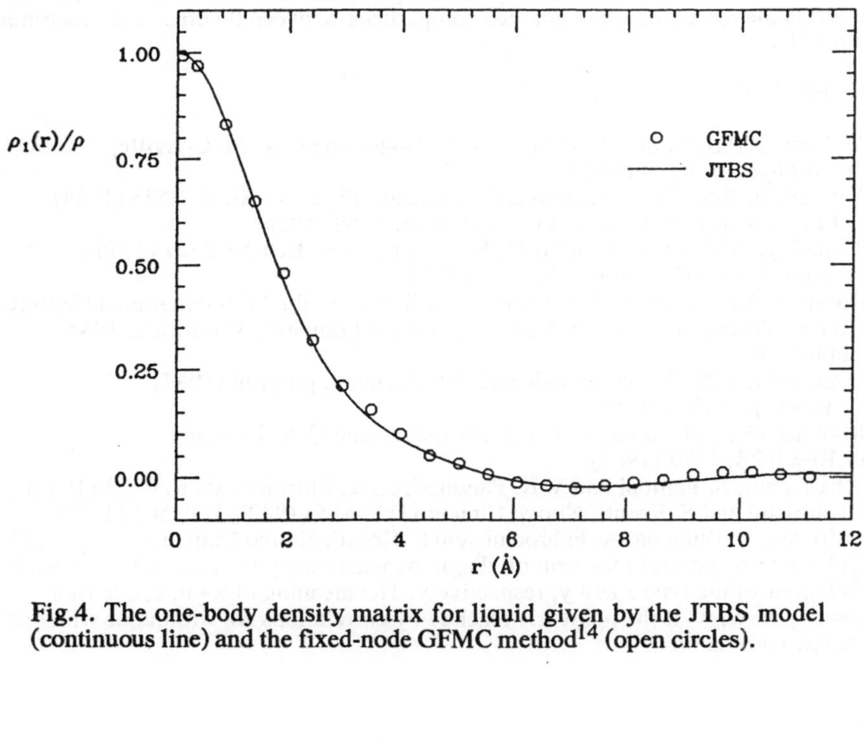

Fig.4. The one-body density matrix for liquid given by the JTBS model (continuous line) and the fixed-node GFMC method[14] (open circles).

Table II. n(k) values calculated at the experimental equilibrium density and in correspondence to different models; the numbers in parentheses are the JTB results of ref.7; the GFMC results are taken from ref.14; the available empirical data[15] are reported in the last column. Z is the discontinuity at $k=k_F$.

k/k_F	J	JT	JTB	JTBS	GFMC	Ref. 15
0.	0.473	0.479	0.484 (0.507)	0.483	0.52	0.48
$1.^-$	0.394	0.390	0.396 (0.398)	0.379	≈0.38	0.41
$1.^+$	0.101	0.099	0.110 (0.123)	0.129	≈0.18	0.12
1.5	0.060	0.061	0.062 (0.062)	0.064	0.07	0.060
Z	0.293	0.291	0.286 (0.275)	0.250	≈0.20	0.29

To illustrate the influence on the momentum distribution of the various correlations included in the wave function, values n(k) correspondent to different models are given for a few k-values in table II, together with the discontinuity Z at $k=k_F$. As it can be seen by inspection of table II, the triplet, backflow and spin-dependent correlations only slightly modify the n(k) evaluated with the Jastrow correlated wave function.

At present, the JTBS model allows for the most refined variational calculation. The agreement of the calculated n(k) with the empirical one is good at k=0 and for large k-values; there are small differences at $k \approx k_F$ which can have different explanations. As an example, CBF corrections[17] could be important in liquid ^{3}He for $k \approx k_F$, or the impulse approximation adopted in ref.15 to obtain n(k) may be not completely justified.

In conclusion, we believe that the important ingredients to obtain n(k) have been correctly introduced in the calculation, but detailed comparisons between theories and experiments are not yet feasible.

REFERENCES

1. R.A. Aziz, V.P.S. Nain, J.S. Carley, W.L. Taylor and G.T. McConville, J.Chem.Phys.**70**, 4330 (1979)
2. M.Viviani, E. Buendia, S.Fantoni and S. Rosati, Phys.Rev.**B38**, 4523 (1988)
3. R.P. Feynman and M. Cohen, Phys.Rev.**105**, 1189 (1956)
4. D. Ceperley, G.V. Chester and M.H. Kalos, Phys.Rev.Lett.**54**,2367 (1977)
5. S. Fantoni, Nuovo Cimento **44A**, 191 (1978)
6. S. Rosati, E. Buendia and M. Viviani, Contribution to the IV International Conference on Recent Progresses in Many-Body Theories, S.Francisco, California, 1985 (unpublished)
7. A. Fabrocini, V.R. Pandharipande and Q.N. Usmani, preprint (1987)
8. M. Viviani, preprint (1989)
9. E. Manousakis, S. Fantoni, V.R. Pandharipande and Q.N. Usmani, Phys.Rev.**B28**,3770 (1983)
10. Q.N.Usmani, S. Fantoni and V.R. Pandharipande, Phys.Rev.**B26**, 6123 (1982)
11. A. Fabrocini and S. Rosati, Nuovo Cimento **D1**, 615 (1982), **D1**, 567 (1982)
M. Viviani, E. Buendia, A. Fabrocini and S. Rosati, Nuovo Cimento **D8**, 561 (1986)

(*) $F_{xy}(1,1')$ corresponds to the sum of diagrams where the points 1 and 1' are reached by correlations of the type x and y, respectively. The meaning of x= d, e, c is well known[12], x=w, w_c if the point 1 is reached by correlations w_2 without or with one exchange correlation.

12.See for example: S. Rosati, Proceedings of the International School of Physics "Enrico Fermi", Course LXXIX, A. Molinari ed., North-Holland Publishing Company, 73 (1981)
13. M.L. Ristig and J.W. Clark, Phys.Rev. **B14**, 2875 (1976)
14. P. Whitlock and R.M. Panoff, Can.J.Phys. **65**, 1409 (1987)
15. J. Carlson, R.M. Panoff, K.E. Schmidt and M.H. Kalos, Phys.Rev.Lett. **55**, 237 (1985)
16. P.E. Sokol, K. Sköld, D.L. Price and R. Kleb, Phys.Rev. **54**, 909 (1985)
17. S. Fantoni and V.R. Pandharipande, Nucl.Phys. **A427**,473 (1984)

OPTIMIZED ^{4}He WAVE FUNCTIONS USING MONTE CARLO INTEGRATION

K.E. Schmidt
Physics Department
Arizona State University
Tempe, Arizona 85287 U.S.A.

Silvio Vitiello
Courant Institute of Mathematical Sciences
New York University
251 Mercer St.
New York, New York 10012 U.S.A.

ABSTRACT

We describe a basis set approach to optimize quantum fluid and solid wave functions. We test the method on ^{4}He using the hypernetted chain approximation and compare to paired-phonon analysis results. We show that only about ten variational parameters are needed to accurately specify an optimized Jastrow correlation. We then give some preliminary results using Monte Carlo integration to optimize these parameters for both the liquid and the solid.

INTRODUCTION

Optimized Jastrow wave functions within the hypernetted chain (HNC) and related approximations have been available for some time.[1-3] (There are of course an enormous number of other references that could mentioned here, many by attendees of this conference.) However, the HNC energy does not agree well with Monte Carlo integrations. Much better agreement is obtained by adding various approximations to the elementary diagrams to the HNC equations.[4-8] Still, these integral equation methods may not be entirely trustworthy, especially when dealing with more complicated correlations and for nonuniform systems and solids. We describe an approach whereby these problems can be solved using the Monte Carlo method.

JASTROW BASIS

As recently shown by Umrigar, Wilson and Wilkes[9] and previously by Conroy[10,11] and Lowther and Coldwell,[12] a highly parameterized variational wavefunction with 20-50 well chosen variational parameters can be optimized using the Monte Carlo

Condensed Matter Theories, Volume 5
Edited by V.C. Aguilera-Navarro
Plenum Press, New York, 1990

method by minimizing a linear combination of the energy and its variance. The variational calculation produces both an upper bound and a lower bound (for example, the Temple[13] or Stevenson[14,15] lower bound theorems) to the ground-state energy. Since the lower bound is almost always the poorest constraint, and is determined by the variance of the energy, minimizing the error from both above and below typically weights the variance much more heavily. In addition, since the variance is the average of a squared quantity, it is inherently easier for Monte Carlo to calculate accurately. The Monte Carlo method produces information about the entire distribution of energies, so other functions of the energy distribution function, in addition to its mean and variance can be used to optimize trial functions.

We parameterize the Jastrow function as

$$f(r) = \sum_n c_n f_n(r), \tag{1}$$

where the $f_n(r)$ are the solutions to the two-body equation

$$-\frac{\hbar^2}{m}\nabla^2 f_n(r)+v(r)f_n(r) = \lambda_n f_n(r), \tag{2}$$

with the boundary conditions

$$f_n(r\geq d) = 1, \tag{3}$$

$$\nabla f_n(r=d) = 0.$$

Note, the method of determining the jastrow function introduced by Pandharipande[16] is just the first term of this expansion with the range d taken to be a variational parameter.

This expansion of f(r) has three main advantages.

1. In the core region where two particles are close together, f(r) will be a solution to the two-body Schroedinger equation as required. Since Monte Carlo calculations do not sample the core region often (the wave function is small there), it is difficult to optimize functions that behave unphysically in this region.
2. The f(r) heals smoothly to 1 at r=d. In Monte Carlo calculations the range of the correlations is restricted by the simulation cell. While Ewald sum methods can be used to sum the correlations from particles in other cells, typically the range of correlations is chosen to be less than L/2 where L is the side of the simulation cell. Our construction automatically truncates f(r) smoothly at r=d, and we take $d \leq L/2$.
3. The basis set is complete for r<d.

HNC TESTS

We have implemented an HNC program using our expansion of f(r), and applied it to liquid ^4He. For a fixed number of terms N in our expansion, the coefficients are varied to minimize the Jackson-Feenberg form for the HNC energy.[17] In figures 1 and 2, we compare our results for d = 7.668 A (this distance corresponds to the range of correlations in simulations of about 100 particles) and equilibrium density to results using paired phonon analysis (PPA)[1,3] for optimizing the HNC energy. We have used the HFDHE2 potential of Aziz *et al.*[18] The energies for various densities and numbers of terms in the expansion are given in Table 1.

The jastrow function using our expansion is in excellent agreement with the fully optimized HNC calculations for r<d. We take this to be a strong confirmation that our expansion is general enough to produce nearly optimal jastrow correlations. The long range phonon tail is energetically unimportant, and well understood. Such structural information can only be obtained by our method by very large Monte Carlo simulations. Since the optimal HNC jastrow correlation can be well represented by approximately 10

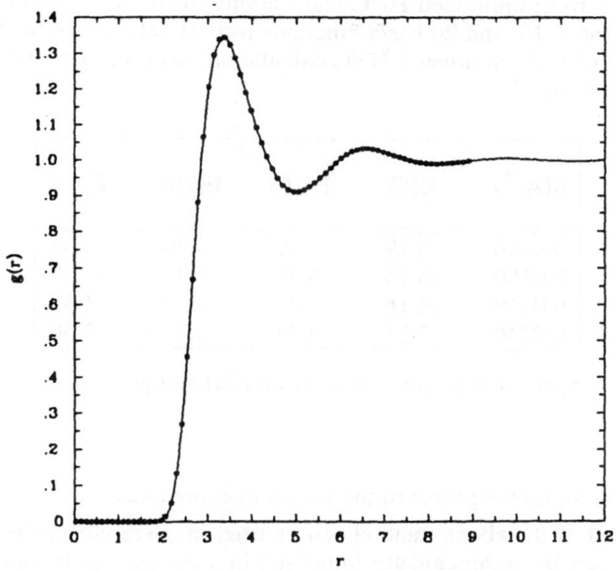

Figure 1. The two-body distribution functions calculated using HNC approximation and paired-phonon phonon analysis (PPA) and using ten basis functions from the two-body equation which heal to 1 at r=7.668A. The solid line is the PPA result; circles are the basis set result.

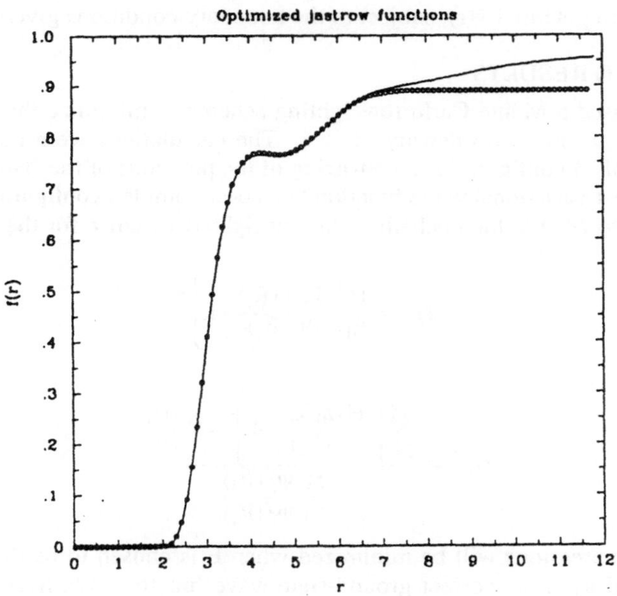

Figure 2. The jastrow factors calculated using HNC approximation and paired-phonon phonon analysis (PPA) and using ten basis functions from the two-body equation which heal to 1 at r=7.668A. The basis set solution was arbitrarily multiplied by 0.89 to display the agreement of the two curves. The solid line is the PPA result; circles are the basis set result.

Table 1. Energies from optimized HNC calculations for liquid ^4He. E(5), E(10), E(20) are the energies for 5, 10, and 20 basis functions respectively which heal to 1 at 7.668A. E_{ppa} is the result of fully optimized HNC calculations with the HFDHE2 potential. All energies are in Kelvin.

$\rho(A^{-3})$	E(5)	E(10)	E(20)	E_{ppa}*
0.0160	-5.29	-5.36	-5.36	-5.37
0.0180	-5.23	-5.36	-5.37	-5.39
0.0219	-4.16	-4.49	-4.50	-4.55
0.0240	-2.91	-3.43	-3.44	-3.50

*private communication from Karl Runge

terms, this is well within the power of the variational method.

An extension of the above method can be used in cases where the original basis is poor. The basis can be systematically improved in a manner analogous to that used by Wilson in his solution of the Kondo problem.[19] After optimization with N basis functions where N can be as small as two, we calculate the potential which gives the current optimized f(r) as its ground state, i.e.

$$\tilde{v}(r) = \frac{\hbar^2}{m}\frac{\nabla^2 f(r)}{f(r)}.$$

We then take as our next basis, the first N solutions of the two-body Schroedinger equation, Eq. (2), with potential $\tilde{v}(r)$, subject to the boundary conditions given by Eq. (3).

MONTE CARLO RESULTS

We have used a Monte Carlo reweighting scheme to minimize the energy variance for one liquid and one solid density of ^4He. The calculations were done, as usual, by sampling a set of M configurations, consisting of the positions of the ^4He atoms, from the square of our best variational wave function ψ_s. These sampled configurations are labeled R_i for i from 1 to M. We then calculate the reweighted variance for the trial wave function by either

$$\sigma_1{}^2 = \sum_{i=1}^{M}\left[\frac{H\psi_t(R_i)}{\psi_t(R_i)} - E\right]^2,$$

or

$$\sigma_2{}^2 = \frac{\sum_{i=1}^{M}\left[\frac{H\psi_t(R_i)}{\psi_t(R_i)} - E\right]^2 \frac{\psi_t^2(R_i)}{\psi_s^2(R_i)}}{\sum_{i=1}^{M}\frac{\psi_t^2(R_i)}{\psi_s^2(R_i)}}.$$

Both of these expressions will be minimized when E is chosen to be the correct ground state energy and ψ_t is the correct ground-state wave function. We have not noticed any significant advantage to using either form. Note that if E is not the correct ground-state energy, we are essentially minimizing a linear combination of the energy average and its variance.

The minimization was done using a Levenberg-Marquardt method to calculate the values of the coefficients c_i in Eq. 1. The quality of the resulting wave function is then tested by calculating the energy expectation value in an independent variational calculation. We repeat the whole minimization process several times with new sampled positions, and with different starting values for c_i.

In Table 2 we give results for ^{4}He liquid at equilibrium density and at one solid density. The trial wave function for these cases are of the form[20]

$$\psi_t = \prod_{i<j} f_{ij} \prod_{i<j<k} f_{ijk}^{(3)} \prod_i \phi_i,$$

where f_{ij} is given by Eqs. 1 and 2, $f_{ijk}^{(3)}$ is a three-body correlation of the form

$$f_{ijk}^{(3)} = \exp(-\frac{\lambda}{2} \sum_{cyc} \xi_{ij}\xi_{ik}\vec{r}_{ij}\cdot\vec{r}_{ik})$$

with

$$\xi=(R_c-r)^3\exp(-(r-r_0)^2/w),$$

cyc means to sum over cyclic permutations, and ϕ_i is one for the liquid and

$$\phi_i=\exp(-\alpha(\vec{r}_i-\vec{R}_i)^2),$$

for the solid, where $\vec{R}_i$ are the lattice positions for a face-centered cubic lattice. Both calculations are with 108 particles. The values of the parameters other than the c_i are taken from our best previous variational results. Clearly, they should be reoptimized as well. The liquid ^{4}He results at equilibrium density are about 0.1K better than our best previous results using a PPA/HNC jastrow factor which was arbitrarily cutoff at L/2. This result agrees with that of Usmani et al.[5] when they used a three-body correlation of the form used here.

The results in the solid phase show an even greater improvement of 0.2K over our best previous variational Monte Carlo calculations. These previous calculations used a

Table 2. Monte Carlo optimization results ^{4}He. The triplet parameters are $\lambda=-14$ for the liquid and $\lambda=-8$ for the solid, $r_0=2.096$A, w=.1.278A, and $R_c=7.668$A for the liquid and $R_c=7.428$A for the solid. The one-body parameter is $\alpha=1.225$A^{-2}.

	Liquid	Solid
E(K)	−6.79±0.01	−4.00±0.03
ρ(A^{-3})	0.0218	0.0329
c_1	0.94365	0.97854
c_2	0.06051	0.01635
c_3	-0.00964	0.01382
c_4	0.00675	-0.01647
c_5	-0.00280	-0.01204
c_6	0.00304	-0.00741
c_7	-0.00190	0.00425
c_8	0.00112	-0.00214
c_9	-0.00153	0.00204
c_{10}	0.00079	-0.00102

Jastrow factor of the McMillan form, $f(r)=\exp(-\frac{1}{2}(b/r)^5)$, with b = 2.89 A. Our calculation shows that we are not limited to the bulk liquid state. Other correlations can be easily investigated in a similar way, as well as more complicated nonuniform and fermi systems.

CONCLUSION

We have shown that by using a basis set given by the solutions of the constrained two-body equation, a jastrow factor can be easily optimized using the Monte Carlo method. This general technique should allow us to parameterize and optimize a variety of correlations in complicated systems.

ACKNOWLEDGEMENTS

We wish to thank Dr. Karl Runge for providing the paired-phonon analysis results shown in Table 1. and the figures. We would also like to thank Profs. M.H. Kalos, M.A. Lee, and P.A. Whitlock for valuable suggestions. K.E.S. is grateful for the hospitality of the Aspen Center for Physics where some of this work was done. This work was supported by the National Science Foundation grant ASC-8715641.

REFERENCES

1. E. Feenberg, *Theory of Quantum Fluids,* Academic Press, New York (1969).
2. L.J. Lantto and P.J. Siemens, *Phys. Lett.* **68B**, 311 (1977).
3. F.J. Pinski and C.E. Campbell, *Phys. Lett.* **79B**, 23 (1978).
4. Q.N. Usmani, B. Friedman, and V.R. Pandharipande, *Phys. Rev.* **B25**, 4502 (1982).
5. Q.N. Usmani, S. Fantoni, and V.R. Pandharipande, *Phys. Rev.* **B26**, 6123 (1982).
6. Y. Rosenfeld and N.W. Ashcroft, *Phys. Rev.* **A20**, 1208 (1979).
7. A. Fabrocini and S. Rosati, *Nouvo Cimento* **1D**, 567 (1982).
8. M. Viviani, E. Buendia, A. Fabrocini, and S. Rosati, *Nouvo Cimento* **8D**, 561 (1986).
9. C.J. Umrigar, K.G. Wilson, and J.W. Wilkins, *Phys. Rev. Lett.* **60**, 1719 (1988).
10. H. Conroy, *J. Chem. Phys.* **41**, 1331 (1964).
11. H. Conroy, *J. Chem. Phys.* **41**, 1336 (1964).
12. R. L. Coldwell and R. E. Lowther, *Int. J. Quant. Chem. Symp.* **12**, 329 (1978).
13. G. Temple, *Proc. Roy. Soc. London* **119**, 276 (1928).
14. A.F. Stevenson, *Phys. Rev.* **53**, 199 (1938).
15. A.F. Stevenson and M.F. Crawford, *Phys. Rev.* **54**, 375 (1938).
16. V.R. Pandharipande and H.A. Bethe, *Phys. Rev.* C7, (4)1312 (1973).
17. H.W. Jackson and E. Feenberg, *Ann. Phys. (N.Y.)* **15**, 266 (1961).
18. R.A. Aziz, V.P.S. Nain, J.S. Carley, W.L. Taylor, and G.T. McConville, *J. Chem. Phys* **70**, 457 (1974).
19. K.G. Wilson, *Rev. Mod. Phys.* **47**, 773 (1975).
20. K.E. Schmidt, M.H. Kalos, M.A. Lee, and G.V. Chester, *Phys. Rev. Lett.* **45**, 573 (1980).

THE NORMAL PHASE OF A CORRELATED BOSE FLUID

G. Senger and M. L. Ristig

Institut für Theoretische Physik
der Universität zu Köln
D-5000 Köln 41, West-Germany

ABSTRACT

The variational density-matrix approach for describing the statistical mechanics of strongly correlated bosons is extended to a treatment of the normal-fluid phase at elevated temperatures. The microscopic formalism is based on an appropriate class of density matrices and employs the Gibbs-Delbrück-Molière minimum principle for the Helmholtz free energy. The model exhibits a Bose-Einstein transition to a bosonic phase with nonzero condensate at sufficiently low temperatures.

Microscopic many-body theories have been highly successful in analyzing spatial correlations in strongly interacting quantum systems. The variational approach or, more generally, the method of correlated basis functions has brought a detailed formal understanding and reliable enumeration of the correlation structure of liquid helium, charged fluids, and of nuclear matter, at zero temperature [1,2]. These ab-initio procedures may be properly adapted and successfully employed for studying ground state correlations generated by various lattice models which are of interest in gauge theories [3] or in solid state physics [4,5]. Moreover, the variational approach may be suitably generalized to deal with the thermal equilibrium properties of strongly correlated many-body systems at elevated temperatures [6,7]. The formalism at nonzero temperatures is based on the Gibbs-Delbrück-Molière minimum principle for the Helmholtz free energy [6-8],

$$F_0 \leq F = Tr\{WH + \beta^{-1} W ln W\}. \tag{1}$$

At constant temperature $T = (k_B\beta)^{-1}$ the trial free energy F is bound from below by the exact free energy F_0 of a N-body quantum system described by a Hamiltonian of the form

$$H = T + V = -\sum_{j}^{N} \frac{\hbar^2}{2m}\Delta_j + \sum_{i<j}^{N} v(r_{ij}). \tag{2}$$

The N-body operator W represents an element of a suitably chosen set of trial density matrices. These matrices must be non-negative and be normalized to unity, $Tr\,W = 1$. In coordinate-space representation the statistical operator W for spinless bosons may be cast into the general form [7]

$$W(\underline{\mathbf{R}}, \underline{\mathbf{R}}') = \frac{1}{I}\, \Phi(\underline{\mathbf{R}}) Q(\underline{\mathbf{R}}, \underline{\mathbf{R}}') \Phi(\underline{\mathbf{R}}'), \tag{3}$$

Condensed Matter Theories, Volume 5
Edited by V.C. Aguilera-Navarro
Plenum Press, New York, 1990

where $\underline{\mathbf{R}} = (\underline{\mathbf{r}}_1, \underline{\mathbf{r}}_2, ..., \underline{\mathbf{r}}_N)$. The incoherence factor $Q(\underline{\mathbf{R}}, \underline{\mathbf{R}}')$ contains no factors which depend only on primed or unprimed coordinates alone and it reduces to unity at zero temperature. The constant I is the generalized normalization integral

$$I = \int d\underline{\mathbf{R}}\, \Phi^2(\underline{\mathbf{R}})\, Q(\underline{\mathbf{R}}, \underline{\mathbf{R}}), \tag{4}$$

which ensures that the trace of operator W is normalized to unity.

For N bosons confined to a box of volume Ω a popular choice of trial functions $\Phi(\underline{\mathbf{R}})$ is the Jastrow ansatz,

$$\Phi(\underline{\mathbf{R}}) = exp \sum_{i<j}^{N} \tfrac{1}{2} u(r_{ij}). \tag{5}$$

At sufficiently low temperatures the correlated bosons will be in a Bose-Einstein condensed phase characterized by a non-vanishing zero-momentum condensate. In this case the incoherence factor $Q(\underline{\mathbf{R}}, \underline{\mathbf{R}}')$ may be suitably represented by the choice [7],

$$Q(\underline{\mathbf{R}}, \underline{\mathbf{R}}') = exp \sum_{i,j}^{N} \gamma(|\underline{\mathbf{r}}_i - \underline{\mathbf{r}}_j'|). \tag{6}$$

Ansatz (6) provides a first adequate step for describing the equilibrium properties in the low temperature region. However, it is not sufficiently flexible for representing the properties of a normal-fluid phase. Instead, an appropriate and simple first choice for approximately representing the density matrix of the normal phase of correlated bosons with zero condensate would be [7],

$$Q(\underline{\mathbf{R}}, \underline{\mathbf{R}}') = \underset{i,j}{Perm}\{\Gamma(|\underline{\mathbf{r}}_i - \underline{\mathbf{r}}_j'|)\}. \tag{7}$$

Expression (7) is a $N \times N$ permanent constructed from the element $\Gamma(r)$ being a variational trial function. Ansatz (7) becomes exact in the high-temperature limit where $\Gamma(r)$ assumes a Gaussian form [7], $\Gamma(r) = exp\{-\pi r^2 \lambda^{-2}\}$, with $\lambda^2 = 2\pi\beta\hbar^2/m$.

The variational density-matrix approach proceeds in two steps. One begins adopting ansatz (3),(5) and (6) or (7) and evaluates the Helmholtz free energy F as a functional of the trial functions $u(r)$ and $\gamma(r)$ or $u(r)$ and $\Gamma(r)$ which characterize the density matrices for the Bose-condensed phase or the normal phase, respectively. In a second step one employs principle (1) and constructs explicit expressions for the associated Euler-Lagrange equations which determine the optimal solutions $u(r)$, $\gamma(r)$ or $u(r)$, $\Gamma(r)$.

A detailed realization of this program has been put forward in Refs. [6,7] analyzing the Bose-condensed phase of the fluid. Here, we shall report on some of our results extending the formalism for exploring the properties of normal boson fluids based on eq. (3) and assumptions (5) and (7). We shall concentrate on the main ideas and analytic steps omitting detailed algebraic manipulations. A forthcoming article will describe our formal results in more detail and will present a numerical study of normal ^{4}He within the density-matrix approach.

We begin with the task to evaluate the internal energy $U = Tr\, H\, W$ and the entropy $T\, S_e = -\beta^{-1} Tr\{W\, ln\, W\}$ and therewith the trial Helmholtz free energy F as functionals of the dynamic quantity $u(r)$ and the statistical function $\Gamma(r)$ which constitute ansatz (3),(5),(7).

Employing the familiar Jackson-Feenberg identity [9] the internal energy of the normal

Bose fluid may be represented in the form,

$$U = \frac{N}{2}\rho\int d\underline{r}\, v^*(r)g(r) + \sum_{\underline{k}} \epsilon_0^*(k)n_{cc}(k). \tag{8}$$

The integral involves the radial distribution function $g(r)$ associated with eqs. (3),(5),(7) at particle number density $\rho = N/\Omega$. It also contains the effective potential [9] $v^*(r) = v(r) - \frac{\hbar^2}{4m}\Delta u(r)$. The second term in eq. (8) collects the effective energy contributions $\epsilon_0^*(k)$ of single-particle states with wave vector $\underline{\mathbf{k}}$ occupied in the average by $n_{cc}(k)$ bosons. Switching off the interparticle interaction the effective energy $\epsilon_0^*(k)$ reduces to the kinetic energy $\epsilon_0(k) = \hbar^2k^2/2m$ of a boson with bare mass m and the statistical weight function $n_{cc}(k)$ approaches the familiar thermodynamic Bose-distribution function $\{exp[\beta\epsilon_0(k) - \beta\mu_0] - 1\}^{-1}$ for a gas of non-interacting bosons with chemical potential μ_0.

In the presence of correlations the function $g(r)$ and the dressed quantities $\epsilon_0^*(k)$ and $n_{cc}(k)$ may be evaluated by employing a generating functional $\Lambda[u(r), \Gamma(r)]$. This quantity is defined by

$$I = \Omega^N exp\{N\Lambda[u, \Gamma]\}. \tag{9}$$

Standard cluster-expansion procedures [9-11] are available for constructing explicit representations of the logarithm of the generalized normalization integral I and therewith of the functional $\Lambda[u, \Gamma]$. Here, we need this quantity in the thermodynamic limit (N and Ω going to infinity while the density ρ is kept constant). The result of such a formal analysis may be represented in terms of generalized Ursell-Mayer graphs [1]. The two- and three-body cluster contributions to the functional $\Lambda[u, \Gamma]$, in this limit, are diagramatically displayed in Fig. 1. We represent the function $u(r)$ by the dynamic bond $\eta(r) = exp\, u(r) - 1$ (a dashed line with two root points) and the statistical function $\Gamma(r)$ assumed to be normalized to unity at $r = 0$ (an oriented line with two root points). We learn from Fig. 1 that the functional may be decomposed into an irreducible portion and a reducible (factorizable) component,

$$\Lambda[u, \Gamma] = \Lambda_{irr}[u, \Gamma] + \Lambda_{red}[u, \Gamma]. \tag{10}$$

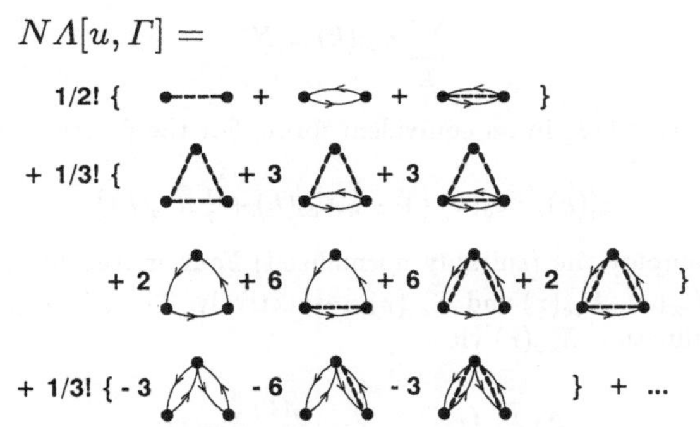

Fig. 1 Two- and three-body cluster contributions to the generating functional $\Lambda[u, \Gamma]$, in the thermodynamic limit. Diagrams in the first and second bracket are irreducible, the diagrams in the third bracket are reducible (factorizable).

We may re-arrange this cluster expansion and perform a partial summation up to infinite cluster order. The result reads

$$\Lambda[u,\Gamma] = \Lambda_{irr}[u,\Gamma_{cc}] + c - 1 - ln\, c. \tag{11}$$

Eq. (11) achieves a dressing

$$\Gamma_{cc}(r) = c\,\Gamma(r) \tag{12}$$

of the statistical bond. The associated renormalization constant c is determined by a sum rule,

$$c\left\{1 + \rho\int d\underline{r}\,\Gamma(r)\frac{\delta\Lambda_{irr}[u,\Gamma_{cc}]}{\delta\Gamma_{cc}(r)}\right\} = 1. \tag{13}$$

The radial distribution function, the energy $\epsilon_0^*(k)$ and the weight function $n_{cc}(k)$ appearing in eq. (8) may be constructed from the variational derivatives,

$$\frac{\delta\Lambda_{irr}[u,\Gamma_{cc}]}{\delta u(r)} = \tfrac{1}{2}g(r), \tag{14}$$

and

$$\frac{\delta\Lambda_{irr}[u,\Gamma_{cc}]}{\delta\Gamma_{cc}(r)} = g_{cc}(r). \tag{15}$$

The function $g_{cc}(r)$ may be graphically represented by a set of circular (cc) diagrams with two root points. The graphs are characterized by a loop of statistical bonds that begins at one root point, connects p intermediate field points ($p = 0,1,2,...$) and ends at the second root point. We may decompose quantity (15) into the dressed statistical component $\Gamma_{cc}(r)$, a nodal (N) and a non-nodal (X) portion [1],

$$g_{cc}(r) = X_{cc}(r) + N_{cc}(r) + \Gamma_{cc}(r). \tag{16}$$

We are now in a position to formulate the explicit expressions which define the ingredients $n_{cc}(k)$ and $\epsilon_0^*(k)$. A detailed formal analysis of the sum appearing in eq. (8) in terms of the variational derivative (15) and the components (16) arrives at

$$n_{cc}(k) = \Gamma_{cc}(k)\{1 - X_{cc}(k) - \Gamma_{cc}(k)\}^{-1} \tag{17}$$

with

$$\sum_{\underline{\mathbf{k}}} n_{cc}(k) = N \tag{18}$$

representing the sum rule (13) in an equivalent form. For the dressed single-particle energies we find

$$\epsilon_0^*(k) = \epsilon_0(k)\left\{1 - \tfrac{1}{2}X_{cc}(k) + \tfrac{1}{2}\widetilde{X}_{cc}(k)\right\}. \tag{19}$$

Eqs. (17) and (19) employ the (suitably normalized) Fourier transforms $\Gamma_{cc}(k)$, $X_{cc}(k)$ and $\widetilde{X}_{cc}(k)$ of functions $\Gamma_{cc}(r)$, $X_{cc}(r)$ and $\widetilde{X}_{cc}(r)$, respectively. The latter quantity is constructed from the non-nodal function $X_{cc}(r)$ via

$$\Delta_1\widetilde{X}_{cc}(r_{12}) = \underline{\nabla}_1 \cdot \underline{\nabla}_1^{(\Gamma)}X_{cc}(r_{12}). \tag{20}$$

The notation $\underline{\nabla}_1^{(\Gamma)}X_{cc}(r_{12})$ indicates the following graphic prescription: In the set of diagrams representing quantity $X_{cc}(|\underline{r}_1 - \underline{r}_2|)$ replace the dressed statistical bond Γ_{cc} that appears at the root point $\underline{r}_1$ by the gradient $\underline{\nabla}_1\Gamma_{cc}$.

The various ingredients of eqs. (8), (16)-(20) may be further studied and evaluated by analyzing the functions $g(r)$ and $g_{cc}(r)$ within a properly adapted hypernetted-chain (HNC) scheme [1]. We decompose the radial distribution function into irreducible direct-direct (dd), direct-'exchange' (de) and 'exchange'-'exchange' (ee) portions,

$$g(r) = 1 + \{X_{dd}(r) + N_{dd}(r)\} + 2\{X_{de}(r) + N_{de}(r)\} + \{X_{ee}(r) + N_{ee}(r)\}. \tag{21}$$

Eq. (21) bears a formal relationship with the analogous FHNC-decomposition of the radial distribution function $g(r)$ associated with a ground state of the Jastrow-Slater type for a system of spatially correlated spinless fermions. This is, of course, apparent since we exactly recover the ground state density matrix of Jastrow-Slater form by converting the permanent (7) into a determinant and by simultaneously replacing function $\Gamma(r)$ by the Slater exchange function $l(r)$. For this reason we use the familiar subscripts dd, de, ee and cc. However, keep in mind that 'exchange' refers in the present case to the statistical bonds $\Gamma(r)$, $\Gamma_{cc}(r)$ and not to function $l(r)$.

The FHNC equations properly adapted to the present thermodynamic problem read,

$$\begin{aligned}
X_{dd}(r) &= e^{u+N_{dd}+E_{dd}} - 1 - N_{dd}, \\
X_{de}(r) &= e^{u+N_{dd}+E_{dd}}\{N_{de} + E_{de}\} - N_{de}, \\
X_{ee}(r) &= e^{u+N_{dd}+E_{dd}}\{N_{ee} + E_{ee} + (N_{de} + E_{de})^2 + (N_{cc} + E_{cc} + \Gamma_{cc})^2\} - N_{ee}, \\
X_{cc}(r) &= e^{u+N_{dd}+E_{dd}}\{N_{cc} + E_{cc} + \Gamma_{cc}\} - N_{cc} - \Gamma_{cc},
\end{aligned} \tag{22}$$

and,

$$\begin{aligned}
N_{dd}(k) &= (X_{dd} + X_{de})(X_{dd} + N_{dd}) + X_{dd}(X_{de} + N_{de}), \\
N_{de}(k) &= (X_{de} + X_{ee})(X_{dd} + N_{dd}) + X_{de}(X_{de} + N_{de}), \\
N_{ee}(k) &= (X_{de} + X_{ee})(X_{de} + N_{de}) + X_{de}(X_{ee} + N_{ee}), \\
N_{cc}(k) &= (X_{cc} + \Gamma_{cc})(X_{cc} + N_{cc} + \Gamma_{cc}),
\end{aligned} \tag{23}$$

$N_{\alpha\beta}(k)$ and $X_{\alpha\beta}(k)$ being, respectively, the dimensionless Fourier transforms of the nodal function $N_{\alpha\beta}(r)$ and the non-nodal function $X_{\alpha\beta}(r)$. The hypernet equations (22) and the chain equations (23) in conjunction with the sum rule (17), (18) may be used for a numerical analysis of a strongly correlated system such as normal liquid ^{4}He.

For evaluating the entropy portion of the Helmholtz free energy (1) we may begin with a study of the traces of powers of the density-matrix operator in analogy to the treatment of Ref. [7] and then continue to evaluating

$$T\,S_e = -\beta^{-1} Tr\{W\, ln\, W\} = -\beta^{-1} \frac{\partial}{\partial\alpha} Tr\{W^{1+\alpha}\}\Big|_{\alpha=0}. \tag{24}$$

We shall report on the realization of this enterprise in a forthcoming publication [12].

Here, we employ a plausible approximation S_0 for the exact entropy (24),

$$T\,S_0 = \beta^{-1} \sum_{\underline{\mathbf{k}}} \{(n_{cc}(k) + 1)\, ln\,(n_{cc}(k) + 1) - n_{cc}(k)\, ln\, n_{cc}(k)\}. \tag{25}$$

This expression is the analogue of the corresponding approximant used in Ref. [7] for the entropy of a Bose-condensed phase based on the separability assumption. Eq. (25) may be interpreted as an exact expression for the entropy of N non-interacting bosons occupying single-particle orbitals with momentum $\hbar\underline{\mathbf{k}}$ and with properly renormalized energy $\epsilon(k)$ and chemical potential μ_ϵ. Their thermodynamic distribution over the available states at temperature T is given by the familiar expression

$$n_{cc}(k) = \{exp[\beta\epsilon(k) - \beta\mu_\epsilon] - 1\}^{-1}. \tag{26}$$

At sufficiently small wave numbers $\underline{\mathbf{k}}$ these elementary excitations are characterized by an effective mass, $\epsilon(k) \simeq \hbar^2 k^2/2m^*$. In general, it depends on the density ρ and the temperature T with $m^* \to m$ in the high-temperature limit.

For the second main step in developing the density-matrix approach of the normal-fluid phase we minimize the trial Helmholtz free energy subject to the sum-rule condition (18). Alternatively, we may minimize the functional

$$F_\lambda = F + \lambda\{ N - \sum_{\underline{\mathbf{k}}} n_{cc}(k) \} \tag{27}$$

without constraints. The optimized functions $u(r)$ and $\Gamma_{cc}(r)$ generating this minimum are solutions of two Euler-Lagrange equations,

$$\frac{\delta F_\lambda[u,\Gamma_{cc}]}{\delta u(r)} = 0, \qquad \frac{\delta F_\lambda[u,\Gamma_{cc}]}{\delta \Gamma_{cc}(r)} = 0, \tag{28}$$

which must also fulfill the sum rule (18). Employing the results (8), (17), (19), (21) and (25) we may derive explicit expressions for the two coupled equations (28).

The first eq. (28) may be most conveniently cast, in $\underline{\mathbf{k}}$-space, into the form

$$\epsilon(k)\{1 - S(k)\} = 2\,\dot{S}(k), \tag{29}$$

familiar from the paired-phonon treatment of Refs. [7,9] and from the formulation of the optimization problem for the ground state of a Fermi system, Ref. [13]. Quantity $S(k)$ is essentially the Fourier transform of the radial distribution function $g(r)$ and may be identified with the static structure function. The function $\dot{S}(k)$ can be written in the form

$$\begin{aligned}\dot{S}(k) = \;& \rho\int d\underline{\mathbf{r}}\,\frac{\delta S(k)}{\delta u(r)} v^*(r) + \tfrac{1}{N}\sum_{\underline{\mathbf{k}}'} \frac{\delta S(k)}{\delta \Gamma_{cc}(k')}\Gamma_{cc}(k')\{\tfrac{1}{2}\epsilon_0(k') - \epsilon(k') + \mu_\epsilon - \lambda\} \\ & + \rho\int d\underline{\mathbf{r}}_{12}\,\frac{\hbar^2}{4m}\{\underline{\nabla}_1\Gamma_{cc}(r_{12})\}\cdot\underline{\nabla}_1^{(\Gamma)}\frac{\delta S(k)}{\delta\Gamma_{cc}(r_{12})}\end{aligned} \tag{30}$$

and may be interpreted as the derivative of a suitably defined generalized structure function $S(k,\alpha)$ with respect to the parameter α. We emphazise that quantity (30) differs distinctly from the analogous function of Ref. [7] derived for an adequate description of a Bose-condensed fluid.

The second eq. (28) leads to the explicit representation

$$\begin{aligned}\epsilon(k) - \mu_\epsilon = \;& \epsilon_0^*(k) + \tfrac{1}{2}\epsilon_0(k)\{\Gamma_{cc}(k) + \widetilde{X}_{cc}(k)\} \\ & -\lambda + \{1 - X_{cc}(k) - \Gamma_{cc}(k)\}\,\dot{S}_{cc}(k).\end{aligned} \tag{31}$$

Eq. (31) determines the optimal functions $\epsilon(k)$, $n_{cc}(k)$ and μ_ϵ. The derivative $\dot{S}_{cc}(k)$ is the circular analogue of quantity (30).

The solutions of eqs. (29) and (31) in conjunction with the sum rule (18) play a crucial role in determining instabilities and phase boundaries of a normal boson fluid. At sufficiently low temperatures $T \leq T_{BE}$ eqs. (29) and (31) do not permit solutions which satisfy the sum rule (18) with $n_{cc}(0) < \infty$. In this temperature region the formalism presented here must be generalized allowing for the presence of a macroscopically occupied zero-momentum state, $n_{cc}(0) \sim N$. This phenomenon is familiar from the statistical mechanics of a system of non-interacting bosons. The critical temperature T_{BE} signals a transition from the normal-fluid phase to the Bose-condensed phase characterized by the appearance of a zero-momentum condensate. Thus, the formalism developed here provides the analytic tools for studying the Bose-Einstein condensation of a strongly correlated boson fluid. Of course, only detailed numerical calculations based upon the HNC equations (22), (23), the Euler-Lagrange equations (29), (31), and the sum rule (18), can determine the accurate location of the Bose-Einstein condensation line $T_{BE}(\rho)$ in the (T, ρ) phase diagram for a particular Bose fluid of interest. Numerical calculations for liquid ^{4}He are in progress.

REFERENCES

[1] J. W. Clark, in *Progress in Particle and Nuclear Physics*, Vol. 2, ed. D. H. Wilkinson (Pergamon, Oxford, 1979).

[2] J. W. Clark and M. L. Ristig, in *Momentum Distributions*, ed. R. N. Silver (Plenum, New York, 1989), in press.

[3] A. Dabringhaus and M. L. Ristig, in *Condensed Matter Theories*, Vol. 4, ed. J. Keller (Plenum, New York, 1989), in press.

[4] S. Fantoni, X. Wang, E. Tosatti and Lu Yu, Physica C **153**-**155**,1255(1988).

[5] M. L. Ristig, submitted to Z. Phys. B.

[6] C. E. Campbell, K. E. Kürten, M. L. Ristig and G. Senger, Phys. Rev. B**30**,3728(1984).

[7] G. Senger, M. L. Ristig, K. E. Kürten and C. E. Campbell, Phys. Rev. B**33**,7562(1986).

[8] A. Huber, in *Methods and Problems of Theoretical Physics*, Vol. 37, ed. J. E. Bowcock (North-Holland, Amsterdam, 1970).

[9] E. Feenberg, *Theory of Quantum Fluids* (Academic, New York, 1969).

[10] T. Morita and K. Hiroike, Prog. Theor. Phys. **23**,1003(1960).

[11] K. Hiroike, Prog. Theor. Phys. **24**,317(1960).

[12] G. Senger, doctoral thesis, and to be published.

[13] J. C. Owen, Phys. Rev. B**23**,2169(1981).

A NEW APPROACH TO EXCITED STATES IN ^{4}He: ROTONS AND VORTICES

S. A. Vitiello

Courant Institute of Mathematical Sciences
New York University, New York, N. Y. 10012, USA

M. H. Kalos

Laboratory of Atomic and Solid State Physics
and Theory Center
Cornell University, New York, N. Y. 14853, USA

L. Reatto

Dipartimento di Fisica
Università Degli Studi di Milano, 20133 Milano, Italy

INTRODUCTION

The shadow wave function has been extensively used to describe quantum solids and liquids.[1,2] The good results so far obtained for the ground state properties of a system of ^{4}He atoms have encouraged us to apply this same formalism in more complex situations. In this paper we investigate the elementary excitation spectrum of liquid ^{4}He and nonuniform states of this system with a vortex line.

In the shadow wave function particles are correlated both directly by a Jastrow term and indirectly via auxiliary variables, one for each particle. An auxiliary variable is a way to represent the quantum delocalization of a given particle. This quantum hole should be present also in the low energy excited states suggesting that a better trial wave function for these states is obtained if the terms representing the excitation are written in terms of these auxiliary variables.

We start by recalling the definition of the shadow wave function for uniform systems. Then we consider a trial wave function able to describe the elementary excitation spectrum of a system of ^{4}He atoms and we give the preliminary results of a numerical simulation of the excitation spectrum. Finally we discuss a model wave function for a system with a vortex line based on the same ideas. We derive the expression of the expectation value for the velocity field and discuss qualitatively the behavior of this velocity field and of its curl. The most important qualitative aspect is that this new wave function gives a distributed vorticity and the density in the core does not vanish.

Condensed Matter Theories, Volume 5
Edited by V.C. Aguilera-Navarro
Plenum Press, New York, 1990

THE SHADOW WAVE FUNCTION

The trial shadow wave function $\Psi_T(R)$ for the ground-state of a system with N ^{4}He atoms of coordinates $R \equiv \{\mathbf{r}_1, \mathbf{r}_2, \ldots, \mathbf{r}_N\}$ can be written as

$$\Psi_T(R) = \psi_r(R) \int dS \prod_i \Theta(\mathbf{r}_i - \mathbf{s}_i)\ \psi_s(S)\ , \tag{1}$$

where S denotes a set of auxiliary variables $S \equiv \{\mathbf{s}_1, \ldots, \mathbf{s}_N\}$, and dS stands for the product $dS \equiv ds_1 ds_2 \ldots ds_N$. In this expression, the factors ψ_i are of the Jastrow form, and thus given by a product of terms of the e^{-u_i} form. In the computations performed so far the pseudopotential for the particles is assumed to be $u_r(|\mathbf{r}_i - \mathbf{r}_j|) = \frac{1}{2}(\frac{b_r}{|\mathbf{r}_i - \mathbf{r}_j|})^5$. The pseudopotential of the auxiliary variables is chosen as $u_s(|\mathbf{s}_i - \mathbf{s}_j|) = (\frac{b_s}{|\mathbf{s}_i - \mathbf{s}_j|})^5$. Both b_r and b_s are variational parameters. The auxiliary variables can be thought as if they were coordinates of fictitious particles - the shadow particles - since their interactions are analogous to those of the real particles. The interaction term between the particles and the shadows in Eq. (1) is assumed to be a Gaussian given by $\Theta(\mathbf{r}_i - \mathbf{s}_i) = e^{-C\ (\mathbf{r}_i - \mathbf{s}_i)^2}$; C being a variational parameter.

The ground state energy of the system described by an Hamiltonian H has an upper bound given by

$$E_0 \le E_T = \frac{\int \Psi_T H \Psi_T dR}{\int |\Psi_T|^2 dR}\ . \tag{2}$$

This equation can be rewritten in terms of Eq. (1) so that it reads:

$$E_T = \iiint \left\{ \frac{H\psi_r(R) \prod_i \Theta(\mathbf{r}_i - \mathbf{s}'_i)}{\psi_r(R) \prod_i \Theta(\mathbf{r}_i - \mathbf{s}'_i)} \right\} p(R, S, S') dR dS dS'\ , \tag{3}$$

where $p(R, S, S')$ is given by

$$p(R, S, S') = \frac{\psi_r^2(R) \prod_i \Theta(\mathbf{r}_i - \mathbf{s}_i)\psi_s(S) \prod_j \Theta(\mathbf{r}_j - \mathbf{s}'_j)\psi_s(S')}{\iiint \psi_r^2(R) \prod_i \Theta(\mathbf{r}_i - \mathbf{s}_i)\psi_s(S) \prod_j \Theta(\mathbf{r}_j - \mathbf{s}'_j)\psi_s(S') dR dS dS'}\ . \tag{4}$$

Therefore the variational energy E_T can be estimated by taking the average of the energy estimator in Eq. (3) with respect to the probability density function $p(R, S, S')$

$$\bar{E}_T = \left\langle \frac{H\Psi_T}{\Psi_T} \right\rangle\ . \tag{5}$$

The integral in Eq. (3) can be evaluated by the Metropolis Monte Carlo algorithm[3] generalized to take into account the shadow coordinates.

THE ELEMENTARY EXCITATION SPECTRUM

The Feynman form[4] of the wave function for an excited state of momentum $\hbar\mathbf{k}$ is simply the product of the density fluctuation of wavevector $\mathbf{k}$ times the ground state. As argued in the Introduction a better wave function should be obtained if this density fluctuation is written in terms of the auxiliary variables. Therefore we write the unnormalized wave function in the form

$$\Psi_k(R) = \psi_r(R) \int dS \prod_i \Theta(\mathbf{r}_i - \mathbf{s}_i)\ \psi_s(S)\ \sigma_k\ , \tag{6}$$

where

$$\sigma_k = \sum_{i=1}^{N} e^{i\mathbf{k}\cdot\mathbf{s}_i}\ . \tag{7}$$

All other factors in (6) are taken to be the same as in the ground state.

It is very easy to see that the wave function Ψ_k is an eigenstate of the linear momentum operator $\mathbf{P} = -i\hbar \sum_i \nabla_{\mathbf{r}_i}$. The operator $\mathbf{P}$ will act only on the terms under the integration sign since ψ_r depends only on the relative coordinates. Let us apply $\mathbf{P}$ on Ψ_k and consider the following change of variables $\mathbf{s}_i \rightarrow \mathbf{r}_i - \mathbf{s}_i$

$$\mathbf{P}\Psi_k(R) = -i\hbar\psi_r(R)\times$$

$$\times \sum_i \nabla_{\mathbf{r}_i} \int dS \prod_i \Theta(\mathbf{s}_i) \prod_{l<m} e^{u_s(|\mathbf{s}_m - \mathbf{s}_l + \mathbf{r}_l - \mathbf{r}_m|)} \sum_{n=1}^{N} e^{i\mathbf{k}\cdot(\mathbf{r}_n - \mathbf{s}_n)}\ . \tag{8}$$

In the above equation we can see that the product of the Jastrow form does not contribute because it depends only on the relative distance of the particles. The only term that makes a contribution to the total linear momentum of the system comes from the last sum; as expected, the result is

$$\mathbf{P}\Psi_k(R) = \hbar N\mathbf{k}\Psi_k(R)\ . \tag{9}$$

The wave function (6) does not contain any new variational parameter so that we can use the configurations generated for the ground state of liquid ^{4}He and the estimate of the excitation spectrum can be reduced to

$$\bar{\mathcal{E}}_k = \frac{\left\langle \sigma_k{}'\sigma_k \frac{H\Psi_T}{\Psi_T} \right\rangle}{\left\langle \sigma_k{}'\sigma_k \right\rangle} - \bar{E}_T\ , \tag{10}$$

where $\bar{E}_T$, the estimate of the ground state energy of Eq.(5) has to be subtracted. The prime on σ_k denotes the expression of Eq. (7) using coordinates S'.

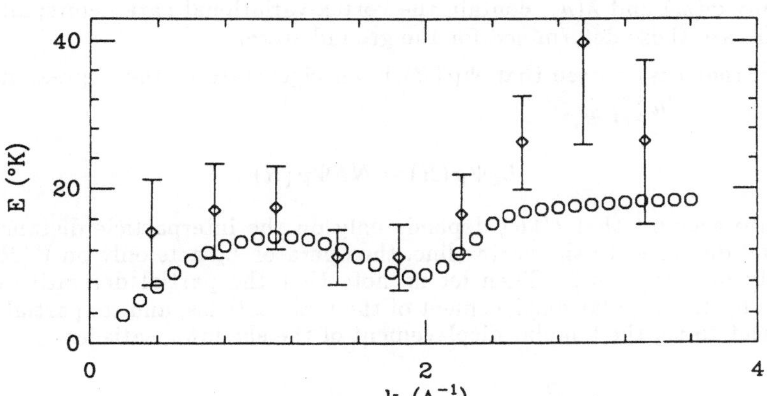

Fig. 1. Elementary excitation spectrum computed for the discrete set of values of the Born-von Karman wavevector consistent with the periodicity of the simulation cell. The experimental data is shown by circles.

Our preliminary results for the excitation spectrum obtained at the equilibrium density and using Eq. (10) is presented at Fig. 1 together with the experimental data.[5] We find that the energy of a roton is 11.2 ± 4 °K. For comparison the usual Feynman form gives an energy of 20 °K and the Feynman-Cohen[6] form gives[7] 14 °K. It is clear that the new wave function represents a significant improvement with respect to the Feynman form but for a more precise conclusion we must reduce the statistical fluctuations that result from the phase factors contained in (10).

A TRIAL WAVE FUNCTION FOR SYSTEMS WITH A VORTEX

The wave function for a vortex line[8] in liquid He which has been studied so far is characterized by a localized vorticity on the axis of the vortex and by a hollow core, i.e. the density of atoms vanishes on the axis. We can easily write a wave function with distributed vorticity if the flow field is associated with the auxiliary variables. More precisely we write

$$\Psi_V(R) = U(R) \int dS \; V(R,S) \; , \tag{11}$$

where U(R) is a real function and V(R,S) a complex one. The factor $U(R)$ in Eq. (11) given by

$$U(R) = \psi_r(R) \prod_{i_1}^{N} \varphi(\rho_i) \; , \tag{12}$$

contains besides the factor $\psi_r(R)$ of the Jastrow form of Eq.(1), a one-body factor $\varphi(\rho_i)$ that depends on the distance ρ_i of particle i to the vortex axis. The factor $V(R,S)$ is

$$V(R,S) = \prod_i \Theta(\mathbf{r}_i - \mathbf{s}_i) \; \psi_s(S) \; \prod_{j=1}^{N} (\mathrm{e}^{i\vartheta_j} \lambda(\varrho_j)) \; . \tag{13}$$

This factor has also a real one-body term $\lambda(\varrho_j)$ which is a function of the distance ϱ_j of the jth shadow particle to the vortex line. The complex term $\mathrm{e}^{i\vartheta_j}$ takes into account the angular displacement ϑ_j of this shadow particle. In the wave function $\Psi_V(R)$ only $\varphi(\rho_i)$ and $\lambda(\varrho_j)$ contain the vortex variational parameters; all the other parameters are those determined for the ground state.

It is rather easy to see that $\Psi_V(R)$ is an eigenstate of the angular momentum operator $L_z = -i\hbar \sum_i \frac{\partial}{\partial \theta_i}$,

$$L_z \Psi_V(R) = N\hbar \Psi_V(R) \; . \tag{14}$$

Taking into account that $U(R)$ depends only on the interparticle distances and on their radial distances to the vortex line, the operator L_z acts only on $V(R,S)$, more specifically on $\Theta(\mathbf{r}_i - \mathbf{s}_i)$. Then let us note that the partial derivative of Θ with respect to θ_i, the angular displacement of the real particles, and its partial derivative with respect to ϑ_i, the angular displacement of the shadows, satisfies

$$\frac{\partial}{\partial \theta_i} \Theta(\mathbf{r}_i - \mathbf{s}_i) = -\frac{\partial}{\partial \vartheta_i} \Theta(\mathbf{r}_i - \mathbf{s}_i) \; . \tag{15}$$

Using the above equality in the left hand side of Eq. (14) and performing an integration by parts we get the result

$$L_z\Psi_V(R) = -i\hbar U(R)\int dS\prod_i \mathrm{e}^{-C\,(\mathbf{r}_i-\mathbf{s}_i)^2}\times$$

$$\times\sum_j\frac{\partial}{\partial\vartheta_j}[\prod_{k<l}\mathrm{e}^{-v(|\mathbf{s}_k-\mathbf{s}_l|)}\prod_{m=1}^{N}(\mathrm{e}^{i\vartheta_m}\lambda(\varrho_m))]\ . \qquad (16)$$

The factors $\mathrm{e}^{-v(|\mathbf{s}_k-\mathbf{s}_l|)}$ do not contribute because they depend only on the relative displacements and the only term that contributes is $\prod \mathrm{e}^{i\vartheta_m}$. Thus the right hand side of Eq. (14) is immediately obtained.

The velocity field $\mathbf{v}(\mathbf{r})$ is defined in terms of the current density $\mathbf{j}(\mathbf{r})$ and of the number density $n(\mathbf{r})$ by

$$\mathbf{v}(\mathbf{r}) = \frac{\mathbf{j}(\mathbf{r})}{n(\mathbf{r})}\ . \qquad (17)$$

Considering the definition of the current density operator,

$$\hat{\mathbf{j}}(\mathbf{r}) = \frac{\hbar}{2mi}\sum_l[\delta(\mathbf{r}-\mathbf{r}_l)\nabla_{\mathbf{r}_l} + \nabla_{\mathbf{r}_l}\delta(\mathbf{r}-\mathbf{r}_l)]\ , \qquad (18)$$

its expectation value can be written as

$$\mathbf{j}(\mathbf{r}) = \mathcal{N}\frac{\hbar\,N}{2mi}\int dR\delta(\mathbf{r}-\mathbf{r}_1)[\Psi_V^*\nabla_{\mathbf{r}_1}\Psi_V - \Psi_V\nabla_{\mathbf{r}_1}\Psi_V^*]\ , \qquad (19)$$

where $\mathcal{N}^{\frac{1}{2}}$ is the normalization factor of the trial wave function. Of course we have

$$\mathcal{N} = \frac{1}{\int dR\,dS\,dS'\,U^2(R)V^*(R,S')V(R,S)}\ . \qquad (20)$$

In terms of the functions $U(R)$ and $V(R,S)$ of Eq. (11) the expression of $\mathbf{j}(\mathbf{r})$ can be rewritten as

$$\mathbf{j}(\mathbf{r}) = \mathcal{N}\frac{\hbar\,N}{2mi}\int dR\,dS\,dS'\,\delta(\mathbf{r}-\mathbf{r}_1)U^2(R)$$
$$\times[\frac{V^*(R,S')V(R,S)}{\Theta(\mathbf{r}_1-\mathbf{s}_1)}\nabla_{\mathbf{r}_1}\Theta(\mathbf{r}_1-\mathbf{s}_1) - \frac{V^*(R,S)V(R,S')}{\Theta(\mathbf{r}_1-\mathbf{s}_1)}\nabla_{\mathbf{r}_1}\Theta(\mathbf{r}_1-\mathbf{s}_1)]\ . \qquad (21)$$

Integrating by parts this equation and taking into account that

$$\nabla_{\mathbf{r}_1}\Theta(\mathbf{r}_1-\mathbf{s}_1) = -\nabla_{\mathbf{s}_1}\Theta(\mathbf{r}_1-\mathbf{s}_1)\ , \qquad (22)$$

the only term that survives, after the subtraction is carried out, is the one that contains the factor $\nabla_{\mathbf{s}_1}\vartheta_1$. The latter is $\nabla_{\mathbf{s}_1}\vartheta_1 = \frac{1}{\varrho_1}\mathbf{u}_{\vartheta_1}$, where $\mathbf{u}_{\vartheta_1}$ is the unit vector in the direction of the angular displacement of shadow 1, and so we can write

$$\mathbf{j}(\mathbf{r}) = \mathcal{N}\frac{\hbar\,N}{m}\int dR\,dS\,dS'\,\delta(\mathbf{r}-\mathbf{r}_1)U^2(R)V^*(R,S')V(R,S)\frac{1}{\varrho_1}\,\mathbf{u}_{\vartheta_1}\ . \qquad (23)$$

Let $n(\mathbf{r},\mathbf{s})$ be the pair distribution function for a particle in $\mathbf{r}$ and its shadow of type S in $\mathbf{s}$. This is a real function given by

$$n(\mathbf{r},\mathbf{s}) = \mathcal{N}N \int dR\,dS\,dS'\,\delta(\mathbf{r}-\mathbf{r}_1)\delta(\mathbf{s}-\mathbf{s}_1)U^2(R)V^*(R,S')V(R,S)\ . \qquad (24)$$

Taking into account this definition of the pair distribution function $n(\mathbf{r},\mathbf{s})$, the expectation value of the current density operator of Eq. (23) can be rewritten as

$$\mathrm{j}(\mathbf{r}) = \frac{\hbar}{m}\int ds\ n(\mathbf{r},\mathbf{s})\frac{1}{\varrho}\ \mathbf{u}_\vartheta\ . \qquad (25)$$

Finally the velocity field $\mathbf{v}(\mathbf{r})$ of Eq. (17) reads:

$$\mathbf{v}(\mathbf{r}) = \frac{\hbar}{m}\int ds \frac{n(\mathbf{r},\mathbf{s})}{n(\mathbf{r})}\frac{1}{\varrho}\ \mathbf{u}_\vartheta\ . \qquad (26)$$

The ratio $\frac{n(\mathbf{r},\mathbf{s})}{n(\mathbf{r})} = f(\mathbf{s}|\mathbf{r})$ has a significance of its own. It is the the conditional probability that a real particle in the position $\mathbf{r}$ has its shadow of type S in $\mathbf{s}$. Of course we have

$$\int ds\, f(\mathbf{s}|\mathbf{r}) = 1\ , \qquad (27)$$

since $n(\mathbf{r}) = \int ds\, n(\mathbf{r},\mathbf{s})$.

The precise behavior of the velocity field depends on the variational parameters of Ψ_V but we have not yet performed this variational computation. In order to get a qualitative idea of the velocity field given by the new wave function we have computed $\mathbf{v}(\mathbf{r})$ and $\nabla \times \mathbf{v}(\mathbf{r})$ for two models of $f(\mathbf{s}|\mathbf{r})$ which in a certain sense are at two extremes. In the first one the only nonuniformity and anisotropy of $n(\mathbf{r},\mathbf{s})$ is assumed to be contained in $n(\mathbf{r})$. In this case we can take $f(\mathbf{s}|\mathbf{r}) = w(|\mathbf{r}-\mathbf{s}|)$ as the ground state correlation function between a particle and its coupled shadow. This correlation is to a good approximation a Gaussian

$$w(|\mathbf{r}-\mathbf{s}|) = \frac{1}{\pi^{3/2}\Delta^3}\mathrm{e}^{-|\mathbf{r}-\mathbf{s}|^2/\Delta^2}\ . \qquad (28)$$

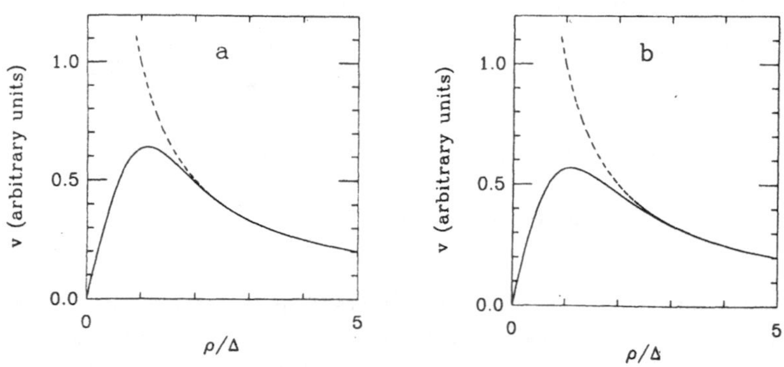

Fig. 2. Velocity field of liquid ^{4}He with a vortex line. The dashed lines represent the classical flow field. In Fig. 2a the shadows are uniform distributed. Fig. 2b shows the result for a distribution of the shadows that have a Gaussian hole around the vortex line, in this plot $\xi = \Delta$.

The parameter Δ, related to the particle-shadow mean distance, is obtained from the computations made for the ground-state and it is of order of 1.3 Å. This model is certainly appropriate if the density profile is mainly determined by the one body factor $\varphi(\rho_i)$ for the particles, and if we have $\lambda(\varrho_i) \approx const.$

The velocity field $\mathbf{v}$ around the vortex axis can be computed immediately. Substituting Eq. (28) in Eq. (26), the integral can be easely performed and we obtain

$$\mathbf{v}(\mathbf{r}) = \frac{\hbar}{m}\frac{1}{\rho}(1 - e^{-\rho^2/\Delta^2})\, \mathbf{u}_\theta \, . \tag{29}$$

A plot of the velocity field $\mathbf{v}(\mathbf{r})$ and of the classical flow field is presented in Fig. 2a. The velocity field is a function only of ρ, the distance to the vortex line in the z-axis. It is null on the vortex axis and has a maximum at a distance of order Δ. Beyond this maximum it rapidly approaches $1/\rho$, the classical flow field.

The curl of $\mathbf{v}$ has just one component along the z-axis in the $\mathbf{u}_k$ direction. Thus the curl of $\mathbf{v}$ is a Gaûssian centered in the z axis and given by

$$\nabla \times \mathbf{v}(\mathbf{r}) = \frac{2\hbar}{m\Delta^2} e^{-\rho^2/\Delta^2}\, \mathbf{u}_k \, . \tag{30}$$

In the second model that we have considered, we take the shadow particles to be nonuniform as well. That is, we assume that the shadow particles have a Gaussian hole of width ξ around the vortex line in the z-axis. So, to model $n(\mathbf{s}) = \int d\mathbf{r}\, n(\mathbf{r}, \mathbf{s})$, the density profile of the shadows around the vortex, we use

$$n(\mathbf{s}) \propto 1 - e^{-\varrho^2/\xi^2} \, . \tag{31}$$

As before, ϱ is the distance from the vortex axis to a shadow particle at $\mathbf{s}$. In this model we will take the following product as the conditional probability that a particle at $\mathbf{r}$ has a particular coupled shadow at $\mathbf{s}$:

$$f(\mathbf{s}|\mathbf{r}) = A(\mathbf{r}) n(\mathbf{s}) w(|\mathbf{r} - \mathbf{s}|) \, , \tag{32}$$

where w is again the Gaussian correlation function between a particle and its coupled shadow as given in Eq. (28), and $A(\mathbf{r})$ is determined in such a way that the condition of Eq. (27) is satisfied. Explicitly, $A(\mathbf{r})$ is given by

$$A(\mathbf{r}) = \frac{1}{1 - \frac{\xi^2}{\xi^2+\Delta^2} e^{-\frac{\rho^2}{\xi^2+\Delta^2}}} \, . \tag{33}$$

The velocity field can be obtained by some lengthy but straightforward computations. The result is

$$\mathbf{v}(\mathbf{r}) = \frac{\hbar}{m}\frac{1}{\rho} \frac{1 - e^{-\frac{\rho^2}{\xi^2+\Delta^2}}}{1 - \frac{\xi^2}{\xi^2+\Delta^2} e^{-\frac{\rho^2}{\xi^2+\Delta^2}}}\, \mathbf{u}_\theta \, . \tag{34}$$

In Fig. 2b we display the velocity field obtained in this situation, *viz*, when the shadows have the nonuniform distribution of Eq. (31). Also this time $\mathbf{v}$ is only a function of ρ, the distance to the vortex axis. This velocity field has a behavior similar to the one of the previous model, but its maximum velocity is at a distance ρ_{Max} approximately equal to

$$\rho_{Max} \approx [(\xi^2 + \Delta^2) ln(1 + \frac{\Delta^2}{\xi^2 + \Delta^2})]^{\frac{1}{2}} .$$

The curl of $\mathbf{v}(\mathbf{r})$ for this model is roughly a Gaussian; more precisely it is given by

$$\nabla \times \mathbf{v}(\mathbf{r}) = \frac{2\hbar}{m} \frac{\Delta^2 e^{\frac{-\rho^2}{\xi^2+\Delta^2}}}{[\Delta^2 + \xi^2(1 - e^{\frac{-\rho^2}{\xi^2+\Delta^2}})]^2} \mathbf{u}_k . \tag{35}$$

In both models that we have analysed, we obtain a vortex with distributed vorticity. The vorticity region is a cylinder with a radius of the order of the particle-shadow mean distance. The velocity field goes to zero in both cases at the vortex axis. This fact comes naturally from our formalism; in neither of the models that we have considered it is necessary to impose that the density of the real particles goes to zero at that line. A variational calculation is under way to determine the numerical predictions of this new wave function.

DISCUSSION

Using the ideas of the shadow wave function we are able to construct trial wave functions that have the necessary properties to describe the elementary excitation spectrum of liquid ^{4}He and the nonuniform states of this system which has a vortex.

Our preliminary simulations of the excitation spectrum show that although we have to deal with strong oscillating phases in the calculations, we can obtain reasonable results. Extending the search in the variational parameter space and doing longer runs we expect a substantial improvement in the agreement with the experimental data.

For liquid ^{4}He with a vortex line we have shown how one can easily construct a wave function with distributed vorticity. Thus for the first time it is possible to carry out a microscopic simulation with this necessary property. The form of the trial wave function is flexible enough to cover a wide range of possibilities.

ACKNOWLEDGEMENTS

We would like to acknowledge many useful discussions with G. V. Chester, K. J. Runge, K. E. Schmidt, and P. A. Whitlock. This work was partially supported by the Condensed Matter Theory Program of the National Science Foundation under Grant DMR-8419083, by the Advanced Scientific Computing Program of the NSF under Grant ASC-8715641, and by the Applied Mathematics subprogram of the Office of Energy Research, U. S. Department of Energy under contract No. DE-AC02-76ER03077. This research was conducted using the Cornell National Supercomputer Facility, a resource of the Center for Theory and Simulation In Science and Engineering at Cornell University, which is funded in part by the National Science Foundation, New York State, and the IBM Corporation. This work was also partially supported by Consiglio Nazionale delle Ricerche under the Italia-US Cooperative Science Program.

REFERENCES

1. Silvio Vitiello, Karl Runge and Malvin H. Kalos, *Phys. Rev. Lett.* **60**, 1970 (1988).

2. S. A. Vitiello, K. J. Runge, G. V. Chester and M. H. Kalos *"Shadow" Wave Function Variational Calculations of Crystalline and Liquid Phases of ^4He*, preprint.

3. N. Metropolis, A. W. Rosenbluth, M. N. Rosenbluth, A. H. Teller and E. Teller, *J. Chem. Phys.* **21**, 1087 (1953).

4. R.P. Feynman, *Phys. Rev.* **94**, 262 (1954)

5. R. A. Cowley and A. D. B. Woods, Can. J. Phys. 49, 177, 1971.

6. R.P. Feynman and M. Cohen, *Phys. Rev.* **102**, 1189 (1956)

7. E. Manousakis and V.R. Pandharipande, *Phys. Rev. B* **30**, 5062 (1984)

8. R.P. Feynman, *Application of Quantum Mechanics to Liquid Helium* in C.J. Gorter ed. *Progress in Low Temperature Physics*, North Holland, Amsterdam, 1955

VIBRATIONAL DENSITY-OF-STATES, ISOTOPE EFFECT, AND SUPERCONDUCTIVITY IN $Ba_{1-x}K_xBiO_3$ CUBIC OXIDES

Marcos H. Degani[1], Rajiv K. Kalia, and P. Vashishta

Materials Science Division
Argonne National Laboratory
Argonne, IL 60439, USA

Vibrational density-of-states of insulating $BaBiO_3$ in orthorhombic phase and superconducting $Ba_{0.6}K_{0.4}BiO_3$ in cubic phase are studied using the molecular dynamics (MD) method. The MD results are compared with the recent inelastic neutron scattering and electron tunneling experiments. The exponent of the oxygen isotope effect is calculated from the first moment of the phonon density-of-states (the weak coupling limit) and from the solution of Eliashberg gap equations. Results are compared with isotope effect experiments. Evidence based on inelastic neutron scattering, tunneling, and isotope effect experiments when combined with the MD calculations suggest that this material is a normal weak coupling BCS superconductor with strong coupling of the carriers to high energy oxygen phonons.

INTRODUCTION

Recent discovery of superconductivity in $Ba_{1-x}K_xBiO_3$ at 30K is of great interest because of the difference in the properties of this material and other high-T_c oxide superconductors. $Ba_{1-x}K_xBiO_3$ is cubic in the superconducting phase, $0.35<x<0.5$, whereas the other high-T_c materials have a distinctly planar (Cu-O) structure. This material has no copper and displays none of the antiferromagnetism common to other high-T_c materials. Measurements by Hinks et al.[1] on $Ba_{0.6}K_{0.4}BiO_3$ indicate a substantial oxygen isotope effect, $\alpha=0.4$, unlike the $YBa_2Cu_3O_{7-\delta}$ which shows a negligible isotope effect. Batlogg et al.[2] find a smaller value $\alpha=0.22$, whereas in a more recent experiment Kondoh et al.[3] have determined $\alpha=0.35$ in $Ba_{0.6}K_{0.4}BiO_3$. Infrared measurements[4]

[1] Permanent address: Instituto de Física e Química de São Carlos, USP
13560 - São Carlos - S.P. - Brasil

Condensed Matter Theories, Volume 5
Edited by V.C. Aguilera-Navarro
Plenum Press, New York, 1990

reveal a superconducting gap (2Δ=8.7 meV) with $2\Delta/k_BT_c$=3.5. Since there is no evidence for magnetic fluctuations and the transition temperature is not so high that one has to invoke an exotic mechanism, it is quite possible that phonon coupling between carriers is responsible for superconductivity in $Ba_{1-x}K_xBiO_3$.[5] Attempts to carry out electron tunneling experiments in superconductor-insulator-superconductor (S-I-S) junctions in $YBa_2Cu_3O_{7-\delta}$ have not been very successful due to the very small coherence length (~ 10Å). In $Ba_{1-x}K_xBiO_3$, however, electron tunneling experiments on S-I-S junctions have recently been carried out by Zasadzinski et al..[6] This is due in part to the fact that the coherence length in this material is of the order of ~50Å. Even though it has not been possible to invert the Eliashberg gap equations using the tunneling data, it is quite clear that there are images of excitations in the tunneling data at high energies in the range of 60 meV. This observation of high energy structure in the tunneling data has important consequences as we shall discuss in the following sections.

The pairing theory[7] (BCS) has given an accurate account of the properties of conventional superconductors as well as superfluid ^{3}He. With the discovery of high-T_c oxide superconductors, the question has been raised whether the pairing theory continues to be applicable but with an exceptionally strong pairing interactions or whether a totally different framework must be developed. It is easier to address some of these questions on cubic, nonmagnetic 30K oxide superconductors $Ba_{1-x}K_xBiO_3$ than on highly complex materials like $YBa_2Cu_3O_{7-\delta}$ with antiferromagnetic phases lingering in the vicinity of superconducting phase.

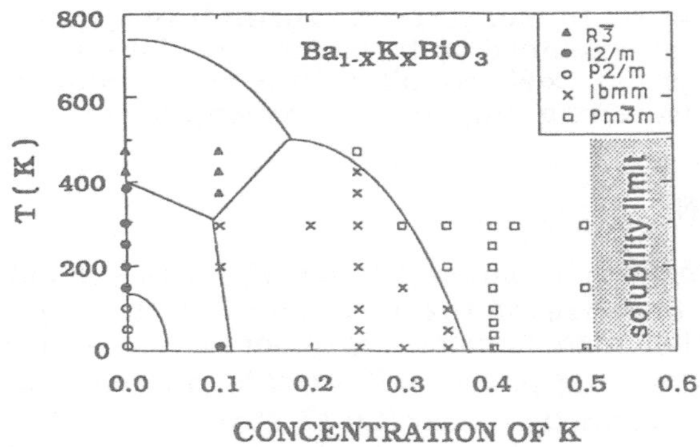

Fig. 1a. Phase diagram for $Ba_{1-x}K_xBiO_3$ as determined from neutron diffraction by Pei et al.[8]

Using neutron diffraction technique Shiyou Pei et al.[8] have determined the structure of five phases in the $Ba_{1-x}K_xBiO_3$ system for potassium concentrations in the range 0.0<x<0.5 below 473K. Only the perovskite phase which exists for x>0.37 exhibits bulk superconductivity. Fig. 1a shows the crystal structures in $Ba_{1-x}K_xBiO_3$, as a function of temperature and potassium concentration. The behavior of the superconducting transition temperature as a function of K concentration is shown in Fig. 1b.

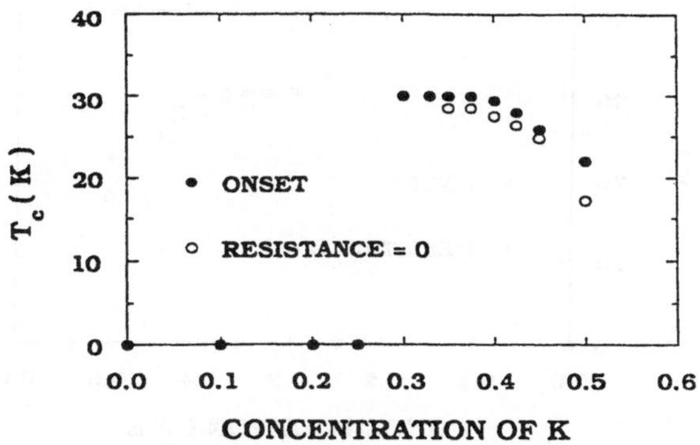

Fig. 1b. Superconducting transition temperature versus concentration.[8]

MOLECULAR DYNAMICS SIMULATION

Effective interparticle interactions were used in the molecular dynamics simulations. The potentials[9] include steric repulsions between ions, Coulomb interactions due to charge transfer effects, and charge-dipole interactions due to large electronic polarizability of O^{--} ions. For $BaBiO_3$ there are six interaction potentials, whereas for $Ba_{0.6}K_{0.4}BiO_3$ there are ten.[10]

Table 1. Experimental data used in the MD simulations.

	$BaBiO_3$	$Ba_{0.6}K_{0.4}BiO_3$
Mass density	7.88 g/cm^3	7.33 g/cm^3
Lattice	orthorhombic	cubic
Structure	a=6.2000Å b=6.1561Å c=4.3474Å	a=4.3160Å

The molecular dynamics[11] calculations were done with 540 particle orthorhombic $BaBiO_3$ and 625 particle cubic $Ba_{0.6}K_{0.4}BiO_3$ at the experimental number densities. Periodic boundary conditions were used. The unit cells of the two systems are shown in Fig. 2. The structural parameters used in the MD simulations are given in Table 1.

superconductivity. Fig. 1a shows the crystal structures in $Ba_{1-x}K_xBiO_3$, as a function of temperature and potassium concentration. The behavior of the superconducting transition temperature as a function of K concentration is shown in Fig.1b.

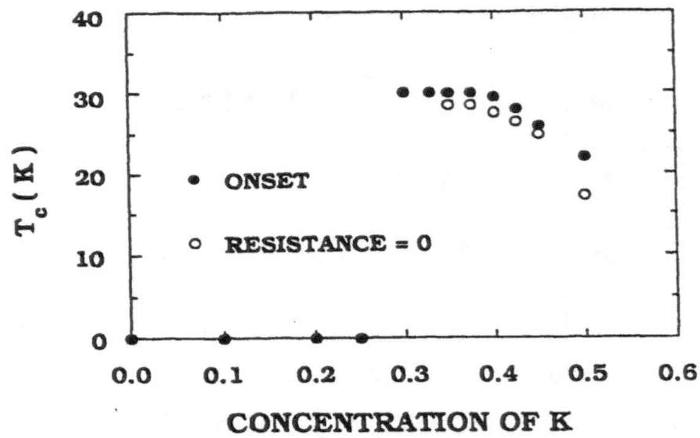

Fig. 1b. Superconducting transition temperature versus concentration.[8]

MOLECULAR DYNAMICS SIMULATION

Effective interparticle interactions were used in the molecular dynamics simulations. The potentials[9] include steric repulsions between ions, Coulomb interactions due to charge transfer effects, and charge-dipole interactions due to large electronic polarizability of O^{--} ions. For $BaBiO_3$ there are six interaction potentials, whereas for $Ba_{0.6}K_{0.4}BiO_3$ there are ten.[10]

Table 1. Experimental data used in the MD simulations.

	$BaBiO_3$	$Ba_{0.6}K_{0.4}BiO_3$
Mass density	7.88 g/cm^3	7.33 g/cm^3
Lattice	orthorhombic	cubic
Structure	a=6.2000Å	a=4.3160Å
	b=6.1561Å	
	c=4.3474Å	

The molecular dynamics[11] calculations were done with 540 particle orthorhombic $BaBiO_3$ and 625 particle cubic $Ba_{0.6}K_{0.4}BiO_3$ at the experimental number densities. Periodic boundary conditions were used. The unit cells of the two systems are shown in Fig. 2. The structural parameters used in the MD simulations are given in Table 1.

The $Ba_{0.6}K_{0.4}BiO_3$ system was obtained from $BaBiO_3$ by randomly replacing 40% of the Ba atoms with K atoms. Before calculating the phonon DOS, it was ensured that the systems were dynamically stable in the appropriate symmetries. The phonon DOS was calculated using three methods: (1) velocity auto-correlation functions, (2) the equation of motion method,[12] and (3) direct diagonalization of the dynamical matrix. The results of all these three calculations are in good agreement with one another.

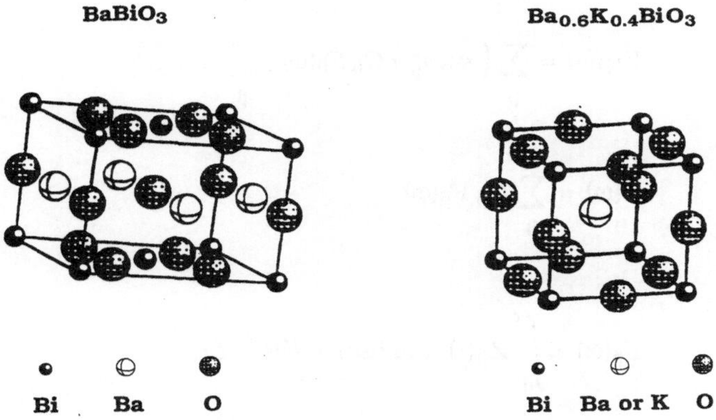

Fig. 2. Unit cell for $BaBiO_3$ and $Ba_{0.6}K_{0.4}BiO_3$.

To establish the dynamical stability of $BaBiO_3$, the system was put in the orthorhombic structure in a fixed volume MD cell. The partial pair distribution function and bond angle distributions were calculated to verify the bond lengths and coordination numbers etc. The system was slowly heated to 600K and thermalized for several thousand time steps. The system was run uninterruptedly for more than 30,000 time steps and various structural correlations were calculated to examine the symmetry. The system at 600K was slowly cooled and then subjected to a steepest descent quench[13](SDQ) which is a mathematically well defined method of examining the underlying mechanically stable structures. The pair correlation functions and bond angle distributions were calculated again to ascertain the symmetry of the MD system. After performing the above mentioned procedure on the 540 particle $BaBiO_3$ system it was determined that the resulting final symmetry was indeed the same as that of the starting orthorhombic structure. The 625 particle cubic $Ba_{0.6}K_{0.4}BiO_3$ system was obtained by randomly replacing

40% of the Ba atoms with K atoms. This system was subjected to the same procedure to ensure the dynamic stability of $Ba_{0.6}K_{0.4}BiO_3$ system.

PHONON DENSITY-OF-STATES

The phonon density-of-states was calculated using three different methods. The first method involves calculating the velocity autocorrelation functions for each species of atoms from the molecular dynamics trajectories. The partial densities of states are obtained by fourier transforming the time correlations.

$$F_N(\omega) = \sum_{\beta} \left(4\pi b_\beta^2 \right) C_\beta F_\beta(\omega) / \sum_{\beta} 4\pi b_\beta^2 C_\beta \tag{1}$$

$$F(\omega) = \sum_{\beta} C_\beta F_\beta(\omega) \tag{2}$$

$$F_\beta(\omega) = \int_0^\tau Z_\beta(t)\ \cos(\omega t)\ e^{-\lambda(t/\tau)^2}\ dt \tag{3}$$

$$Z_\beta(t) = \frac{\left\langle \sum_{i(\beta)} \mathbf{v}_i(t)\cdot\mathbf{v}_i(0) \right\rangle}{\left\langle \sum_{i(\beta)} \mathbf{v}_i^2(0) \right\rangle} \tag{4}$$

where $Z_\beta(t)$ is the normalized velocity autocorrelation function for β^{th} species and its fourier transform $F_\beta(\omega)$ is the partial density-of-states normalized to the total number of particles, N, of the system. The total density-of-states is obtained by summing the concentration weighted partial densities-of-states. Additional weighting with the coherent neutron cross-sections is required to obtain the neutron density-of-states, $F_N(\omega)$, for comparison with the generalized density-of-states, $G(\omega)$, obtained in the neutron scattering experiment.[5]

The second method involves calculating the displacement autocorrelation function using the equation of motion method.[12] To implement this method it is essential to bring the system to the local minimum by carrying out the steepest descent quench which guarantees that the force and the velocity of each particle is zero. Each particle is then given a random displacement and the system is allowed to evolve according to the classical equations of motion.

$$F(\omega) = \frac{4}{\pi \delta_m^2} \int_0^{\tau} f(t) \cos(\omega t)\, e^{-\lambda (t/\tau)^2}\, dt \tag{5}$$

$$f(t) = \sum_{i\mu} \delta r_{i\mu}(t)\, \delta r_{i\mu}(0) \tag{6}$$

$$\delta r_{i\mu}(t) = r_{i\mu}(t) - r_{i\mu 0} \tag{7}$$

$$\delta r_{i\mu}(0) = \delta_m \cos\theta_{i\mu} \tag{8}$$

where $r_{i\mu}(0)$ and $r_{i\mu}(t)$ are the μ^{th}(x, y, or z) component of the displacement of the ith particle, $r_{i\mu 0}$ are the equilibrium positions of the atoms, δ_m is the amplitude of the initial displacement, and $\theta_{i\mu}$ are random angles distributed uniformly between 0 and 2π.

Finally, the phonon density-of-states can be calculated by direct diagonalization of the dynamical matrix. As with the equation of motion method it is essential to bring the system in the correct symmetry with zero force on each particle. This is accomplished by applying the SDQ to the MD configuration. Since we know the interaction potential functions under the influence of which the system is dynamically stable, we can calculate the elements of the dynamical matrix by numerically calculating the x, y, and z derivatives of the force on each particle.

$$F_{j\nu} = - \frac{\partial \phi}{\partial r_{j\nu}} \tag{9}$$

$$D_{ij}^{\mu\nu} = - \frac{1}{\sqrt{M_i M_j}} \frac{\partial F_{j\nu}}{\partial r_{i\mu}} \tag{10}$$

$$\omega^2 u_{i\mu}^{(0)} = \sum_{j=1}^{N} \sum_{\nu=1}^{3} D_{ij}^{\mu\nu}\, u_{j\nu}^{(0)}. \tag{11}$$

Using standard numerical methods the 1620x1620 dynamical matrix for $BaBiO_3$ and the 1875x1875 matrix for $Ba_{0.6}K_{0.4}BiO_3$ were diagonalized to obtain the eigenvectors and eigenvalues from which the partial and total density-of-states can be obtained.

For $BaBiO_3$, Fig. 3 displays the phonon densities-of-states from neutron scattering and MD simulation. The experiment shows prominent peaks at 35, 43, 63, and 71 meV. In addition, there is an indication of a shoulder at 24 meV and weak features in the region of 50-58 meV. In the MD results, we find peaks at 25, 32, 37, 40, 45, 51, 60, 66, and 74 meV. Because of limited resolution of the neutron measurements, the MD peaks at 32 and 37 meV are observed experimentally as a single peak at 35 meV. Similarly, the MD peaks at 60 and 66 meV appear as a single peak at 63 meV in the neutron experiment. Note that there are two additional peaks at 11 and 16 meV in the MD results. These peaks are not observed in the neutron measurements because of the low-energy cutoffs.

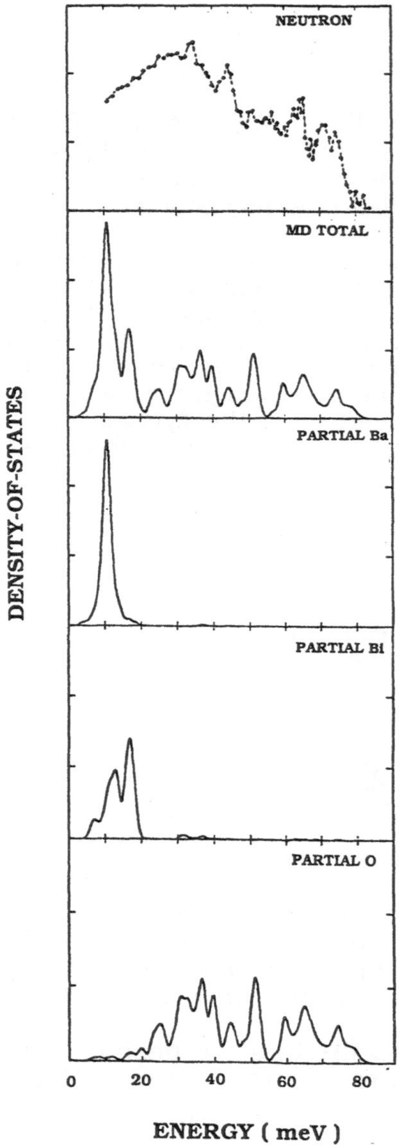

Fig. 3. Phonon densities-of-states for $BaBiO_3$ from experiment and MD simulation.

For $Ba_{0.6}K_{0.4}BiO_3$ the phonon densities-of-states are shown in Fig. 4. Neutron measurements reveal that K doping broadens the peaks due to disorder caused by random substitution of K on Ba sites and also shifts the density-of-states toward lower energies. Because of broadening, only two bands at 30 and 60 meV remain in $Ba_{0.6}K_{0.4}BiO_3$. An overall broadening of the peaks and a shift toward lower energies are

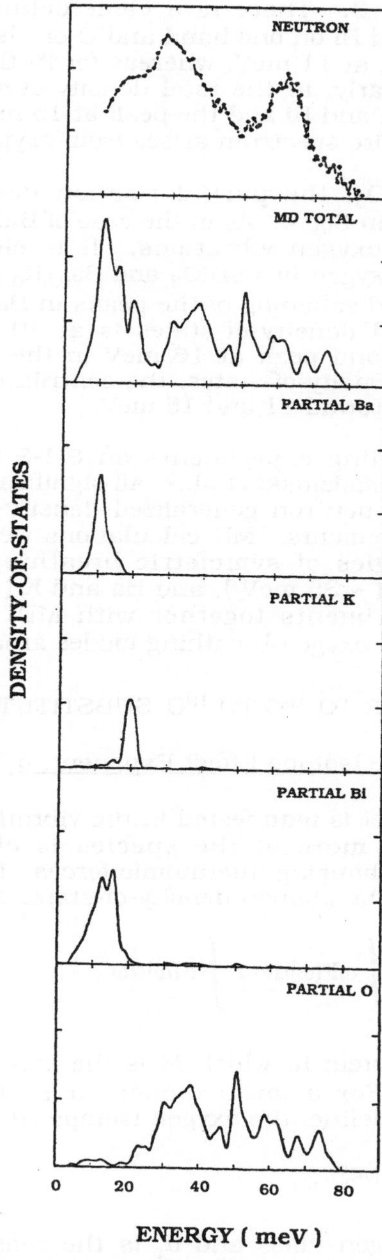

Fig. 4. Phonon densities-of-states for $Ba_{0.6}K_{0.4}BiO_3$ from experiment and MD simulation.

also evident from the MD results shown in Fig. 3 and Fig. 4: (1) a band extending from 25 to 37 meV. (2) a peak around 51 meV, (3) a band between 54 and 65 meV, and (4) small peaks at 67 and 73 meV. In addition, there are small features at 25 and 46 meV. The simulation also shows peaks at 11 and 15 meV, whereas only a shoulder at 16 meV is visible experimentally. Higher-resolution neutron measurements may reveal the additional features observed in the simulation.

To understand the origin of the peaks in the density-of-states, we examine the MD results for the partial phonon density-of-states for $BaBiO_3$ shown in Fig. 3. There is a clear delineation in the peaks associated with Ba and Bi on one hand and O on the other. For Ba there is only one main peak at 11 meV, whereas for Bi there are two peaks at 12 and 17 meV. Clearly, in the total density-of-states the peak at 11 meV is due to both Ba and Bi and the peak at 15 meV is due to Bi alone. Above 20 meV the entire spectrum arises from oxygen vibrations.

For $Ba_{0.6}K_{0.4}BiO_3$, the partial phonon density-of-states from simulation are shown in Fig. 4. As in the case of $BaBiO_3$, the peaks above 20 meV are due to oxygen vibrations. It is clear from the partial density-of-states for oxygen in $BaBiO_3$ and $Ba_{0.6}K_{0.4}BiO_3$ that there is an overall broadening and softening of the peaks in $Ba_{0.6}K_{0.4}BiO_3$. The main peak in the K partial density-of-states is at 20 meV and it strongly overlaps with the second peak at 16 meV in the Bi partial density-of-states. In the total density-of-states, the contributions from Ba, K, and Bi give rise to peaks around 11 and 15 meV.

Electron tunneling experiments on S-I-S and S-I-N junctions were carried out by Zasadzinski et al..[6] All significant features in the MD density-of-states and neutron generalized density-of-states are seen in the tunneling measurements. MD calculations were also carried out to determine the energies of symmetric breathing modes for oxygen vibrations around Bi (~ 35 meV), and Ba and K (~ 60 meV). Neutron and tunneling experiments together with MD simulations suggest coupling of carriers to oxygen breathing modes around 35 and 60 meV.

ISOTOPE EFFECT DUE TO ^{16}O TO ^{18}O SUBSTITUTION

Reference Value of the Isotope Effect Exponent, α_r

The isotope effect is manifested in the vibrations of a lattice when the mass of one or more of the species is changed by isotopic substitution without changing interatomic forces. Let us define the first frequency moment of the phonon density-of-states as,

$$< \omega > = \int \omega F(\omega) d\omega \Big/ \int F(\omega) d\omega \, . \tag{12}$$

For a monoatomic system in which M is the mass of each atom, $<\omega>$, behaves as, $M^{-1/2}$. For a multicomponent systems such as oxide superconductors, we define the oxygen isotope effect exponent as,

$$< \omega > \approx M_O^{-\alpha_r} \, , \tag{13}$$

where M_O is the oxygen mass and α_r is the reference isotope effect exponent which is generally smaller than 1/2 depending on the masses of other atoms in the unit cell.

When superconductivity is due to electron-phonon coupling and the strong coupling effects are included, the isotope effect of the lattice is reflected through the superconducting transition temperature,

$$T_c \equiv \langle\omega\rangle\, e^{-1/N(0)V} \equiv \langle\omega\rangle\, e^{-1/(\lambda-\mu^*)} \equiv \langle\omega\rangle\, e^{-f(\lambda,\dots,\mu^*)} . \qquad (14)$$

In an oxide superconductor the oxygen isotope effect in T_c can be expressed as,

$$T_c \approx M_O^{-\alpha}, \quad \alpha = \alpha_r - \delta\alpha, \qquad (15)$$

where α_r arises from $\langle\omega\rangle$ and $\delta\alpha$ from f in the exponent. Clearly, for a monoatomic system α_r is 1/2 and $\delta\alpha$ is a measure of strong coupling corrections. For most oxide superconductor α_r is around 0.4.[10] This is due to heavy masses of other atomic species compared to the oxygen mass. For $Ba_{0.6}K_{0.4}BiO_3$ the reference isotope effect exponent was calculated from $^{16}\langle\omega\rangle$ and $^{18}\langle\omega\rangle$. Results of these calculations are summarized in Table 2.

Table 2. MD results for the first moment, in meV, and α_r for $Ba_{0.6}K_{0.4}BiO_3$.

$^{16}\langle\omega\rangle$	$^{18}\langle\omega\rangle$	α_r
33.50	31.91	0.41

Isotope Effect Exponent, α, from Eliashberg Gap Equation.

Implications of the strong coupling effects are studied within the Eliashberg theory[14] with a model of $\alpha^2(\omega)F(\omega)$ for ^{16}O, using λ=1 and $^{16}T_c$=29.5K, which is consistent with the electron tunneling experiments.[6] Using the $^{18}F(\omega)$ and the same model for $\alpha(\omega)$ and the same value of μ^*(=0.12) as for ^{16}O, the gap equations were solved to obtain $^{18}T_c$. Results of these calculations are summarized in Table 3.

Table 3. α from T_c for $Ba_{0.6}K_{0.4}BiO_3$ using Eliashberg equations.

$^{16}T_c$ (K)	$^{18}T_c$ (K)	α
29.5	28.2	0.38

CONCLUSIONS

MD simulations in conjunction with the neutron, tunneling, infrared, and isotope effect measurements reveal that $Ba_{0.6}K_{0.4}BiO_3$ is a weak coupling superconductor in which the carriers have a strong matrix element to high frequency oxygen vibrations.

ACKNOWLEDGEMENTS

This work was supported by the U.S. DOE Basic Energy Sciences-Materials Science, under contract No. W-31-109-ENG-38. MD simulations were done on the Energy Research Cray Supercomputer at the National MFE Computing Center(Livermore). M.H.D would like to thank Fundação de Amparo à Pesquisa do Estado de São Paulo (FAPESP), Brazil for a research fellowship.

REFERENCES

1. D. G. Hinks, D. R. Richards, B. Dabrowski, D. T. Marx, and A. W. Mitchell, Nature 335:419 (1988).

2. B. Batlogg, R. J. Cava, L. W. Rupp, Jr., A. M. Mujsce, J. P. Remeika, W. F. Peck, Jr., A. S. Cooper, and G. P. Espinosa, Phys. Rev. Lett. 61: 1670 (1987).

3. S. Kondoh, M. Sera, Y. Ando and M. Sato, Physica C 157:469 (1989).

4. Z. Schlesinger, R. T. Collins, J. A. Calise, D. G. Hinks, A. W. Mitchell, Y. Zheng, and B. Dabrowski, N. E. Bickers, and D. J. Scalapino, submitted to Phys. Rev. B.

5. C.-K. Loong, P. Vashishta, R. K. Kalia, M. H. Degani, D. L. Price, J. D. Jorgensen, D. G. Hinks, B. Dabrowski, A. W. Mitchell, D. R. Richards, and Y. Zheng, Phys. Rev. Lett. 62:2628 (1989).

6. J. F. Zasadzinski, N. Tralshawala, D. G. Hinks, B. Dabrowski, A. W. Mitchell, D. R. Richards, Physica C 158:519 (1989).

7. J. Bardeen, L. N. Cooper, and J. R. Schrieffer, Phys. Rev. 108:1175 (1957)

8. S. Pei, J. D. Jorgensen, B. Dabrowski, D. G. Hinks, D. R. Richards, A. W. Mitchell, J. M. Newsam, S. K. Sinha, D. Vaknin, and A. J. Jacobson, submitted to Phys. Rev. B.

9. P. Vashishta, and A. Rahman, Phys. Rev. Lett. 40: 1337 (1978), and P. Vashishta, R. K. Kalia, and I. Ebbsjö, Phys. Rev. B 39:6034 (1988).

10. M. H. Degani, R. K. Kalia, and P. Vashishta, to be published.

11. A. Rahman, and P. Vashishta, in The Physics of Superionic Conductors, edited by J. W. Perram (Plenum, New York, 1983), and J. P. Hansen, and I. R. McDonald, Theory of simple liquids (Academic Press, 1976).

12. D. Beeman and R. Alben, Adv. in Phys. 26:339 (1977).

13. R. Fletcher, Practical Methods of Optimization (Wiley, New York, 1980), and F. H. Stillinger, and T. A. Weber, Science 225:983 (1984).

14. G. M. Eliashberg, Zh. Eksp. Teo. Fiz. 38:966 (1960).

Variational Monte-Carlo Study of Superconductivity and Magnetism in the Two-Dimensional Hubbard Model

T. Giamarchi*and C. Lhuillier

Laboratoire de Physique Théorique des Liquides
4, Place Jussieu
75252 PARIS Cedex 05
FRANCE

Abstract

We use a variational Monte-Carlo technique to study the ground-state of the two dimensional Hubbard model on a square lattice. We introduce a trial wave-function which allows a continuous description of the paramagnetic, antiferromagnetic and superconducting phases, as well as the coexistence of these phases, with no a priori constraint on double occupancy. The phase diagram is given for intermediate coupling ($U = 10t$). We find that except at half-filling a pure antiferromagnetic phase does not exist but is always in coexistence with a *d*-wave superconducting phase, for doping less than $\delta \sim 0.2$. A pure *d*-wave superconducting phase extends to dopings as large as $\delta \sim 0.3$. The staggered magnetization and the superconducting gap are measured, and a comparison with other analytical or numerical results is made.

1 Introduction

Soon after the discovery of High-Tc superconductivity [1], it was pointed out by Anderson that the two-dimensional Hubbard model can be of some relevance to these compounds [2]. Another model which was also proposed was the so-called t-J model which was first introduced as the strong coupling limit of the Hubbard model [3] and was latter proved to be an efficient model to describe the two-band structure of the High-Tc compounds [4]. Intensive efforts have thus been made to understand the nature of the ground state of these models. One question of crucial interest is the existence of superconductivity and the competition with the usual antiferromagnetic phase that one can expect for repulsive interactions.

A first approach is by Quantum Monte Carlo simulations (QMC) [5,3,6]. This

*Permanent address: Laboratoire de Physique des Solides, U.P.S. Bât 510, 95400 ORSAY, FRANCE

Condensed Matter Theories, Volume 5
Edited by V.C. Aguilera-Navarro
Plenum Press, New York, 1990

method has the advantage to give direct access to the physical quantities of the model without a-priori knowledge on the system, except the Hamiltonian, and to provide a true numerical "experiment". It, nevertheless, still suffers from some numerical problems like the necessity to extrapolate to zero temperature, and instabilities related to fermion statistics. Some of these problems can be partially cured [7,8]. It is also particularly time-consuming. Another possibility is the Monte-Carlo variational calculations [9,10,11,12]. This numerical method has the drawback to require an a-priori idea of the wave-function but has the advantage to provide exact variational upper-bounds for the ground state energy as well as reasonable computing times. It is usually much more simple to deal with than QMC, and suffers from no intrinsic numerical instability.

By using this method on the t-J model Gross, Yokoyama and Shiba [13,14] were the first to prove that close to half-filling the antiferromagnetic phase was higher in energy than a *d*-wave superconducting phase. These results are consistent with small repulsion renormalization calculation which also find a *d*-type pairing away from half filling [15,16,17].

Some questions still remain open. Although the most popular candidate for the origin of superconductivity in the Hubbard model is the antiferromagnetic fluctuations, the possibility of a coexistence of superconductivity and antiferromagnetism has been little considered. Such a coexistence was pointed out theoretically [18] and was checked in the t-J model by projecting the variational superconducting (RVB) function to simulate the presence of a staggered magnetic field [19]. It was found that the coexistence phase was stable close to half-filling and was followed by a very quick disappearance of the long-range antiferromagnetic order at doping close to 5%.

Moreover nearly all the variational results concerning superconductivity in this field have been performed on the t-J model [14,13,19]. This model has the numerical advantage to give simplified variational wave-functions, and therefore much simpler calculations. Although the t-J model can be considered as an independent model [4], it can be interesting to compare with the results obtained for the Hubbard model. This can provide insights in the validity of the strong coupling limit at intermediate U, and on the efficiency of the variational wavefunctions used when applied to the Hubbard model.

We therefore perform here simulations on the Hubbard model, and look for the existence of a coexistence between superconductivity and antiferromagnetism in this model. As for the Hubbard model the antiferromagnetic wavefunction seems to be particularly suitable [20], instead of taking a wavefunction similar to [19], we treat antiferromagnetism and superconductivity on an equal footing. Some simulation with a pure *d*-wave superconducting wavefunction will also be made in order to compare with the t-J model.

2 Model and trial Wavefunction

We use the two-dimensional Hubbard model on a square lattice, with hopping restricted to nearest neighbors

$$H = -t \sum_{\langle i,j \rangle, \sigma} c^\dagger_{i,\sigma} c_{j,\sigma} + U \sum_i n_{i,\uparrow} n_{i,\downarrow} \ , \tag{2.1}$$

where $\langle\rangle$ stands for nearest neighbors, $c_{i,\sigma}, c^\dagger_{i,\sigma}$ respectively destroy (create) an electron with spin σ at site i and $n_{i,\sigma} = c^\dagger_{i,\sigma} c_{i,\sigma}$. U is the on-site Hubbard repulsion ($U > 0$)

and t the hopping parameter. In the following we will take $t = 1$ which gives for $U = 0$ a bandwidth of 8, and express all energies in units of t.

We compute, by the usual Monte-Carlo integration technique [9,10,11,12], the average value of H with a trial wave-function $|\psi\rangle$

$$E = \langle\psi|H|\psi\rangle/\langle\psi|\psi\rangle \tag{2.2}$$

In order to take into account the fact that the doubly occupied sites are unfavored by the Hubbard repulsion U and to have a tractable function, $|\psi\rangle$ is usually taken of the Jastrow-Gutzwiller type [21,22,23]

$$|\psi\rangle = P|\psi_0\rangle \tag{2.3}$$

where P is a projector which reduces the weight of doubly occupied sites and $|\psi_0\rangle$ is a model wavefunction which insures the fermionic antisymmetry.

We will take here the simplest form for the prefactor $P = g^{N_d}$ [22,23], where g is a variational parameter and N_d the number of doubly occupied sites. More refined prefactors have been studied by Yokoyama [12] but will not be considered here for simplicity. The effects of such prefactors will be discussed latter.

The nature of $|\psi_0\rangle$ depends on the expected long-range behaviour. Two functions were mainly studied: a commensurate antiferromagnetic wave function [20,11] with a wavevector $K = (\pi,\pi)$, and a d-wave superconducting BCS type wavefunction [13,14]. The former describes a state with a staggered magnetization and no superconductivity, whereas the later describes a superconducting state with no long-range magnetic order although it exhibits some short range antiferromagnetic correlations. It is possible to include these two states in a unique variational space if instead of pairing two free electrons of opposite spin, we pair together two quasiparticles describing the excitations of the antiferromagnetic phase. We therefore have

$$|\psi_0\rangle = \prod_k (u_k + v_k d^\dagger_{k,\uparrow} d^\dagger_{-k,\downarrow})|0\rangle \ , \tag{2.4}$$

where the u_k, v_k, are the usual BCS coefficients. Up to a nonimportant normalization factor we get

$$v_k/u_k = \Delta_k/(\epsilon_k - \mu + \sqrt{(\epsilon_k-\mu)^2 + |\Delta_k|^2}) \tag{2.5}$$

Δ_k is the BCS variational parameter and is taken to be $\Delta_k = \Delta(\cos(k_x) - \cos(k_y))$ for d-wave superconductivity. μ is the chemical potential, both Δ and μ are taken as variational parameters in our wave function. Note that such a form for Δ_k is not mandatory, and any function can in principle be taken into account.

The $d^\dagger$ are the operators diagonalizing the antiferromagnetic Hartree-Fock Hamiltonian

$$d^\dagger_{k,\sigma} = \alpha_k c^\dagger_{k,\sigma} + \sigma\beta_k c^\dagger_{k+K,\sigma} \qquad d^\dagger_{k+K,\sigma} = -\sigma\beta_k c^\dagger_{k,\sigma} + \alpha c^\dagger_{k+K,\sigma} \ , \tag{2.6}$$

where $K = (\pi,\pi)$ is the commensurate perfect nesting vector, and k is limited to half of the Brillouin zone by $\epsilon_k < 0$. α_k and β_k are the usual Hartree-Fock antiferromagnetic coefficients

$$\begin{aligned} \alpha_k &= [(1 - \epsilon_k/\sqrt{\epsilon_k^2 + D^2})/2]^{1/2} \ , \\ \beta_k &= [(1 + \epsilon_k/\sqrt{\epsilon_k^2 + D^2})/2]^{1/2} \ . \end{aligned} \tag{2.7}$$

The gap D is related to the antiferromagnetic staggered magnetization, whereas the gap Δ is related to the superconducting order parameter. If $D = 0$ our function gives the d-wave type wavefunction, whereas if $\Delta \to 0$ the usual antiferromagnetic phase is recovered.

To be treated numerically, the function $|\psi\rangle$ is projected on a subspace with a fixed number of particles [24,13], and Δ, D, μ, g, are taken as variational parameters and fixed by minimizing the energy E.

3 Simulation and Results

3.1 Simulation

All the calculations were made on a square lattice with periodic-antiperiodic conditions in order to avoid degeneracy of the Fermi surface at half-filling [12,25]. We have mainly studied 8×8 lattices for $U = 10.0$. We have taken an intermediate U because the intermediate coupling region is the region of physical interest in application to High Tc superconductivity [2,26]. This also allows us to qualitatively compare our results with those obtained for the t-J model. In order to check the size dependency we have also performed simulations on 6×6 and 10×10 systems for selected values of the band filling, but we will not perform any systematic extrapolation to infinite size.

The variational parameters were determined by using a method proposed by Umrigar et al. for the study of atomic systems [27]. A set of configurations is generated and then used to minimize the energy. This offers both the advantage of good computing time performances and of using **correlated** measurements, which allows us to compare energies that differ by much less than the statistical error bars on uncorrelated samples. We have used at least 5 independent simulations each of $4\,10^5$ Monte-Carlo steps (Mcs) to determine the minimum energy and parameters and the error bars. Other quantities such as the superconducting order parameter, the staggered magnetization, the kinetic and potential energy, were measured over independent samples of $8\,10^5$ Mcs. The order of magnitude of the time needed to get one minimum is between 1 and 2 hours on a Cray-2. The different physical quantities are much easier to compute and a set of measures takes around 1 hour of Cray-2.

3.2 Results

The optimal parameters for an 8×8 system at $U = 10$ as a function of the band filling are indicated in table 1. By looking at table 1 one can see that, close to half filling, the superconducting parameter **and** the antiferromagnetic parameter are both nonzero indicating a coexistence of superconductivity and antiferromagnetism. For doping larger than $\delta_c = 0.2$ the commensurate antiferromagnet is no more stable but the phase seems to exhibit d-wave superconductivity up to doping of about 0.3. Such a phase diagram is in good qualitative agreement with the mean-field analytical result [18] but in contradiction with the two-hole result [28] which predicted no attraction for two holes in an antiferromagnetic background. We have therefore checked that our result was not due to an artificial finite-size effect by looking at the size dependence of the superconducting variational parameter at half filling, and around $\delta \sim 0.1$ where Δ seems to be maximum. The results are reported in Fig. 1. One clearly sees on this figure the strong decrease of the Δ parameter at half-filling and its probable extrapolation towards a zero value. We can therefore expect, in the thermodynamical limit, a pure antiferromagnetic phase at half filling, which is in agreement with the

Table 1. Optimal parameters for $L = 8 \times 8$ system at $U = 10$. N is the number of particles, δ the deviation from half-filling $(L - N)/L$, D the antiferromagnetic parameter, Δ the superconducting one, μ the chemical potential, g the Gutzwiller prefactor and N_d is the number of doubly occupied sites.

N	δ	D	Δ	μ	g	N_d
64	0.0	1.55 ± 0.04	0.04 ± 0.02	-0.00 ± 0.02	0.45 ± 0.01	2.33 ± 0.02
60	0.06	0.80 ± 0.05	0.11 ± 0.02	-0.19 ± 0.03	0.34 ± 0.01	2.39 ± 0.04
52	0.19	0.05 ± 0.03	0.05 ± 0.02	-0.55 ± 0.06	0.25 ± 0.03	1.93 ± 0.02
44	0.31	0.00 ± 0.03	0.02 ± 0.01	-0.69 ± 0.05	0.29 ± 0.02	1.30 ± 0.05
36	0.44	0.00 ± 0.03	0.00 ± 0.03	-0.87 ± 0.10	0.34 ± 0.02	1.00 ± 0.05

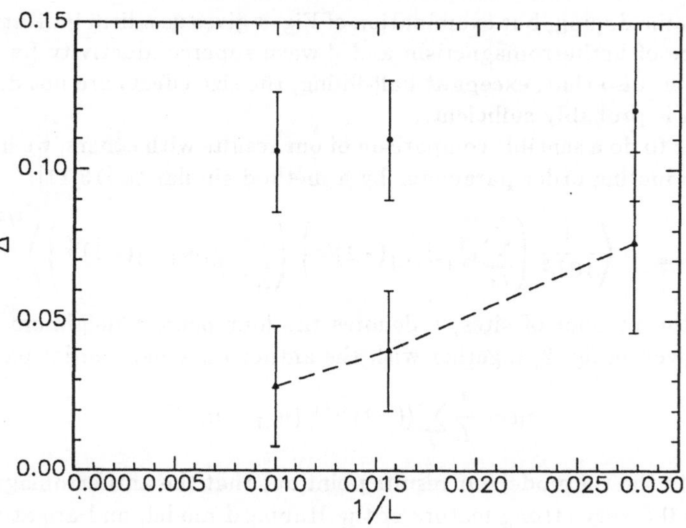

Figure 1: Size dependency of the variational parameter Δ. Triangles correspond to half-filling, whereas bullets stand for systems with 32 (6×6), 60 (8×8), 44 (10×10) particles. These three points correspond to slightly different δ ($\delta = 0.11$, $\delta = 0.0625$, $\delta = 0.12$ respectively) and therefore no extrapolation to infinite size can directly be made.

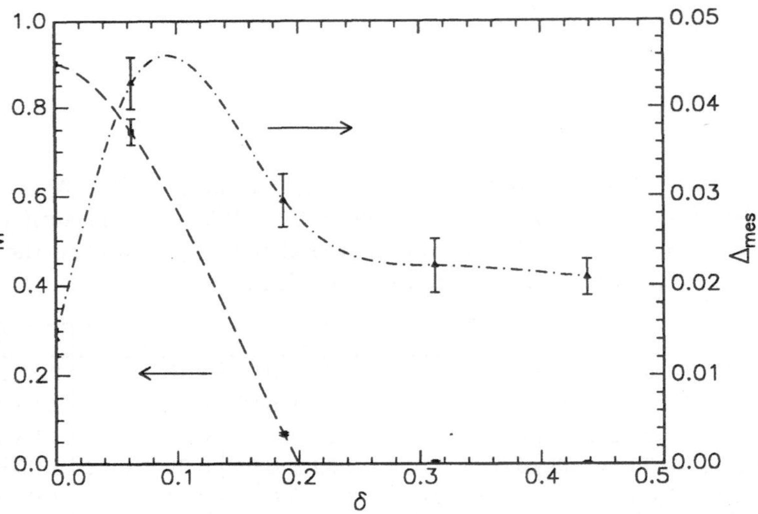

Figure 2. Measured superconducting order parameter (triangles) and staggered magnetization (squares) as function of the band filling for a 8×8 system at $U = 10$. The lines are simply a guide to the eyes.

previously known results [20,19,11,5]. The results for a doped sample do not exhibit the same size effect†. True extrapolation is difficult as the different samples have not exactly the same doping, but examination of Fig. 1 gives us a firm indication, at $T = 0$, of coexistence of antiferromagnetism and d-wave superconductivity for $\delta < \delta_c = 0.2$. Fig. 1 indicates also that, except at half-filling, the size effects are not drastic and our 8×8 system is probably sufficient.

In order to do a sensible comparison of our results with others, we have measured the superconducting order parameter by a method similar to [13,11]

$$\Delta_{\text{mes}} = \left\langle \frac{1}{16L^2} \left(\sum_{i,\tau} c^\dagger_{i,\uparrow} c^\dagger_{i+\tau,\downarrow} (-1)^{\tau_y} \right) \left(\sum_{j,\tau'} c_{j,\downarrow} c_{j+\tau',\uparrow} (-1)^{\tau'_y} \right) \right\rangle^{1/2} , \quad (3.1)$$

where L is the number of sites, τ denotes the four nearest neighbors vectors. The results are given in fig. 2, together with the antiferromagnetic order parameter

$$m = \frac{1}{L} \sum_i \langle (-1)^{x_i + y_i} (n_{i,\uparrow} - n_{i,\downarrow}) \rangle \ , \quad (3.2)$$

In contrast to the t-J model our results point out that the antiferromagnetic instability is at $T = 0$ a very strong feature of the Hubbard model, and are at variance with the QMC conclusions [5,29]. An explanation of the discrepancy could be that VMC mainly gives information on the fundamental state of the system, whereas QMC is intrinsically a $T \neq 0$ method. The critical doping for the disparition of antiferromagnetism ($\delta_c = 0.2$) is in good quantitative agreement with the mean-field results

†As already pointed out, at half filling, if all doubly occupied sites are projected out ($g = 0$) a wave function with a nonzero variational parameter does not exhibit superconductivity [13,14]. In our case since $g \neq 0$, a nonzero Δ implies superconductivity. Since the number of doubly occupied sites is quite small on the smallest systems, ($N_d = 2.3$ for an 8×8 system) both solutions ($\Delta \neq 0$ and $\Delta = 0$) are certainly very close in energy which explains the important size effect at half-filling.

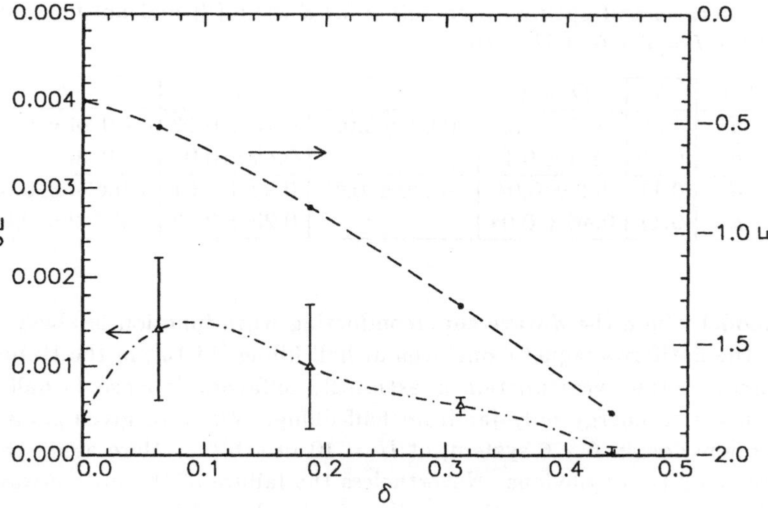

Figure 3. Energies for the optimal parameters given in table 1, and the difference δE with the same phase but with no superconducting gap.

[18] and is nearly identical to the values found for the disparition of a pure antiferromagnetic phase [30,31]. This indicates that, at least for this kind of wavefunction, antiferromagnetism is not drastically affected by the possibility of superconducting pairing. The critical δ found here is much greater than the value of Lee and Feng [19]. This discrepancy comes probably from the fact that their wavefunction is mainly a superconducting wave function in which antiferromagnetism is put only in the projector, whereas in our wavefunction antiferromagnetism and superconductivity are treated on an equal footing, thus naturally giving a wider range of stability for antiferromagnetism. Of course more detailed studies are needed to determine which solution is optimal [32].

We have indicated in fig. 3 the difference in energy between the phase of coexistence and the same phase but with no superconducting gap ($\Delta = 0$). One can see that even in the most favorable cases the difference is extremely small. Such a small value of the pairing energy could explain the absence of visible superconductivity in Quantum Monte-Carlo results [6].

Between $\delta \sim 0.2$ and $\delta \sim 0.3$ a pure *d*-wave superconducting phase is found, as clearly seen on table 1 and Fig. 3[‡]. The pure superconducting phase extends much further in our diagram than in [18], or for the existence of a pure superconducting phase in [33]. This is surprising since close to half-filling the *d*-wave superconducting phase is not specially favorable as will be seen below. So far the interpretation of this discrepancy is not clear.

Let us come back to the comparison with the t-J model. The first approximation for the t-J model, a paramagnetic *d*-wave superconducting wavefunction [13,14] is a poor approximation for the Hubbard model. At the opposite to what happens in

---

[‡]Note that on fig. 2 the measured order parameter Δ_{mes} always keeps a finite value even in the paramagnetic phase. This is an artefact of the formula 3.1 which gives only the off-diagonal long range order parameter in the limit $L \to \infty$.

Table 2. Energies and optimal parameters for d-wave (d) and antiferromagnetic (a) functions for $L = 6 \times 6$ at $U = 10$.

W.F.	δ	D or Δ	μ	g	E
d	0.00	0.5 ± 0.2	0.00 ± 0.02	0.16 ± 0.03	-0.24 ± 0.03
a	0.00	1.5 ± 0.1		0.45 ± 0.02	-0.40 ± 0.02
d	0.11	0.2 ± 0.04	-0.39 ± 0.05	0.22 ± 0.01	-0.634 ± 0.002
a	0.11	0.46 ± 0.04		0.28 ± 0.02	-0.642 ± 0.002

the t-J model where the d-wave superconducting wave function is always extremely close the the antiferromagnetic one even at half filling [14,13], in the Hubbard model the superconducting wavefunction is extremely unfavorable close to half filling and becomes lower in energy only far from half-filling. We have given some values for both functions for an 6×6 system, at $U = 10$ in table 2. Here again the origin of the discrepancy is not obvious. Nevertheless the failure of the pure d-wave function to describe a good approximation to the Hubbard model is manifestly due to an insufficient description of the spin-spin antiferromagnetic correlations in this wave function. Whether this unpleasant feature can be corrected by the use of more refined Gutzwiller-like prefactors like in [12], and a good approximation of the fundamental of the Hubbard model could be obtained for a pure pairing of plane waves is an open question but seems doubtful. The results [19] have shown that the paramagnetic d-wave function was not the best solution of the t-J model and became unstable with the inclusion of a staggered magnetization. Therefore even if the first approximations for the trial wavefunctions considerably differ in the two models (an antiferromagnetic wave function for the Hubbard model and a d-wave superconducting wave function for the t-J model), the more refined ones seem to converge towards the same phase diagram: a coexistence phase followed by a pure superconducting phase.

The order of magnitude of the superconducting order parameter seems to be in good agreement with the values obtained formerly for the t-J model [14]. The staggered magnetization is larger in our wavefunction than it is in Lee and Feng's, but it is well known that a Hartree-Fock wave-function overestimates the magnetization for the Hubbard model [25,5]. On the opposite, according to QMC calculations [5,29], the Lee and Feng function underestimates the magnitude of the staggered magnetization.

The present quality of the wave-function does not seem sufficient to investigate the effects of the neglected terms in the strong coupling approximation which leads to the t-J model. These terms can be still important for values of U comparable to the bandwidth. Note that the number of doubly occupied sites although small is not negligible (see Table 1).

4 Conclusion

In this paper we have presented a variational Monte-Carlo study of the coexistence of superconductivity and antiferromagnetism in the Hubbard model on a square lattice. We have introduced a variational wave-function which allows a continuous description of the paramagnetic, antiferromagnetic, superconducting phases, as well as a coexistence between antiferromagnetism and superconductivity.

We have found that at intermediate coupling a pure antiferromagnetic phase seems to exist only at half filling. Below a certain doping $\delta \sim 0.2$ the most stable phase contains antiferromagnetism and superconductivity. The energy gained in the pairing compared to those of the pure antiferromagnetic phase is very weak. Above this threshold the antiferromagnetism disappears and the system becomes purely superconducting. Finally above $\delta \sim 0.3$ the most stable phase is the paramagnetic one.

We have also shown that contrary to what happens in the t-J model a pure *d*-wave superconducting phase is not energetically favorable close to half filling, compared to a pure commensurate antiferromagnetic phase. This difference can perhaps be cured by more refined projectors, but some quantitative differences are always to be expected. As soon as antiferromagnetism is included the qualitative agreement between the two models becomes rather good.

Clearly the theory is much too crude to be related to High Tc experiments. To the best of our knowledge no coexistence of long-range order has been observed so far. Nevertheless if, as in our wavefunction, the system possesses an antiferromagnetic gap at points where the superconducting gap vanishes (i.e. $k_x = k_y$), all the unusual aspects of *d*-wave superconductivity linked to the presence of zeroes of the gap on the Fermi surface will not be observable and the experimental distinction between *s* and *d*-wave superconductivity will become more difficult [34].

Some extension of the study seems suitable. First the same method can be applied to the t-J model and a comparison with [19] has to be done [32]. Here we also have only considered the case of a commensurate antiferromagnetic phase. The extension to the coexistence of superconductivity and incommensurate antiferromagnetic phases, although difficult, is necessary due to the strong stabilisation of the antiferromagentic phase by incommensurate effects [35,36,37,38,39,40] It is very unlikely that a pure superconducting phase will remain energetically favorable. The possibility of pairing of such states is still open.

It could be added that, within the present study, the optimized superconducting state appears to be a weak superconductor, which only involves the pairing of a small fraction of the electrons around the Fermi surface. Whether a true R.V.B. state [41,42] describing a disordered spin liquid could be a better approximation for the fundamental of the Hubbard model remains to be checked.

Acknowledgments

It is a pleasure to thank H. J. Schulz for his constant interest in this work, many interesting discussions and useful remarks, and M. Gabay for interesting discussions. We would like to thank the Ecole Normale Supérieure, the Centre de Calcul Vectoriel pour la Recherche for grants of computer time. This work has been supported by the D.R.E.T. under grant n° 881342.

References

[1] J. G. Bednorz and K. A. Müller, Z Phys. B **64**, 189 (1986).
[2] P. W. Anderson, Science **235**, 1196 (1987).
[3] J. E. Hirsch, Phys. Rev. Lett. **54**, 1317 (1985).
[4] F. C. Zhang and T. M. Rice, Phys. Rev. B **37**, 3759 (1988).
[5] J. E. Hirsch, Phys. Rev. B **31**, 4403 (1985).

[6] J. E. Hirsch and H. Q. Lin, Phys. Rev. B **37**, 5070 (1988).
[7] J. E. Hirsch, To be published in Phys. Rev. B (1989).
[8] S. Sorella, S. Baroni, R. Car, and M. Parinello, Europhys. Lett. **8**, 663 (1989). And references therein.
[9] W. L. Mac Millan, Phys. Rev. **138**, A442 (1965).
[10] D. Ceperley, G. V. Chester, and K. H. Kalos, Phys. Rev. B **16**, 3081 (1977).
[11] C. Gros, Ann. Phys. **189**, 53 (1989).
[12] H. Yokoyama. *Variational Monte-Carlo Studies of Hubbard Model.* Thesis, Institute for Solid State Physics, University of Tokyo, 1988.
[13] C. Gros, Phys. Rev. B **38**, 931 (1989).
[14] H. Yokoyama and H. Shiba, J. Phys. Soc. Jpn. **57**, 2482 (1988).
[15] I. Dzyaloshinskii, Sov. Phys. JETP **66**, 848 (1987).
[16] I. Dzyaloshinskii and V. M. Yakovenko, Sov. Phys. JETP **67**, 844 (1988).
[17] H. J. Schulz, Europhys. Lett. **4**, 609 (1987).
[18] M. Inui, S. Doniach, P. J. Hirschfeld, and A. E. Ruckenstein, Phys. Rev. B **37**, 2370 (1988). And the erratum Phys. Rev. B **39** 7300 (1989).
[19] T.K. Lee and S. Feng, Phys. Rev. B **38**, 11809 (1988).
[20] H. Yokoyama and H. Shiba, J. Phys. Soc. Jpn. **56**, 3582 (1987).
[21] R. Jastrow, Phys. Rev. **98**, 1479 (1955).
[22] M. C. Gutzwiller, Phys. Rev. Lett. **10**, 159 (1963).
[23] M. C. Gutzwiller, Phys. Rev. **137**, A1726 (1965).
[24] J. P. Bouchaud, A. Georges, and C. Lhuillier, J. Phys. (Paris) **49**, 553 (1988).
[25] H. Yokoyama and H. Shiba, J. Phys. Soc. Jpn. **56**, 1490 (1987).
[26] P. W. Anderson. *50 Years of the Mott phenomenon: insulators, magnets, solids, and superconductors as aspects of the strong repulsion theory.* Lecture series at Varenne Summer School. "Frontiers and Borderlines in Many Particle physics", 1987.
[27] C. J. Umrigar, K. G. Wilson, and J. W. Wilkins, Phys. Rev. Lett. **60**, 1719 (1988).
[28] C. Gros, R. Joynt, and T. M. Rice, Z Phys. B **68**, 425 (1987).
[29] J. E. Hirsch and S. Tang, Phys. Rev. Lett. **62**, 591 (1989).
[30] H. Yokoyama and H. Shiba, J. Phys. Soc. Jpn. **56**, 3570 (1987).
[31] E. Kaxiras and E. Manousakis, Phys. Rev. B **37**, 656 (1988).
[32] T. Giamarchi and C. Lhuillier, work in progress.
[33] H. Shiba. Some aspects of strongly correlated electronic systems - variational monte-carlo studies. In *Two dimensional Strongly Correlated Electronic systems*, Gordon and Breach, 1988.
[34] We are grateful to H. J. Schulz to have pointed out this point.
[35] H. J. Schulz, Orsay, preprint.
[36] H. J. Schulz, Orsay, preprint.
[37] D. Poilblanc and T. M. Rice, Phys. Rev. B **39**, 9749 (1989).
[38] B. I. Shraiman and E. D. Siggia, Phys. Rev. Lett. **61**, 467 (1988).
[39] B. I. Shraiman and E. D. Siggia, Phys. Rev. Lett. **62**, 1564 (1989).
[40] T. Giamarchi and C. Lhuillier, communication at Stat. Phys. 17 (Rio de Janeiro 1989).
[41] J. P. Bouchaud and C. Lhuillier, Europhys. Lett. **3**, 1273 (1987).
[42] S. Liang, B. Douçot, and P. W. Anderson, Phys. Rev. Lett. **61**, 365 (1988).

FINITE-TEMPERATURE MANY-BODY PERTURBATION THEORY FOR SUPERCONDUCTING FERMION SYSTEMS

Donald H. Kobe*

Instituto de Física Teórica
Rua Pamplona, 145
01405-São Paulo-SP-Brazil

ABSTRACT

For a many-fermion system, Bogoliubov's Principle of Compensation of Dangerous Diagrams (PCDD) to determine the coefficients in a canonical transformation to quasiparticles is derived from the variational principle that the number of quasiparticles in the system is a minimum. The PCDD states that the sum of the diagrams going from a two-quasiparticle state to the vacuum is zero. When the PCDD is used with the quasiparticle self energy, both in first order of finite-temperature perturbation theory, the finite-temperature Hartree-Fock-Bogoliubov theory is obtained. Corrections to the quasiparticle energy can be calculated systematically by going to second or higher orders in both the PCDD and self energy, and solving the equations self consistently.

I-INTRODUCTION

The theory of superconductivity developed by Bardeen, Cooper, and Schrieffer (BCS)[1] has been very successful in explaining the properties of superconductors. However, the recent discovery of high transition temperature superconductors indicates that an attractive interaction due to the exchange of phonons may not be adequate, and that a new mechanism for the attractive interaction may be required[2]. Whether the structure of BCS theory with pairing between electrons of equal but opposite momentum and spin will be adequate remains to be seen.

The BCS theory was given an elegant mathematical formulation by Bogoliubov[3], who made a canonical transformation to quasiparticles for which the BCS ground state was the vacuum state. The coefficients in the canonical transformation were determined by minimizing the energy in the quasiparticle vacuum state. When this was done the terms in the Hamiltonian which create or annihilate a pair of quasiparticles were zero. Bogoliubov generalized this condition by stating his Principle of Compensation of Dangerous Diagrams (PCDD)[4], which says that the sum of all diagrams which go from the vacuum to a two-quasiparticle state should be set equal to zero. His reason for stating the PCDD was to eliminate divergences in the perturbation expansion of the ground-state energy. This principle has been

*Permanent address: Department of Physics, University of North Texas, Denton, Texas 76203, U.S.A.

derived from variational principles for both fermions[4] and bosons[5,6].

In this paper we derive a finite temperature version of the PCDD for fermion systems from a variation principle which minimizes the number of Bogoliubov quasiparticles in the system. It is formulated in terms of finite-temperature, many-time causal Green's functions[7], for which an expansion in terms of finite-temperature Feynman diagrams can be made. The self-energy of the Bogoliubov quasiparticle (QP) can also be calculated from the single QP propagator[8,9]. When the first-order dangerous diagrams are used along with the first-order self-energy diagrams, the finite-temperature Hartree-Fock-Bogoliubov (HFB) theory is obtained for the QP energy[6]. However, it is not necessary to stop at the HFB theory, but second-order diagrams can be considered both in the PCDD and in the self energy. In principle these equations can be solved self-consistently to give the effect of dressing in second order on the quasiparticle energy and the energy gap. Since the calculation is difficult, it will be done in another paper.

In Sec. II the Hamiltonian is given. The PCDD is derived in Sec. III from a variational principle. Finite temperature, many-time, causal Green's functions are given in Sec. IV. The self-consistent perturbation expansion in first order is given in Sec. V. Finally, the conclusion is given in Sec. VI.

II- HAMILTONIAN

The Hamiltonian for a system of spin one-half fermions interacting with a two-body potential V is

$$H = \sum_{1} e_1 a_1^{+} a_1 + \frac{1}{2}\sum_{1234} \langle 1,2|V|3,4\rangle a_1^{+} a_2^{+} a_3 a_4 , \tag{2.1}$$

where the sum is over $i = (\vec{k}_i, \sigma_i)$, $\vec{k}_i$ is the wave vector, and σ_i is the z-component of the spin (i= 1,2,3,4). The single particle energy e_k is

$$e_k = (k^2/2m) - \mu, \tag{2.2}$$

where natural units are used such that $\hbar = 1$. The chemical potential μ is determined by setting the average number of particles equal to N, the number of particles in the system. The matrix element in Eq. (2.1) is antisymmetrized, so

$$\langle 12|V|34\rangle = -\langle 21|V|34\rangle = -\langle 12|V|43\rangle = \langle 21|V|43\rangle . \tag{2.3}$$

The creation operator a_1^{+} and annihilation operator a_1 satisfy the usual fermion anticommutation relations

$$\{a_1, a_2^{+}\} = \delta_{12}, \quad \{a_1, a_2\} = 0, \quad \{a_1^{+}, a_2^{+}\} = 0. \tag{2.4}$$

A canonical transformation can be made to Bogoliubov quasiparticles (QP)

$$a_1 = u_1 \gamma_1 + v_{-1} \gamma_{-1}^{+} , \tag{2.5}$$

where the quasiparticle creation and annihilation operators, γ_1^{+} and γ_1 , respectively, satisfy the same fermion anticommutation relations as in Eq. (2.4). The coefficients u_1 and v_1 therefore must satisfy

$$u_1^2 + v_1^2 = 1, \quad u_{-1} = u_1, \quad v_{-1} = -v_1, \tag{2.6}$$

and can be taken to be real without loss of generality. The vacuum state $|\Phi_0\rangle$ for the quasiparticles is defined such that

$$\gamma_1 |\Phi_0\rangle = 0. \tag{2.7}$$

When Eq. (2.5) is substituted into Eq. (2.1), and the operators are put in normal order, with the creation operators on the left and the annihilation operators on the right, the Hamiltonian becomes[8]

$$H = H_{00} + H_{02} + H_{11} + H_{20} + H_{04} + H_{13} + H_{22} + H_{31} + H_{40}. \tag{2.8}$$

The term H_{jk} in the Hamiltonian has j quasiparticle creation operators and k quasiparticle annihilation operators

$$H_{jk} = \sum_{1,2,\cdots,j+k} h_{jk}(1,2,\cdots,j,j+1,\cdots,j+k)\,\gamma_1^{\dagger}\cdots\gamma_j^{\dagger}\gamma_{j+1}\cdots\gamma_{j+k}, \tag{2.9}$$

where the coefficient h_{jk} depends on the u_k, v_k, and the matrix elements of the potential. Since the Hamiltonian is Hermitian, we have

$$H_{jk}^{\dagger} = H_{kj}. \tag{2.10}$$

The term H_{00} in Eq. (2.8) is the ground-state energy of the system in the quasiparticle vacuum state

$$H_{00} = \langle\Phi_0|H|\Phi_0\rangle = \sum_1 \left\{ e_1 + \sum_2 \langle 12|V|21\rangle v_2^2 \right\} v_1^2 + \frac{1}{2}\sum_{1,2} \langle 1,-1|V|2,-2\rangle u_1 v_1 u_2 v_2. \tag{2.11}$$

The coefficient h_{11} in the term H_{11} of Eq. (2.8) is the "kinetic energy" of the Bogoliubov quasiparticle, which is

$$h_{11}(1,1) = U_1^0 (u_1^2 - v_1^2) + \Delta_1^0\, 2u_1 v_1, \tag{2.12}$$

where the zero-temperature Hartree-Fock single-particle energy is

$$U_1^0 = e_1 + 2\sum_2 \langle 1,2|V|2,1\rangle v_2^2, \tag{2.13}$$

and the zero-temperature energy gap is

$$\Delta_1^0 = \sum_2 \langle 1,-1|V|2,-2\rangle u_2 V_2 .$$

(2.14)

The term H_{20} in Eq. (2.8), which describes the creation of two quasiparticles from the vacuum, has a coefficient

$$h_{20}(1,-1) = \frac{1}{2}\{ U_1^0 2u_1 v_1 - \Delta_1^0 (u_1^2 - v_1^2)\} .$$

(2.15)

The other terms in Eq. (2.8), in which the sum of the creation and annihilation operators is four, describe quasiparticle interactions. The coefficients in these terms have previously been given[8,9].

When the ground-state energy in Eq. (2.11) is minimized with respect to u_k and v_k, subject to the constraint in Eq. (2.6), the result is that Eq. (2.15) is equal to zero, so $H_{20} = 0$, $H_{02} = 0$. When this condition is used along with Eq. (2.6), the coefficients in the canonical transformation are determined. The QP energy in Eq. (2.12) then becomes

$$h_{11}(1,1) = E_1^0 = [(U_1^0)^2 + (\Delta_1^0)^2]^{1/2},$$

(2.16)

which is the zero temperature result.

III-PRINCIPLE OF COMPENSATION OF DANGEROUS DIAGRAMS

Bogoliubov[3] formulated the Principle of Compensation of Dangerous Diagrams (PCDD) to obtain the coefficients in the canonical transformation for fermion systems. He said that to eliminate divergences in the perturbation expansion of the ground-state energy the sum of all the diagrams that lead from the vacuum to the two quasiparticle state should be set equal to zero. Here we derive the PCDD at finite temperature from a variational principle that has been used for boson systems[6].

The average number of quasiparticles in the system is

$$N_{QP} = \sum_1 \langle \gamma_1^+ \gamma_1 \rangle ,$$

(3.1)

where the average is calculated using the density matrix ρ,

$$\langle \cdots \rangle = \mathcal{T}r\{\rho \cdots\} .$$

(3.2)

The density matrix in the grand canonical ensemble is

$$\rho = Z^{-1} \exp(-\beta H) ,$$

(3.3)

where the grand partition function is

$$Z = \mathcal{T}r\{\exp(-\beta H)\} .$$

(3.4)

The parameter $\beta = (k_B T)^{-1}$, where T is the absolute temperature and k_B is Boltzmann's constant.

The average number of quasiparticles in the system can be minimized with respect to the parameters u_k and v_k, with the constraint in Eq. (2.6). When Eq. (2.5) and its Hermitian conjugate are used, we can solve for the quasiparticle annihilation operator

$$\gamma_1 = u_1 a_1 + v_1 a_{-1}^{\dagger} . \tag{3.5}$$

The particle creation and annihilation operators, $a_1^{\dagger}$ and a_1, respectively, do not depend on the parameters u_1 and v_1. When Eq. (3.5) and its Hermitian conjugate are substituted into Eq. (3.1), and the expression minimized with respect to u_1 and v_1, using the constraint in Eq. (2.6), we obtain

$$\mathrm{Re} \langle \gamma_k^{\dagger} \gamma_{-k}^{\dagger} \rangle = 0. \tag{3.6}$$

When this amplitude is expressed in terms of Feynman diagrams, it states that the sum of the diagrams going from the vacuum to the two quasiparticle state is zero.

When the average number of quasiparticles in the system in Eq. (3.1) is minimized, the quasiparticle density is decreased and they behave more like an ideal gas of quasiparticles. The interactions between the quasiparticles should be less important, and perturbation theory should converge more rapidly.

IV-FINITE-TEMPERATURE, MANY-TIME, CAUSAL GREEN'S FUNCTIONS

In order to use Feynman diagrams, it is necessary to introduce the Green's functions of quantum field theory. The general finite-temperature, many-time, causal Green's function[4] for the quasiparticles[8] is defined as

$$\mathcal{G}_{nm}(1,2,\cdots,n+m) = -i \langle T\{\gamma_1 \cdots \gamma_n \gamma_{n+1}^{\dagger} \cdots \gamma_{n+m}^{\dagger}\} \rangle , \tag{4.1}$$

where the variables in $\mathcal{G}_{nm}$ are in terms of the states and the times. The time-ordering operator T puts the later times on the left and the earlier times on the right with a plus (minus) sign for an even (odd) number of interchanges of the original order. The creation operators $\gamma_{n+1}^{\dagger}$ or annihilation operators γ_1 are in the Heisenberg picture, as for example

$$\gamma_1 = \gamma_1(t_1) = \exp(iHt_1)\, \gamma_1(0) \exp(-iHt_1). \tag{4.2}$$

The average in Eq. (4.1) is taken with respect to the grand canonical ensemble in Eqs. (3.2)-(3.4). The quasiparticles are not conserved, so the number of creation operators m in Eq. (4.1) is not necessarily equal to the number of annihilation operators n. The Green's function in Eq. (4.1) describes n quasiparticles going into a box and m quasiparticles coming out of the box[8].

The equation of motion for the Green's function in Eq. (4.1) can be obtained by differentiating with respect to one of the time variables[7,8]. To obtain Feynman diagrams in terms of frequency, the Fourier transform of

g_{nm} is taken to give G_{nm} .

In terms of the Green's function G_{02} the PCDD in Eq. (3.6) can be written as

$$\mathrm{Im} \iint_{-\infty}^{\infty} d\omega_1 \, d\omega_2 \; G_{02}(1,2) = 0, \tag{4.3}$$

where the variables in G_{02} are in terms of the state and frequency. Equation (4.3) is shown in Fig. 1, in which the sum of all the diagrams which go from the vacuum to a two quasiparticle state are equal to zero. It is understood that the Green's function in Fig. 1 is integrated over frequency and the imaginary part is taken, as in Eq. (4.3).

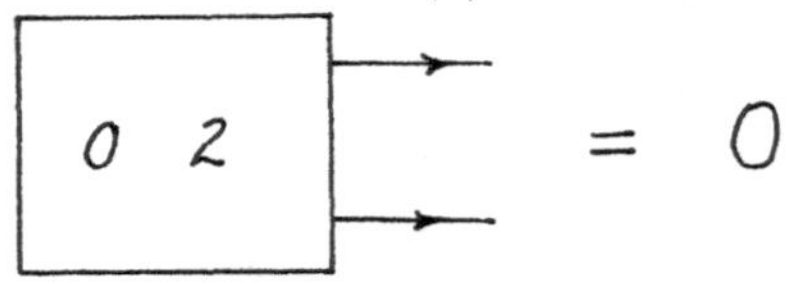

Fig. 1. The PCDD in Eq. (4.3) in terms of Feynman diagrams. All possible processes can take in the box.

V-SELF-CONSISTENT FINITE-TEMPERATURE PERTURBATION THEORY OF THE QUASIPARTICLE ENERGY

In this section we use a perturbation expansion of the PCDD along with the self-energy of the quasiparticle (QP) to obtain its self-consistent, finite temperature Hartree-Fock-Bogoliubov (HFB) energy.

A-PERTURBATION EXPANSION OF THE PCDD

The equation of motion for G_{02} can be obtained and decoupled from the higher Green's functions[9]. When this is done and only the irreducible dangerous diagrams are considered, we get the perturbation expansion shown in Fig. 2. In Fig. 2(a) the vertex h_{20} contributes. In Fig. 2(b) the vertex h_{31} is involved, since one quasiparticle goes into the vertex and three go out. In the second-order diagram in Fig. 2(c) the vertices h_{40} and h_{13} are involved. The internal lines are the bare finite-temperature QP propagators[7],

$$G^{0}_{11}(1,2) = 2\pi\,\delta(1,2)\left\{\frac{1-n_1}{\omega_1 - E_1 + i0} + \frac{n_1}{\omega_1 - E_1 - i0}\right\}, \tag{5.1}$$

where $\delta(1,2) = \delta_{12}\,\delta(\omega_1+\omega_2)$ and E_1 is the QP energy, which is yet to be determined. The average fermion QP occupation number n_1 is

$$n_1 = [\, \exp(\beta E_1) + 1 \,]^{-1}, \tag{5.2}$$

where $\beta^{-1} = k_B T$. At zero temperature $n_1 = 0$ and the diagram in Fig. 2(b) does not contribute.

The diagrams in Fig. 2(a) and (b) give

$$h_{20}(1,2) + 3\sum_{3} h_{31}(1,2,3,3)\, n_3 = 0 . \tag{5.3}$$

When the functions h_{20} in Eq. (2.15) and h_{31} are used in Eq. (5.3), we obtain

$$U_1\, 2u_1 v_1 - \Delta_1 (u_1^2 - v_1^2) = 0 . \tag{5.4}$$

The finite-temperature Hartree-Fock energy is

$$U_1 = e_1 + 2\sum_{2} \langle 12|V|21\rangle [v_2^2 + (u_2^2 - v_2^2) n_2] , \tag{5.5}$$

and the finite-temperature energy gap Δ_1 is

$$\Delta_1 = \sum_{2} \langle 1,-1|V|2,-2\rangle u_2 v_2 \tanh(\beta E_2/2) . \tag{5.6}$$

Equations (5.5) and (5.6) generalize their zero temperature counterparts in (2.13) and (2.14), respectively.

When Eq. (5.4) and the constraint in Eq. (2.6) are solved for the coefficients, we obtain

$$u_1^2 - v_1^2 = U_1 (U_1^2 + \Delta_1^2)^{-1/2} \tag{5.7}$$

and

$$2u_1 v_1 = \Delta_1 (U_1^2 + \Delta_1^2)^{-1/2} . \tag{5.8}$$

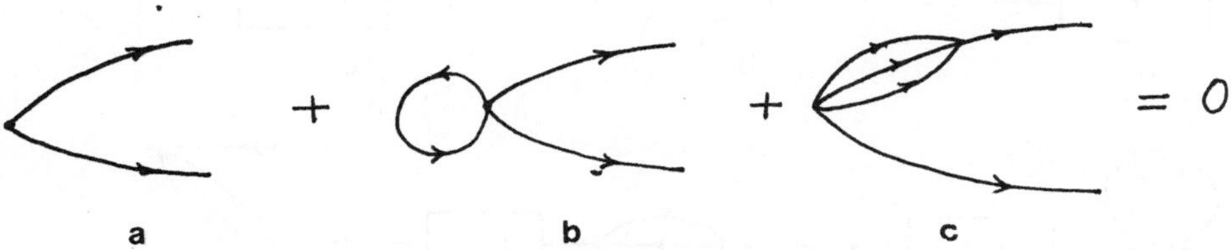

Fig. 2. The perturbation expansion of the PCDD.

B-QUASIPARTICLE SELF-ENERGY

The single-quasiparticle propagator G_{11} has an equation of motion for which the perturbation expansion is given in Fig. 3[9]. The external potential in Fig. 3(c), denoted by $\times$, which is added and subtracted to the original Hamiltonian, can be chosen to cancel the other self-energy diagrams. If we choose Fig. 3(c) to cancel Fig. 3(d), then we obtain for the quasiparticle energy

$$E_1 = h_{11}(1,1) + 4\sum_2 h_{22}(1221)n_2 . \tag{5.9}$$

If the function h_{11} in Eq. (2.12) and h_{22} are substituted into Eq. (5.9), we obtain

$$E_1 = U_1 (u_1^2 - v_1^2) + \Delta_1 (2u_1v_1), \tag{5.10}$$

where U_1 and Δ_1 are defined in Eqs. (5.5) and (5.6), respectively.

When Eqs. (5.7) and (5.8) are used in Eq. (5.10) for the quasiparticle energy, we obtain

$$E_1 = [U_1^2 + \Delta_1^2]^{1/2} . \tag{5.11}$$

Equation (5.11) is the finite temperature energy for the Bogoliubov quasiparticle, which is obtained from finite-temperature perturbation theory using only the dangerous diagrams in Figs. 2(a) and (b) and the self-energy in Fig. 3(c) and (d). Equation (5.11) is the finite temperature generalization of Eq. (2.16).

When Eq. (5.8) is used in Eq. (5.6), we obtain

$$\Delta_1 = \sum_2 \langle 1,-1|V|2,-2\rangle (\Delta_2/2E_2)\tanh(\beta E_2/2), \tag{5.12}$$

which is the finite temperature integral equation for the energy gap[10].

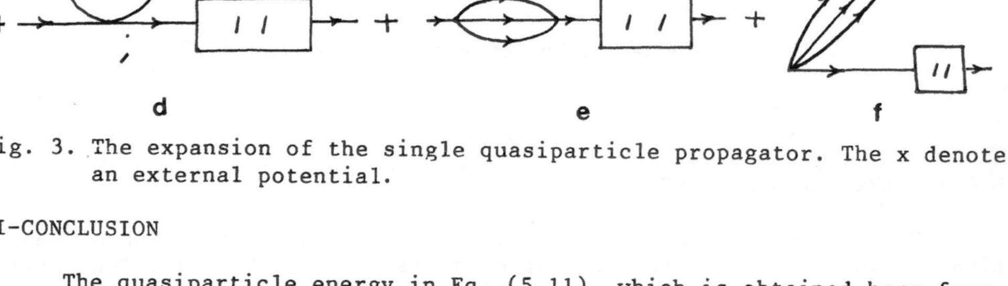

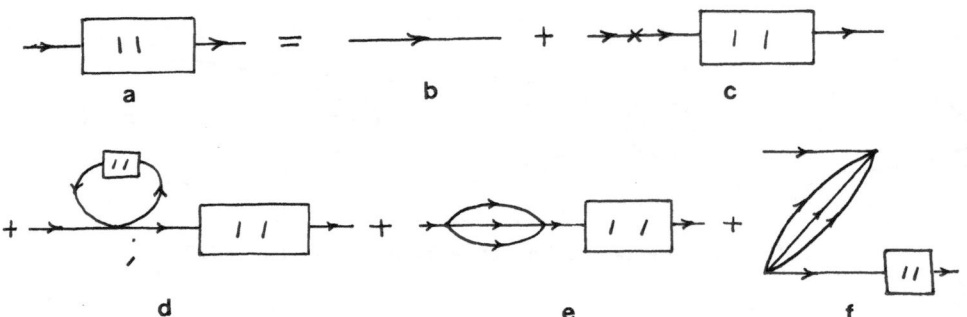

Fig. 3. The expansion of the single quasiparticle propagator. The x denotes an external potential.

VI-CONCLUSION

The quasiparticle energy in Eq. (5.11), which is obtained here from self-consistent, finite temperature perturbation theory, is the same as obtained by making the replacement of the quasiparticle vacuum expectation values $<\phi_0| a_k^\dagger a_k |\phi_0>$ and $<\phi_0| a_k a_{-k} |\phi_0>$ by their averages at finite temperature, $<a_k^\dagger a_k>$ and $<a_k a_{-k}>$ 10. However, this replacement is only an ad hoc recipe to obtain the finite temperature expressions, and does not show how to extend the theory to calculate higher-order corrections. On the other hand, the self-consistent finite-temperature perturbation theory based on the PCDD shows what diagrams are needed to go beyond the first order HFB theory, which is calculated here. Clearly the calculation of the second-order corrections to the PCDD and the self-energy would be worthwhile, to find the effect on the energy gap in Eq. (5.12). Work on this problem is in progress.

ACKNOWLEDGMENTS: I would like to thank Professor R. Aldrovandi, the director of IFT, the other faculty, the staff, and the students at the Instituto de Física Teórica for their hospitality to me this year. To the Fulbright Commission in Brazil for a fellowship and to the University of North Texas for a Faculty Development Grant I would like to express my gratitude.

REFERENCES

[1] - J.Bardeen, L.N. Cooper, and J.R.Schrieffer, Phys.Rev. 108, 1175(1957)

[2] - See the review: C.P. Enz, Helv.Phys.Acta 61, 741(1988)

[3] - N.N. Bogoliubov, Zh. Eksperim. Teor.Fiz. 34, 58(1958) [English Transl.: Sov.Phys. - JETP 7, 41(1958)] Usp.Fiz. Nauk SSSR 67, 549(1959) [English Transl.: Sov.Phys. - Usp. 2, 236(1959)] N.N. Bogoliubov, V.V. Tolmachev, and D.V. Shirkov, A New Method in the Theory of Superconductivity (Academy of Sciences of the USSR Press, Moscow, 1958) [English Transl.: Consultants Bureau, New York, 1959]

[4] - D.H. Kobe, Phys.Rev. 140, A825(1965); J.Math.Phys. 8, 1200(1967)

[5] - D.H. Kobe, J.Math.Phys. 9, 1779, 1795(1968)

[6] - D.H. Kobe and G.W. Goble, J.Math.Phys. 15, 1835(1974)

[7] - D.H. Kobe, Ann.Phys. (N.Y.) 75, 9(1973)

[8] - D.H. Kobe and W.B. Cheston, Ann.Phys. (N.Y.) 20, 279(1962)

[9] - D.H. Kobe, Ann.Phys. (N.Y.) 28, 400(1964) J.R. Schrieffer, Nucl.Phys. 35, 363(1962)

[10]- J.R. Schrieffer, Theory of Superconductivity (Benjamin, New York, 1964), pp. 53-56

ABNORMAL OCCUPATION, TIGHTER-BOUND COOPER PAIRS and HIGH T_c SUPERCONDUCTIVITY

M. de Llano*

Physics Department
North Dakota State University
Fargo, North Dakota, 58105

J. P. Vary

Physics Department
Iowa State University
Ames, Iowa 50011

The classic Cooper electron-pair problem is generalized via a Fermi sea having *two* concentric surfaces, rather than the familiar sphere of a perfect (interactionless) fermion gas. Substantially tighter-bound pairs are obtained for fixed phonon coupling with the BCS model interaction, no matter how weak. This will admit increased transition temperatures for superconductivity in the BCS theory and beyond, and suggests that *designing* materials having multiply-connected Fermi surfaces with maximal *interior area* will yield larger T_c values.

INTRODUCTION

The Cooper electron pair problem[1] is vitally central to the Bardeen-Cooper-Schrieffer (BCS) microscopic theory of superconductivity[2]. It consists in a Schrödinger equation for two particles of opposite momenta and spins being scattered by a particular attractive interaction *only* into one-electron plane wave (PW) states just above the spherical Fermi sea occupied by the PW orbitals of the other N - 2 background electrons in the N-electron system. PW states are used for simplicity, instead of Bloch states. The well-known resulting eigenvalue equation for the pair energy E_o is then

$$1 = V \sum_{\mathbf{k}}{}' \frac{1}{2E(\mathbf{k}) - E_o} \longrightarrow V \int_{E_F}^{E_F+\hbar\omega_D} d\mathcal{E} \frac{g(\mathcal{E})}{2\mathcal{E} - E_o} , \tag{1}$$

*A travel grant from US Army Research Office is gratefully acknowledged.

where $E(\mathbf{k}) = \hbar^2k^2/2m$ are the unperturbed single-particle energies and $V > 0$ is the strength of the effective attractive electron-electron interaction induced by the electron-phonon coupling. This is nonzero only within a very thin shell of thickness $\hbar\omega_D$ above the Fermi surface of energy $E_F = \hbar^2k_F^2/2m$, k_F being the Fermi sphere radius; it is the so-called "BCS model interaction". Here, $\hbar\omega_D$ is the maximum energy possible for an ionic lattice phonon, and is typically 10^{-3} to 10^{-2} times smaller than E_F. The prime over the summation sign means restriction to those PW states such that $E_F < E(\mathbf{k}) < E_F + \hbar\omega_D$, i.e., to states that are unoccupied. The integral involves the free-electron density of states $g(\mathcal{E}) \propto \mathcal{E}^{1/2}$, which can in turn be factored out from the integral as a constant $g(\mathcal{E}_F)$ due to the smallness of $\hbar\omega_D/E_F$. This leaves an elementary integral to be performed that gives a logarithm. Solving for the eigenvalue E_o then yields

$$E_o = 2E_F - \frac{2\hbar\omega_D}{e^{2/g(E_F)V} - 1} \equiv 2E_F - \Delta_o \xrightarrow[V\to 0]{} 2E_F - 2\hbar\omega_D e^{-2/g(E_F)V}. \tag{2}$$

Putting $\epsilon_o \equiv E_o/2E_F$ and $\nu \equiv \hbar\omega_D/2E_F$, the first equation can be rearranged to read

$$e^{-2/g(E_F)V} = \frac{\epsilon_o - 1}{\epsilon_o - 1 - 2\nu}. \tag{3}$$

The lhs (a constant for fixed coupling and density of states) and rhs of this equation are displayed, for the special but typical case $2\nu = 10^{-3}$, in Figure 1 as function of $\epsilon \equiv 1 - \Delta_o/2E_F$, where Δ_o is the pair binding energy. We see that ϵ_o differs *very little* from unity (and hence $\Delta_o/2E_F$ from zero) for all but the largest values of the coupling parameter $\lambda \equiv g(E_F)V/2$. Since (3) results from solving a *two*-body Schrödinger equation (in momentum representation), its validity is *not* limited to weak coupling, unless regarded as a Bethe-Goldstone equation (ladder approximation) treatment of the many-fermion problem, which is exact only for small $k_F\lambda$.

Also restricted to weak coupling is the BCS *many*-electron theory[2]. In this formalism a temperature-dependent energy gap $\Delta(T)$ emerges, which for $T = 0$ is found to be

$$\Delta(0) \equiv \Delta_0^{BCS} = \frac{\hbar\omega_D}{\sinh[1/2\lambda]} \xrightarrow[\lambda\to 0]{} 2\hbar\omega_D\, e^{-1/2\lambda} , \tag{4}$$

i.e., *similar* to the weak-coupling limit of the Cooper pair binding energy Δ_o in (2). The "normal metal" to "superconductor" transition temperature

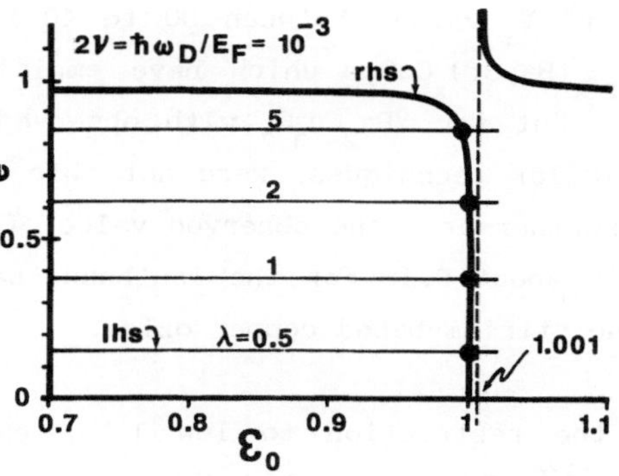

Fig. 1. Graphical solution of Eq. (3) for $2\nu = 10^{-3}$ and several values of the coupling parameter λ as indicated.

T_c is then determined by the vanishing of $\Delta(T)$, or $\Delta(T_c) = 0$, whose solution is the celebrated relation

$$\Delta_0^{BCS} = \pi e^{-\gamma} kT_c \simeq 1.76kT_c \tag{5}$$

where $\gamma \simeq 0.577$ is the Euler constant. Combining (4) and (5) we have

$$T_c = 1.13\, \Theta_D\, e^{-1/2\lambda}, \tag{6}$$

where $\Theta_D \equiv \hbar\omega_D/k$ is the Debye temperature. For elemental superconductors measured T_c values range from a very low 3.25×10^{-4} K for Rh to 9.26 K for Nb. Since Θ_D is proportional to an appropriately averaged sound speed $\bar{c}$ for the lattice, and since $\bar{c} \propto M_{ion}^{-1/2}$, with M_{ion} the mass of the lattice ions, we see that $T_c \propto M_{ion}^{-1/2}$. This is known as the *isotope effect*, observed for many elemental superconductors but *not* for the new copper-oxide high-T_c materials. Since $\Theta_D \sim 10^2$ K, (6) severely limits T_c to a few degrees K with acceptable values of λ. Other T_c formulas[3], beginning with the MacMillan[4] formula based on strong-coupling Migdal-Eliashberg theory[5], give values of T_c as high as around 40 K. Indeed, a recent realistic tight-binding calculation with the Eliashberg equations gave Weber[6] T_c values between 30 to 40 K for the copper-oxide superconductors $La_{2-x}(Ba,Sr)_xCuO_4$, which have empirical T_c values in the range 30 to 36 K. But for $YBa_2Cu_3O_7$ with observed T_c = 95 K Weber and Mattheiss, using similar techniques, were not able to extract a T_c larger than about 30 K. Furthermore, the observed value of the exponent α in the formula $T_c \propto M_{ion}^{-\alpha}$ is about 0.18 for the lanthanum-based copper oxide just cited, and 0 for the yttrium-based copper oxide.

We show how the restriction to low T_c's, exemplified by (6) and traceable to the $e^{-1/\lambda}$ factor in the Cooper pair problem (2), can be surmounted through *a generalized Fermi sea*. And, moreover, a negligibly weak isotope effect is then possible, even with a phonon-mediated interaction.

The fact that (quasi) electron-pair binding occurs for arbitrarily weak attraction---as well as the non-perturbative (essential singularity in V) property of the energy Δ_o needed to break up a Cooper pair---occurs even for one-body potential well problems[8] in two-dimensions (2D). Recalling that $g(\mathcal{E})$ = const in 2D, it is then clear why the Cooper pair problem behaves like a 2D quantum system, a fact which may be pivotal in understanding the superconductivity of the *layered* copper-oxide ceramics.

ABNORMAL OCCUPATION

The all-important Fermi sea assumed in the Cooper problem is appropriate to the perfect Fermi gas Slater determinant ground state wave function for N particles enclosed in a volume Ω,

$$\Phi = (N!)^{-1/2}\det_{n_{\mathbf{k}}^{o}}\left[\Omega^{-1/2}e^{i\mathbf{k}_i\cdot\mathbf{r}_j}\right], \tag{7}$$

with $i,j = 1,2,\ldots N$, and $n_{\mathbf{k}}^{o} = \theta(k_F - k)$ the unit step function, $\theta(x) \equiv \frac{1}{2}[1+\mathrm{sgn}(x)]$. For an interacting system the most general occupation scheme consistent with the Pauli principle, however, is merely

$$n_{\mathbf{k}} = 0 \text{ or } 1, \qquad \sum_{\mathbf{k}} n_{\mathbf{k}} = N. \tag{8}$$

We have raised the general question[9] of what the optimum scheme might be for a *non-ideal*, fully-interacting many-fermion assembly in any dimension. After all, Fermi surfaces for many materials are known which contrast drastically with the familiar Fermi sphere. Overhauser[10], for example, has considered multiply-connected Fermi seas associated with both charge and spin density wave states. These states break the translation symmetry and will not concern us here. Indeed, as early as 1950 Fröhlich[11] had already contemplated a departure from the simple Fermi sphere, with ordinary PW orbitals. He took a spherical shell concentric upon, but disjointed from, an interior sphere, of occupied electronic states, and used this with his new electron-phonon hamiltonian in (second-order) perturbation theory to find a lower energy state, *if* the coupling exceeded a certain critical value. This behavior, however, sharply differs from the empirical fact that superconducting critical temperatures can be arbitrarily small, a drawback eventually circumvented by the (infinite-order) treatment[1] of Cooper with the standard Fermi sea. More recently, in an infinite meson-nucleon system a transition from a Fermi sphere to a Fermi "shell" [cf. below, Eq. (10)] distribution has been interpreted as a possible phase transition from "nuclear" to "quark" matter within a relativistic Hartree-Fock (HF) treatment[12]. The (first-order) transition is signalled by a large jump at high densities in the resulting low-temperature specific heat. Finally, abnormal occupation in a finite nucleus has been discovered in ^{24}Mg, in constrained HF calculations[13] (with good total angular momentum and its projection), using a realistic effective two-nucleon interaction based on the Reid soft-core potential.

Within the (nonrelativistic) plane-wave Hartree-Fock (PWHF)

approximation with many-fermion hamiltonian H we originally showed[9] that for a simple one-dimensional system under a sufficiently attractive (but non-collapsing in the thermodynamic limit), short-ranged, two-body interaction v_{12}, *lower* HF total energies

$$\mathcal{E}_{HF}[n_k] \equiv < \Phi|H|\Phi > \\ = \sum_k E(k)n_k + \frac{1}{2}\sum_{k_1k_2} \{<k_1k_2|v_{12}|k_1k_2> - <k_1k_2|v_{12}|k_2k_1>\} n_{k_1}n_{k_2} \tag{9}$$

emerge for some particle densities $N/\Omega = k_F^3/3\pi^2$, with a generalized Fermi sea defined by

$$n_k = \theta(k - \beta k_F)\theta(\gamma k_F - k) \\ \gamma^3 - \beta^3 = 1, \quad \gamma > \beta \geq 0, \quad \gamma \geq 1. \tag{10}$$

This corresponds to a spherical shell in k-space of inner and outer radii βk_F and γk_F, respectively. In other words, it was established that $\mathcal{E}_{PWHF}[n_k] < \mathcal{E}_{PWHF}[n_k^o]$ for a range of densities, and in a manner reminiscent of a (first-order) gas-liquid phase transition. Any lowering of the total HF energy at fixed density comes entirely from the last (exchange) term in (9), since the first (kinetic energy) term can only increase for any n_k other than a spherical Fermi sea, while the second (direct) term is unchanged if the interaction is local, as then that matrix element is independent of the summation indices for PW states. The search for lower-energy, abnormally occupied Slater PW determinants for a wide variety of pair-interaction cases was subsequently extended[14] to three dimensions (3D), and to a much larger class of abnormal occupation schemes n_k. Moreover, starting from a very-many-shell structure modeling the general distribution (8), random-search and random-walk numerical techniques[14] established the "single-shell" distribution (10) as the optimum one in numerous cases. In still further work[15] it was proved, for example, that *any* hard-core plus square-well two-body potential prefers abnormal occupation, no matter how weak the attraction. This potential has been employed[16] to model liquid-^{3}He semi-realistically. Finally, several many-boson systems were also found[17] which prefer abnormal occupation.

Summarizing, even at the PWHF level of approximation, Fermi seas more general than the usual spherical sea are favored for sufficiently strong interparticle coupling. A possible physical rationale[18] for this is the appearance, as coupling is increased, of "particle clusterings" of some kind, since abnormal occupation precludes *small*-k states, meaning suppression of particle orbits with *large* spatial extensions.

TIGHTER-BOUND COOPER PAIRS

Robustly tighter bound Cooper pairs are possible, for *any* coupling strength, with abnormal occupation. We suggest that when suitably incorporated into the BCS-Bogoliubov[19]-Anderson[20]-Gor'kov[21]-Migdal -Eliashberg[5] formalisms, this may yet provide a comprehensive understanding of *both* low- and high-T_c superconductivity, perhaps in terms of the phonon mechanism alone. The term "normal" is here used in the sense of Landau's Fermi liquid theory, since a key assumption for a first-principles derivation by Klein[22] of Landau's theory is that the lowest-energy single-particle states be occupied, i.e., that the familiar finite-temperature Fermi distribution holds for the single-particle spectra, instead of some *other*, more general, distribution.

Guided by our PWHF studies[9,14,17] we employ the model (10) merely as an illustration. This particular occupation scheme is a definite step beyond the perfect Fermi gas picture, and suffices to uncover an instability in the Fermi-sphere-induced Cooper pair. The two surfaces in (10) are situated at energies $E_1 \equiv \beta^2 E_F \equiv (\gamma^3 - 1)^{2/3} E_F \geq 0$ and $E_2 = \gamma^2 E_F \geq E_F$. The parameter γ is presumably characteristic of the fully-interacting *many*-body system, and can later be fixed, within a BCS-like theory, e.g., variationally, with respect to the total superconducting ground-state energy. The integral in (1) now becomes

$$1 \simeq V \left\{ g(E_1) \int_{E_1 - \hbar\omega_D}^{E_1} \frac{d\mathcal{E}}{2\mathcal{E} - E} + g(E_2) \int_{E_2}^{E_2 + \hbar\omega_D} \frac{d\mathcal{E}}{2\mathcal{E} - E} \right\}, \tag{11}$$

provided that E_1 is significantly larger than $\hbar\omega_D$ (as was verified *post hoc*). Scattering is now allowed by the BCS model interaction in the vicinity of *both* surfaces. Performing the integrals makes solving for $e^{-2/g(E_F)V}$ easier than for the new eigenvalue E. Putting $E/2E_F \equiv \epsilon$ and $\hbar\omega_D/2E_F \equiv \nu$, Eq. (11) leads to a transcendental equation for ϵ which, in the 3D application presented here, is

$$e^{-1/\lambda} \equiv e^{-2/g(E_F)V} = \left[\frac{\epsilon - (\gamma^3-1)^{2/3} + 2\nu}{\epsilon - (\gamma^3-1)^{2/3}}\right]^{(\gamma^3-1)^{1/3}} \left[\frac{\epsilon - \gamma^2}{\epsilon - \gamma^2 - 2\nu}\right]^{\gamma}. \quad (12)$$

Both quantities in the square brackets must be non-negative to ensure that the lhs of (12) be real.

We have carried out a numerical search for values of ϵ and γ satisfying (12), such that $\gamma > 1$ and $\epsilon_o > \epsilon \equiv 1 - \Delta/2E_F$, with Δ positive, for a wide range of (λ, ν) values. However, a better feeling for the solutions of our model is obtained from Fig. 2, where both sides of Eq. (12)

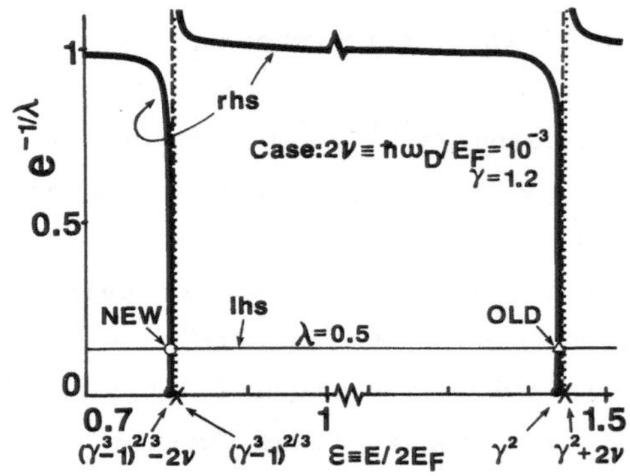

Fig. 2. Typical case illustrating graphical solution of Eq. (12).

are plotted for the case specified by $\lambda = 0.5$, $2\nu = 10^{-3}$, and $\gamma = 1.2$. The lhs then equals $\simeq 0.135$. The rhs of (12) gives rise to the full thick curves shown; the dashed curves correspond to negative quantities inside both square brackets of (12) and so are not relevant. The two zeros (poles) are designated by dots (crosses) on the ϵ-axis. They are labeled by their general values, for any $\gamma \geq 1$ and 2ν, which for the specific case just cited are respectively equal to 0.808259, 0.809259, 1.440 and 1.441.

The dot-to-cross distances have been exaggerated somewhat in the figure, for clarity. The lowest energy eigenvalue occurs from the projection of the intersection marked "NEW" and indicated by an open circle: it is a (quasi) bound state since $\epsilon < 1$ and moreover, by inspection, $\epsilon < \epsilon_o$ (= $1 - 10^{-3} = 0.999$). The intersection marked "OLD" and indicated by an open triangle is now an *unbound* level in the pair continuum, and becomes the ordinary Cooper bound state (2) or (3) as $\gamma \rightarrow 1$, when the open circle intersection is pushed leftwards outside the $\epsilon > 0$ region. The reason for tighter-bound Cooper pairs with abnormal occupation is almost trivial: an interior Fermi surface allows for *lower* kinetic energy electron pairs with roughly the same potential energy. As in the usual Cooper pair case, the new, tighter-bound Cooper pair solution *will survive, no matter how weak the coupling* V, since smaller coupling merely lowers the horizontal line marked "lhs". More interesting, however, is that these tighter-bound Cooper pairs can lead to transition temperatures T_c with a *robust* contribution of order T_F (~ 10^4 to 10^5 K), in addition to a much weaker "isotope-effect" contribution of order Θ_D (~ 10^2 K), as we now discuss.

HIGH-TEMPERATURE SUPERCONDUCTIVITY

In the weak-coupling limit ($\lambda \equiv g(E_F)V/2 << 1$) the intersection marked "new" in Fig. 2 will correspond to

$$1 - \Delta/2E_F \equiv \epsilon = (\gamma^3 - 1)^{2/3} - 2\nu - \eta, \tag{13}$$

with $0 < \eta << 1$. Instead of working with this, we obtained the transcendental equation for ϵ for the Fröhlich abnormal Fermi distribution

$$\begin{gathered} n_k = \theta(\alpha k_F - k) + \theta(k - \beta k_F)\,\theta(\gamma k_F - k) \\ 0 \le \alpha \le \beta \le \gamma, \quad \alpha \le 1, \quad \gamma \ge 1 \\ \alpha^3 - \beta^3 + \gamma^3 \equiv 1 \end{gathered} \tag{14}$$

which becomes the normal one when all equalities hold, in particular $\alpha = 1$. Instead of (13) one then has $\alpha^2 - \eta$ on the rhs. Inserting the last member of this into the new transcendental equation and expanding about $\eta = 0$ gives $\eta \simeq 2\nu e^{-1/\alpha\lambda}$. Recalling (2), (4) and (5), this leads to

$$T_c \simeq 1.13[(\Theta_D T_F)(1 - \alpha^2 + \frac{\Theta_D}{T_F} e^{-1/\alpha\lambda})]^{1/2} + \ldots, \tag{15}$$

where α is the radius (in units of k_F) of the *smallest* surface of the abnormal distribution (14). For $\alpha \longrightarrow 1$, (15) becomes the BCS normal occupation result (6). Otherwise, T_c *values well in excess of the "phonon barrier" (~ 40 K) are possible without invoking stronger electron-phonon coupling, nor unconventional interaction mechanisms.* The latter, however, are not ruled out in principle. We also note that from (15) T_c now depends on $M_{ion}^{-1/4}$, instead of $M_{ion}^{-1/2}$, so that the isotope effect will be weaker.

CONCLUSIONS

Tighter-bound Cooper pairs will arise by generalizing the assumed Fermi sea for the occupied background electrons, *without* invoking either stronger electron-phonon coupling in the BCS interaction nor unconventional interaction mechanisms.

Multiple-surfaced Fermi topologies can lead to T_c's scaling both as the Debye temperature Θ_D (~ 10^2 K) as in normal-occupation BCS theory, as well as with the Fermi temperature T_F (~ 10^4 to 10^5 K), thus much higher T_c values are possible.

If more were known experimentally about multiply-connected Fermi surfaces with *different energies* and how to manipulate them, perhaps one could design materials (most probably compounds) having *interior* surfaces which, allowing for Cooper pairs with smaller kinetic energies, would result in higher T_c values.

ACKNOWLEDGEMENTS

This work was supported in part by the U.S. Department of Energy under Contracts Nos. DE-FG02-87ER40371, Division of High Energy and Nuclear Physics. We wish to acknowledge discussions with Professors O. Civitarese, J.R. Clem, D. Cox, Christina Keller, M. Luban, H.G. Miller, S.A. Moszkowski, O. Rojo, M. Saraceno, B.D. Serot, V.V. Tolmachev and R.N. Zitter.

REFERENCES

1. L.N. Cooper, Phys. Rev. **104**, 1189 (1956)
2. J. Bardeen, L.N. Cooper and J. Schrieffer, Phys. Rev. **108**, 1175 (1957)
3. P.B. Allen and B. Mitrović, Sol. State Phys. **37**, 1 (1982)
4. W.L. McMillan, Phys. Rev. **167**, 331 (1968)
5. A.B. Migdal, Sov. Phys.-JETP **7**, 996 (1958); G. M. Eliashberg, *ibid* **11**, 696 1960)
6. W. Weber, Phys. Rev. Lett. **58**, 1371 (1987)
7. W. Weber and L.F. Mattheiss, Phys. Rev. **B 37**, 599 (1988)
8. L.D. Landau and I.M. Lifshitz, *Quantum Mechanics*, (Pergamon, London, 1977) p. 163
9. M. de Llano and J.P. Vary, Phys. Rev. **C 19**, 1083 (1979)
10. A.W. Overhauser, in *Highlights of Condensed-Matter Theory,* Proc. of the Intl. School of Phys. "E. Fermi", Course LXXXIX, ed. by F. Bassoni *et al* (North-Holland, Amsterdam, 1985)
11. H. Fröhlich, Phys. Rev. **79**, 845 (1950)
12. C.J. Horowitz and B.D. Serot, Phys. Lett. **109B**, 341 (1982)
13. R.M. Quick and H.G. Miller, Phys. Rev. **C 34**, 1458 (1986)
14. M. de Llano, A. Plastino and J.G. Zabolitzky, Phys. Rev. **C 20**, 2418 (1979)
15. V.C. Aguilera-Navarro, R. Belehrad, M. de Llano, M. Sandel and J.P. Vary, Phys. Rev. **C 22**, 1260 (1980)
16. T.W. Burkhardt, Ann. Phys. (N.Y.) **47**, 516 (1968)
17. V.C. Aguilera-Navarro, R. Barrera, J.W. Clark, M. de Llano and A. Plastino, Phys. Lett. **80 B**, 327 (1979); Phys. Rev. **C 25**, 560 (1982); M.C. Cambiaggio, M. de Llano, A. Plastino and L. Szybisz, Rev. Mex. Fís. **28**, 91 (1981)
18. S.A. Moszkowski, priv. comm.
19. N.N. Bogoliubov, Sov. Phys. (J.E.T.P.) **34**, 41 (1958); N.N. Bogoliubov, V.V. Tolmachev and D.V. Shirkov, *A New Method in the Theory of Superconductivity* (Consultants Bureau, Inc., N.Y., 1959)
20. P.W. Anderson, Phys. Rev. **112**, 1900 (1958)
21. L.P. Gor'kov, Sov. Phys. JETP **9**, 1364 (1959)
22. A. Klein, Phys. Rev. **121**, 957 (1961)

ON THE ROLE OF ELECTRON-MEDIUM COUPLING IN HIGH TEMPERATURE SUPERCONDUCTORS*#

F. B. Malik

Physics Department, Southern Illinois University
Carbondale, Illinois, 62901, U.S.A.

ABSTRACT

The effective Hamiltonian for a pair of Fermion derived from the basic many body Hamiltonian allows for the interaction of a pair with its medium in a number of ways e.g., once via the mean field and another via the pairing part of the interaction. The two state model developed earlier has been solved in a better approximation. The model has then been applied to calculate pairing energy, critical temperature, coherence length and critical field for $YBa_2Cu_3X_{7-\delta}$ with X = O, S, Cl, F, Br, I. The model can account for the general trend and magnitude of these observables, particularly the very large value of the critical field. Using a model Hamiltonian of Frölich type, the electron-phonon coupling strength is estimated. The electron-phonon coupling strength for our particular system is more than an order of magnitude larger than the one in normal superconductors. The Eliashberg dimensionless coupling parameter λ is found to be about 2.0 to 3.0.

1. INTRODUCTION

Transition from a normal state, be it a conductor or a semiconductor or a perovskite, to a superconducting state involves a phase transition. Whereas, electrons in a normal state behave like incoherent Fermions, the formation of Boson-like coherent pairs is a necessary criterion for a superconducting state. Near critical temperature, T_c, one has, therefore, two types of states of a system, one representing incoherent Fermion and the other coherent quasi boson like pairs. A set of coupled equation commensurate with this situation has been derived[1,2] and the zeroeth order solution has been successful in predicting the trend of T_c in Y-Ba-Cu-O-X with X = O, S, Cl and F.

*A travel grant from the U. S. Army Research Office is thankfully acknowledged.

#Dedicated to Professor Don Lichtenberg in appreciation of his human qualities and contribution to physics.

The key to the pairing has been assumed to be the formation of doubly ionized negative ion X^{2-}. Qualitatively, one, therefore, expects that the substitution by Br and I both of which form negative ions should give rise to superconducting behavior and this has recently been confirmed[3]). In sections 2, we present a better approximation of the two state model and calculate T_c, energy gap, coherence length and the critical field. In section 3 we discuss the relation of the calculated energy gap to electron-phonon coupling constant using an interaction similar to the one proposed by Frölich. Hamiltionian containing this interaction has been diagonalized using the method of Huang and Rys[5]. We estimate the electron-phonon coupling strength in a simple model and present there a discussion of the Eliasberg coupling constant.

2. THE THEORY

As noted in ref. 1, the equation for a pair of Fermion in state α in a medium is given by

$$[\theta(i) + \theta(j) + V(i,j) - \varepsilon_\alpha]\ \phi_\alpha(i,j) + (1/2)\sum_{\beta=1}^{n} \int dkdl\phi_\beta^*(k,l)\ W\phi_\alpha(i,j)\phi_\beta(k,l) = 0 \qquad (1)$$

where V(i,j) is the effective two-electron interaction obtained by integrating over all lattice coordinates R, of the bare electron-electron and electron-lattice interaction h_{ee} and h_{e-x}, respectively.

$$V(i,j) = \langle h_{ee}(ij) + h_{e-x}(R,ij)\rangle_R \qquad (2)$$

θ(i) is the effective one body operator which could be bare kinetic energy or an effective one-body interaction. W is given by

$$W = [V(i,k)\ (1-X(i,k)) + V(i,l)\ (1-X(i,l)) + V(j,k)\ (1-X(j,k)) + V(j,l)\ (1-X(j,l))] \qquad (3)$$

X(i,j) is the exchange operator transposing all coordinates i and j. ϕ is an antisymmetric function. W is the average potential generated by pairs other than ϕ_α. In case all the pair functions are replaced by 2x2 determinants of single particle orbitals, (1) reduces to the set of Hartree-Fock equation. Above critical temperature the total electronic wave function ψ_F is given by properly ordered permutation P of of Fermion type of pair function

$$\psi_F = (1/\sqrt{2n!}) \sum_P \varepsilon^P \phi_{\alpha_1}(12) \ldots\ldots \phi_{a_n}(2n-1,2n) \qquad (4)$$

Below critical temperature ψ_s is a linear combination of Schafroth pair functions[6]) and leads to condensation necessary for superconductivity

$$\psi_s = C_n \sum_P (-1)^P)[\phi(12)\phi(34)\ldots\ldots\phi(2n-1,2n)] \qquad (5)$$

Because of the difference between these two types of wavefunctions, the average potential seen by a pair i.e., the terms containing W in (1), differs significantly in two cases. That might result in a change in total energy signaling a phase transition. Near critical temperature we expect significant components of both types of states to be present and the total wavefunction may be written as

$$\chi = A_1\psi_F + A_2\psi_S \tag{6}$$

Thus, near critical temperature, one is to diagonalize the matrix

$$\begin{pmatrix} H_{11} & H_{12} \\ H_{21} & H_{22} \end{pmatrix} \tag{7}$$

where $H_{ii}(i=1,2)$ contains diagonal matrix elements in states 1, identified as Fermion like states and in state 2, identified as quasi-Boson like states. $H_{12} = H_{21}^*$ are the coupling terms. H_{ii} is a sum of two terms, ε_i containing effective average field seen by a pair and ε_{ii}, the interaction between a pair. The coupling term H_{ij} contains at least the effective two body interaction V(i,j) but might also take into account the difference in mean field in two cases.

The diagonalization of (7) leads to

$$E_\pm = \frac{1}{2}(H_{11} + H_{22}) \pm \frac{1}{2}\sqrt{(H_{11} - H_{22}^2) + 4|H_{12}^2|} \tag{8}$$

E_-, being lower in energy, is to be identified with the superconducting state and E_+ with the normal state. Since $(H_{11} - H_{22})^2$ and $|H_{12}|^2$ are,respectively, of the order of a few eV and a few tens of meV, it is reasonable to assume $4|H_{12}|^2 << |H_{11}-H_{22}|^2$. Then

$$E_- \cong H_{22} - \frac{|H_{12}|^2}{H_{11}-H_{22}} = H_{22} - \frac{|H_{12}|^2}{\varepsilon_1-\varepsilon_2+\varepsilon_{11}-\varepsilon_{22}} \tag{9}$$

$\varepsilon_{11}-\varepsilon_{22}$ is usually much less than $(\varepsilon_1-\varepsilon_2)$. Therefore,

$$E_- \cong \varepsilon_2+\varepsilon_{22} - \frac{|H_{12}|^2}{\varepsilon_1-\varepsilon_2} \tag{10}$$

∴ Pairing energy in the superconducting state ΔE

$$\Delta E = (E_- - \varepsilon_2) = \varepsilon_{22} - \frac{|H_{12}|^2}{\varepsilon_1-\varepsilon_2} \tag{11}$$

Both ε_{22} and $|H_{12}|^2$ involve integration over all lattice coordinates and represent a coupling between electrons and the medium. These depend on various parameters of media or lattice and are collectively denoted by κ.

Since this is a zero-temperature theory one can relate ΔE to T_c using the relation[7,8]

$$(\pi/\gamma)kT_c = \Delta E \tag{12}$$

(Here k and γ are, respectively, Boltzmann and the exponential of Euler constants and $\pi/\gamma \cong 1.764$). Assuming $|H_{12}|^2$ to depend on mass of ion X as $1/\sqrt{M}$, we may write

$$T_c = a(\kappa) - b(\kappa)/\sqrt{M}(\varepsilon_2-\varepsilon_1) \tag{13a}$$

$$\equiv a(\kappa) - b(\kappa)\ \Delta W \tag{13b}$$

Using T_c = 93°K and 90°K for X = O and S, respectively, we find $a(\kappa)$=31°K and $b(\kappa) \cong$ 2521°K. Since ΔW for O and S are very close, we expect them to have similar T_c which is, indeed, the case. As noted in refs. 1 and 2, the superconducting state in this system is identified with the localization of electron pairs in forming doubly ionized negative ion, X^{2-} and the normal state with singly ionized negative ion, X^{1-}. In table 1 we have compared calculated T_c with the observed ones. For X = Br and I, we provide theoretical estimates using a range of electron affinities of Br^- and I^-. The theory can reproduce the trend.

There is some uncertainty about the validity of (12) in this type of superconductors. Nevertheless, we have used it to calculate the gap parameter and tabulated them in table 2. The measurements for the oxygen case is in accord with the calculated number but indicate that the constant of proportionality between Δ and T_c is about twice the one in (12).

Coherence length ζ_o is related gap parameter by

$$\zeta_o = \hbar V_F/\pi\Delta \tag{14}$$

where V_F is the velocity near Fermi surface. Assuming the V_F in this type of superconductors is close to those in type II superconductors, we may determine this from Nb_3Sn(T_c = 18°K, ζ_o = 50 Å). Calculated values are tabulated and compared with data in table 2.

Since Δ and ζ_o for this type of superconductors are, respectively, substantially larger and smaller than those in type I and II superconductors, we may call these type III superconductors. The large Δ and small coherence length are indicative of very strong coupling between pairs and the environment and the coherence is more localized. Thus, the BCS approximation of a constant weak coupling between a pair of electron via phonons is invalid for these materials.

The difference between the total energy of the normal and superconducting states is related to critical field H_c by $H_c^2/8\pi$ = energy difference per unit volume

$$H_c^2/8\pi = \frac{1}{2}\rho \times \sqrt{(H_{11}-H_{22})^2 + 4|H_{12}|^2} \tag{15}$$

where ρ is the number of electrons per unit volume. We have used (8) to get the energy difference per pair between the normal and the superconducting state.

Assuming that $(\varepsilon_{11}-\varepsilon_{22}) \ll (\varepsilon_1-\varepsilon_2)$ we get

$$\begin{aligned}(H_c^2/8\pi)\ (2/\rho) &\cong \left|(\varepsilon_1-\varepsilon_2) + \frac{2|H_{12}|^2}{\varepsilon_1-\varepsilon_2}\right| \\ &= |(\varepsilon_1-\varepsilon_2) - 2(\pi/\gamma)k\ b(\kappa)\ \Delta W| \\ &\cong |\varepsilon_1-\varepsilon_2|\end{aligned} \tag{16}$$

Table 1. Calculated T_c (col. 5), are compared to observed one (col. 6). Electron affinities in col. 2 and for O and S in col. 3 are taken from ref. 18. Electron affinities for F^- and Cl^- are theoretical calculation and Br and I are estimation. (a), (b), (c), (d), (e) and (f) are refs. 19-23 and 3, respectively.

X	e.a(x)(eV)	ea(x^-)eV	ΔW	T_c(cal.)°K	T_c(expt)°K
O	1.47	-8.73	-0.02451	93	93[a]
F	3.45	-4.81	-0.02778	102	159[b] 140[c] 80 to 89[d]
S	2.07	-5.51	-0.02332	90	90[e]
Cl	3.61	-6.11	-0.01739	75	72[e]
Br	3.36	-6.00 -5.00 -4.00	-0.01202 -0.01346 -0.01528	61 65 70	~ 75[f]
I	3.06	-6.00 -5.00 -4.00	-0.00979 -0.01100 -0.01259	56 59 63	~ 50[f]

Table 2. Calculated gap Δ (col.2), coherence length ζ_o (col. 4), critical field (col. 6) are compared with respective observed values, in columns 3, 5 and 6.(a), (b), (c), and (d) are, respectively, ref. 24-27.

X	Δ(cal.) MeV	Δ(expt) MeV	ζ_0(calc.)Å	ζ_0(exp)Å	Hψ(cal.)G	H_c(exp)O_e
O	14	15-23[a)] 32[b] ~50[c)]	10	~22[d)]	25 x 10^3	(10±2[d])10^3
F	16	-	9	-	-	-
S	14	-	10	-	-	-
Cl	11	-	12	-	-	-
Br	9-11	-	13-15	-	-	-
I	9-10	-	15-17	-	-	-

The last approximation is valid since the second term is of the order of a few tens of meV but $|(\varepsilon_1-\varepsilon_2)|$ is about 5 to 10eV. Eq. (16) is an important result and is very different from the expression obtained in the Schafroth or BCS theory. In latter two cases instead of (15), one has[9] $(H_c^2/8\pi)(2V/\rho_E) \simeq \Delta^2$. Thus, the theory predicts that critical field in this types of superconductors is much larger than that in type I superconductors. We may estimate ρ from normal superconductors e.g., tin which has a $T_c = 3.73°K$ and $H_c = 306$ Gauss. Noting the ρ per meV = ρ_E/V, we get $\rho = 10^{18}/cm^3$ from the tin data. Calculated critical field using this, is noted and compared to the observed value in table 2. The theory can account for the large critical field seen in high T_c superconductors.

3. EFFECTIVE COUPLING STRENGTH IN A FROHLICH MODEL

Describing the medium and the electrons, respectively, by Boson operators, b_k and Fermion operators a_k we may write the Hamiltonian for a linear coupling between the two as follows

$$H = \sum_\lambda \varepsilon_\lambda a_\lambda^+ a_\lambda + \sum_{\vec{k}} h\omega(\vec{k})\,[b_k^+ b_k + \tfrac{1}{2}]$$

$$+ \sum_{\lambda,\vec{k}} a_\lambda^+ a_\lambda\,[v(\vec{k},\lambda) b_k + v(\vec{k},\lambda) b_k^+] \tag{17}$$

This is basically the Hamiltonian proposed by Fröhlich[4] and used in the Eliashberg[11] formalism. Taking $v(\vec{k},\lambda)$ to be the matrix element of $(\hat{\varepsilon}.\nabla V)$ where $\hat{\varepsilon}$ and V are, respectively, unit polarization vector and potential at lattice sites, one may define a dimensionless coupling parameter λ within the framework of Eliasberg formalism. The averaged electron-phonon interaction $\alpha^2(\omega)$, averaged electron-phonon matrix elements $\langle g^2\rangle$ and the dimensionless coupling constant λ are related as follows:

$$\lambda = 2\int \alpha^2(\omega)F(\omega)d\omega/\omega \tag{18}$$

$$\frac{N(0)\langle g^2\rangle}{2M} = \int_0^\infty \omega\alpha^2(\omega)F(\omega)d\omega \tag{19}$$

In the above $F(\omega)$ is phonon density of states, and $N(o)$ is electronic density of state at the Fermi energy and M is the atomic mass. In case α^2 = constant, as suggested by McMillan[14]

$$\lambda = \frac{N(0)\langle g^2\rangle}{M\,\langle\omega^2\rangle} = 2\alpha^2\int F(\omega)\omega^{-1}\,d\omega \tag{20}$$

$$\text{with } \langle\omega^2\rangle \equiv \int F(\omega)\omega d\omega \Big/ \int F(\omega)\omega^{-1}\,d\omega \tag{21}$$

Evaluation of α^2 and $\langle g^2\rangle$ require a knowledge of λ and either $\langle\omega^2\rangle$ or $\langle\omega^{-1}\rangle$.

λ can be determined from observed T_c. For small and moderate values of λ, McMillan's relation holds[14]

$$T_c = K \exp[-\frac{1.04(1+\lambda)}{\lambda-\mu^*(1+0.62\lambda)}] \tag{22}$$

Here μ^* is McMillan's Coulomb parameter and K is either logarithmic average phone frequency $\omega_{log}/1.2$ or Debye temperatue $T_D/1.45$.

For very large λ i.e.$\lambda>$ five times (phonon frequency)^2, Allen and Dyne's[13] relation holds

$$T_c = 0.180(1+2.60\ \mu^*)^{-1/2}\ \lambda^{1/2} \tag{23}$$

Implication of (23) is that there is no upper limit of T_c. λ, is not expected to be very large for the 1-2-3 superconductors. Since ω_{log} is not very well known, we can estimate λ using $K = T_D/1.45$. For X = oxygen $T_c = 93$, and $T_D = 812$ (ref. 16). Calculated values for λ are 1.38, 1.89, 2.09, 2.45 and 2.81 for $\mu^* = 0$, 0.10, 0.13, 0.18 and 0.22, respectively. Thus, λ is estimated to be between 2 and 3. Both α^2 and $\langle g^2\rangle$ can be evaluated once phonon spectra become available.

One may, however, follow a different approach noting that the Hamiltonian (17) can be diagonalized exactly following the method of Huang and Rys[5,12]. The expression for diagonalized energy E is given by

$$E_\lambda(n,k) = \varepsilon_\lambda - \Delta_\lambda + \sum_{\vec{k}} \hbar\omega(\vec{k})\ [n_k + \tfrac{1}{2}] \tag{24}$$

with

$$\Delta_\lambda = \sum_R |v(\vec{k},\lambda)|^2/\hbar\omega(\vec{k}) \tag{25}$$

$v(\vec{k},\lambda)$ could take various forms, including the matrix element of $(\varepsilon.\nabla V)$. We can, however, make a simple estimation of the electron-phonon coupling strength by setting $v(\vec{k},\lambda) = g\ \hbar\omega(k)$. In this case[28]

$$\Delta_\lambda = g_\lambda^2 \sum_k \hbar\omega(k) \quad g_\lambda^2\ k\ T_D \tag{26}$$

Using the observed values of $\Delta \cong 32$ meV[15] and 50 meV[17] and a Debye temperature[16] $T_D = 812°$ K, we get $g^2 \cong 0.46$ and 0.71, respectively.

Both (23) and (25) in conjunction with (12) imply that there is, in principle, no limit of T_c. The key to high T_c is strong electron-phonon coupling as suggested by Fröhlich[4].

Since the model should be valid for type I and II superconductors, we can calculate g^2 for those cases. In Table 3, we have compared effective coupling strength g^2 for type I, II and III superconductors. The electron-phonon coupling in high T_c superconductors is two orders of magnitude larger than that in type I superconductors. It is about an order of magnitude larger than that in type II superconductors.

Table 3. Coupling strength g^2 in some type I, II and III superconductors.

Substance →	Al	Re	Sn	Nb	Y-Ba-Cu-0
T_c in °K	1.16	1.69	3.72	9.22	93.00
k x debye Temperature (meV)	36.9	35.8	17.20	23.90	70.06
g^2	0.003	0.004	0.019	0.037	0.46 to 0.71

4. CONCLUDING REMARKS

Although in this analysis, particularly in section 3, one has mentioned electron-phonon coupling, the theory outlined in section 2 does not restrict superconductivity to effective pairing caused by electron-phonon coupling case only. IN FACT, ANY MECHANISM THAT MAKES EFFECTIVE INTERACTION (2) ATTRACTIVE AND ALLOWS TO FORM SCHAFROTH'S PAIR COULD GIVE RISE TO SUPERCONDUCTIVITY. A coupled electron pair in a singlet state and with opposite momenta is usually the lowest energy state in many potential wells. Hence, it is natural to use this coupling scheme along with the coherence state condition for superconducting states. One can form such a state in many ways e.g., by surrounding a pair of electrons by positively charge ions. Given the right condition, that system could also exhibit superconducting behavior, even though no electron-phonon coupling is directly involved. There is no upper limit of T_c and a strong electron-phonon or electron-environment interaction is a necessary requirement for high T_c, as noted by Fröhlich[4]).

REFERENCES

1. P. Haapakosi and F. B. Malik, Condensed Matter Theories, ed. J. Keller (Plenum Publishing Corp. 1989) Vol. 4., p. 311.
2. P. Haapakoski, A. Kallio and F. B. Malik, Phys. Lett A. (submitted).
3. Yu. A. Osip'yan, O. V. Zharikov, G. V. Novikov, N. S. Sidorov, V. I. Kulakov, L. V. Sipavina, R. K. Nikolaer and A. M. Gromov, Pis'ma Zh. Eksp. Teor. Fiz 49, 61 (1989) [Eng. Trans. 49, 73 (1989)].
4. H. Fröhlich, Proc. Roy. Soc. (Lond.) A215, 291 (1952).
5. K. Huang and A. Rys, Proc. Roy. soc. (Lond.) A204, 406 (1950).
6. M. R. Schafroth, Phys. Rev. 96, 1442 (1954).
7. J. Bardeen, L.N. Cooper and J. R. Schrieffer, Phys. Rev. 108, 1175 (1957).
8. B. Mühlschlegel, Z. Physkik 155, 313 (1959).
9. J. M. Blatt, Theory of Superconductivity (Academic Pr4ss, New York 1964).
10. R. J. Cana et al. Phys. Rev. Lett. 58, 1676 (1987).
11. G. M. Eliashberg, Zh. Eksp. Teor. Fiz., 38, 966 (1960) (Eng. JETP 11, 696 (1960)) and 39, 1437 (1960) (Eng. JETP 12, 1000 (1961)).
12. C. B. Duke and G. D. Mahan, Phys. Rev. 139A, 1965 (1965).
13. P. B. Allen and R. C. Dynes, Phys. Rev. B 12, 905 (1975).
14. W. L. McMillan, Phys. Rev. 167, 331 (1968).
15. Z. Schlesinger, R. T. Collins, D. L. Kaiser and F. Holtzberg, Phys. Rev. Lett. 59, 1958 (1987).
16. M. De Llano, Private Communication.
17. M. D. Kirk et al. Phys. Rev. B 35, 8850 (1987).
18. Robert C. Weast ed., Handbook of Chemistry and Physics (Chemical Rubber Co., Cleveland. 60th edition 1979).
19. M. K. Wu et al. Phys. Rev. Lett. 58, 908 (1987).
20. R. N. Bhargara, S. P. Herko and W. N. Osborne, Phys. Rev. Lett. 59, 1468 (1987).
21. X. R. Meng et al. Solid State Comm. 64, 325 (1987).
22. P. K. Davies et al. Solid Sate comm. 64 1441 (1987).
23. I. Felner et al. Phys. Rev. B 36, 3923 (1987).
24. J. R. Kirtley et al. Phys. Rev. B 35, 8846 (1987).
25. Z. Schlesinger, R. T. Collins, D. L. Kaiser and F. Holtzberg, Phys. Rev. Lett. 59 (1987).
26. M. Kirk et al. Phys. Rev. B 35, 8850 (1987).
27. R. J. Cara et. al. Phys. Rev. Lett. 58, 1676 (1987).
C. B. Duke, F. B. Malik and F. W. K. Firk, Phys. Rev. 157, 879 (1987).

CORRELATED SPIN-DENSITY-WAVE THEORY

X. Q. Wang,[1)] S. Fantoni,[2)] E. Tosatti,[1,3)] and Lu Yu[3)]

[1)]International School for Advanced Studies
34014 Trieste, Italy
[2)]Department of Physics, University of Lecce
73100 Lecce, Italy
[3)]International Center for Theoretical Physics
34100 Trieste, Italy

Abstract

An Fermi Hypernetted Chain (FHNC) method is devised to calculate the pair distribution function and the one body density matrix of a correlated spin-density-wave (SDW) model on a lattice. Preliminary results for the 1D Hubbard model at half filling are presented and discussed.

1. INTRODUCTION

The Hubbard model,[1] given by

$$H = -t \sum_{<i,j>,\sigma} c^{\dagger}_{i\sigma} c_{j\sigma} + U \sum_{i} n_{i\uparrow} n_{i\downarrow}, \tag{1}$$

where i, j are sites of a D-dimensional lattice, t is the electron hopping energy, and U is the on-site repulsion, is a well-known homework problem of strongly interacting electrons. More recently, interest in the Hubbard model has been revived by Anderson's suggestion[2] that non-Fermi liquid behavior away from half filling might be at the origin of high T_c superconductivity. In $D = 1$ dimension, the Hubbard model has been exactly solved by Lieb and Wu.[3] However, the underlying physics and the nature of strong electron correlations remain still an open question. Recently developed methods of numerical solutions[4,5] and variational calculations[6-9] have been proven to be extremely powerful in the study of the Hubbard model. The variational approaches, in particular, have been very useful to understand the role played by pairing correlations [as postulated by the resonating-valence-bond state of Anderson] and correlations between empty and doubly occupied sites.[10]

Expectation values of the ground state energy and other physical quantities of interest can be calculated by the Monte Carlo or FHNC method.[11,12] The latter calculation can be done at the outset in the thermodynamic limit. Therefore it is not faced with any finite size problem, which instead hamper the stochastic approaches. Moreover, working with higher dimensions is not a serious problem in FHNC, in contrast with Monte Carlo based techniques. Although the FHNC diagrammatic expansion at present cannot be expressed in a fully closed form (which makes this approach intrinsically approximate), it has been proved to be very accurate for continuous models such as nuclear matter and liquid ^{3}He,[13,14] and more recently also for discrete lattice models as the Hubbard model.[9]

The variational ground state is taken of the form

$$|\Psi >= G|\Phi >, \tag{2}$$

where $|\Phi >$ is a reference state and G is a correlation operator which builds the important correlations in that state. However, such correlations cannot be so strong as to change the symmetries and the character of $|\Phi >$ in order for the FHNC scheme to converge. In particular if $|\Phi >$ has a Fermi surface the approximate correlations will invariably conserve a remnant of the Fermi liquid character. Such a character however is known to be incorrect at least in $1D$ Hubbard model.[2,15] It is therefore necessary to consider reference states like $|\Phi >= |BCS >$, or $|\Phi >= |SDW >$, where the Fermi-liquid character is not present at all, and compare the equation of state, the momentum distribution function $n(\mathbf{k})$ and the excitation spectrum $e(\mathbf{k})$ obtained in the various correlated models.

The FHNC formalism to treat the correlated BCS has already been developed[17] and its application to the 1D Hubbard model has been recently presented.[9] In this contribution we focus our attention to a band antiferromagnetic ground state with spatial periodicity π/k_F, which is referred to as SDW state. In this case the Fermi surface is destroyed by single particle Bragg scattering. In section II we formulate the FHNC treatment to be used in connection with the correlated SDW model, in the case of the Jastrow ansatz $G = \prod_{i<j=1}^{A} f(\mathbf{r}_{ij})$. Preliminary results of the ground state energy and $n(\mathbf{k})$ in the 1D case are presented and discussed in the last section.

2. MODEL AND FHNC TREATMENT

The reference state $|\Phi_{SDW} >$ of the correlated SDW function is a Slater determinant of single particle wave functions[6,16]

$$\varphi_{k\sigma}(\mathbf{r}_i) = (u(\mathbf{k})e^{i\mathbf{k}\mathbf{r}_i} + sgn(\sigma)v(\mathbf{k})e^{i(\mathbf{k}+\mathbf{Q})\mathbf{r}_i})\xi_\sigma(i), \qquad k < k_F \tag{3}$$

with the normalization condition $u^2(\mathbf{k}) + v^2(\mathbf{k}) = 1$. sgn($\sigma$) takes either the value +1 or -1 according to whether $\sigma =\uparrow$ or $\downarrow$, and $\mathbf{Q}$ is the wavevector characterizing the extra periodicity due to the antiferromagnetic ordering. In principle, $\mathbf{Q}$ should be determined variationally. Here, we choose a simple two-sublattice commensurate antiferromagnetic structure. $\mathbf{Q}$ is then defined by the condition that $e^{i\mathbf{Q}\mathbf{t}} = -1$ for all translations $\mathbf{t}$ which transform a sublattice into the other. Thus $\mathbf{Q}$ is π for 1D chain and (π, π) for $2D$ square lattice.

The orthogonality condition requires that for $k > k_F$

$$\varphi = (v(\mathbf{k})e^{i\mathbf{k}\mathbf{r}_i} - sgn(\sigma)u(\mathbf{k})e^{i(\mathbf{k}+\mathbf{Q})\mathbf{r}_i})\xi_\sigma(i), \tag{4}$$

so that the single particle spectrum has a gap 2Δ at $\mathbf{k} = \mathbf{k}_0 = \pm\mathbf{Q}/2$.

The function $u(\mathbf{k})$ has to be determined variationally. In the uncorrelated model ($f = 1$) one has $u_k = \{[1 + \epsilon_k/E_k]/2\}^{1/2}$, $E_k = (\epsilon_k^2 + \Delta^2)^{1/2}$, where $\epsilon_k = -2t\cos ka$ (for 1D) and $\epsilon_k = -2t(\cos k_x a + \cos k_y a)$ (for the 2D square lattice). At half filling, the single particle spectrum is given by[16] E_k for $k < k_F$ and $-E_k$ for $k > k_F$. The gap Δ turns out to be given by

$$\Delta = -Um/2, \tag{5}$$

where m is the spin polarization per particle in the z direction. In our correlated model we keep $u(\mathbf{k})$ of the same form as above and consider Δ as a variational parameter. The ordinary paramagnetic correlated ground state is a special case of this function, with $\Delta = 0$.

The expectation value of the hamiltonian is given by[8]

$$< H > /N = -2t\rho n(a) + U\rho^2 g(0)/2, \tag{6}$$

where N are the lattice points separated by a and $\rho = A/N$. The functions $g(\mathbf{r}_{ij})$ and $n(\mathbf{r}_{ij})$ are the pair distribution function and the one body density matrix respectively.

The explicit derivations of FHNC equations for $n(\mathbf{r})$ and $g(\mathbf{r})$ are presented in Ref. 18 and will not be repeated here. The FHNC diagrams are built in terms of the dynamical correlations $h = f^2 - 1$ and $\xi = f - 1$ and of the uncorrelated density matrix

$$\begin{aligned} \rho(1,2) &= \frac{1}{N}\sum_{k,\sigma} \varphi^*_{k\sigma}(1)\varphi_{k\sigma}(2) \\ &= \sum_{\sigma} \xi^*_{\sigma}(1)\xi_{\sigma}(2)[(l_u(\mathbf{r}_{12}) + e^{i\mathbf{Q}\mathbf{r}_1} l_v(\mathbf{r}_{12}) sgn(\sigma)] \\ &= \sum_{\sigma} \xi^*_{\sigma}(1)\xi_{\sigma}(2)\hat{\rho}_{\sigma}(\mathbf{r}_{12}), \end{aligned} \tag{7}$$

where

$$l_u = \frac{1}{N}\sum_{k<k_F} \{u^2(\mathbf{k})e^{-i\mathbf{k}\mathbf{r}_{12}} + v^2(\mathbf{k})e^{-i(\mathbf{k}+\mathbf{Q})\mathbf{r}_{12}}\}, \tag{8}$$

$$l_v = \frac{1}{N}\sum_{k<k_F} u(\mathbf{k})v(\mathbf{k})e^{-i\mathbf{k}\mathbf{r}_{12}}(1 + cos(\mathbf{Q}\mathbf{r}_{12})). \tag{9}$$

One can easily verify that $\rho = \rho(1,1) = A/N$. The FHNC equations for dd, de and ee functions have the conventional form, whereas, due to the phase factor $e^{i\mathbf{Q}\mathbf{r}_{ij}}$, it is convenient to consider ρ and consequently X_{cc}, N_{cc} as two component vectors: $\hat{Z} = (Z_u, Z_v)$. The components of $\hat{\rho}$ are given by l_u and l_v respectively. The convolution integral is then given by:

$$Z = (G|F), \tag{10}$$

$$\begin{aligned} Z_u &= \rho\int d\mathbf{r}_3 F_u(\mathbf{r}_{13})G_u(\mathbf{r}_{32}) + \rho\int d\mathbf{r}_3 F_v(\mathbf{r}_{13})cos(\mathbf{Q}\mathbf{r}_{13})G_v(\mathbf{r}_{32}), \\ Z_v &= \rho\int d\mathbf{r}_3 F_v(\mathbf{r}_{13})G_u(\mathbf{r}_{32}) + \rho\int d\mathbf{r}_3 F_u(\mathbf{r}_{13})cos(\mathbf{Q}\mathbf{r}_{13})G_v(\mathbf{r}_{32}). \end{aligned} \tag{11}$$

which can be put into a convenient matrix form when Z_u and Z_v are calculated in Fourier space[18].

The product between vector quantities is a scalar quantity given by

$$\hat{F}_{12} \bullet \hat{G}_{12} = Z_{12} = 2F_u(12)G_u(12) + 2cos(\mathbf{Q}\mathbf{r}_{12})F_u(12)G_v(12), \tag{12}$$

where the factor 2 is due to spin summation. The parallel connection of a scalar quantity with a vector one is a vector quantity $Z(12) = F(\mathbf{r}_{12})\hat{G}(1,2) = (F(\mathbf{r}_{12})G_u(\mathbf{r}_{12}), F(\mathbf{r}_{12})G_v(\mathbf{r}_{12}))$. In doing the cc-type chain summation one has to keep track of the ordering of various cc-elements, due to the asymmetrical structure of the convolution integral (11). It follows that the cancellation process amongst cc-diagrams which is

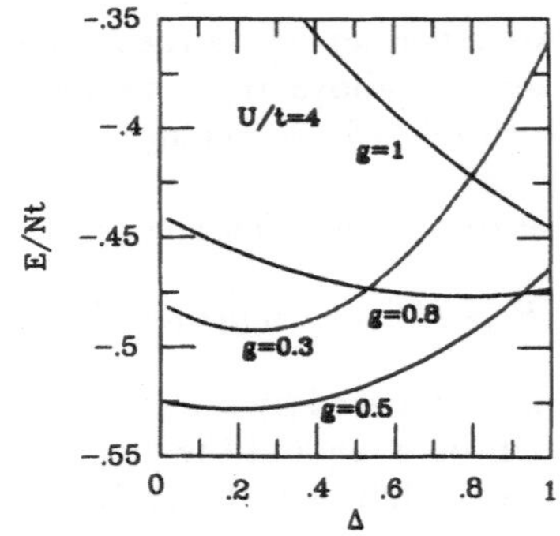

Figure 1. Energy expectation values of the correlated SDW model for the 1D Hubbard hamiltonian as a function of the two variational parameters Δ and g.

present in the standard FHNC theory, is destroyed here, so that the FHNC equation for $\hat{N}_{cc}$ results to be[18]

$$\hat{N}_{cc} = (\hat{X}_{cc} + \hat{N}_{cc} - \hat{\rho}|\hat{X}_{cc} - (\hat{X}_{cc}|\hat{\rho})) - (\hat{X}_{cc}|\hat{\rho}). \tag{13}$$

This equation reduces to the conventional FHNC equation when $l_v = 0$. The equations for $X_{\alpha\beta}$ totally coincide with the conventional ones,[9] with the *proviso* of correctly calculating the products between vector quantities (like for instance $\hat{N}_{cc} \bullet \hat{N}_{cc}$ in X_{ee}) and of a scalar with a vector quantity.

Similarly, in the calculation of the density matrix the nodal equations which sum the diagrams having an open exchange loop take care of the ordering discussed above.

3. RESULTS AND DISCUSSION

We present results of FHNC/0 calculations for the 1D chain. The correlation function $f(\mathbf{r}_{ij})$ has been taken of the following simple form

$$f(x_{ij}) = \begin{cases} g & x_{ij} = 0, \\ g_1 & |x_{ij}| = a, \\ 1 & otherwise. \end{cases} \tag{14}$$

In the case of $g_1 = 1$, $\prod_{i<j} f(\mathbf{r}_{ij})$ corresponds to the Gutzwiller correlation operator. Fig. 1 shows the results of $< H > /tN$ for various values of the variational parameters Δ and g at $U/t = 4$. The optimal value of g_1 has been always found around 1. Values of $g_1 \approx 0.95 \sim 1$ lower the energy upper-bound only of a few parts per thousand.

The correlated SDW obviously includes the correlated paramagnetic ($\Delta = 0$) and the uncorrelated SDW ($g = g_1 = 1$) models. Results for the last two models are represented by the curve labelled $g = 1$ and that on the vertical axis ($\Delta = 0$), respectively. The lowest energy at $U/t = 4$ is found to be -0.53 at $\Delta = 0.24$ and $g = 0.5$ in quite a good agreement with the variational Monte Carlo results obtained in Ref. 6, $E/Nt = -0.54$ and $\Delta = 0.3$. The convergence of the FHNC treatment is fast even

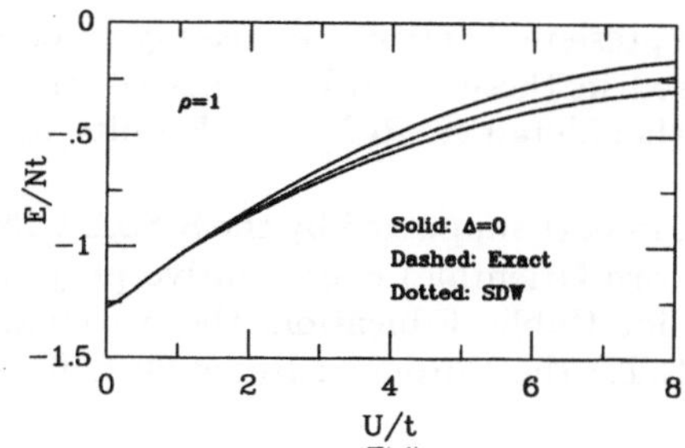

Figure 2. Energy expectation values as a function of U/t. The results of the correlated paramagnetic model[6,9] (upper curve) and those of correlated SDW model (middle curve) are compared with the exact results of Lieb and Wu (lower curve).

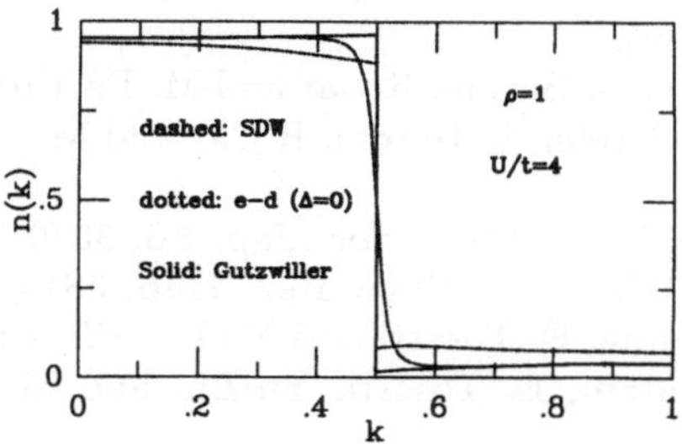

Figure 3. Momentum distribution functions for the correlated SDW model ($\Delta = 0.24$, $g = 0.5$).

at large value of U/t. The contribution from bridge diagrams in the case considered has always been found to be less than 5%. The optimal values of the parameter Δ are considerably lower than those in the uncorrelated model ($g = 1$). Such a feature implies that the spin polarization is reduced when dynamical correlations are taken into account. A detailed analysis of these results will be given elsewhere.[18]

In Fig. 2 the energy of the correlated SDW state is represented as a function of U/t and compared with exact results and the results of the correlated paramagnetic model.

The momentum distribution function at $U/t = 4$ is shown in Fig.3 and compared with that obtained with the correlated paramagnetic model with and without $e - d$ correlations.[9] One can see that, except for the expected change at $k = k_F$, the correlated SDW $n(\mathbf{k})$ has a shape similar to that of the correlated paramagnetic $n(\mathbf{k})$. This clearly indicates that the effect of "$e - d$" correlations is only partly included (at the mean-field level) in the present correlated SDW state.

Interesting quantities to be calculated are the single particle spectrum, the spin polarization in the z direction and the magnetic structure function. They can be obtained within our theory and work in this direction is in progress.

The FHNC method presented here can be readily extended to higher dimensions. The equations to be used in these cases have exactly the same structure as those described here and further detail in Ref. 18. Results of such calculation will be presented elsewhere.[18]

This work has been in part supported by the SISSA-CINECA (Centro di Calcolo Elettronico dell'Italia Nord-Orientale) collaborative project, under the sponsorship of the Italian Ministry for Public Education, the National Institute for Physics of Matter (INFM), and (E.T.) the European Research Office, U.S. Army, contract # 6280-PH-01.

REFERENCES

1. J. Hubbard, *Proc. Roy. Soc.* **A276**, 238 (1964); **A285**, 542 (1965).
2. P. W. Anderson, *Science* **235**, 1196 (1987).
3. E. H. Lieb and F. Y. Wu, *Phys. Rev. Lett.* **20**, 1445 (1968).
4. J.E. Hirsh, *Phys. Rev.* **B31**, 4403 (1985); J.E. Hirsh and H.Q. Lin, *Phys. Rev. Lett.* **31**, 4403 (1989);
5. S. Sorella, E. Tosatti, S. Baroni, R.Car and M. Parrinello, *Intern. J. Mod. Phys.* **B1**, 993 (1988); S. Sorella, S. Baroni, R.Car and M. Parrinello, *Europhys. Lett.* **8**, 663 (1989);
6. H. Yokoyama, H.Shiba, *J. Phys. Soc. Jap.* **56**, 3570 (1987); 1490 (1987);
7. C. Gros, R.Joynt, T.M. Rice, *Phys. Rev.* **B36**, 381 (1987).
8. S. Fantoni, X.Q. Wang, E. Tosatti and Yu Lu, *Physica C* **153-155**, 1255(1988).
9. X.Q. Wang, S. Fantoni, E. Tosatti, Yu Lu, and M. Viviani, *Phys. Rev.* **B**, (1990) in press.
10. T.A. Kaplan, P. Horsh and P. Fulde, *Phys. Rev. Lett.* **49**, 889 (1982); P. Fazekas and K. Penc, *Intern. J. Mod. Phys.* **B1**, 1021 (1988).
11. S. Fantoni and S. Rosati, *Nuovo Cim.* **A25**, 593 (1975).
12. E. Krotscheck and M.L. Ristig, *Nucl. Phys.* **A242**, 389 (1975).
13. J.W. Clark, in *Progress in Particle and Nuclear Physics*, ed. D.H. Wilkinson (Pergamon, Oxford, 1979), Vol.2.
14. M. Viviani, E. Buendia, S. Fantoni, S. Rosati, *Phys. Rev.* **B38**, 4523, (1988) and references therein.
15. A. Parola, S. Sorrela, E.Tosatti, M. Parrinello (to be published).
16. J.R. Schrieffer, X.G. Wen, and S.C. Zhang, *Phys. Rev.* **B39**, 11663 (1989).
17. S.Fantoni, *Nucl. Phys.* **A363**, 381 (1981).
18. X.Q. Wang, S. Fantoni, E. Tosatti, and Yu Lu, in *Recent Progress in Many Body Theory*, Vol.2. Y. Avishai *et al.* eds. Plenum Press, London.

Effective Dielectric Response of Composites: A New Diagrammatic Approach

Rubén G. Barrera^{+}, **Cecilia Noguez**$^{+*}$ **and Enrique V. Anda**$^{++}$

$^{+}$*Instituto de Física, UNAM, México, D.F., México*
**Facultad de Ciencias, UNAM, México, D.F., México*
$^{++}$*Instituto de Física, UFF, Niterói, Brazil*

INTRODUCTION

The interest in the dielectric response of composites has been renewed by the application of a wide variety of mathematical techniques borrowed from other fields of physics. The use of these materials as selective absorbers in solar energy devices [1] and the study of fluids in rocks and porous materials for oil exploration [2], has also contributed to the revival of the actual research in composites. It has been recognized that the topology of the composite plays a crucial role in the response of the system to an external perturbation [3]. Here we treat the dielectric response of a system composed by spherical inclusions located at random in an otherwise homogeneous matrix. Although the problem was posed more than a century ago [4] only until recently, theories beyond the mean field approximation started to be developed. Multiple scattering theory [5], cluster expansions [6], lattice gas models [7], numerical simulations [8], homogenization theory [9], renormalization [10] and diagrammatic techniques [11] have been the main ingredients of the recently developed theories. Comparison with experiment has been troublesome because the experiments have been done in samples with a poorly characterized microstructure. Also, the generalization of the theories applicable to models which described better the actual experimental conditions, is not straightforward. This situation requires theoretical work in two different directions. First, the development of a theory which retains the main aspects of the problem but that is simple enough to be extendable to more complicated situations *i.e.* spheres with a given distribution of radii, the inclusion of multipolar interactions, the effects of clustering and dimensionality and systems of particles with different shapes. On the other hand, it is also necessary to develop, even in the simplest model, a systematic approach which allows to make

Condensed Matter Theories, Volume 5
Edited by V.C. Aguilera-Navarro
Plenum Press, New York, 1990

an adequate comparison of the different types of calculations and which generates a scheme for obtaining better and better approximations. Here we will present a new diagrammatic approach for the calculation of the effective dielectric response of a composite which encompasses, in a certain way, both of these directions. It can also be viewed as a reformulation of the diagrammatic approach developed in Ref. 11

Formalism

Lets consider an homogeneous an isotropic ensemble of $N \gg 1$ identical spheres, with radius a_o and dielectric functions ϵ_s, located at random positions $\{\mathbf{R}_i\}$ within a homogeneous matrix with dielectric function ϵ_h. The system is in the presence of a position dependent external field $\mathbf{E}^{ex}(\mathbf{r},\omega)$ oscillating at frequency ω and with wavelength much larger than a_o and the typical separation between spheres. The dipolar moment $\mathbf{p}_i$ induced at the i-th sphere is then given by

$$\mathbf{p}_i = \alpha[\mathbf{E}_o(\mathbf{R}_i) + \sum \overleftrightarrow{\mathbf{t}}_{ij} \cdot \mathbf{p}_j] \tag{1a}$$

where $\mathbf{E}_o$ is the electric field in the absence of the spheres,

$$\alpha = a_o^3(\epsilon_s - \epsilon_h)/(\epsilon_s + 2\epsilon_h) \tag{1b}$$

is the effective polarizability of each sphere,

$$\overleftrightarrow{\mathbf{t}}_{ij} = (1 - \delta_{ij})\nabla_i\nabla_j(1/R_{ij}) \tag{1c}$$

is the dipole-dipole interaction tensor in the quasi-static approximation and we omit the explicit dependence on ω.

The induced average polarization per unit volume (polarization field)

$$\langle\sum_i \mathbf{p}_i\delta(\mathbf{r} - \mathbf{R}_i)\rangle \equiv n\mathbf{p}(\mathbf{r}) \tag{2}$$

is related to the macroscopic (or effective) dielectric response of the system ϵ_M by [10]

$$\frac{\epsilon_h(\omega)}{\epsilon_M(\omega)} = 1 - 4\pi\epsilon_h(\omega)\chi^{ex,\ell}(\mathbf{q} \to 0, \omega) \tag{3}$$

or by

$$\frac{\epsilon_M(\omega)}{\epsilon_h(\omega)} = 1 + 4\pi\chi^{ex,t}(\mathbf{q} \to 0, \omega) \tag{4}$$

where n is the number density of spheres, $\langle ... \rangle$ means ensemble average and $\overleftrightarrow{\chi}^{ex}(q,\omega)$ is the Fourier transform of the external susceptibility, defined through

$$n\mathbf{p}(\mathbf{q},\omega) = \overleftrightarrow{\chi}^{ex}(\mathbf{q},\omega) \cdot \mathbf{E}^{ex}(\mathbf{q},\omega). \tag{5}$$

Here $\mathbf{p}(\mathbf{q})$ is the spatial Fourier transform of $\mathbf{p}(\mathbf{r})$ and the superscripts ℓ and t mean longitudinal and transverse projections, respectively. We have used the fact that the $\mathbf{q} \to 0$ limit of ϵ_M^ℓ and ϵ_M^t of a system which is homogeneous, isotropic and invariant under inversions, are identical.

For the purpose of calculation we choose Eq. (3). Furthermore, since there is no macroscopic coupling between transverse and longitudinal fields, due to the symmetry properties of the ensemble, it is sufficient to consider only a single Fourier longitudinal component of the external field, that is

$$\mathbf{E}^{ex}(\mathbf{r}) = \hat{\mathbf{q}} E^{ex} e^{i\mathbf{q}\cdot\mathbf{r}}, \tag{6}$$

where $\mathbf{q}$ is the wavevector, $\hat{\mathbf{q}} = \mathbf{q}/q$ and the explicit dependence on ω has been omitted. Then by substituting Eq. (6) into Eq. (1a) one obtains

$$\mathbf{P}_i(\mathbf{q}) = \alpha[\hat{\mathbf{q}} E^{ex}/\epsilon_h + \sum_j \overleftrightarrow{\mathbf{T}}_{ij}(\mathbf{q}) \cdot \mathbf{P}_j(\mathbf{q})] \tag{7a}$$

where

$$\mathbf{P}_i(\mathbf{q}) = \mathbf{p}_i e^{-i\mathbf{q}\cdot\mathbf{R}_i}, \tag{7b}$$

$$\overleftrightarrow{\mathbf{T}}_{ij}(\mathbf{q}) = \overleftrightarrow{\mathbf{t}}_{ij} e^{-i\mathbf{q}\cdot(\mathbf{R}_i - \mathbf{R}_j)}, \tag{7c}$$

have been defined only in order to remove the trivial exponential factors. The Fourier transform of the polarization field is given by

$$\mathbf{p}(\mathbf{q}) = \langle \mathbf{P}_i(q) \rangle \equiv \langle \mathbf{P} \rangle \tag{8}$$

where we have assumed that the volume and ensemble averages are identical. Also, the ensemble average of $\mathbf{P}_i$ is independent of i due to the translational symmetry of the ensemble.

We now add and substract in the rhs of Eq. (7a) the term

$$\sum_j \overleftrightarrow{\mathbf{T}}_{ij} \cdot \langle \mathbf{P}_j \rangle \equiv N \langle \mathbf{T} \rangle \cdot \langle \mathbf{P} \rangle \text{and write}$$

$$\mathbf{P}_i = \alpha[\mathbf{E}_L + \sum_j \Delta\overleftrightarrow{\mathbf{T}}_{ij} \cdot \mathbf{P}_j], \tag{9a}$$

where

$$\mathbf{E}_L \equiv \hat{\mathbf{q}}E^{ex}/\epsilon_h + N\langle\mathbf{T}\rangle \cdot \langle\mathbf{P}\rangle \tag{9b}$$

is called the Lorentz field,

$$\Delta\overleftrightarrow{\mathbf{T}}_{ij} \equiv \overleftrightarrow{\mathbf{T}}_{ij} - \langle\overleftrightarrow{\mathbf{T}}\rangle, \tag{9c}$$

the relation $\langle\overleftrightarrow{\mathbf{T}}\rangle = (1/N)\sum_j \overleftrightarrow{\mathbf{T}}_{ij}$ has been used and the explicit dependence of $\mathbf{q}$ has been omitted. Here N is the total number of spheres.

The formal solution of Eq. (7) is immediately given by

$$\mathbf{P}_i = \alpha \sum_j (\overleftrightarrow{\mathbf{V}}^{-1})_{ij} \cdot \mathbf{E}^L, \tag{10a}$$

where $(\overleftrightarrow{\mathbf{V}}^{-1})_{ij}$ is the ij-th element of the inverse operator $\overleftrightarrow{\mathbf{V}}$, whose elements are defined by

$$\overleftrightarrow{\mathbf{V}}_{ij} = \overleftrightarrow{\mathbf{1}}\delta_{ij} - \alpha\Delta\overleftrightarrow{\mathbf{T}}_{ij}. \tag{10b}$$

We now take ensemble average and longitudinal projection of Eq. (10a) in order to obtain

$$n\langle P\rangle^\ell = n\alpha\langle\sum_j (V^{-1})^\ell_{ij}\rangle E^\ell_L \equiv \chi^{L,\ell}E^\ell_L, \tag{11}$$

where $\overleftrightarrow{\chi}^L$ is called the Lorentz susceptibility. It can be shown [10] that

$$\overleftrightarrow{\mathbf{E}}_L = \hat{\mathbf{q}}(E^{ex}/\epsilon_h - \frac{8\pi}{3}n\langle\mathbf{P}\rangle), \tag{12}$$

and using this result it can be easily proved that

$$\chi^{ex,\ell}(\mathbf{q},\omega) = \frac{\chi^{L,\ell}(\mathbf{q},\omega)}{1 + \frac{8\pi}{3}\chi^{L,\ell}(\mathbf{q},\omega)}. \tag{13}$$

We define a renormalized polarizability α^* as

$$n\alpha^*(\omega) \equiv \chi^{L,\ell}(\mathbf{q} \to 0, \omega) \tag{14}$$

and substituting Eqs. (13) and (14) into Eq. (3) we finally get

$$\frac{\epsilon_M - \epsilon_h}{\epsilon_M + 2\epsilon_h} = f\hat{\alpha}^*, \tag{15}$$

where $\hat{\alpha}^* \equiv \hat{\alpha}/a_o^3$ and $f \equiv n4\pi a_o^3/3$ is the volume fraction occupied by the spheres. We want to emphasize that Eq. (15) is an it exact expression. It has the same functional form as the Maxwell-Garnett mean field approximation (MGT) [12] but with a renormalized polarizability $\hat{\alpha}^*$ instead of the bare polarizability $\hat{\alpha} \equiv \alpha/a_o^3$.

We now use Eq. (11) and a series representation of its inverse in order to write

$$\sum_j (\overleftrightarrow{\mathbf{V}}^{-1})_{ij} = \overleftrightarrow{\mathbf{1}} + \alpha \sum_j \Delta\overleftrightarrow{\mathbf{T}}_{ij} + \alpha^2 \sum_{jk} \Delta\overleftrightarrow{\mathbf{T}}_{ik} \cdot \Delta\overleftrightarrow{\mathbf{T}}_{kj} + ... \tag{16}$$

Then we take the longitudinal projection and the ensemble average of Eq. (16) using the following simplifying assumption

$$\rho^{(m)}(\mathbf{R}_1, ..., \mathbf{R}_m) = \Pi_{i,j}\rho^{(2)}(R_{ij}), \tag{17}$$

where $\rho^{(m)}$ is the m-particle distribution function as defined in Ref. 11 and the product is over all the open sequential pairs. The result can be expressed in the language of diagrams through the definitions introduced in Ref. 11, *i.e.*

 $\equiv \lim_{\mathbf{q}\to 0} \int \hat{\mathbf{q}} \cdot \overleftrightarrow{\mathbf{T}}_{12} \cdot \overleftrightarrow{\mathbf{T}}_{21} \cdot \hat{q}\rho^{(2)}(R_{12}) d^3R_2.$ (18)

We obtain

$\xi \equiv \frac{\alpha^*}{\alpha} = \sum_r \sum_s L(r,s) =$ 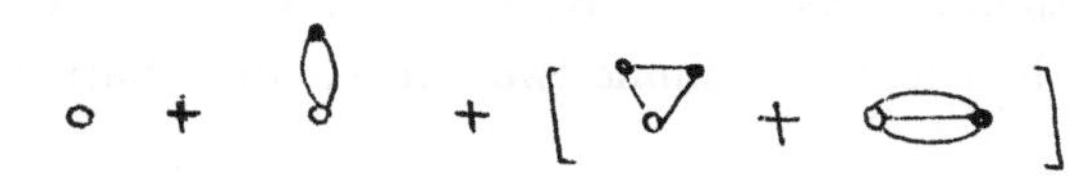

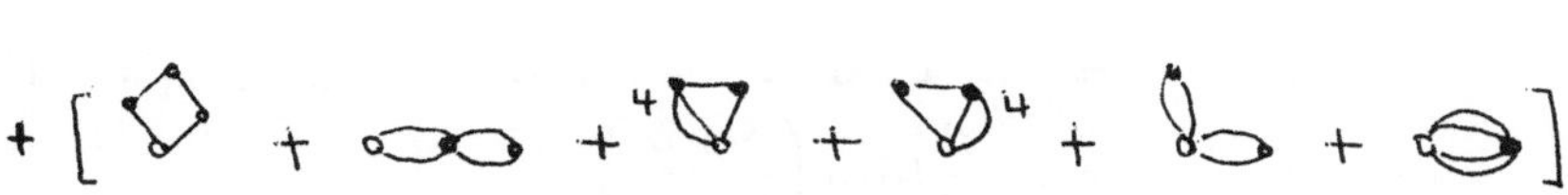 $+ \cdots$ (19)

where $L(r,s)$ are the sum of all renormalized graphs. That is all graphs with r lines and s black dots which can be drawn using the same rules given in Ref. 11 but omitting all graphs that can be disjoint into two separate graphs by cutting a single line between two dots. For example, the following graphs are omitted in $L(r,s)$:

 (20)

Lets recall that each graph is proportional to $\hat{\alpha}^r f^s$, thus the graphs in square brackets in Eq. (1a) correspond to an expansion in powers of $\hat{\alpha}$.

If now we take

(i)$\xi =$ ∘ $\equiv 1$ (21)

we recover MGT.

(ii)$\xi =$ 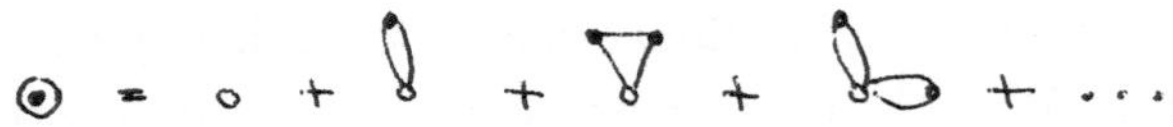 (22)

that is the sum of all simply-connected rings, we recover the results of Ref. 11. We recollect that a cruder approximation can be constructed by taking ξ as the solution of

 (23)

which yields $\hat{\alpha}^*$ in a simple analytical form [11]

$$\frac{\hat{\alpha}^*}{2} = 1 - \frac{\sqrt{1 - f_e\hat{\alpha}^2}}{f_e\hat{\alpha}} \tag{24}$$

This approximation has also an intuitive interpretation [10] which allows a straightforward generalization to system of spheres with a given distribution of radii [13] an also to more complicated systems.

Right now we are analyzing two new summations of specific classes of diagrams using the excluded volume two-particle distribution function:

(i)

$\xi =$  (25)

$$= 1 + \frac{2}{3} f \hat{\alpha} \ell n \left(\frac{8 + \hat{\alpha}}{8 - 2\hat{\alpha}} \right).$$

This expression can be considered a low density approximation because it takes the lowest number of black dots (one) for a given number of lines. It also agrees with Eq. (18) of Ref. 11 where a similar summation was also performed, and it should be now compared with some new results recently reported [14].

(ii) $\xi =$ 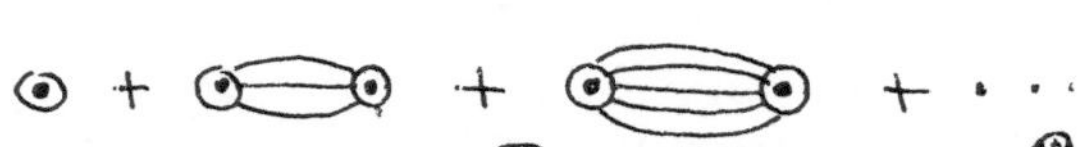 (26a)

$\Delta \equiv$ 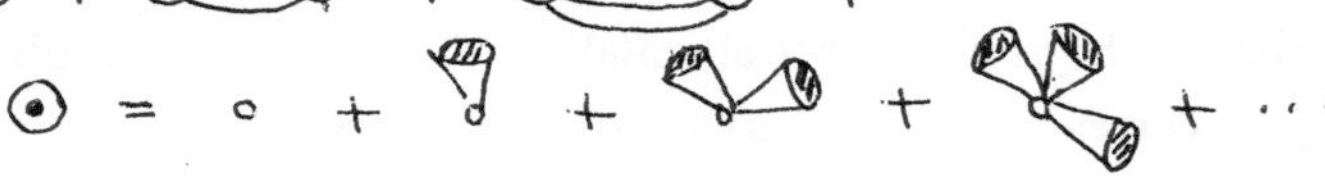 (26b)

(26c)

which leads to

$$\xi = \Delta + \frac{1}{3} f \hat{\alpha} \Delta^2 \left[\ell n \frac{64 - \hat{\alpha}^2 \Delta^2}{64 - 4\hat{\alpha}^2 \Delta^2} \right] \tag{27a}$$

where Δ is the self-consistent solution of the following equation:

$$\Delta = \left[1 - \frac{1}{3} f \hat{\alpha} \sqrt{\Delta} \ell n \frac{(4 + \hat{\alpha}\sqrt{\Delta})(8 + \hat{\alpha}\sqrt{\Delta})}{(4 - \hat{\alpha}\sqrt{\Delta})(8 - \hat{\alpha}\sqrt{\Delta})} \right]^{-1} \tag{27b}$$

If we now compare Eq. (26) with Eqs. (25) and (27) of Ref. 11, we can see that Eq. (31) of this same reference is obtained by taking $\xi = \Delta$. This approximation can be considered to be valid in the intermediate density regime, because the graphs included in Eq. (26b) have more black dots, for a given number of lines, than other possible graphs. Since in [Eq. (26)] we are also considering graphs of the type shown in Eq. (25), we expect that this new approximation should be good for densities in the whole range, from low to intermediate. The numerical results as well as the discussion of this new approximations will be reported elsewhere.

In conclusion, we have developed a new diagrammatic approach for the calculation of the effective dielectric response ϵ_M of a system of identical spheres embedded in a homogeneous matrix (within the dipolar long-wavelength approximation). We obtain an expression for ϵ_M which has the same analytical form as MGT but with a renormalized polarizability $\hat{\alpha}^*$ instead of the bare polarizability $\hat{\alpha}$. We showed that this renormalized polarizability can be expressed as an infinite sum of irreducible diagrams and previously reported results can be immediately derived as partial infinite summations of specific classes of diagrams. The main advantage of this new approach, as compared with the one developed in Ref. 11, is that the diagrammatic series for ξ contains only irreducible graphs. Finally we showed two specific ways of carrying out new type of summations applicable for low and intermediate concentrations.

Acknowledgements

Two of us (R.G.B. and E.V.A.) would like to acknowledge the kind hospitality of the Instituto de Fisica of Universidade Federal Fluminense (Brazil) and the Instituto de Física, Universidad Autonoma de México (México) where part of this work was done. Illuminating discussions with W. Luis Mochán are also acknowledged. This work was partially supported by CNPq (Brazil), FINEP (Brazil), TWAS (Trieste, Italy) and ICTP (Trieste, Italy).

References

1. See for example: A.J. Sievers in "Solar Energy Conversion: Topics in Applied Physics", Vol. 31 ed. by B.D. Seraphin Springer, Berlin 1979.
2. See for example: F. Browers, A. Ramsamugh and V.V. Dixit, *J. Mat. Sci.* **22** 2759 (1987).
3. See for example: J.M. Ziman, "Models of Disorder", Cambridge Univ. Press, N.Y. 1979.
4. J.C. Maxwell, "Treatise on Electricity and Magnetism", Vol. 1 (Reprint) Dover, N.Y. 1954. Sec. 314,p. 440.
5. L. Tsang and J.A. Kong, *J. Appl. Phys.* **53** 7162 (1982).
6. B.U. Felderhof and R.B. Jones, *Phys.* **B62** 231 (1986); **62** 225 (1986); **62** 215 (1986); **62** 43 (1985).
7. A. Liebsch and P. Villaseñor-González, *Phys. Rev.* **B29** 6907 (1984).
8. W. Lamb, D.M. Wood and N.W. Aschcroft. *Phys. Rev.* **B21** 2248 (1983); J.M. Gerardy and M. Ausloos, *Phys. Rev.* **B26** 4703 (1982).
9. D.J. Bergman, *J. Phys.* **C12** 4947 (1979); *Phys. Rev.* **B19** 2359 (1979); *Phys. Rep* **43** 337 (1978).
10. R.G. Barrera, G. Monsivais and W.L. Mochán, *Phys. Rev.* **B38** 5371 (1988).
11. R.G. Barrera, G. Monsivais, W.L. Mochán and E.V. Anda, *Phys. Rev.* **B39** 9998 (1989).
12. J.C. Maxwell Garnett, *Philos. Trans. R. Soc. London* **203** 3;85 (1904).
13. R.G. Barrera, P. Villaseñor-González, W.L. Mochán, M. del Castillo-Mussot and G. Monsivais, *Phys. Rev.* **39** 3522 (1989).
14. B. Cichocki and B.U. Felderhof, *J. Stat. Phys.* **43** 499 (1988); B.U. Felderhof and R.B. Jones, *Phys. Rev.* **B39** 5669 (1989).

THE TRAJECTORIES OF MAGNETIC FIELD LINES IN TOKAMAKS WITH HELICAL WINDINGS

I.L. Caldas, M.V.A.P. Heller, S.J. Camargo, M.C.R. Andrade

Instituto de Física, Universidade de São Paulo
C.P. 20.516
01498 - São Paulo - SP, Brazil

ABSTRACT: The distribution of the magnetic lines in a tokamak plasma perturbed by resonant helical windings is considered. A Hamiltonian perturbation theory is applied to investigate the break-up of magnetic surfaces inside the plasma for large aspect-ratio tokamaks. This effect is also investigated integrating numerically the line equations, obtaining a Poincaré mapping of these non-integrable fields in order to analyse the extent of satellite magnetic islands and chaotic regions. Average invariants are also obtained to describe approximately the field lines distribution inside the plasma.

1. INTRODUCTION

The concept of plasma confinement in tokamaks is based upon the idea that the magnetic field lines lie on a set of closed, nested, toroidal magnetic surfaces. For these axial symmetric configurations the existence of such surfaces is proved theoretically and verified experimentally[1]. However, these surfaces can be partially destroyed by resonant helical fields due to plasma oscillations[2] or external perturbations created by helical windings[3]. The superposition of an helical perturbation upon an equilibrium with axial symmetry brokes both symmetries and the perturbed field is no more integrable[4]. In tokamaks, the amplitudes of these perturbations are so small that the field can be considered an almost integrable system[5].

After the work on the Pulsator[3], resonant helical windings have been used in other tokamaks to generate helical perturbations. Experiments with these external coils showed that plasma oscillations could be controlled and the nature of disruptive instability could be investigated. In fact, there are experimental evidences that disruptive instabilities in tokamaks could be triggered by the overlapping of magnetic islands, inside the plasma, created by helical windings[2,3].

In this paper we report theoretical work that have been done at the University of São Paulo to investigate the field line trajectories of the perturbed fields and their topological properties in the tokamak TBR-1[6], which can be modified by the helical winding current, a relevant control parameter in the mentioned experiences. Typical parameters of this tokamak are used in the numerical applications.

Condensed Matter Theories, Volume 5
Edited by V.C. Aguilera-Navarro
Plenum Press, New York, 1990

2. ALMOST INTEGRABLE FIELD

In this section we consider a large aspect-ratio tokamak with circular cross section, represented by a periodical cylinder with length $2\pi R$, and assume the tokamak scaling.

$$B_{o\theta}/B_{oz} \approx na/(mR) << 1 \tag{1}$$

where R and a are respectively the major and minor plasma radii, m and n are the poloidal and toroidal wave numbers and $B_{o\theta}$ and B_{oz} are the equilibrium magnetic field components in cylindrical coordinates (Fig. 1). The magnetohydrodynamic (mhd) equilibrium is determined by the external uniform field B_{oz} and the plasma current density

$$\vec{j} = j_o\ (1 - r^2/a^2)^{\gamma}\ \hat{z} \tag{2}$$

where j_o and γ are constants[7]. In this case the equilibrium magnetic surfaces comprise a system of nested concentric cylinders and are characterized by the rotational transform of their lines

$$\iota(r) = 2\pi\ R\ B_{o\theta}/(r\ B_{oz}) \tag{3}$$

The helical perturbation on equilibrium depends on the coordinates r and $u = m\theta - nz/R$ and is created by currents I flowing in m pairs of helical windigs, equally spaced, with radius b wounded on a circular cylinder (corresponding to a large aspect-ratio tokamak). The ratio m/n characterizes the winding field helicity. We consider currents I flowing in opposite direction in adjacent windings (Fig. 2) . The field created by these windings is[7]:

$$\vec{B}_1\ (r,\ u) = \nabla\Phi \quad ,\ \Phi = (\mu_o I/\pi)(r/b)^m\ \sin u \tag{4}$$

In this article we consider the linear superposition of the unperturbed field $\vec{B}_o$ with the perturbation $\vec{B}_1$:

$$\vec{B}\ (r,\ u) = \vec{B}_o\ (r) + \vec{B}_1(r,\ u) \tag{5}$$

where $B_o/B_1 << 1$, since the current I on the windings is much lower than the plasma current I_p ($I/I_p \sim 1 \times 10^{-2}$). This approximation is not valid for marginally stable states, for which the plasma response should not be neglected.

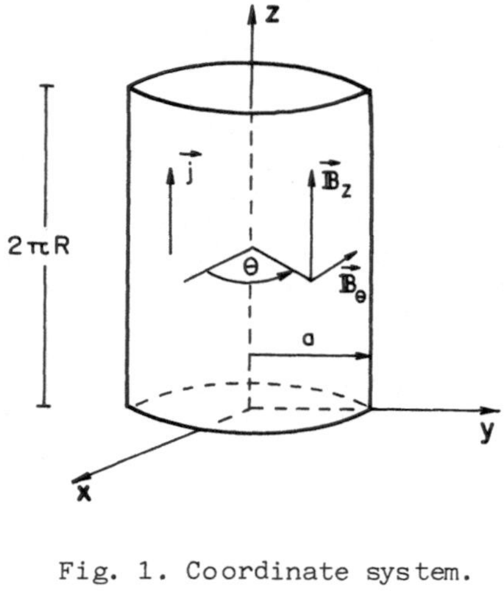

Fig. 1. Coordinate system.

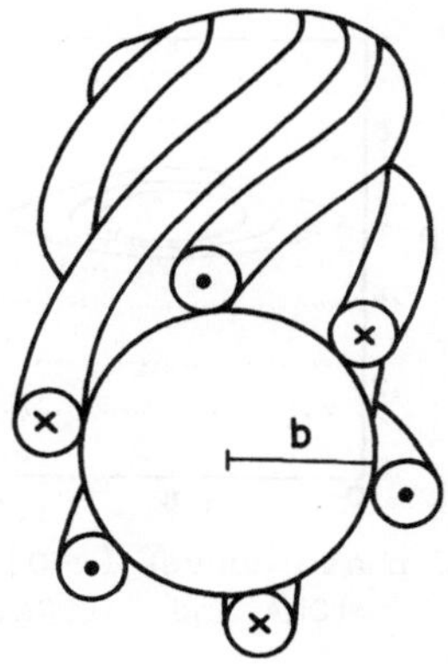

Fig. 2. Outline of the helical windings on the toroidal camera.

The equations for the field lines have the following form:

$$dr/B_r = rd\theta/B_\theta = dz/B_z \tag{6}$$

where the field components are known in terms of the coordinates r and u.

Introducing a new variable $\rho = r^2/2$, it is possible to show that eqs. (6) are equivalent to the following Hamilton equations:

$$\frac{\partial H}{\partial \rho} = -\frac{d\theta}{dt} \quad , \quad \frac{\partial H}{\partial \mu} = \frac{d\rho}{dt} \tag{7}$$

where H is the Hamiltonian and t the "time" given by

$$H = nB_{oz}\, r^2/(2R) - m \int B_{o\theta}\, dr + (\mu_o Im/\pi)(r/b)^m \cos u \ , \tag{8}$$

$$dz/dt = B_z\ (r,\ u) \tag{9}$$

ρ and u are the canonical variables and H is invariant along the field lines[8]

$$\vec{B}.\nabla H = 0 \tag{10}$$

The phase trajectories in the $(\rho,\ u)$ plane are determined by the equation

$$d\rho/du = -(\partial H/\partial u)/(\partial H/\partial \rho) \tag{11}$$

and are intersections of the magnetic surfaces with the plane t=const. Fig. 3 shows some trajectories for $\iota(a) = 2\pi/5$, $\iota(0) = 2\pi$ and I=100A. For low intensity helical currents the island width is proportional to $\sqrt{I}$ [7].

Using standard procedures, we can introduce action-angle variables(J,α)[5]

$$J = 1/2\pi \oint \rho du \qquad\qquad \alpha = \frac{\partial}{\partial J}\int_0^u \rho du' \tag{12}$$

In terms of these variables, the field line equations can be easily integrated, once we neglected the tokamak toroidal curvature. If we consider corrections on $\vec{B}_o$ due to this effect, the symmetry is broken and the field $\vec{B}$ (r, θ, u) is not integrable anymore. This correction can be considered by multiplying the constant B_{oz} by the factor $[1 + (r/R)\cos\theta]^{-1}$ in the field

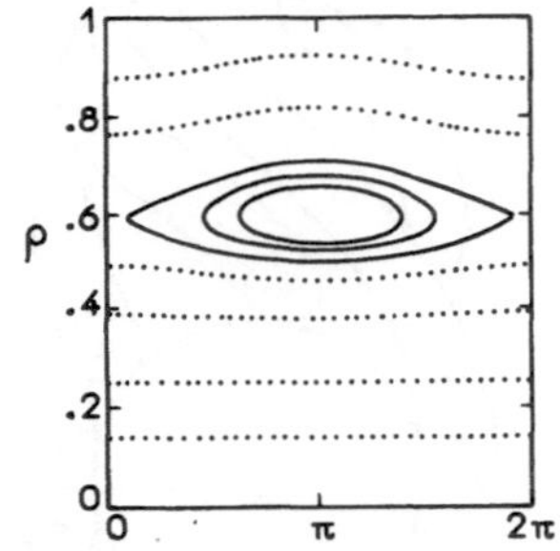

Fig. 3. Computed phase curves (z=0) for $\iota(a) = 2\pi/5$, $\iota(0)=2\pi$, I_p=10kA and I=100A.

line equations (6). For small but finite aspect-ratio $\varepsilon = r/R$ it is possible to use the perturbation theory to obtain[8]:

$$dI/dt = -(\varepsilon r/\omega)(nzB_{oz}/R - mB_{o\theta}/r)(\partial\Phi/\partial r) \quad (13)$$

$$d\alpha/dt = \omega-\varepsilon\omega(mr^2\ \partial\Phi/\partial\theta)\ [-nB_{oz}/R + (m/r^2)(rB_{o\theta} + \partial\Phi/\partial\theta)]^{-1}$$

where $\omega = dH/dJ$.

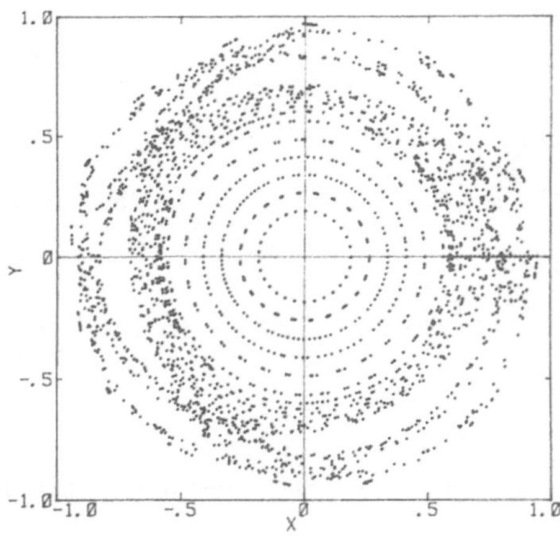

Fig. 4. Field line intersections on the poloidal plane z=0 for the same parameters of Fig. 3.

3. FIELD LINE TRAJECTORIES

Considering the toroidal correction ($\varepsilon \neq 0$), we integrated numerically the differential equations (6) for different equilibria and various perturbations strenghts. Fig. 4 shows the intersections of the magnetic field trajectories with a poloidal plane z=0 for an equilibrium characterized by $\iota(a) = 2\pi/5$ and $\iota(0) = 2\pi$, perturbed by m=3/n=1 helical windings with current I=100A. The major effect of the toroidal correction is the appearence of m±1 satellite islands (with width $\Delta\alpha \; \varepsilon \; \sqrt{I}$) on the surfaces with $\iota = 2\pi n/(m\pm 1)$ and a chaotic region between the islands. This chaotic dis - tribution of magnetic line intersections outside the remaining magnetic surface intersections depends upon the amplitude I, $\iota(a)$ and $\iota(0)$. This dependence is usually indicated by the stochasticity parameter S defined as

$$S = \frac{\Delta_{m,n} + \Delta_{m'n'}}{2(r_{m,n} - r_{m',n'})} \qquad q(r_{m,n}) = m/n \tag{14}$$

Thus S or I can be considered as the important control parameter in the analysis of the Poincaré map of this almost integrable field ($S \sim 1$).

The radial excursions of the line intersections in the considered map, generated by each magnetic line were also investigated. A computer programme provides, for an initial point (r_o, μ_o), the successive intersections in the plane z=0. Fig. 5 shows the radial departure of 100 field line intersections after every toroidal turn, for r_o=6,62cm and μ_o=0 (a point in the chaotic region) for I=10A and I=80A. In the first case, in contrast with the last one, it is still possible to recognize some regularity in the observed oscillations. The oscillations amplitudes increase and become more irregular with the increasing of I (or S). While for I=10A the oscillation is quasi-periodical, for I=80A different periods can be identified in the oscillation spectrum (Fig. 6).

The effect of the mentioned toroidal correction on the magnetic islands can be observed in Fig. 7, obtained for an initial point with coordinates r_o = 6cm and $\theta_o = 80^o$ (inside a 3/1 island). Comparing the curves of the Fig. 7, we can conclude that the introduction of this correction causes the quasi-regular oscillation seen in the curve B, in contrast with the periodical oscillation represented by curve A (obtained considering ε=0). The lost of regularity can be associated to the contraction of the islands and to the distortion of their separatrices due to the islands superposition[5].

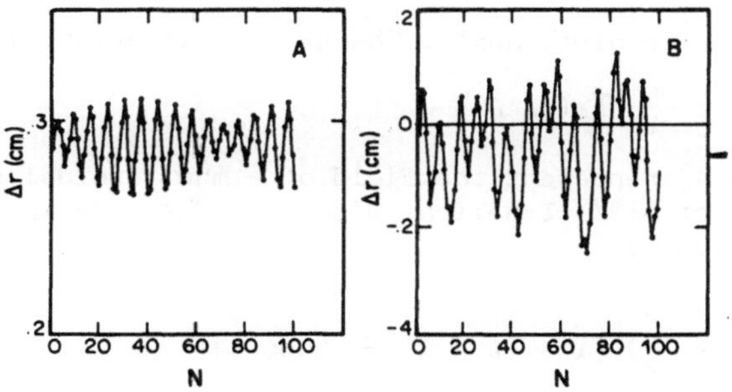

Fig. 5. Radial excursion of successive intersections of a line crossing the plane at r_o=6,62 cm μ_o=0, on the chaotic region of the map obtained for $\iota(a)$= $2\pi/5$, $\iota(0)$= 2π, I_p=10kA and I=10A (curve A) or I=80A (curve B).

4. AVERAGE MAGNETIC SURFACES

The field line equation[6] can be derived from a variational principle using the vector potential $\vec{A}$ of the magnetic field $\vec{B}$ [10]. Applying the Noether's theorem to the Lagrangian of the considered system, it can be concluded that the component A_i of the vector potential is an invariant if $\vec{A}$ is independent of the coordinate x_i. Since the considered field has no symmetry, an average vector potential $\overline{A}j$ was defined in order that an average

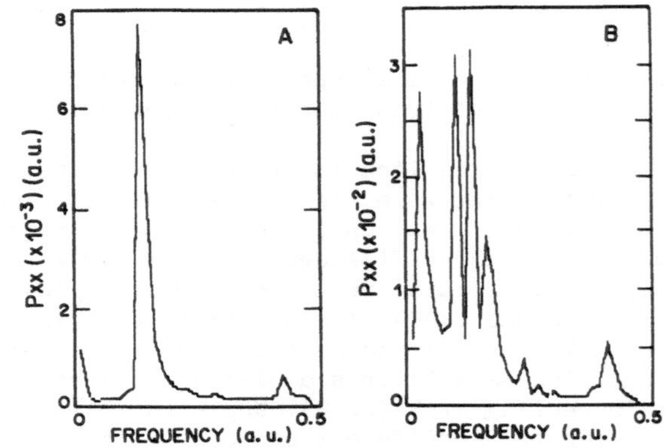

Fig. 6. Power spectra of the oscillations shown in Fig. 5.

invariant could be used to describe average magnetic surfaces. The vector $\vec{A}$, defined through an average on the poloidal angle θ on a line with u constant, is given by

$$\overline{A}j = \frac{1}{2\pi} \int_0^{2\pi} d\theta \; Aj \; (\rho \; , \; \theta, \; \phi) \tag{15}$$

where $\phi = z/R$ is the toroidal angle. The average invariant is given by

$$\overline{\Psi} = \overline{A}_\theta \; (\rho \; , \; u) + \frac{m}{n} \; \overline{A}_\phi(\rho, \; u) \tag{16}$$

The components Aj represent the field of a mhd toroidal equilibrium perturbed by a set of m=3/n=1 helical windings. The invariant of the axial symmetric equilibrium is

$$\psi = \frac{\mu_o \; I_p R}{4\pi} \; (1 - \frac{r^2}{a^2}) \; [1 - (\Lambda + 1) \; \frac{r}{R} \; \cos \theta] \tag{17}$$

and was obtained by solving the Grad-Shafranov equation[11]. Λ is a parameter determined by the kinetic and magnetic pressures and the plasma inductance.

Fig. 8 shows the intersections of the average magnetic surfaces with the poloidal plane $\Phi=0$ for $I_p=18kA$, $\Lambda=0,28$, $B_\phi=0,4T$ and $I=100A$[12].

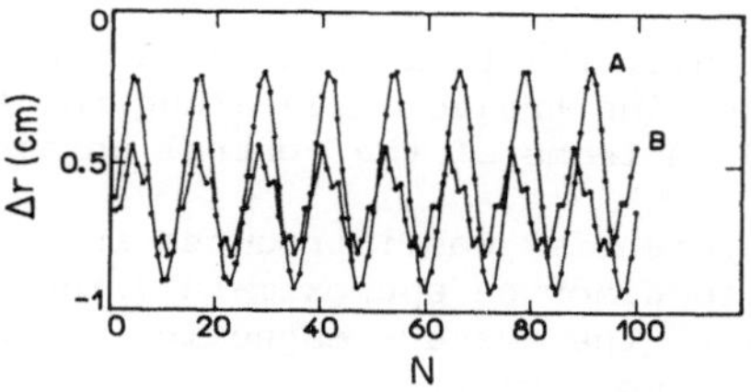

Fig. 7. Radial excursion of successive intersections of a line crossing the plane at r_o=6.0cm, μ_o=80° inside a 3/1 island on the map, obtained for $\iota(a)= 2\pi/5$, $\iota(0)= 2\pi$, I_p=10kA and I=125A. A and B were obtained neglecting and considering the toroidal correction.

Through the analysis of the average surfaces, it was concluded that the average magnetic islands go nearer the plasma centre and that their width become smaller, as the pressure increases. It must be mentioned that to consider the pressure of the plasma is equivalent to consider the displacement of the magnetic axis in the toroidal geometry, because for low tokamak plasma pressure the magnetic axis coincides with the geometric axis[12].

5. CONCLUSION

A perturbation theory was applied to almost integrable fields produced by the perturbation of the large aspect-ratio tokamak equilibrium by resonant helical windings. Action-angle variables were introduced to describe the field line trajectories. The changes in these trajectories due to the toroidal correction were calculated in terms of non-perturbed functions.

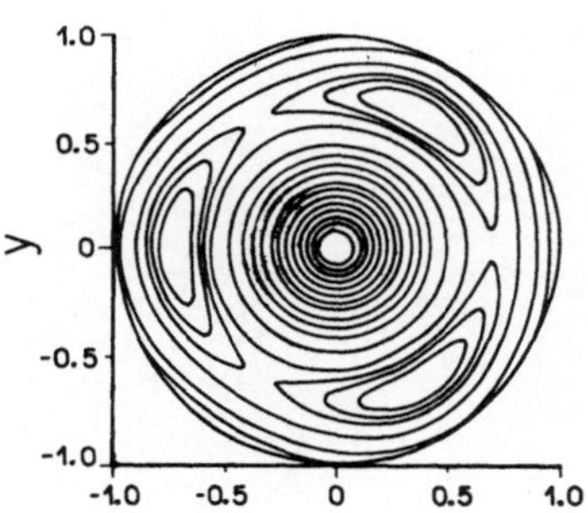

Fig. 8. Intersections of the average toroidal magnetic surfaces with a poloidal plane ϕ=0 for I_p=10kA, $B_\phi \simeq$0,4T, Λ=0,28 and I=100A.

Integrating numerically the field line equations, a Poincare map was used to investigate the line trajectories along the torus. The transition to chaos was discussed in terms of the control parameter S.

An average vector potential was introduced in order that an average invariant could be used to describe approximately the field lines distribution. As the pressure increases the average magnetic islands turn to be smaller and dislocate to the plasma center.

REFERENCES

1. J. P. Freidberg, Ideal Magnetohidrodynamic Theory of Magnetic Fusion System, Rev. Modern Physics 54:801 (1982).
2. D. C. Robinson, Ten Years of Results from Tosca Device, Nuclear Fusion 25:1101 (1985).
3. F. Garger and O. Kluber, The Pulsator Tokamak, Nuclear Fusion 25:1059 (1985).
4. H. Grad, Theory and Applications of the Nonexistence of Simple Toroidal Plasma Equilibrium, Int. J. Fusion Energy 3:33 (1985).
5. A. J. Lichtenberg and M. A. Liberman, "Regular and Stochastic Motion", Springer Verlag, New York (1983).
6. A. Vannucci, I. C. Nascimento and I. L. Caldas, Disruptive Instabilities in the Discharges of the TBR-1 Small Tokamak, Plas. Phys. Control. Fus. 31:147 (1989).

   A. Vannucci, O. W. Bender, I. L. Caldas, I. H. Tan, I. C. Nascimento and E. K. Sanada, Influence of Resonant Helical Windings on the Mirnov Oscillations in a Small Tokamak, Il Nuovo Cimento 10D:1193 (1988).
7. A. S. Fernandes, M. V. A. P. Heller, I. L. Caldas, The Destruction of Magnetic Surfaces by Resonant Helical Windings, Plas. Phys. Contr. Fus. 30:1203 (1988).
8. M. C. R. Andrade, "Formalismo Hamiltoniano para Superfícies Magnéticas", Master of Science Thesis, IFUSP (1989), to be published.
9. M. V. A. P. Heller, I. L. Caldas, work in development.
10. J. R. Cary, Construction of Three-dimensional Vacuum Magnetic Fields with Dense Nested Flux Surfaces, Phys. Fluids 27:119 (1984).
11. V. Okano, Master of Science Thesis, in preparation (IFUSP).
12. S. J. Camargo, "Average Magnetic Surfaces", Master of Scicence Thesis, IFUSP (1989), to be published.

SPIN-SPLITTED PHASE TRANSITION IN THE QUANTIZED HALL EFFECT IN NARROW-GAP $Hg_{(1-x)}Cd_xTe$ INVERSION LAYERS

Marta Silva dos Santos
Departamento de Física, Universidade do Amazonas
69000, Manaus, AM, Brazil

Gilmar Eugenio Marques
Departamento de Física, Universidade Federal de São Carlos
13560, São Carlos
SP, Brazil

Abstract: The scattering of electrons by impurities in spin-splitted states on the inversion layer of narrow-gap semiconductors is studied in the presence of a quantizing magnetic field. The electron-electron exchange interaction is also included. Depending on the system, it is shown that the calculated longitudinal magneto-conductivity, as a function of the magnetic field B, presents discontinuities when the ratio of the average exchange potential to the average electron-impurity potential increases.

I. INTRODUCTION

The quantum Hall effect (QHE) occurs at low temperature and is only observed in systems presenting a completely quantized two-dimensional electron gas (2DEG). The quantization is due to an one-dimensional spatial potential, say in the z-direction, which gives a free 2DEG in a plane perpendicular to the one-dimension, the xy-plane, plus a magnetic field applied perpendicular to the xy-plane, which will quantize the free motion of the electrons in the sample[1,2]. The effect is very sensitive to scattering processes in the sample and to the interface properties which are, mainly, due to the crystal growth conditions. The 2DEG can be produced in silicon[1] MOS or in II-VI MOS[3] samples and also in III-V

Condensed Matter Theories, Volume 5
Edited by V.C. Aguilera-Navarro
Plenum Press, New York, 1990

heterostructures[2,3] grown by MBE techniques. The QHE has been used as an important characterization probe to the layers of the semiconductors where the 2DEG is formed. In this work we will address to a striking difference[3] in the quantized Hall effect observed in a II-VI narrow-gap inversion layer on p-type $Hg_{0.78}Cd_{0.22}Te$ MISFET, grown by solid state recrystallization technique and in the MBE grown III-V heterostructure, made of $Ga_{0.7}Al_{0.3}As$ - GaAs, with the barrier n-type doped with Si.

II.. SUBBAND STRUCTURE

In narrow-gap semiconductors the surface electric field, due to charge redistribution in the inversion layer, is extremely large (about 10^5 eV/cm) and the charge layer is very thin (about 50 Å). Therefore, if the band gap is smaller than 600 meV, one must use a multi-band effective mass approximation[4], instead of a single-band effective mass approximation[5] to describe the motion of electrons in the surface states. In each case, one must use properly derived boundary conditions for the one component (single-band [5]) or for all components (multi-band[4]) of the wave function at the oxide-semiconductor interface.

The quantized electronic structure of inversion layers in MISFETs of narrow-gap semiconductors, such as InSb or Hg(Cd)Te, exhibits large spin-splitting as was theoretically demonstrated by Marques and Sham[4] and experimentally confirmed by Kotthaus et al.[6] and, more recently, by Koch et al.[7]. The splitting is an important feature of the electronic subband structure and has

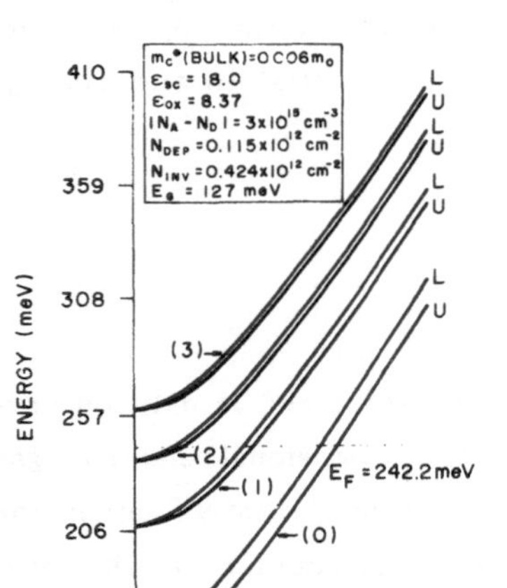

Fig.1 The self-consistent subband of an inversion layer on p-type $Hg_{.78}Cd_{.22}Te$. The electron density, N_{inv} = 0.424 10^{12} e/cm^2, is close to the experimental value. Notice the large spin-splitting and the number of occupied subbands.

permitted a neat interpretation of the optical absorption of the 2DEG in InSb-MOSFET[4] as due to the transitions between spin-splitted subbands with different spin-states. For any given subband number, the states with lower energy were called U-states and those with higher energy were called L-states. This spin-splitting of subbands is mainly due to the derived boundary conditions at the oxide-semiconductor interface and to the self-consistent internal electric field due to the charge redistribution in the inversion layer [4].

Fig.[1] shows the self-consistent electronic subband structure of an in-version layer in $Hg_{0.78}Cd_{0.22}Te$ MISFET , in the absence of the magnetic field. At the used Cadmium concentration, x=0.22, the band-gap energy is $E_g = 127$ meV and the effective-mass for electrons in the conduction band is $m_c^* = 0.006\ m_o$. All the other relevant parameters are given in the upper part of the figure. It should be noticed the large spin-splitting between the U-states and the L-states and the large number of occupied subbands. The splitting is proportional to the internal electric field and this explains why the smaller index subbands have larger splitting than the greater index ones. These features are common to inversion layers in direct-gap semiconductors with small energy gap. The derived boundary conditions for the envelope wave functions in the absence of applied magnetic field[4] remain unchanged in the case of applied field[8], therefore, the effects due to the applied magnetic field (quantization of the xy-plane) will compete with the effects due to the internal electric field plus boundary conditions (quantization of the z-direction).

III. MAGNETO-TRANSPORT

The magnetic field B, applied to the heterostructure, will quantize the spin-splitted states, U and L, into series of Landau levels. For small magnetic field, the effect due to the internal electric field dominates the Landau levels. At high magnetic fields the opposite is observed. In fig.[2], we show the calculated Landau levels as a function of the applied magnetic field. The strong interaction between Landau levels gives the extra splitting but this effect is more important from medium to high magnetic field. Notice also that, at small values of B, there are many occupied Landau levels and, at high values, only the few lower ones are occupied.

Let us next discuss the aspects of the motion of electrons in the xy-plane. In the absence of scattering of electrons by any mechanism, the Landau levels display density-of-states (DOS) very close to delta functions[9,10]. The degeneracy, D, of a given Landau level depends of the value of B and of the area of the sample in the xy-plane. It can be shown that D=eB/ħc which gives the area associated with the flux quanta[9] in the sample.

In the presence of scattering mechanisms, such as due to electron-electron, electron-hole, electron-impurity or electron-phonon interactions, the DOS will show a broadening and it will assume a Gaussian-like function[10]. The states bordering the peak of the DOS are referred as "extended states" and they will determine the mobility edge for the sample. All other are referred as "localized states" and they represent immobile electrons in the magnetic field, therefore they will not contribute to the magneto-transport. In general, only a small fraction[9] (about 5%) of the total number of electronic states in a given Landau

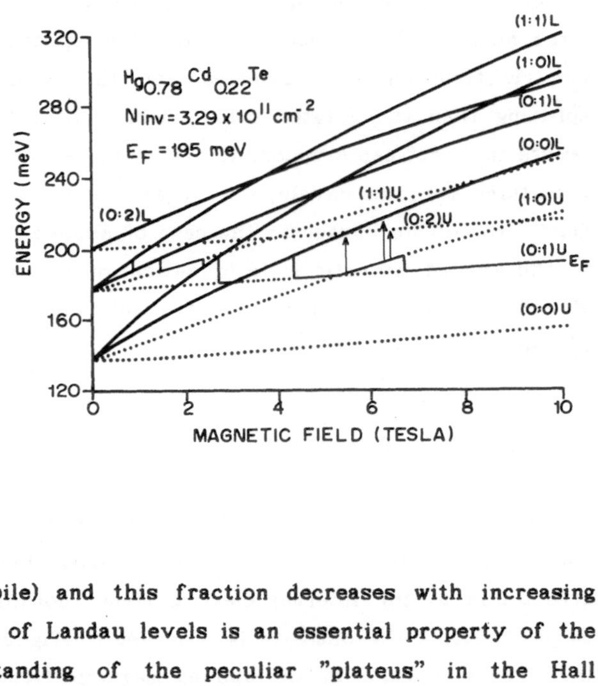

Fig.2 The Landau levels for N_{inv} of fig.1. The local Fermi level, E_F, changes from Landau levels as B increases, and their crossing is due to the strong interaction. The arrows show possible "crossing" of peaks in the magneto-optical absorption observed in ref.[6].

level are extended states (mobile) and this fraction decreases with increasing magnetic field. The broadening of Landau levels is an essential property of the DOS and permits the understanding of the peculiar "plateus" in the Hall resistivity and their corresponding "zeroes" in the longitudinal resistivity. The plateus and zeroes ocurr for integer as well as for fractional occupation of the Landau levels. In this work we will be dealing with the integer Hall effect only, and we will address now to the fine features (kinks) observed in the magneto-conductivity of a n-type inversion layer on a p-type MISFET of Hg(Cd)Te[3]. In order to attain it let us describe the model to be used.

First of all, for simplicity of calculation, we will restrict ourselves to the case of high field, B, where only a subband is occupied. Therefore only the lowest Landau level of each spin state is important and is shown in the fig. 3a. In this sence our single-valley two-branch system is structurally equivalent to the two-valley single-branch Si-MOSFET case studied by Gummich and Sham[11].

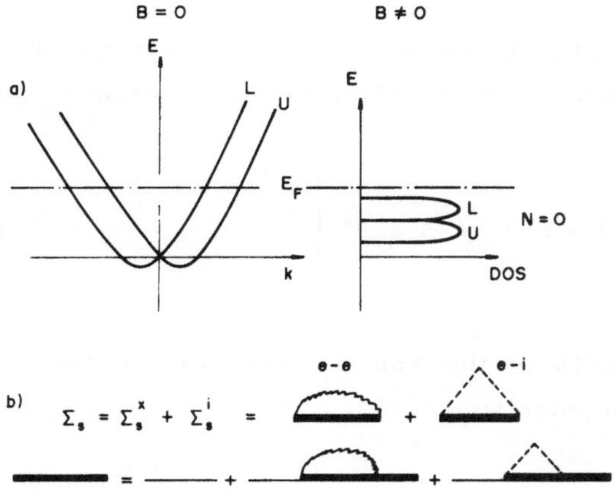

Fig.3 - a) The single-valley spin-splitted model used in this work. b) The exchange and electron-impurity contributions to the self-energy and the diagramatic Dyson equation for the fully interacting 2D- electron gas.

Next, we will write the Hamiltonian for the electronic system only as a sum of the Hartree-Fock energy plus the e-impurity interaction and the intra-branch electron-electron exchange interaction. The neglected electron-electron correlation and the inter-branch exchange will give extremely small contributions and will not add any new qualitative aspect to the final result. Thus, we have

$$H = H_o + H_i + H_{xc} \quad . \tag{1}$$

Here, the free motion is given by

$$H_o = \sum_{k,s} E_o(k) \; a^+_{k,s} \; a_{k,s} \quad , \tag{2}$$

the electron-impurity interaction, for an impurity placed at the position z_o away from the interface, is given by

$$H_i = \sum_{k,\,k',s} < k, s|V^i(\vec{r} - \vec{r}', \; z_o)| \, k', s > a^+_{k,s} \, a_{k',s} \quad , \tag{3}$$

and, finally, the electron-electron exchange interaction is given by

$$H_{xc} = \frac{1}{2} \sum_{\text{all } k,s} V(k, k'; k'', k''') \; a^{+}_{k,s} \; a^{+}_{k'', s'} \; a_{k''', s'} \; a_{k',s} \; . \tag{4}$$

In the above expressions, $E_o(k)$ is the energy of the first occupied level of Landau, and $a_{k,s}$ and $a^{+}_{k,s}$ are respectively the annihilation and creation operators of the electronic state of momentum k and spin s. The electron-impurity interaction potential, in eq.(3), for a semiconductor with dielectric constant ϵ_{sc}, is

$$V'(\vec{r} - \vec{r}', z_o) = \frac{e^2}{\epsilon_{sc}} \left[(\vec{r} - \vec{r}')^2 + (z_o)^2 \right]^{-\frac{1}{2}} \tag{5}$$

and, finally, the function V, in eq.(4), is the Fourier transform of the two-dimensional electron-electron Coulomb potential

$$V(\vec{r} - \vec{r}') = \int dz \int dz' \, | F_o(z) |^2 \, | F_o(z') |^2 \, V(\vec{r} - \vec{r}'; z - z'), \tag{6}$$

$F_o(z)$ being the envelope function for the two electrons in the lowest occupied state in the inversion layer. The potential inside the integral in eq.(6) is the same as in eq.(5) after replacing the impurity coordinate, z_o, by the difference between the z-coordinates of the two interacting electrons, (z-z').

We will assume that the impurity concentration is small compared to the electron concentration. Therefore the broadening of the Landau level, caused by the electron-impurity scattering, can be treated in the self-consistent Born approximation[10]. In this approximation, the contribution to the electron self-energy, due to the electron-impurity scattering (see fig.3.b) is[10]

$$\sum\nolimits_s^{i} = \frac{1}{4} \, \Gamma^2 \, G_s(E), \tag{7}$$

where the broadening of the levels, Γ^2, is the average of the electron-impurity potential, $< k, s \mid V^i \mid k', s >$, in the impurity configuration. For short-range potentials, the broadening average can be associated to the mobility, μ, and to the relaxation time ,τ, as

$$\Gamma = \sqrt{\frac{4}{2\pi R_c^2}\;\frac{e^2\hbar^3}{m_c^2\;\mu}} = \sqrt{\frac{2}{\pi}\;\hbar\omega_c\;\frac{\hbar}{\tau}}\;. \tag{8}$$

The contribution to the self-energy, due to the exchange potential will be treated in the Hartree-Fock approximation. The product of four operators in eq.(3), is considered only for the pairing , k = k" and k"" = k, then we are not allowing the change of branch. Therefore, we get for the self-energy (see fig.3b)

$$\sum\nolimits_s^{xc} = -\,\alpha\,\upsilon_s\;, \tag{9}$$

where, $\upsilon_s = N_s/D$, is the fractional filling of the occupied Landau level in the s-branch and α was taken as the average of the exchange potential, in the Hartree-Fock approximation,

$$\alpha = \left\langle \sum_{k,k'} V(\,k,\,k''\;;\,k'',\,k) \right\rangle\;. \tag{10}$$

The Green function for a given branch, s = L or U, is given by the Dyson equation

$$G_s(E) = G_s^{(o)}(E) + G_s^{(o)}(E)\left[\sum\nolimits_s^{i} + \sum\nolimits_s^{xc}\right]G_s(E), \tag{11}$$

which can be easily solved to give the value

$$G_s(E) = \frac{2}{\Gamma^2}\left\{(\,E - \epsilon_o - \sum\nolimits_s^{xc}\,) - i\left[\;\Gamma^2 - (E - \epsilon_o - \sum\nolimits_s^{i}\,)^2\,\right]^{\frac{1}{2}}\right\} \tag{12}$$

The above Green function, is used to calculate the total energy of the electronic system. Here it will be assumed the Fermi distribution for electrons in the limit T=0 K, since all the experiments are done at extremely low temperature. Let $\tilde{E}$ be the total energy of the two-dimensional electron gas divided by ratio of the area of de sample (L^2) to the area of a cyclotron orbit ($2\pi R^2$). Let us define also $\epsilon_s = (\,E_F - \epsilon_o + \alpha\,\upsilon_s)/\Gamma$ as the energies measured from the Fermi level, E_F . One gets then

$$\tilde{E} = \sum_{s=L,U} [\epsilon_o \upsilon_s - \tfrac{1}{2} \alpha \upsilon_s^2 - \frac{2\Gamma}{3\pi} (1 - \epsilon_s^2)^{\frac{3}{2}}] \tag{13}$$

where, the first term is the kinetic energy, the second term is the exchange energy and the third term is the contribution of the electron-impurity energy.

The total energy can now be minimized with respect of the total fractional occupation , $\upsilon = (Ninv/D) = (\upsilon_L + \upsilon_U)$. There are two possibilities which assure a minimum energy: a) If the non-interacting energy, characterized by the Landau level width Γ, dominates over the exchange energy, characterized by α, then a minimun occurs if the branches are filled simultaneously, i.e., $\upsilon_L = \upsilon_U = \upsilon/2$. Call this the P-phase. b) If the opposite occurs, then a minimum is founded if the branches are filled sequentially, i.e. i) $\upsilon_L = \upsilon$ and $\upsilon_U = 0$ or ii) $\upsilon_L = 0$ and $\upsilon_U = \upsilon$. Call this the F-phase. This is the phase transition for the electronic system and it should be noticed that the P-phase is similar to the paramagnetic, whereas the F-phase is similar to the ferromagnetic phases in normal metals.

These two phases depend on the ratio $\gamma = \Gamma/\alpha$. For $\gamma > \gamma_c$ only the P- phase exists, because the kinetic term dominates over the other two terms. However, below this critical ratio, there exist two values of the fractional occupation where the phase transition can be observed. For decreasing magnetic field B, or increasing occupation of the Landau levels, it is found a first occupation, υ_1 where the system undergoes a transition from the P-phase to the F-phase (high-B). By continuing to decrease the field B, later on, at υ_2, the system will undergo the next transition back to the P-phase. In order to see this feature let us calculate the longitudinal magneto-conductivity for the system under applied magnetic field.

For the sake of simplicity, we will use the same model described above. Therefore, the longitudinal conductivity can be related to the calculated Green's function as

$$\sigma_{xx}(\upsilon) = \frac{e^2}{2\hbar\pi^2} \sum_{s=L,U} \int dE \left(\frac{df}{dE} \right) \frac{[G_s^2(E)]^2}{\{ [G_s^1(E)]^2 + [G_s^2(E)]^2 \}}, \tag{14}$$

where f(E) is the Fermi distribution function and G_s^1 and G_s^2 are the real and imaginary parts of G_s (E). For T= 0° K, the derivative of f(E) becomes a delta function.

In fig.(4) we show the calculated conductivity, for an inversion layer in p-type MISFET on $Hg_{0.78}Cd_{0.22}Te$, with a total density Ninv=0.424 10^{12} el/cm^2.

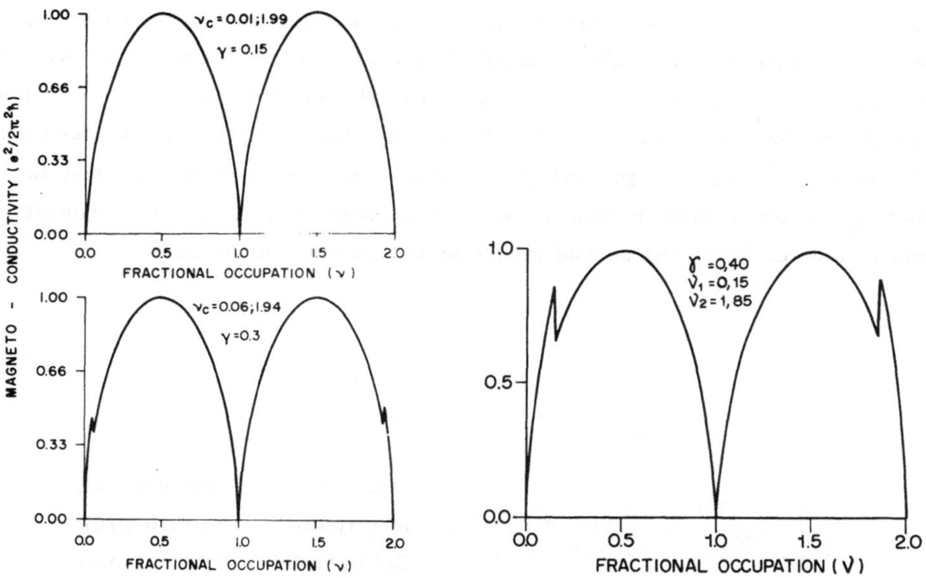

Fig.4 **The calculated longitudinal magneto-conductivity. a) - For a system with low impurities, γ=0.15, no visible kink can be seen. b) - As γ increases, two kinks, due to the described phase-transition, become apparent. c) - Finally, for γ=0.4, we can observed two transitions at fractional fillings υ_1 = 0.15 and υ_2 = 1.85.**

Notice that the conductivity is proportional to the DOS (since the imaginary part of G gives the DOS). At very small impurity concentration, i.e. very small values of the ratio γ, no visible effect is observed. For larges values of γ, two kinks can be seen and they are due to the mentioned two phase transitions at the values of υ_1, υ_2. These critical fractional concentrations correspond to two values of the applied magenetic field, B_1(high) and B_2(low). Also, the size of each kink is an increasing function of γ.

In fig.(5) we show the experimental Hall resistivity and longitudinal resistivity obtained by Kirk[(3)] et al. The solid lines refer to an inversion layer in HgCdTe MISFET, grow by recrystallization, and the dotted lines to the inversion layer on a Ga(Al)As-GaAs heterojunction grow by MBE. Both results were obtained simultaneously on the same equipment, which assures that the striking difference in the QHE in each sample cannot be assigned to any instrumental effect. Notice that the heterostructure shows no visible kink whereas the Hg(Cd)Te sample presents kinks at the places where the extended states give a non-zero longitudinal conductivity or the Hall resistivity is changing between two consecutive Hall plateaus. According to our results the difference can be understood since the Hg(Cd)Te MISFET sample has low mobility (μ=5,0 10^4 $cm^2/V.sec$) and very small effective mass for the conduction band (m_c= 0.006 m_o) and, on the other side, the GaAs heterojunction has high mobility (μ=1.9 10^5 $cm^2/V.sec$) and much larger effective mass for the conduction band (m_c = 0.0665 m_o). These values correspond to a large value of γ for the first sample and a small value of γ for the second sample as can be seen in the eq.(8).

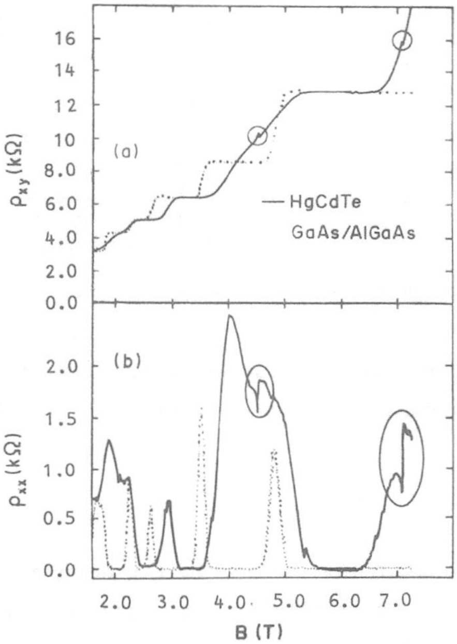

Fig.5 The experimental Hall and longitudinal resistivities for Ga(Al)As-GaAs heterostructure (dotted lines) and for Hg(Cd)Te MOSFET (solid lines) taken from ref.[3]. The second sample shows clearly the kinks due to the phase-transition described in the text.

Also, the samples grown by MBE technique have higher mobility since the doping impurities are very far away from the electronic channel whereas, in an MOSFET sample, the impurities are placed in the same region where the inversion layer is formed. Therefore, the III-V heterostructure in the experiment, can be assumed to have very small impurity concentration which effectively can contribute to the scattering, and therefore, a very small value of Γ. Thus the large observed kinks in HgCdTe are mostly due to the very small effective mass at the concentration x = 0.22 % of Cd and to the higher concentration of scattering impurities , i.e. , $|N_A - N_D|$ = 3.0 10^{16} cm^{-3}.

IV. ACKNOWLEDGMENT

The authors are indebted with N. Studart and U. Gummich for valuable discussions on the system.

VI. REFERENCES

1] - K. von Klitzing, G. Dorda and M. Pepper, Phys. Rev. Lett. 45, 494 (1980).

2] - H. L. Störmer, J. P. Eisenstein, A. C. Gossard, W. Wiegmann and K. Baldwin, Phys. Rev. Lett. 56, 85, (1986).

3] - W. R. Kirk and P. S. Kobiela, R. A. Schiebel and M. A. Reed, J. Vac. Sci. Technol. A4 (4), 2132, (1986).

4] - G. E. Marques and L. J. Sham, Surface Sci. 113, 131 (1982).

5] - L. J. Sham and M. Nakayama, Phys. Rev.B20, 734 (1979).

6] - M. Horst, U. Merkt and J. P. Kotthaus, Surface Sci. 113, 315 (1982).

7] - R. Sizmann, J. Chu, R. Wollrab, F. Koch, J. Ziegler and H. Maier, 19th Int. Conf.on Phys. Semiconductors,(ed. W. Zawadski) Warsaw, vol.1, p.471 (1988).

8] - S. Lamari and L. J. Sham, Surface Sci. 196, 551 (1988).

9] - K. von Klitzing, Rev. of Modern Phys. 58, 519 (1986).

10] - T. Ando and Y. Uemura, J. Phys. Soc. Jpn. 36, 959 (1974).

11] - U. Gummich and L. J. Sham, Phys. Rev. B26, 5611 (1982).

HIGH MAGNETIC SUSCEPTIBILITY LIQUID METALS

Carmen de Teresa and Jaime Keller

Div. de Ciencias Básicas, F.Q.
Universidad Nacional Autónoma de México
Apartado 70-528, 04510 México, D.F., México

We discuss the magnetic susceptibility and the magnetic moment formation in high magnetic susceptibility pure liquid rare earth and transition metals. After an analysis of an obvious case, the magnetic susceptibility of liquid gadolinium, we apply a local-density-functional, linear response, definition of the local enhanced spin susceptibility, and a local Stoner-like stability criterion to liquid iron to show how a localized magnetic moment develops and the resulting magnetic susceptibility.

PACS. 75.10.-b, 75.50.Bb, 71.25.Lf, 75.10.Lp

I. INTRODUCTION

For many years the study of liquid metals was restricted to low melting temperature elements and alloys, mercury, gallium, etc., for technical reasons, mainly because liquid metals corrode most containers of the type commonly used in the determination of the electronic and magnetic properties of liquids and, secondly, because the high temperatures needed to melt transition metals (TM) and rare-earth metals (RE) could not be achieved in the standard equipment. Finally at the beginning of the 70's both problems were solved in the physics and chemistry laboratories: first for the lower melting point alloys of the TM and the RE and second for the higher melting point alloys and pure metals[1].

As expected the liquid transition and rare earth metals showed a high susceptibility compared to the Pauli enhanced magnetic susceptibility of the normal metals, up to two orders of magnitude higher, notably in the case of liquid cobalt. For the case of liquid rare earth metals where a localized f magnetic moment was expected. The explanation was taken for granted but for the liquid transition metals explanation of this high susceptibility was not given in terms of the orientation of localized atomic moments (see Levin, Bass and Bennemann for example[2]) but in terms of a very high density of states at the Fermi level, arising from virtual bound states, and an exchange enhancement similar then to that of the normal metals magnetic susceptibility, but much stronger.

Nevertheless there were two points which were poorly or not explained by those theoretical approaches a) the positive temperature coefficient of the inverse susceptibility exactly similar to that expected for a material with a localized atomic moment, following a Curie-Weiss law and b) the fact that in order to explain the slope of the susceptibility-temperature curve, a magnetic moment equal or even higher for that of the corresponding ferromagnetic solids was needed which was far from the analysis given by enhanced spin susceptibilities as above.

In relation to these liquid metal systems a considerable number of interesting effects can be studied by means of a real space approach to the calculation of their electronic structures. An approach we have frequently used consists in the analysis of a cluster of atoms embedded in a medium with the appropriate boundary conditions. In fact, this approach has been recognized as a suitable technique for dealing with disordered, liquid and amorphous materials. We have enlarged the use of this approach to include also the case of long-range order, where traditional band based techniques can be used. One of the main reason for using this technique is all the conceptual information it brings, mainly in relation to local order and the short range of chemical bonding. See Section II.

An additional advantage comes from the evidence that in many experiments the measurements correspond to local properties. This is true very often even in periodic solids because the measurement itself breaks the translational symmetry of the environment. Thus a real space calculation is useful since we obtain a better understanding of the electronic distribution as a function of $\underline{r}$ and the effect of the local environment, from it we can obtain different local response functions of the material, knowing which regions will be more susceptible to an external potential. This is what we have done in the definition of a local spin susceptibility (in the sense of the diagonal $\chi_s(r,r)$ part of the non local in general $\chi(r,r'))$ in Section III.

Several different materials have been used as examples of the information we may obtain with the present approach. In some previous papers[3] we have presented our results for paramagnetic materials (niobium and vanadium), ferromagnetic solids (Bcc iron or HCP cobalt or the cobalt-gadolinium alloy) and itinerant magnetism system ($TiBe_2$). In Section IV we discuss our central example: liquid iron.

II. ELECTRONIC STRUCTURE CALCULATION

Although for several of our previous examples (V, Nb, Bcc, Fe, Co, $GdCo_5$ and $TiBe_2$) band structure calculations are the more accurate method for the analysis of the electronic structure, in this paper we use a real space cluster in condensed matter approach to the calculation of the electronic structure. We use multiple scattering techniques[4] for a finite cluster whose atoms are immersed in an average potential corresponding to that of the solid. This effective potential is constructed as the sum of a coulombic part, from the overlapping of relativistic self-consistent atomic charge densities, through Poisson equation, and an exchange correlation part from a local density functional ($X_{\alpha\beta}$ with universal parameters)[5]. Long-range order is then not included explicitly. The potential is required to reach convergence, with respect to the number of atomic shell considered that is, to be unchanged by the inclusion of additional shells of atoms. The space is partitioned in non overlapping cells centered in the atomic sites. Within the spheres enclosing given cells, the potential is spherically symmetric and it is constant in the regions outside the cell, the interstitial region[6].

The effect of the potential enters the multiple scattering process through the phase-shifts $\delta^i_\ell(E)$ produced by the potential in each cell i, where ℓ is an angular momentum index and E is the energy. With these quantities and the Friedel sum rule it is possible to calculate the single site density of states, $n^{i\ell}_{ss}(E)$, for the site i. as a function of E that is: free electron like ($n(E) = A\sqrt{E}$) for the s and p electrons and a single, pronounced peak for the d and f electrons, corresponding to their larger atomic character.

The multiple scattering technique will now be briefly summarized[3] The potential enters through the phase-shifts, which are used to construct the reactance k-matrix, for an angular independent potential in each region

$$k_\ell(E) = -\tan \delta_\ell(E)/\kappa \; ; \quad \kappa^2 = E \tag{1}$$

Structural information is contained in the free electron propagator for the cluster, $G_o^+(E)$, whose elements are of the form

$$[G_o^+(E)]_{L\ i,L\ j} = \sqrt{E} \sum_{L_1} 4\pi i^{\ell_1}\ h_{\ell_1}^+\left(\sqrt{E}|\underline{r}_i-\underline{r}_j|\right)\ C_{L',L}^{L_1}\ Y_{L_1}(\hat{r}_{ij}) \tag{2}$$

where the angular momentum composite index $L = (\ell, m)$, the $h_\ell^+\ (\sqrt{E}|\underline{r}_i-\underline{r}_j|)$ are spherical Hankel functions, the $C_{L',L''}^{L}$ are Gaunt numbers (integrals of a product of three spherical harmonics, products of Clebsch-Gordan coefficients) and the Y_L are spherical harmonics. The total Green's function $G^+(E)$

$$G^+(E) = G_o^+(E) + G_o^+(E)\ k(E)\ G^+(E) \tag{3}$$

where $G^+(E)$ is the $(L\ i,\ L'j)$ supermatrix corresponding to the one particle propagator for the cluster embedded in the solid's potential, and $k(E)$ a matrix of single site k-matrices for all i.

The density of states is given in terms of this propagator by

$$N(E) = -\frac{1}{\pi}\ \mathrm{Im}\ \mathrm{Tr}\ G^+(E). \tag{4}$$

We define the multiple scattering ratios, (MSR's) by

$$r_L^i(E) = \frac{\mathrm{Im}\ G^+(E)_{iL,iL}}{\mathrm{Im}\ g^+(E)_{iL,iL}} \tag{5}$$

the denominator being in fact proportional to the (in our calculation m independent) single site density of states $n_{ss}^{i\ell}(E) = -\ \mathrm{Im}\ g^+(E)_{iL,iL}/\pi$ we factor the density of states as

$$N_i(E) = \sum_L N_L^i(E) = \sum_L r_L^i(E) n_{ss}^{iL}(E) \tag{6}$$

and, remembering that $n_{ss}^{i\ell}(E)$ is a one peak curve, it follows that all the detailed structure in the actual density of states curve (peaks, gaps, "shoulders", plateaus) must come from the MSR's curve, mainly from the effects of the local symmetry and bonding-antibonding conditions.

In the actual series of calculations we have used the density functional theory approach with the universal $X_{\alpha\beta}$ exchange approximation ($\alpha = 2/3$ and $\beta = 0.025$) for the potential V_X . The use of the β terms are important because even if the average value of the gradient terms for correlation partially cancels the average value of the gradient terms of exchange, this cancellation is not the case for the <u>local</u> values of V_x and V_{corr}. Our experience has shown that the $X_{\alpha\beta}$ approximation, even if more difficult to use, gives good results in practice.

III. LOCAL SUSCEPTIBILITY

A criterion for the stability of the system against a magnetic transition, the so-called Stoner criterion, states that the system will be unstable if $D = 1 - I\ N(E_F) \leq 0$, is related to the idea of enhancement, that is the case when D, being positive but smaller than 1.0, enhances the free electron value χ_o for the susceptibility, in such a way that $\chi = \chi_o/D$. In these equations I is the Stoner parameter or exchange integral and $N(E_F)$ is the electronic density of states at the Fermi level.

More recently[7] this kind of approach has been combined with the local density functional formalism to obtain formulae for the calculation of I. Different approximations to the exchange correlation potential have been tested resulting in similar values for I. So, within present day choices of the particular band structure (or other kind of) method for the calculation of $N(E_F)$ and of the exchange-correlation functional, the calculation of the spin susceptibility has become standard.

The theoretical approach starts from the consideration that the magnetization of a paramagnet in the presence of an homogeneous external field $H_e(\underline{r})$ is given by the integral equation

$$m(\underline{r}) = \int d\underline{r}'\chi(\underline{r},\underline{r}')[H_e(\underline{r}') - K(\underline{r}')m(\underline{r}')] \qquad (7)$$

where $\chi(\underline{r},\underline{r}')$ is the unenhanced real space susceptibility and the term $K(\underline{r}')m(\underline{r}')$ is a local exchange-correlation enhancement of the field obtained from the second derivative of the exchange-correlation energy $E_{xc}(\rho, m(\underline{r}))$ with respect to the magnetization[7].

The solution to this equation is achieved in two steps. First defining the relative local electronic density of states $\gamma(\underline{r})$ by

$$\gamma(\underline{r}) = \sum_{i,L} \frac{r_L^i(E_F)|\psi_{i,\ell}(\underline{r},E_F)|^2}{N(E_F)} \qquad (8)$$

where $r_L^i(E_F)$ are MSR's giving the actual amount of L symmetry electrons (of a given spin) in the i-site due to the effect of the crystal potential (through multiple scattering interactions), and $\psi_{i,\ell}(\underline{r},E_F)$ are the solutions of the effective Schrödinger like equation for a given spin in the atomic sites (remember that we use spherically independent potentials in each cell to compute the $\psi_{i,\ell}$). All quantities evaluated at the Fermi energy.

Once we have the electron density, a local linear response ansatz is made. This is

$$m(\underline{r}) = C\gamma(\underline{r}) \tag{9}$$

where $\gamma(\underline{r})$ must correspond to that spin with the lower density of states at the Fermi level. Determining C by means of boundary conditions one obtains an expression for the enhanced spin susceptibility of a region

$$\chi = \frac{\int m(\underline{r})d\underline{r}}{H_e} = \frac{N(E_F)}{1+N(E_F)\int dr'\gamma^2(\underline{r}')K(\underline{r}')} = \frac{N(E_F)}{1-N(E_F)I} \tag{10}$$

which has the Stoner form. The global enhancing effect is the result of the total exchange-correlation energy changes weighted by the spin densities within the atomic cell. Thus, large contributions to the enhancement will come from regions where the spin density is large, or where the exchange energy gain in aligning spins is large or both.

Using the above theory in a very small region actually reduced to a point we can define the local spin susceptibility by

$$\chi(\underline{r}) = \frac{N(E_F)\gamma(\underline{r})}{1-I(\underline{r})\gamma(\underline{r})N(E_F)} \tag{11}$$

with $I(\underline{r}) = \gamma^2(\underline{r})|K(r)|$. This is an approximation to $\chi(\underline{r},\underline{r})$, the boundary conditions being obtained from the calculation procedure. In this way we have the local effects explicitly included and retain the Stoner-like criterion for the stability of a region towards the formation of a magnetic state. This local criterion could detect unstable regions that would not necessarily appear when $K(\underline{r}')$ is averaged throughout the atomic cell in the global criterion given by (10).

The following examples will extend this point. The fact that impurities can acquire permanent moments or add their large enhanced spin susceptibility contribution to the total susceptibility indicates that local criteria should be used for these quantities[3].

IV. HIGH MAGNETIC SUSCEPTIBILITY LIQUID METALS

a) Magnetic Properties of Liquid Metals.

The experimental magnetic susceptibility of normal metals, of atomic number Z and valency Z', is typically of the order of 10^{-5} emu/g-atom [which is, in a first approximation, to be considered the sum of the (exchange enhanced) Pauli spin susceptibility χ_s of the Z' conduction electrons

$$\chi_s \approx (Z')^{1/3} 10^{-5} \text{emu/g-atom} \qquad (12)$$

plus the, negative, free electron diamagnetic susceptibility χ_{dia}, which in a first approximation is given by

$$\chi_{dia} \approx -(1/3)(Z')^{1/3} 10^{-5} \text{emu/g-atom} \qquad (13)$$

and the diamagnetic contribution of the core electrons χ_{core}, of the order

$$\chi_{core} \approx -(Z - Z') \, 10^{-5} \text{emu/g-atom}.$$

From the approximate values and formulae it can be estimated that it is implied for the normal metals a density of electronic states $N(E_F)$ at the Fermi level E_F , for free electrons

$$N_o(E_F) = \frac{\Omega}{2\pi^2} E^{1/2} \approx 1 \text{ to } 10 \text{ electron states/Ry} \qquad (14)$$

Above we have used in the formulae shown the value for the Bohr magneton $\mu_B = e\hbar/2mc = 9.27\text{x}10^{21}$ erg/gauss. We can in these cases express the experimental susceptibility as

$$\chi_{exp} = \chi_{dia} + \chi_{core} + \chi_s \qquad (15)$$

The electronic density of states at the Fermi level should be comparable with the free electron value of, for example, a monovalent metal with a density similar to that of copper or in the extreme case of a normal metal with that of a trivalent metal with an electronic density of states similar to that of aluminum. More accurate values can be obtained if diamagnetic contribution of the core electrons can be computed from a knowledge of the average value of the square of the electronic radii

$$\chi_{core} = - e^2 \langle r^2 \rangle / 6mc^2 \quad \text{with} \quad \langle r^2 \rangle = \sum_{i=core} \langle r_i^2 \rangle, \qquad (16)$$

With the core electrons diamagnetic susceptibility of the order of $-3\text{x}10^{-6}$emu/g-atom and the conduction electrons diamagnetic susceptibility of the order of $-1\text{x}10^{-6}$emu/g-atom. The enhancement mechanism for the Pauli susceptibility

$$\chi_s = \chi_s^o/(1-N(E_F)I) \qquad (17)$$

which can be large for materials with a $N(E_F)$ of the order of 30 or more electron states/Ry because the exchange enhancement integrals I are of the order of 0.02-0.07 Ry/state. Those materials are the transition metals, where $N(E\ ) \approx$ 10-30 electron states/Ry.

Then $\chi_s \approx N(E_F)\ \frac{\mu_B^2}{4}\ /(1-N(E_F)I)$ can be in practice one or two orders of magnitude larger for the transition metals (where the enhancement can be as large as $1/(1-N(E_F)I) \approx$ 10-30).

Most of these contributions are very slightly temperature dependent except the spin susceptibility, then additional information should be obtained from the χ-T studies. For the Pauli spin susceptibility

$$\chi_s(T) = \chi_s^o\ [\frac{1+F(T)}{1-I\ N(E_F)\ F(T)}] \qquad (18)$$

For molten normal metals, the expected temperature coefficient of the spin susceptibility, $1+F(T) \approx 1$ due, to the nearly free electron behaviour of the energy density of electron states.

Surprisingly some pure metals and alloys show a large, positive, temperature coefficient of the experimental susceptibility. The observed behavior is well known otherwise to correspond to the Curie-Weiss susceptibility χ_{CW} law:

$$\chi_{CW} = \chi_{LM} = \frac{A}{T-T_c} = \frac{g^2\mu_B^2(J(J+1))}{3k_B(T-T_c)} \qquad (19)$$

this term can also, properly, be called the localized moments susceptibility χ_{CW} because it depends on the magnitude of some moments to be ordered, $p_{eff}^2 = g^2\mu_B^2 J(J+1)$, and on their interaction through the Curie temperature $T_C \approx - - \frac{\left(m_J + 1\ \frac{\Delta E}{DF}\right)}{3k_B}$, where ΔE is the energy to reverse the direction of the magnetization of one localized moment in the average interaction potential of all other magnetic moments. The local magnetic order is frequently incorporated through the topological factor DF(DF $\approx$ 1 for FCC lattices and similar closed packed structures).

We will first analyze an example of a material, liquid gadolinium, where the largest contribution to χ can safely be attributed to a localized moments term arising from a half filled f shell.

The quantities for the relations mentioned above are obtained from a cluster electron density of states calculation.

Considering now the total susceptibility as a sum of (independent) contributions $\chi = \chi_{LM} + \chi_D + \chi_s$, where χ_{LM} is a localized moment contribution, χ_D is the diamagnetic contribution and χ_s the spin enhanced susceptibility.

χ_s can be obtained from the density of states and an average exchange enhancement integral as discussed above

$$\chi_s = \frac{g^2\mu_B^2}{4} N(E_F)/(1 + N(E_F) \int d\underline{r}\ \gamma^2(\underline{r})\ K(r)) \qquad (20)$$

where $2K(\underline{r})\delta(\underline{r}-\underline{r}') = \partial^2 E_{xc}/\partial m(\underline{r})\partial m(\underline{r}')|_{m=m_o}$ is the second derivative of the exchange correlation $F_{xc}[\rho]$ energy with respect to magnetization (see also 8). The diamagnetic contribution, arising from at least two terms $\chi_D = \chi_{DO} + \chi_{DC}$; the first, from core electrons χ_{DO}, is obtained, as mentioned, from the integrals $\chi_{DO} = -e^2\langle r^2\rangle/6mc^2$. The diamagnetic contribution from conduction electrons χ_{DC} can be approximated from Landau's formula $\chi_{DC} = -2m\mu_B^2\ (2mE_F)^{1/2}/12\pi^2$.

The localized moments contribution, once the moment itself has been evaluated selfconsistently, is given by a Curie-Weiss law

$$\chi_{LM} = \frac{A}{T - T_C}\ ; \quad A = g^2\mu_B^2 J(J + 1)/3k_B = 0.125\ p_{eff}^2 \qquad (21)$$

T_C can be obtained to a very good first approximation from a RKKY interaction analysis[9], or from a calculation of the difference of total energy ΔE per atom when the surrounding atoms have the same direction of magnetization or the opposite[10], for total spin S the simplest approximation is

$$k_B T_C = - \frac{(S + 1)}{3} \Delta E \qquad (22)$$

Calculated magnetic susceptibilities for liquid Gd are given in Table 1.

The different contributions to χ are not really independent and additional, here negligible, terms arise from core and band polarizations and from Van Vleck core and band terms χ_{VV}.

The above relationships for the susceptibility of a metal, using the simplest approximations for each one of the expressions and the assumption of the additivity of the different contributions to the susceptibility.

Normal metals, from the point of view of magnetic susceptibility, present nearly free electrons behaviour (in practice: Cu, Ag, Au, Na, Al, etc.). But if we consider transition metals where d, or f bond enhanced s and p conduction bands are found near the Fermi level, with an electronic density of states at the Fermi level of the order of $N(E_F) \approx$ 30 electron-states/Ry, then the exchange enhancement can be as large as to increase the susceptibility by orders of magnitude, or even as large as driving the formation of spontaneous magnetic moment. This has been the subject of a vast literature especially devoted to this problem, see for example: Moment Formation in Solids, Buyers 1984[11].

One of the central problems being the discussion of the possible permanent magnetic moment formation, which should be discussed without the use of standard band structure theory and formalisms based on this approach. The Anderson model is one of the best known examples of the type of approach which can be followed.

In the case of metals where a permanent magnetic moment is not formed, it is when the magnetic susceptibility is not changed to a Curie-Weiss law, the paramagnetic susceptibility and the other several contributions to the experimentally observed susceptibility are almost temperature independent. The core diamagnetic susceptibility is practically unaffected by temperature because thermal energies do not appreciably promote core electrons to empty electronic states. The conduction electrons diamagnetic susceptibility is practically unaffected by temperature because, unless a structural phase transition is driven by temperature, the diamagnetic susceptibility depends only on the average density of conduction electrons. A temperature dependence of the enhanced Pauli susceptibility can however be expected from the broadening found at the Fermi level of the Fermi-Dirac distribution law with increasing temperature. In this case, if the second energy derivative of the electronic density of states at the Fermi level is not negligible in units of $(kT)^{-2}$, it can change the spin susceptibility (see for example Levin, Bass and Bennemann[2]), as mentioned above, to

$$\chi_s(T) = \chi_s^o \left(\frac{1 + F(T)}{1 - I\, N(E_F)\, F(T)} \right) \qquad (23)$$

with $F(T) = a(kT)^2 \partial^2 N(E)/\partial E^2$

for molten transition metals ($T \approx 1000K$) the experimental temperature

Table 1. Magnetic Susceptibility of Liquid Gadolinium (emu/mol).

E_F	$N(E_F)$	T_C	A	χ_{DO}	χ_{DC}	χ_D	χ_s
0.297	23	160	7.49	-60×10^{-6}	-17×10^{-6}	-77×10^{-6}	216×10^{-6}
$\chi_{Total} = 7.49\ (T - 160)^{-1} + 0.000139$ (emu/mol)							

coefficient of the spin susceptibility is again small, that is $1 + F(T) \approx 1$, with the second energy derivative of the density of states $D^2 = \partial^2 N(E)/\partial E^2$ being of the order of $D^2 \approx -N(E)/4E^2$. In order to explain a noticeable temperature dependence of the susceptibility the actual values of the D^2 should be 4 orders of magnitude larger. Then a localized moments formation should be explored.

b) Cluster Calculation of the Magnetic Moment of Fe in Liquid Iron.

In a form similar to what has been described in the previous examples[3] the electronic structure of a cluster of four Fe atoms was computed in a series of calculations:

a) No magnetic moment assumed.

b) All atoms with an assumed magnetic moment of: 1, 2, 2.2, 2.5, 2.75, 3.21, 3.5 and 3.74 spins per atom, ferromagnetically coupled.

c) The cluster of four atoms in a tetrahedral configuration was assumed to have no net magnetic moment but each atom had a magnetic moment of 1, 2, 2.2, 2.5, 2.75, 3.21, 3.5 and 3.74.

In all cases the local magnetic spin susceptibility was computed and it was found that it diverges except for the tetrahedral cluster with antiferromagnetic coupling when each atom is assumed to have a magnetic moment of 3.5 spin per atom. Because of this series of approximations involved in the actual calculations we do not consider our method to be fine enough to allow a direct calculation through the use of the spin susceptibility, for differences in moment per atom smaller than 4-5%. Then we think that given an interpolation between 3.21 and 3.74 indicating that the stable moment (in the antiferromagnetic coupling) was to be found with 3.5 spins per atom and this value being confirmed by direct calculation, we have reached the limits of the accuracy of our method. We should remember that we are representing a liquid by mere four atoms in the medium, that we have selected the tetrahedral cluster because there is only one non magnetic, one ferromagnetic and one antiferromagnetic configuration for this cluster and that our magnetization consistency procedure has shown us that differences in magnetization smaller than 2-3% are not predicted with higher accuracy than, say, changes in the number of atoms or changes in the higher ℓ-partial wave would admit. Then for the purpose of the rest of our analysis we will take m = 3.5 spa to be a reasonable theoretical value.

In Fig. 1 we show the radial criteria (as given by equation 11) of magnetic stability and in Fig. 2 the computed density of states per atom, for molten transition metals ($T \approx 1000K$) the experimental temperature for the cases of 2.75, 3.21, 3.5 and 3.74 spin per atom magnetic moment

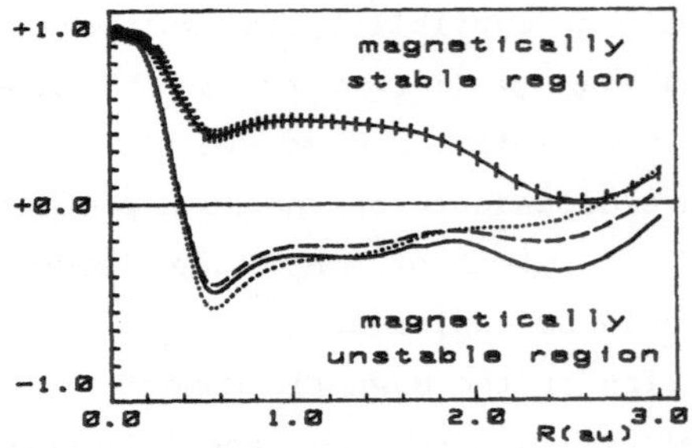

Fig. 1. Radial criteria of magnetic stability of liquid iron for the cases: 2.75 (·····), 3.21 (– – –), 3.5 (┼┼┼┼┼) and 3.74 (———) spin per atom magnetic moment and antiferromagnetic coupling between atoms.

and antiferromagnetic coupling between the atoms. In the case of the antiferromagnetic configuration a moment larger than 3.74 spa cannot be induced in the cluster.

We can now look for a fit to the magnetic susceptibility χ of liquid iron, as reported by Güntherodt, Meier, Hauser, Künzi and Müller[1], with the general formula useful for the localized moments part of the susceptibility (as in the case of liquid Gd above)

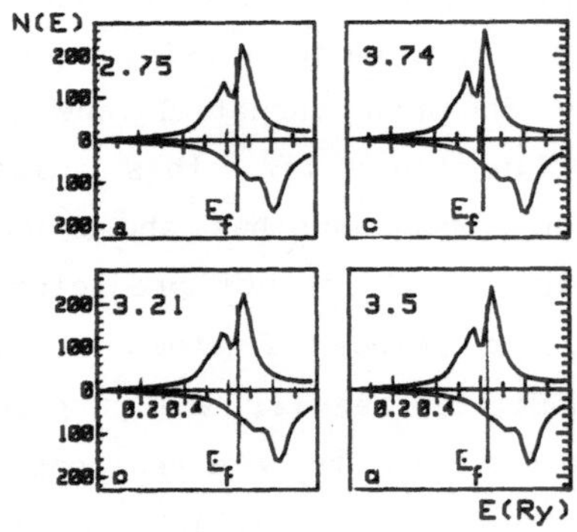

Fig. 2. Computed density of states per atom of liquid iron for 2.75 (———), 3.21 (·····), 3.5 (-·-·-·) and 3.74 (– – –) spin per atom.

$$\chi \approx \chi_{LM} = \frac{A}{T-T_c} \tag{24}$$

where the constant A for iron, in our calculation for a value of the spin S = 3.5, will be A = 2.4.

As a first approximation let us assume $T_c \approx 0$, then

$$\chi \approx \chi_{LM} \approx \frac{A}{T} = \frac{2.4}{T} \; \frac{\text{emu}}{\text{g-atom}} \tag{25}$$

and for $T \approx 1200K$ we would have

$$\chi \approx 2 \times 10^{-3} \; \frac{\text{emu}}{\text{g-atom}} \tag{26}$$

in a very good first agreement with the experimental value of 1.75×10^{-3} emu/g-atom.

We now search for a reasonable value of T_c that could reproduce the inverse susceptibility curves reported by Güntherodt et al., Figs. 3a and 3b. We find that $T_c \approx -200K$ will both, reproduce the inverse susceptibility versus temperature curve and correct for the calculated value of the susceptibility at the melting point to obtain

$$\chi_{LM} \; (T = 1200 \text{ K, computed}) = 1.7 \times 10^{-3} \; \frac{\text{emu}}{\text{g-atom}}$$

From our calculations we found that the magnetic moments in liquid iron tend to couple antiferromagnetically, but we can not estimate the Neel temperature as we did for the liquid gadolinium example.

We have of course tried to obtain this value of the susceptibility from other approaches, for example a highly enhanced spin susceptibility with corrections for temperature dependence and we have found that none of those approaches could be consistent with the magnetic susceptibility of liquid iron being intermediate between that of bcc and fcc solid iron extrapolated to higher temperatures and to a temperature dependence similar to that of those systems, with a T_c much smaller in value than the Curie temperature of the solids (our actual estimation being negative (Neel) $T_c = -200K$).

We have found, in conclusion, that the magnetic susceptibility of liquid iron is best described by that of a collection of magnetic moment carrying Fe atoms, coupled (between nearest neighbors in the liquid state) antiferromagnetically. This is consistent with both the high value of the observed magnetic susceptibility and with its temperature dependence; other terms which would be slightly temperature dependent are obscured by the uncertainties in the actual values of the moment per atom ($m \approx 3.5$ spa) and of the antiferromagnetic critical temperature

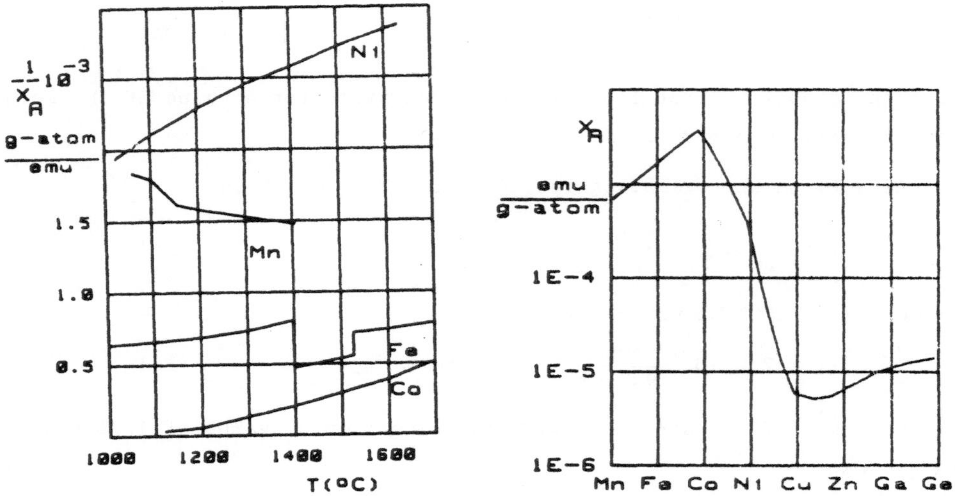

Fig. 3. The magnetic susceptibility of liquid transition metals. a) As a function of temperature. b) Values at the melting point (Taken from Güntherodt H.J., Künzi H.,U. and Müller, H. A. in ref. 1)

$T_c \approx -200K$. This additional contributions are nevertheless small enough not to change the results of our analysis, as they would be one to two orders of magnitude smaller and tend to cancel among themselves.

DISCUSSION

We have presented a method to evaluate the magnetic stability of an assumed electronic structure of a material, useful for both non-magnetized and magnetized electronic structures. The more widely used methods for electronic structure calculations are able either to calculate a consistent (between assumed magnetization and found magnetization) magnetization per atom or in the best approaches a totally selfconsistent magnetization per atom. Some methods can even allow the calculation of the total energy associated with those states, then that state of minimal total energy (total free energy) would be, in principle, the more favorable one; but none of those methods can assure that the electronic structure thus calculated is stable against either lattice deformations or further magnetic transitions. Lattice deformation can, in principle, be studied by changing the lattice structure and if either lower minima or degenerate structures are found we know that the distortion would occur. In the case of the magnetic stability we have to use criteria such as the present one or equivalent.

ACKNOWLEDGEMENTS

We would like to thank many useful discussions with H.J. Güntherodt, H.U. Künzi, L. Schlapbach, R. Müller and H.A. Meier on the subject of the properties of liquid metals. This work was supported in part by the Program "Estructura Electrónica de Materiales de Alta Tecnología" of the Consejo Nacional de Ciencia y Tecnología, México and by the Sistema Nacional de Investigadores (SNI) of México. The technical assistance of Mrs. Irma Aragón is also gratefully acknowledged.

References

1. H.J. Güntherodt and H.A. Meier, Phys. Kondens. Materie 16, 25 (1973); L. Schlapbach, Phys. Kondens. Materie 18, 189 (1974); H.J. Güntherodt, H.U. Künzi and R. Müller, in "Amorphous and Liquid Normal Transition and Rare Earth Metals", J. Keller, ed., U.N.A.M., Mexico, (1975) 186-200; H.J. Güntherodt, E. Hauser and H.U. Künzi, in "Amorphous and Liquid Normal Transition and Rare Earth Metals", J. Keller, ed., U.N.A.M., Mexico, (1975). 356-385.
2. K. Levin, R. Bass and K.H. Bennemann, Phys. Rev. B6, (1972) 1865; R. Bass and K.H. Bennemann, Bull. Am. Phys. Soc. 18, (1973), 417; R. Bass, J. of Phys. F4 (1974) 1256.
3. Keller J., J. of Molecular Structure 93, 93 (1983; Keller J., Amador C. and de Teresa C., Physica 130B, 37 (1985); Pisanty A., Orgaz E., de Teresa C. and Keller J., Physica 102B, 78 (1980); Keller J., Amador C., de Teresa C., in "Future Trends in Material Sciences", J. Keller, ed., World Scientific Advanced Series in Surface Science V2, (1987), 161-227.
4. Keller J, J. Phys. C4, 3143 (1971); Lloyd P and Smith P.V., Adv. in Phys. 21, 69 (1972); Keller J. and Evans R., J. Phys. C Solid St. Phys. 4, 3133-45 (1971); Eyges L., Phys. Rev. 111 683 (1958); Johnson K.H., J. Chem. Phys. 44, 3085 (1966); Johnson K.H., in Advances in Quantum Chemistry, P.P. Löwdin, ed., Vol. 7, 143, Academic, New York 1973; Johnson K.H. and Smith F.C., Phys. Rev. B5, 831 (1972); Keller J., Int. J. Quantum Chem. 9, 583 (1975); Keller J., in Computational Methods for Large Molecules and Localized States in Solids, f. Herman, A.D. McLean and R.K. Nesbet, eds., 341-56, Plenum Press 1973 and J. Physique 33, C3 241 (1972); Keller J., Hyperfine Interact. 6, 15 (1979); Keller J., Fritz J. and Garritz A., J. Physique 35, C4 379 (1974); Keller J. and Fritz J.,

in "Cluster Method Calculation of Density of States", J. Stuke, E. Brenig, eds., Taylor & Francis, 975-80 (1973); Garritz A. and Keller J., in "Amorphous and Liquid Normal, Transition and Rare Earth Metals", J. Keller, ed., University of Mexico, 236-250 (1973).

5. Herman F., van Dyke J.P. and Ortenburger I.B., Phys. Rev. Letters 22, 807 (1969).
6. Loucks T.L., Augmented Plane Wave Method, Benjamin Inc. (1967); Castro M., Keller J. and Rius P., Hyperfine Interactions 12 261 (1982).
7. Vosko S.H. and Perdew J.P., Can. Journal of Phys. 53, 1385 (1975); Gunnarsson O., J. Phys. F 6 587 (1976); Janak J.F., Phys. Rev. 16, 255 (1977); Keller J., Amador C. and de Teresa C., Rev. Mex. Fis. 30, 447 (1984).
8. Janak J.F., Phys. Rev. B16, 255 (1977).
9. Xavier R.M., Da Silva X.A. and Baltensperger W., Phys. Lett. 15, (1965) 126.
10. Keller J. and Garritz A., Int. Conf. Physics of Transition Metals, Toronto (1977). Inst. Phys. Conf. Ser., 39, 372 (1978).
11. Moment Formation in Solids, W.J.L. Buyers Ed. NATO ASI Series B: Physics Vol. 117, Plenum Press (New York) (1984).

TRANSLATIONALLY-INVARIANT COUPLED CLUSTER THEORY APPLIED TO THE ^{4}He NUCLEUS

R.F. Bishop and M.F. Flynn
Department of Mathematics, UMIST
P.O.Box 88,Manchester M60 1QD, England
and
M.C. Boscá, E. Buendía and R. Guardiola
Departamento de Física Moderna, Universidad de Granada
18071 Granada, Spain

Abstract

In this work we discuss a series of calculations for the ^4He nucleus which have been motivated by the coupled cluster method. For pedagogical reasons we restrict ourselves to the case of pure Wigner nucleon-nucleon interactions. All numerical work is done in the standard harmonic oscillator basis and with an exact treatment of the centre-of-mass motion. Particular emphasis is placed on elucidating the meaning of the coupled cluster wave function and its coordinate-space representation, as well as the relation with a variational-like use of the coupled cluster ansatz.

1. Introduction

Since the original proposal of Coester and Kümmel [1,2] a large amount of work has been carried out within the framework of exp(S) or coupled cluster (CC) theory. A full description of CC theory may be found in the article of Kümmel, Lührmann and Zabolitzky [3] that has become a classical reference in this field. On the other hand, in order to acquire a flavour of the wide domain of applications and the success of CC theory it is suggested that the reader consult the recent pedagogical review given in Ref. [4].

In spite of the many well-recognized successes of CC theory, it is still perhaps true that many of its underlying features remain obscure. As is often the case in many-body theories, the basic ansatz is rather simple. Its physical content is quite clearly established in its starting form for the wave function. However, when going to practical applications, simplicity is lost. The algebraic details of applying the underlying formalism tend to make the theory more and more obscure. Of course, this should be expected, given the richness and rather universal applicability of the CC ansatz. To quote a particular example, the $SUB(2)$ approximation, i.e., considering only $2p-2h$ excitations, in an infinite medium gives for the basic two-body amplitude a non-linear integral equation which contains not only

Condensed Matter Theories, Volume 5
Edited by V.C. Aguilera-Navarro
Plenum Press, New York, 1990

Table 1. The ground-state energy of the ^4He nucleus using the MTV interaction, and as obtained by means of various theoretical methods.

Method	Reference	g.s. energy (MeV)
Green Function Monte Carlo	[10]	-31.3 ± 0.2
Coupled Cluster	[11]	-31.36
ATMS (Amalgamation of two-body corr...)	[12]	$-32.8 \leq E \leq -31.3$
Jastrow full Euler-Lagrange variational	[13]	-31.35
Jastrow variational second-order	[14]	-31.19 ± 0.05
Hyperspherical harmonics	[8]	$-31.22 \leq E \leq -30.48$
Yakubowsky equations	[15]	-29.6
Configuration interaction ($10\hbar\omega$) space	[16]	-18.31

such individual contributions as the RPA terms, with Pauli exchanges, the various ladder contributions (*hh*, *pp* and *ph*), the particle and hole potential insertions, among other terms, but also the self-consistent union of *all* such terms iterated simultaneously [5].

The aim of this work is to present with clarity the underlying structure of the CC wave function, as well as to study the characteristics of the energies obtained by solving the CC equations. In order to expose these essential features it is convenient to deal with a rather simplified problem. We found particularly adequate for this purpose the study of the ^4He nucleus where the nucleons interact by means of Wigner-type two-body potentials.

Ignoring the spin and isospin dependence of the two-body interaction undoubtedly means moving relatively far away from the real physical structure of ^4He. For example, we know that the physical nucleon-nucleon interaction has a very important tensor component. Nevertheless, the advantages of working in these simplified conditions are considerable. First of all, we may view the ^4He nucleus under these circunstances as a system of four bosons, with a fully space-symmetric ground-state wave function. This will permit the determination of exact reference or benchmark values, by means of Monte Carlo integration of the many-body equation [6]. In addition, all of the ensuing Fock algebra will deal with bosonic (i.e., obeying commutation relations) creation and anihilation operators.

The second reason is of a didactic character. Our four-nucleon system will be simple enough to permit its study in both Fock and coordinate space. In this manner we will be able to show explicitly the way CC theory describes the nuclear wave function in coordinate space. In turn, this will permit us to connect CC theory with such other theories as the Jastrow variational theory [7] or the hyperspherical harmonics integro-differential equation method [8], which are entirely described in coordinate space.

The last motivation for dealing with this simplified problem is the existence of a large amount of work for the ^4He using the MTV potential of Malfliet and Tjon [9] , and employing both many-body and few-body methods, with which we may compare. A selected set of results concerning the ^4He nucleus with MTV forces is shown in Table 1.

2. Nucleon-nucleon interactions

We have considered four different forms for the two-body nucleon-nucleon interaction which between them cover a wide range of *complexity*. We use this word in the common many-body sense, i.e., interactions with a strong repulsion at short distances are considered *complex*, whereas interactions without core or a quite small one are said to be *simple*. Our aim is then to show the performance of CC theory over as wide a range of problems as our simple model permits.

The simplest potential we use was introduced by Malvin Kalos [17] in a pioneering paper on the application of stochastic methods to quantum systems. It is a fully attractive interaction given by

$$V_K = -72.2\exp\{-(r/1.191555)^2)\}, \tag{1}$$

and it is included here largely for historical reasons.

Our second potential in order of increasing complexity is the Wigner part of the Brink and Boeker $B1$ interaction [18],

$$V_B = 389.5\exp\{-(r/0.7)^2\} - 140.6\exp\{-(r/1.4)^2\}, \tag{2}$$

which has received very much attention in nuclear structure calculations.

The above two interactions are *effective* interactions, which have no direct relation with two-body nucleon-nucleon scattering data. We have also considered two other *realistic* interactions, namely the Wigner part of the $S3$ interaction of Afnan and Tang [19],

$$V_S = 1000\,e^{-3r^2} - 163.35\,e^{-1.05r^2} - 83\,e^{-0.8r^2} - 21.5\,e^{-0.6r^2} - 11.5\,e^{-0.4r^2} \tag{3}$$

and the already-mentioned MTV interaction [9],

$$V_{MTV} = 1458.27\exp(-3.11r)/r - 578.18\exp(-1.55r)/r. \tag{4}$$

These last two potentials correspond to the Wigner part of interactions fitted to the $\ell = 0$ two-body phase shifts and to the deuteron binding energy. In each of equations (1-4) the potential is measured in MeV and the internucleon distance r is measured in fm. In order to fix all details of our calculations, we also note that our value for the nucleon mass m has been chosen so that $\hbar^2/m = 41.5\ MeV\ fm^2$.

With the exception of the MTV potential, all of the other interactions are combinations of gaussians. Given that we will work exclusively in the harmonic oscillator basis, all calculations regarding these gaussian interactions can be carried out by means of semi-numerical algorithms. In addition to speeding up the computations, our corresponding calculations will be free of most numerical approximation errors.

3. An irritating question: the centre-of-mass spuriosity

In a light system, like ^4He, the quantum mechanical description may be explicitly written down in terms of a set of intrinsic, translationally-invariant coordinates, such as the Jacobi set of coordinates. The price that has to be paid is that to impose the Bose or the Fermi statistics one has to translate simple nucleon exchange permutations into some much more complicated transformations among the intrinsic coordinates. On the other hand, in CC theory, as well as in shell-model theory, all states are most readily expressed in terms of the individual nucleon coordinates, referred to some external origin, and because of this the wave function is not translationally-invariant. It is possible to insert by hand some constraint to remove the spurious centre-of-mass coordinate [20], or even to try to remove it optimally [21]. Nevertheless, both of these two approaches will result in tying all of the nucleon coordinates together, thereby converting the calculation of matrix elements of one- or two-body operators into a calculation which involves all of the nucleon coordinates.

There is, however, a special case where the effects of the spurious centre-of-mass coordinate may be removed without paying the heavy price of an intrinsically A-body description. It corresponds to wave functions constructed in a shell model with single particle wave functions from within a harmonic oscillator potential. It is well-known that such a suitably (anti)symmetrized non-interacting many-body harmonic oscillator wave functions ($HOWF$) in which the single particle levels are filled up in order of increasing energy, so that a new major shell is only started when all lower energy shells are completely filled, factorizes into the product of two terms, one depending only on the centre-of-mass (CM) coordinate $\mathbf{R} = \sum_i \mathbf{r}_i/A$, and the other depending only on internucleon distances

$$\Psi_{HOWF}(\mathbf{r}_1, ..., \mathbf{r}_A) = \Psi_{CM}(\mathbf{R}) \prod_{i=1}^{A} \Psi(\mathbf{r}_i - \mathbf{R}), \tag{5}$$

with the centre of mass being in the $0s$ state of a scaled harmonic oscillator. This factorization property permits us to work directly with Ψ_{HOWF}, and all spurious contributions due to Ψ_{CM} may be removed quite easily at the end of the calculations. In this way one can still work with the individual nucleon coordinates, rather than with the much more complicated and less symmetric (under permutations) intrinsic coordinates.

Unfortunately, this property is in general lost when considering the $1p-1h$, $2p-2h$, ... states which are necessary to describe the physical correlations induced in the system by the interparticle forces. There is still a way of removing the spurious CM effect but which is very costly in terms of computational effort [22]. It necessarily involves considering the full space corresponding to a given number of excitation quanta in top of the model non-interacting $HOWF$. In other words, it corresponds to considering all $np-nh$ states with a total harmonic oscillator energy less than or equal to a given energy $N_{max}\hbar\omega$. The disadvantage of this approach is that many of the states which result after the diagonalization of the hamiltonian matrix are not physical, because they correspond to excitations of the centre-of-mass. Moreover, the dimension of the space grows very rapidly with N_{max} [16,23].

We have found a method which produces factorizable wave functions, i.e. which maintains the centre of mass in the $0s$ state. For simplicity of presentation we concentrate on the present ^4He case to discuss the technicalities. The starting state is the non-interacting $HOWF$ in a $(0s)^4$ configuration,

$$\Psi_0 = (4!)^{-1/2} \, (a_0^\dagger)^4 \, |0\rangle \tag{6}$$

where the subindex 0 represents the $0s$ orbital. Excitations with respect to Ψ_0 which maintain the basic factorization and preserve the symmetry of the wave function must correspond, in coordinate space, to at least two-body operators depending only on relative coordinates, i.e., something like $\mathbf{S} = \sum_{i<j} S(r_{ij})$. We note furthermore that $S(r_{ij})$ depends only on the distance between particles i and j in order to maintain the zero angular momentum of the starting wave function. The Fock representation of this operator will be of the type $\mathbf{S} = \frac{1}{2} \sum_{ijkl} S_{ijkl} a_i^\dagger a_j^\dagger a_k a_l$, where each of the indices represents the three quantum numbers necessary to label the single particle wave functions, namely $\{n\ell m\}$. The symmetrized amplitudes S_{ijkl} correspond formally to the matrix element

$$S_{ijkl} = \langle \phi_i(r_1)\phi_j(r_2) + \phi_j(r_1)\phi_i(r_2)|S(r_{12})|\phi_k(r_1)\phi_l(r_2)\rangle/2.$$

The important point here is the strong simplifications which arise in our specific ^4He problem: thus k and l must correspond to $0s$ states, and i and j must couple to zero angular momentum. Finally, when transforming $\phi_i(r_1)\phi_j(r_2)$ to centre-of-mass and relative coordinates, the centre of mass *must* remain in the $0s$ state. Putting all of these constraints together, there results for the operator $\mathbf{S}$ the form

$$\mathbf{S} = \sum_{n>0} S_n \sum_{n_i n_j \ell} \langle n\,0\,0\,0\,0|n_i\,\ell\,n_j\,\ell\,0\rangle [a_{n_i\ell}^\dagger \times a_{n_j\ell}^\dagger]^0\, a_0\, a_0, \tag{7}$$

where the amplitudes $\{S_n\}$ are arbitrary. For a given value of n one has to consider *all* possible sets of indices $\{n_i\, n_j\, \ell\}$ which are compatible with the restriction implied in the Brody-Moshinsky bracket $n = n_i + n_j + \ell$. A given value of n thus corresponds to $2n\hbar\omega$ excitation energy and there is hence only *one* $2p-2h$ state with this energy which respects the restriction of zero angular momentum, while maintaining the $0s$ centre-of-mass motion.

We also note that in eq. (7) there appear also terms where $n_i = n$, $n_j = 0$ and $\ell = 0$, as well as $n_i = 0$, $n_j = n$ and $\ell = 0$, which effectively correspond to $1p-1h$ excitations. So, even if we started from a (formally) two-body operator, we have ended up with a mixture of $1p-1h$ and $2p-2h$ operators. Finally, this discussion could also be extended in principle to consider $3p-3h$ and $4p-4h$ excitations from Ψ_0, but for present purposes we restrict ourselves to the $1p-1h$ and $2p-2h$ excitations considered above.

4. Translationally-invariant coupled cluster theory (TICC)

Standard coupled cluster theory assumes for the ground-state wave function the general form

$$\Psi = \exp\{\mathcal{S}_1 + \mathcal{S}_2 + \mathcal{S}_3 + \ldots\}\Psi_0, \tag{8}$$

where each $\mathcal{S}_n$ corresponds to $np-nh$ excitations with respect to some suitable model or reference state Ψ_0. We have shown in the previous section that this ansatz will spoil the centre-of-mass factorizability property and that at least operators $\mathcal{S}_1$ and $\mathcal{S}_2$ must appear in a special combination. Restricting ourselves to $SUB(2)$ approximation, i.e. only up to $2p-2h$ operators, our $TICC$ ansatz will be

$$\Psi = \exp\{\mathbf{S}\}\Psi_0, \tag{9}$$

with $\mathbf{S}$ given by eq. (7). Given that ^4He has only four particles, eq. (9) is equivalent to

$$\Psi = (1 + \mathbf{S} + \mathbf{S}^2/2)\Psi_0, \tag{10}$$

and the question now arises as to wether the quadratic terms will also respect the factorization property. The answer is that *they do not*, and that the $TICC$ ansatz must be correspondingly modified by adding a normal-ordering prescription to the exponential of the operator, namely

$$\Psi_{TICC} =: \exp\{\mathbf{S}\} : \Psi_0. \tag{11}$$

A particular way of checking this statement is to examine the coordinate representation of Ψ_{TICC}, as will be described in detail elsewere [24]. One has to project over the Fock space field operators [25], and after various Brody-Moshinsky transformations there results the expression

$$\Psi_{TICC}(\mathbf{r}_1\mathbf{r}_2\mathbf{r}_3\mathbf{r}_4) = \frac{\alpha^6}{\pi^3}\{1+2\sum_n S_n \sum_{i<j} \mathcal{F}_n(r_{ij}) + 2\sum_{nm} S_n S_m \sum_{i<j} \mathcal{F}_n(r_{ij})\mathcal{F}_m(r_{kl})\} e^{-\alpha^2 \sum_{p=1}^4 r_p^2/2} \tag{12}$$

where

$$\mathcal{F}_n(r) = \{\frac{2^n n!}{(2n+1)!!}\}^{1/2} L_n^{(1/2)}(\alpha^2 r^2/2), \tag{13}$$

and where $L_n^{1/2}(x)$ is the usual associated Laguerre polynomial. The last term in the curly bracket of eq. (12) has to be understood in a special way, namely the pair $\{kl\}$ represents the remaining two particles once the pair $\{ij\}$ has been selected. This wave function is not normalized althoug it does obey the so-called intermediate normalization condition, $\langle\Psi_0|\Psi_{TICC}\rangle = 1$. Finally, in these equations we have assumed a harmonic oscillator parameter $\alpha = (m\omega/\hbar)^{1/2}$. We note in addition that the quantities $\mathcal{F}(r)$ represent a relative n, s-wave motion for the pair of particles.

The equation equivalent to (12) but in terms of Fock operators is

$$\Psi_{TICC} = (4!)^{-1/2}\{(a_0^\dagger)^4 + 12\sum_n S_n \Omega_n\, a_0^\dagger a_0^\dagger + 12\sum_{nm} S_n S_m \Omega_n \Omega_m\}|0\rangle, \tag{14}$$

with

$$\Omega_n = \sum_{n_i n_j \ell} \langle n\,0\,0\,0\,0|n_i\,\ell\, n_j\,\ell\,0\rangle [a^\dagger_{n_i\ell} \times a^\dagger_{n_j\ell}]^0. \tag{15}$$

In conclusion, the $SUB(2)$ ansatz for a translationally-invariant coupled cluster theory of ^4He has three kinds of terms. The first is the non-interacting harmonic oscillator ground state corresponding to the first term inside the curly brackets in eq. (12). Then we have terms with only one function $\mathcal{F}_n$ corresponding to the excitation of a pair of particles in all possible ways to a state with relative n and $\ell = 0$ motion, and finally there are other terms in which one pair is excited to $n, \ell = 0$ and the other to $m, \ell = 0$. The unknowns of the problem are the c-number coefficients $\{S_n\}$ which have to be determined by solving the Schrödinger equation.

The problem is thus reduced to solving a set of non-linear algebraic equations for the coefficients $\{S_n\}$ and the energy eigenvalue. There is however another way of viewing eq. (12) which consists of introducing a two-body *correlation function*

$$g(r_{ij}) = \sum_n S_n\, \mathcal{F}_n(r_{ij}), \tag{16}$$

in terms of which we may rewrite the wave function as

$$\Psi_{TICC} = \frac{\alpha^6}{\pi^3}\{1 + 2\sum_{i<j} g(r_{ij}) + 2\sum_{i<j} g(r_{ij})g(r_{kl})\}\exp\{-\alpha^2\sum_p r_p^2/2\}. \tag{17}$$

In this form we have only an unknown function $g(r_{ij})$ to be determined. For this function it is possible to write down an integro-differential equation following the CC evaluation method [26] or by means of an Euler-Lagrange variational approach, as in the case of Jastrow correlations [13], or, finally, by following a method similar to the hyperspherical harmonics method of describing nuclei [8]. Actually, if the quadratic terms were omitted from eq. (17) we would have a form for the trial wave function very similar to the one used in the hyperspherical harmonics theory (e.g., see eq. (5) of ref. [27]). For present purposes we have decided to work in the harmonic oscillator basis, by using directly eqs. (12) or (14). In the first case the calculation were carried out by using specific properties of the generating function of the associated Laguerre polynomials. In the second option, eq. (14), we have simply used standard shell-model machinery. Details of the calculations will be published elsewhere [24].

There are several ways of actually using the parametrization given by eqs. (12, 14) which we briefly mention. The first is to consider the linear approximation in which the terms quadratic in the coefficients $\{S_n\}$ are neglected. By minimization of the resulting expectation value $\langle\Psi_{TICC}|H|\Psi_{TICC}\rangle/\langle\Psi_{TICC}|\Psi_{TICC}\rangle$ with respect to the set of coefficients $\{S_n\}$ we will end up with a matrix eigenvalue problem. This is equivalent to a configuration-interaction calculation using a selected set of basis states. The calculations performed in this way will be referred as $LTICC$, L standing for *linear.*

Secondly, we may determine the amplitudes $\{S_n\}$ in the standard coupled cluster way, i.e. by projecting the Schrödinger equation for the full wave function eq. (14) onto the uncorrelated state $(a_0^\dagger)^4|0\rangle$ and onto our special $2p-2h$ states $\Omega_n\, a_0^\dagger\, a_0^\dagger|0\rangle$. Once a maximum value for $n = n_{max}$ has been assumed, we will thereby end up with $n_{max}+1$ non-linear equations involving the ground-state energy E and the unknown amplitudes $\{S_n\}$. This method will be referred to as $TICC$, namely the standard, translationally-invariant coupled cluster approach.

Thirdly, once the amplitudes $\{S_n\}$ have been determined as above, we can compute the expectation value of the hamiltonian with the already known approximate wave function. This method will be referred to as $\langle TICC\rangle$. The difference between this way of determining the energy and the $TICC$ form gives a measure of the goodness of the coupled cluster approximation.

Fourthly and lastly, one may generalize eqs.(12, 14) by the formal replacement of $S_nS_m \rightarrow C_{nm}$, so that the amplitudes of the quadratic terms are no longer tied to the amplitudes of the linear terms. This is again a configuration-interaction calculation with a larger basis space than in the $LTICC$ approach. This kind of calculation will be referred to as the $QTICC$ method, with Q standing for *quadratic.*

The four approaches have been mentioned in order of computational complexity. For example, at $n_{max} = 29$, which corresponds to $58\hbar\omega$ excitation, the first method requires the computation of $30X30$ matrix elements, whereas the $QTICC$ requires the calculation of

Table 2. The ground-state energy in MeV of ^4He for the four interactions considered. The results shown correspond respectively to a variational calculation with respect to the oscillator parameter α in an uncorrelated wave function given by eq. (6) ($HOWF$), linear CC theory ($LTICC$), standard coupled cluster theory ($TICC$), diffusion Monte Carlo (DMC) and the Jastrow variational method ($JASTROW$). The asterisks indicate results not fully converged.

	METHOD				
POTENTIAL	HOWF	LTICC	TICC	DMC	JASTROW
Kalos	−23.15	−28.74	−28.79	-29.25 ± 0.05	−29.11
B1	−28.16	−37.80	−37.85	-38.5 ± 0.10	−36.44
S3	−5.89	−25.29*	−25.47*	-26.9 ± 0.20	−24.29
MTV	−6.40	−26.77*	−27.06*	-31.5 ± 0.20	−29.48

465×465 matrix elements. In any case, these numbers are still very much smaller than the dimensions corresponding to a full configuration interaction calculation [23] in the complete $n_{max}\hbar\omega$ space.

5. Results and discussion

The results of our calculations are shown in Tables 2 and 3, and in Figures 1 and 2. The two figures analyze the convergence of the $LTICC$ calculation in terms of the number of basis states and also in terms of the harmonic oscillator parameter α. We were very surprised by the results displayed in these two figures, as well as the equivalent plots for Fig. 1 in the case of the other three interactions. In order to get convergence it is necessary to go up to $n_{max} \simeq 15$ for the mild interactions (K and $B1$), but even $n_{max} = 30$ is not high enough to get a stabilized result for either the $S3$ or the MTV potentials. Naive (and unjustified) extrapolations to $n \rightarrow \infty$ suggest that the ground-state energy will reach values around $-26\,MeV$ for $S3$ and around $-28\,MeV$ for MTV. Unfortunately we could not go even further in our calculations, because it was already necessary at an appreciably lower level to move to quartic precision (128 bits) in order to avoid rounding errors associated with the Laguerre polynomials of correspondingly high orders. In other words, it is clearly not appropriate to work in the harmonic oscillator basis for these model problems. Instead it would have been more convenient to work directly in the coordinate representation. The second property to be stressed is the dependence of the energy on the harmonic oscillator parameter α. We have shown only the case of the $S3$ interaction (see Fig. 1) where there is only a small region around the minimum which can be considered flat. The same happens in the case of the MTV interaction, and also in the case of the smooth interactions K and $B1$, even if in these two cases the flat region is wider. The usual received wisdom in such generalized shell-model calculations is that one expects the results to be independent of the basis, once convergence is reached, but this is clearly not the case for our results. In other words, the $3p-3h$ and $4p-4h$ states which are lacking in our calculation must be responsible for making the comparable plots to those of Fig. 1 flat. Formulated in yet a different way, it is dangerous to extract conclusions about the relative importance of various clusters unless the

Table 3. The ground-state energy in MeV of ^4He for the three interactions of gaussian shape. The calculations correspond to the quadratic configuration-interaction method ($QTICC$), coupled cluster theory ($TICC$), and the expectation value of the energy for the coupled cluster wave function ($\langle TICC \rangle$). In these calculations $n_{max} = 12$ so the $S3$ results are very far from convergence.

	METHOD		
POTENTIAL	QTICC	TICC	⟨TICC⟩
Kalos	−28.873	−28.791	−28.802
B1	−37.193	−37.178	−37.182
S3	−20.216	−20.186	−20.193

proper harmonic oscillator parameter is used. We note that the two main conclusions coming from the analysis of Figs. 1 and 2, namely the need for very large bases, and the residual dependence on the harmonic oscillator parameter, are by no means exclusive to our coupled cluster approach. They also apply to more general configuration-interaction calculations.

The numerical results of our calculations are also shown in Table 2. The column labelled $HOWF$ corresponds to the optimal value for a $(0s)^4$ configuration, and the $LTICC$ and $TICC$ columns correspond respectively to the linear and to the general translationally-invariant coupled cluster calculation. The results are compared with the exact (within statistical errors) results of a diffusion Monte Carlo calculation (column DMC) [24,28] and with the results of a variational calculation for a simple trial Jastrow function depending on three parameters [28]. We see from this Table that the dominant contribution to the energy comes from the linear part of the coupled cluster wave function, the contribution of quadratic terms being a rather small fraction of $1\,MeV$. Focussing attention on our fully converged results, we observe that we are less than $1\,MeV$ from the exact results, thus indicating that $2p-2h$ excitations are by far the most important contribution to the correlated wave function.

Finally, Table 3 is concerned with the goodness of the coupled cluster form of solving the Schrödinger equation. In this Table we show in the column labelled $QTICC$ the results computed as discussed at the end of the preceeding Section, i.e., by decoupling the quadratic terms from the linear terms, and diagonalizing the hamiltonian in our special $2p-2h$ and $4p-4h$ basis with pairs of particles coupled to zero angular momentum. In addition, the third column labelled $\langle TICC \rangle$ corresponds to first performing the standard coupled cluster theory (with translation invariance incorporated, as in column $TICC$) and then taking the expectation value of the energy for the wave function so determined. These three columns should in principle be the same in the case of an exact calculation in which no truncations in the coupled cluster basis are made. Actually they are very close, the difference being less than $0.1\,MeV$, and this clearly implies that the $SUB(n)$ truncation scheme is a very appropriate one. We note that the calculations needed to compute this Table are much more time-consuming than the corresponding calculations shown in Table 2. For this reason, we could not compute with such a very large value of n_{max}, being only able to reach the value of 12. However, the same behaviour was observed at smaller values of n_{max} so that one should

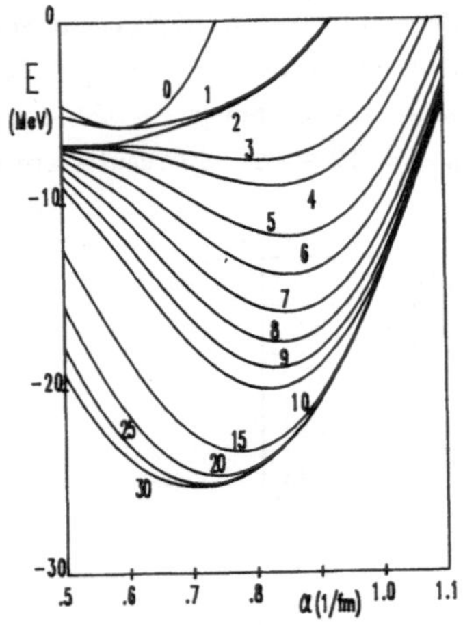

Figure 1: The convergence of the *LTICC* method for the *S*3 interaction as a function of the harmonic oscillator parameter α for different values of n_{max}. The curves are labelled by the value of n_{max}.

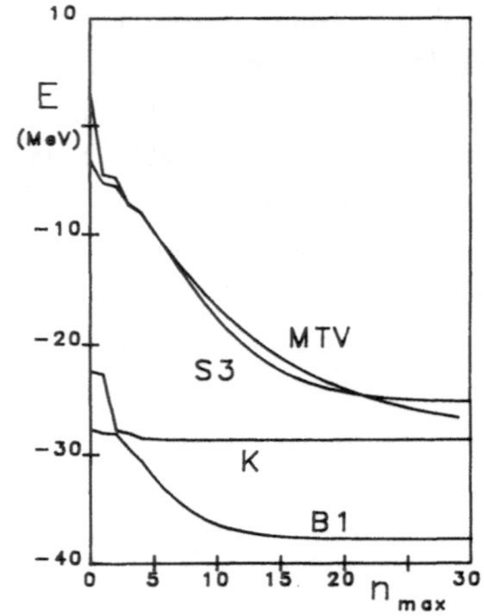

Figure 2: The convergence of the *LTICC* calculations in terms of n_{max} for the four interactions considered. The harmonic oscillator parameter is the optimal value (for $n_{max} \simeq 30$) in each case.

not expect singnificant changes at the higher values of n_{max} needed to attain fully converged results.

The main conclusion of our work is that for this light system the most important part of the wave function may be written in the form $\{\sum_{i<j} f(r_{ij})\}|(0s)^4\rangle$. This form is very similar to that used in the hyperspherical harmonics approximation. Furthermore it corresponds to a configuration-interaction calculation in a special basis and it is also equivalent to a small-correlation expansion of the Jastrow form. Thus, we see rather clearly that all of these theories are very efficient to describe light systems like ^4He. Conversely, the very slow an rather non-uniform approach to convergence of all of our calculations in the harmonic oscillator basis, even when carried out to such virtually unprecedently high levels of excitation energy as the $60\hbar\omega$ reported here must cause grave concern about the efficacy of the standard implementations of the nuclear shell model which aim to go beyond an effective-interaction level of approximation.

Acknowledgements

This work was developed under an *Acciones Integradas* programme between Spain and the United Kingdom. The authors acknowledge the financial support from the corresponding joint committee. M.C.B., E.B. and R.G. are also supported by the *Comisión Interministerial de Ciencia y Tecnología, Spain* under contract 969/87. R.F.B. also acknowledges the support of a research grant from the *Science and Engineering Research Council* of Great Britain.

References

[1] F. Coester, *Nucl. Phys.* 7 (1958) 421.

[2] F. Coester and H.G. Kümmel, *Nucl. Phys.* 17 (1960) 477.

[3] H. Kümmel, K.H. Lührmann and J.G. Zabolitzky, *Phys. Rep.* 36C (1978) 1.

[4] R.F. Bishop and H. Kümmel, *Physics Today* 40 (1987) No. 3, p. 52.

[5] R.F. Bishop and K.H. Lührmann, *Phys. Rev.* B17 (1978) 3757

[6] D.M. Ceperley and M.H. Kalos, *Quantum Many Body Problems* in *Monte Carlo Methods in Statistical Physics*, K. Binder editor (Springer Verlag, New York 1979).

[7] R. Jastrow, *Phys. Rev.* 98 (1955) 1479.

[8] M. Fabre de la Ripelle, H. Fiedeldey and A. Sofianos, *Phys. Rev.* C38 (1988) 449.

[9] R.A. Malfliet and J.A. Tjon, *Nucl. Phys.* A127 (1969) 161.

[10] J.G. Zabolitzky, K.E. Schmidt and M.H. Kalos, *Phys. Rev.* C25 (1982) 1111.

[11] J.G. Zabolitzky, *Phys. Lett.* 100B (1981) 5.

[12] Y. Akaishi, *Nucl. Phys.* A416 (1984) 409.

[13] L. Bracci, S. Rosati and M. Viviani, in *Proceedings of the Secondo Convegno su Problemi di Física Nucleare Teorica*, L. Bracci, P. Christillin, A. Fabrocini, S. Fantoni, G. Fiorentini, A. Kievsky, M. Rosa-Clot, S. Rosati, S. Servadio and M. Viviani editors. (ETS editrice, PISA 1988), p. 34.

[14] J. Carlson and V.R. Pandharipande, *Nucl. Phys.* A371 (1981) 301.

[15] J.A. Tjon, *Phys. Rev. Lett.* 40 (1978) 1239.

[16] R. Ceuleneer and P. Vandepeutte, *Phys. Rev.* C31 (1985) 1528.

[17] M.H. Kalos, *Phys. Rev.* 128 (1962) 1791.

[18] D.M. Brink and E. Boeker, *Nucl. Phys.* A91 (1967) 1.

[19] I.R. Afnan and Y.C. Tang, *Phys. Rev.* 175 (1968) 1337.

[20] H.J. Lipkin, *Phys. Rev.* 110 (1958) 1395.

[21] C.M. Vincent, *Phys. Rev.* C8 (1973) 929.

[22] J.B. McGrory and B.H. Wildenthal, *Phys. Lett.* 60B (1975) 5.

[23] R. Ceuleneer, P. Vandepeutte and C. Semay, *Phys. Lett.* 196 (1987) 303.

[24] R.F. Bishop, M.F. Flynn, M.C. Boscá, E. Buendía and R. Guardiola, in preparation.

[25] S.S. Schweber, *An Introduction to Relativistic Quantum Field Theory*, Harper and Row (New York, 1964), chap. 6.

[26] J.G. Zabolitzky, *Nucl. Phys.* A228 (1974) 272.

[27] M. Fabre de la Ripelle, *C. R. Acad. Sci. Paris* 306, serie II (1988) 1313.

[28] R.F. Bishop, M.F. Flynn, M.C. Boscá, E. Buendía and R. Guardiola, in Proceedings of the NATO school *The Nuclear Equation of State*, W. Greiner editor (Plenum Press, New York, to be published).

ELECTRON CORRELATIONS IN ATOMS

C.E. Campbell†, Tao Pang†, and E. Krotscheck‡
† School of Physics and Astronomy
University of Minnesota
Minneapolis, Minnesota 55455, U.S.A.

‡ Theoretical Physics Institute
School of Physics and Astronomy, University of Minnesota
Minneapolis, Minnesota 55455
and
Center for Theoretical Physics, Department of Physics
Texas A & M University College Station, Texas 77843, U.S.A.

INTRODUCTION

Theories of electron correlat.ıns in condensed matter systems with a high degree of density variation and/or electron localization may be tested in many-electron atoms and ions, where nearly exact correlation energies are known. We have adapted the inhomogeneous, non-local Feenberg-Jastrow Euler-Lagrange theory developed by Krotscheck and his collaborators[1,2] to apply to atomic systems. Correlation energies for four electron and ten electron closed shell atoms (Be and Ne, respectively) are in good agreement with the known results. We also calculate the pair correlation functions, which are found to exhibit an extreme sensitivity to the location of the pair of electrons within the atom.

In the next section we define the theory and the approximations which are necessary within the theory. A more complete discussion of this theory in the context of other inhomogeneous quantum fluids is given by Krotscheck *et al.* elsewhere in this volume.

The third section contains the results of this work for Be and Ne. We conclude with a brief discussion.

FEENBERG-JASTROW EULER-LAGRANGE THEORY OF INHOMOGENEOUS QUANTUM FLUIDS

The Hamiltonian for an inhomogeneous system is

$$H = -\frac{1}{2}\sum_i^N \nabla_i^2 + \sum_i^N v_1(\mathbf{r}_i) + \sum_{i<j} v_2(r_{ij}), \tag{1}$$

Condensed Matter Theories, Volume 5
Edited by V.C. Aguilera-Navarro
Plenum Press, New York, 1990

where $v_1(r)$ is the external potential, $v_2(r_{ij})$ is the two-body potential, and N is the number of electrons in the system. In the present case, $v_1(r) = -Z/r$ where r is the distance to the nucleus, and $v_2(r) = 1/r$, where we use atomic units ($\hbar = e = m = 1$), and the nucleus is taken to have an infinite mass.

The Jastrow-Feenberg wave function space is defined by:

$$|\Psi\rangle = \exp\left\{\frac{1}{2}\left[\sum_i u_1(\mathbf{r}_i) + \sum_{i<j} u_2(\mathbf{r}_i, \mathbf{r}_j)\right]\right\}|\Phi\rangle \tag{2}$$

where $|\Phi\rangle$ is a Slater determinant of one body orbitals which are mutually orthonormal, and $u_1(\mathbf{r})$ and $u_2(\mathbf{r}_i, \mathbf{r}_j)$ are the one- and two-body Jastrow pseudopotentials. Each of these functions is determined by functional variation to minimize the energy expectation value, E:

$$\frac{\delta E}{\delta u_n} = \frac{\delta}{\delta u_n}\frac{\langle\Psi|H|\Psi\rangle}{\langle\Psi|\Psi\rangle} = 0, \qquad (n = 1, 2) \tag{3}$$

The optimal $u_1(\mathbf{r})$ depends solely on the choice of the single- particle orbitals ψ_i which which are solutions of the correlated Hartree-Fock equation:

$$[-\frac{1}{2}\nabla^2 + U_{ext}(\mathbf{r}) + V_H(r)]\psi_i(r) + \int d^3r' V_g(\mathbf{r}, \mathbf{r}')\rho_1(\mathbf{r}, \mathbf{r}')\psi_i(\mathbf{r}') = \epsilon_i\psi_i(\mathbf{r}), \tag{4}$$

where $V_H(\mathbf{r})$ is the generalized Hartree potential, $V_g(\mathbf{r}, \mathbf{r}')$ is the exchange/correlation interaction, and $\rho_1(\mathbf{r}, \mathbf{r}')$ is the one body density matrix of the wave function.

The two-body Euler-Lagrange equation involves the two-body Jastrow pseudopotential $u_2(\mathbf{r}_i, \mathbf{r}_j)$ and the pair distribution function $g(\mathbf{r}, \mathbf{r}')$, defined by

$$g(\mathbf{r}_1, \mathbf{r}_2)\rho_1(\mathbf{r}_1)\rho_1(\mathbf{r}_2) = N(N-1)\frac{\int d^3r_3 \ldots d^3r_N |\Psi(\mathbf{r}_1, \ldots, \mathbf{r}_N)|^2}{\int d^3r_1 \ldots d^3r_N |\Psi(\mathbf{r}_1, \ldots, \mathbf{r}_N)|^2}, \tag{5}$$

where $\rho_1(\mathbf{r})$ is the one-body density of the system, as well as other diagrammatically defined two-body and three-body functions.

The chief remaining task in the formal theory is to find a tractable relationship between the ingredients of the wave function (Φ, $u_1(\mathbf{r})$, and $u_2(\mathbf{r}_i, \mathbf{r}_j)$) and the derived quantities such as the density, pair distribution function, exchange/correlation energy, etc. For highly correlated systems, this is achieved using the fermion version of the hypernetted chain resummation (FHNC), which then requires an approximation for manageable calculations. In this work we use the minimal acceptable approximation, FHNC//0, whose chief ingredient is the "direct" two-point distribution function $g_{dd}(\mathbf{r}, \mathbf{r}')$, which is one plus the sum of all direct two-point Born-Mayer type diagrams where there is no exchange between the two external points $(\mathbf{r}, \mathbf{r}')$ with any internal points. Thus $g_{dd}(\mathbf{r}, \mathbf{r}')$ is a renormalized version of $\exp(u_2(\mathbf{r}, \mathbf{r}'))$, and $\Gamma_{dd}(\mathbf{r}, \mathbf{r}') = g_{dd}(\mathbf{r}, \mathbf{r}') - 1$ can be viewed as the dimensionless correlation hole around a particle at $\mathbf{r}'$ as a function of $\mathbf{r}$. In the FHNC//0 approximation, the pair distribution function is given by

$$g(\mathbf{r}, \mathbf{r}') = [1 + \Gamma_{dd}(\mathbf{r}, \mathbf{r}')]\left\{g_F(\mathbf{r}, \mathbf{r}') + \frac{1}{\sqrt{\rho_1(\mathbf{r})\rho_1(\mathbf{r}')}}\left[S_F * \tilde{\Gamma}_{dd} * S_F - \tilde{\Gamma}_{dd}\right](\mathbf{r}, \mathbf{r}')\right\}, \tag{6}$$

where $S_F(\mathbf{r},\mathbf{r}')$ and $g_F(\mathbf{r},\mathbf{r}')$ are the structure factor and pair distribution function of the uncorrelated state $|\Phi\rangle$. $[A * B](\mathbf{r},\mathbf{r}')$ means the convolution integral of the two two-body functions, and the tilda above a function is defined by $\tilde{A}(\mathbf{r},\mathbf{r}') = \sqrt{\rho_1(\mathbf{r})}A(\mathbf{r},\mathbf{r}')\sqrt{\rho_1(\mathbf{r}')}$.

RESULTS FOR Be AND Ne

The energy of atomic systems is almost entirely accounted for by Hartree Fock. Correlations in electronic systems are significant only for low density regimes, which means the outer part of the atomic systems. The correlation energy E_c is defined as the difference between the total energy E_{TOT} and the Hartree Fock energy E_{HF}:

$$E_c = E_{TOT} - E_{HF} \tag{7}$$

This correlation energy consists of two parts: a positive contribution which comes from the fact that the single particle orbitals ψ_i are solutions of the *correlated* Hartree-Fock equation, and thus are not the best *uncorrelated* Hartree-Fock orbitals; and the remaining contributions which arise primarily from the difference between the particle-hole interaction and the bare Coulomb interaction, as well as the differences between the fully correlated $g(\mathbf{r},\mathbf{r}')$ and the uncorrelated $g_F(\mathbf{r},\mathbf{r}')$. This latter negative quantity exceeds the positive Hartree Fock shift, giving a total negative value for the correlation energy.

Table. 1. Correlation energies for Neon and Beryllium atoms in atomic units. a: experimental data[3]; b: present work; c: LSD results[4]; d: generalized gradient expansion of LM[5]; and e: generalized gradient expansion of Perdew[6].

	a	*b*	*c*	*d*	*e*
Be	-0.0944	-0.096	-0.224	-0.099	-0.094
Ne	-0.39	-0.33	-0.74	-0.41	-0.39

This correlation energy is of the order of 1% or less of the total energy. Nevertheless, Clementi and Veillard[3] have obtained this energy from experiments on four and ten electron atomic systems. Their results, shown in the Table, have been corrected for center of mass energy and relativistic effects, and thus can be compared directly to our results for non-relativistic atoms with an infinite mass nucleus. It is seen from the table that our results are in good agreement with these experiments; some differences should be expected from the FHNC//0 apprximation, which produces approximately 10% errors in the case of jellium.

The density profile of these atoms is very close to the Hartree Fock densities, although the small differences which appear in the low density tail of the atoms is a significant effect in the correlation energies.

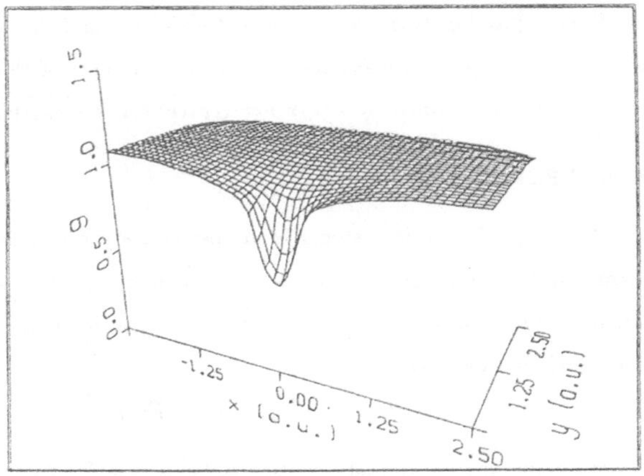

Fig. 1a. $g(\mathbf{r},\mathbf{r}')$ in Neon for an electron located at the origin, i.e. $r' = 0$ a.u. $x = r\hat{\mathbf{r}}\cdot\hat{\mathbf{r}}'$ is the projection of $\mathbf{r}$ on $\mathbf{r}'$ and $y = r\left[1-(\hat{\mathbf{r}}\cdot\hat{\mathbf{r}}')^2\right]^{\frac{1}{2}}$. The atomic center is at the origin. Atomic units are used.

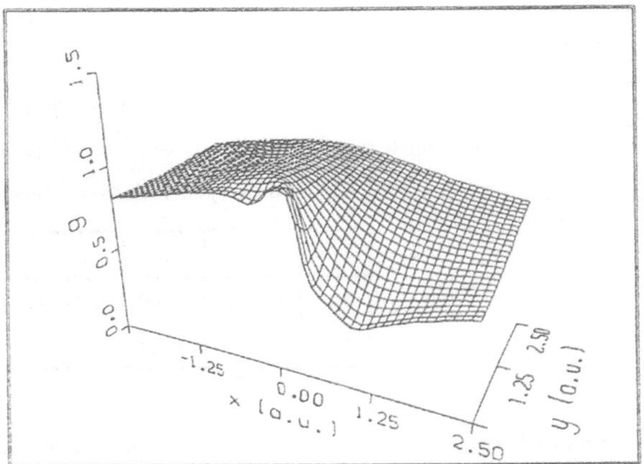

Fig. 1b. Same as Fig. 1a for a particle located a distance 1 a.u. from the nucleus

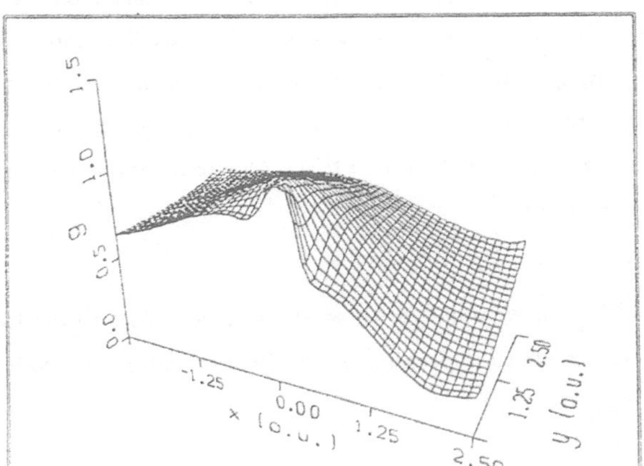

Fig. 1c. Same as Fig. 1a for a particle located at a distance 2 a. u. from the nucleus.

Fig. 2a. Same as Fig. 1b for the Hartree-Fock approximation $g_F(\mathbf{r},\mathbf{r}')$.

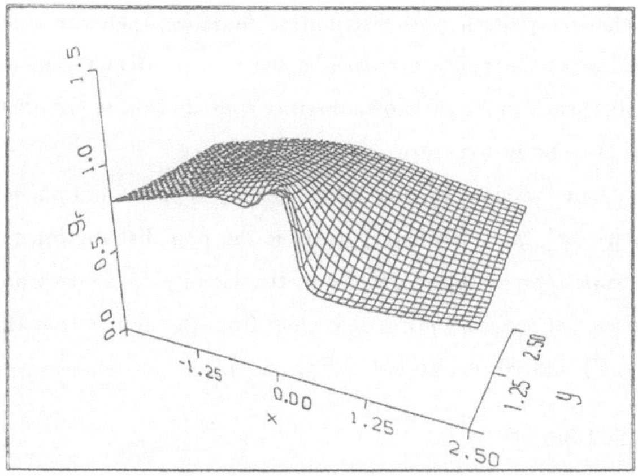

Fig. 2b. Same as Fig. 1b for the correlation contribution $g_{dd}(\mathbf{r},\mathbf{r}') = 1 + \Gamma_{dd}(\mathbf{r},\mathbf{r}')$.

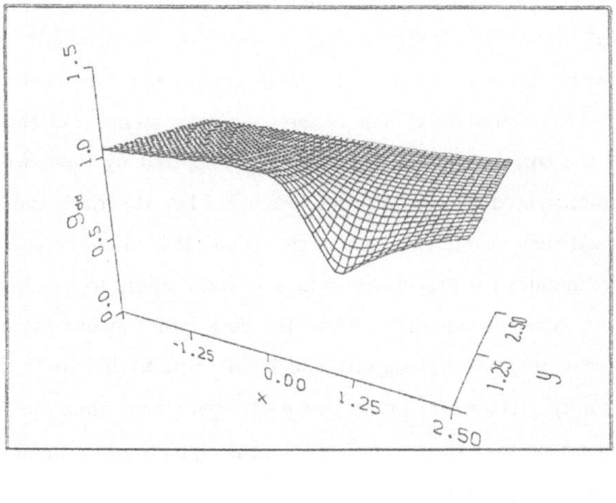

The pair distribution function is a familiar tool for examining two-body correlations in classical and quantum fluid systems, and gives a similar but more complex view of these correlations in inhomogeneous systems. The spherical symmetry of closed shell atomic systems simplifies the exhibition of this structure somewhat. Thus $g(\mathbf{r},\mathbf{r}')$ depends only on the distances r and r' from the origin and the angle between the two vectors. In Figure 1 we present this function as a function of $\mathbf{r}$ for fixed $\mathbf{r}'$, where the origin (the nucleus) and $\mathbf{r}'$ define the x axis, and $g(\mathbf{r},\mathbf{r}')$ is plotted above the plane defined by $\mathbf{r}$ and $\mathbf{r}'$. (For the spherical atom, all such planes are equivalent). It can be seen in Figure 1 that the exchange-correlation hole (the depression of the value of $g(\mathbf{r},\mathbf{r}')$ as $\mathbf{r}$ approaches $\mathbf{r}'$) becomes increasingly important as $\mathbf{r}'$ increases into the low density tail of the atom. Of course even the Hartree Fock result must contain an exchange hole by the Pauli principle, so the question of how much of the exhange-correlation hole can be attributed to exchange (i.e., statistical correlations) and how much to correlation (i.e., dynamical correlations) requires further examination. Fortunately, as Eq. 6 shows, these effects factor to a very good approximation. Thus $g(\mathbf{r},\mathbf{r}')$ is approximately the product of $g_{dd}(\mathbf{r},\mathbf{r}')$ and $g_F(\mathbf{r},\mathbf{r}')$, where the former is dominated by dynamic correlations and is 1 if the two-body interaction vanishes, and the latter is the free particle pair distribution function, therefore containing only statistical correlations. These two quantities and g are each shown in a particular case in Figure 2, where it is seen that these two factors each make their own distinct contribution to the exchange-correlation hole as well as the rest of the two- body structure.

One version of the local density approximation is obtained by approximating $g(\mathbf{r}, \mathbf{r}')$ by $g_J(|\mathbf{r}-\mathbf{r}'|, \langle\rho\rangle_{r,r'})$, where $g_J(r,\rho)$ is the pair distribution function for jellium at density $\langle\rho\rangle_{r,r'}$, which is some representative density in the vicinity of the two points $\mathbf{r}, \mathbf{r}'$, e.g., the geometric or arithmetic average of $\rho(r)$ and $\rho(r')$. It is clear from the figures that these approximations are not appropriate for $g(\mathbf{r}, \mathbf{r}')$; indeed, it also fails for $g_F(\mathbf{r}, \mathbf{r}')$.

DISCUSSION

Small atoms provide one of the most stringent tests of a theory of electron correlations, since the density falls very rapidly to zero away from the nucleus. Consequently our results for neon and beryllium are very encouraging. Preliminary results for other closed shell four and ten electron systems (Ne^{+6}, Mg^{+2}, and Ca^{+10}) are similarly encouraging. In principle, the theory can also be applied to open shell systems; however the lack of spherical symmetry will require additional effort.

It is abundantly clear in these simple systems that the local density approximation is inadequate for the correlation energies. This was expected by those who introduced this approximation[7] and was demonstrated by comparisons between LDA theories[4] and experiments for the four and ten electron closed shell atoms as is seen in the Table. It is also seen there that adjusted gradient expansions[5,6] give agreement with experiment which is comparable to our theory. However it was necessary in each of these gradient expansions to fit the data either at one atom (Perdew[6] adjusts his theory to fit Ne) or to several systems (Langreth and Mehl[5], who fit Be, Ne^{+6}, and Ne); an unadjusted gradient expansion actually gives worse agreement with experiment than the LDA. Thus we conclude that the Jastrow Euler-Lagrange theory is a very promising method for including none-local density effects in atomic and molecular systems.

Since this theory is also a theory of the pair distribution function, it provides much information about the two-body structure of the system, and permits a fairly clean separation between dynamic correlations and statistical correlations.

ACKNOWLEDGEMENTS

We would like to thank Professor V. Aguilera-Navarro for organizing this stimulating workshop, and the Army Research Office for travel support in conjunction with this workshop. This research was supported in part by the National Science Foundation, the Robert A. Welch Foundation , the Theoretical Physics Institute of the University of Minnesota, the Graduate School of the University of Minnesota and the Minnesota Supercomputer Institute. One of us (EK) would like to thank the Theoretical Physics Institute of the University of Minnesota for warm hospitality.

REFERENCES

1. E. Krotscheck, Phys. Rev. **B31**, 4258 (1985).
2. E. Krotscheck and W. Kohn, Phys. Rev. Lett. **57** (1986) 862; E. Krotscheck, G.-X. Qian, and W. Kohn, Phys. Rev. **B32** (1985) 5693.
3. E. Clementi, J. Chem. Phys. **38** (1963) 2248; A. Veillard and E. Clementi, J. Chem. Phys. **45** (1968) 2415.
4. Y. Tong and L. J. Sham, Phys. Rev. **144** (1966) 1.
5. D. C. Langreth and M. J. Mehl, Phys. Rev. **B28** (1983) 1809.
6. J. Perdew, Phys. Rev. **B33** (1986) 8822.
7. W. Kohn and P. Vashishta, in *Theory of the inhomogeneous electron gas*, p. 79 ff, edited by S. Lundqvist and N. H. March (Plemum, N.Y. 1983).

THE FOUNDATION OF THE NUCLEAR SHELL MODEL†

W. H. Dickhoff and P. P. Domitrovich

Department of Physics, Washington University
St. Louis, Missouri 63130, USA

A. Polls and A. Ramos

Departament d'Estructura i Constituents de la Matèria
Universitat de Barcelona, E-8028 Barcelona, Spain

Abstract: The nuclear shell model exists now for forty years. The notion of independent particle motion which is at the basis of this simple picture of nuclei has recently been challenged by new experimental data. Nevertheless, a considerable amount of experimental data has successfully been interpreted over the years in terms of this picture. On a microscopic level, the success of the shell model has been linked to the strong short-range repulsion in the nuclear interaction by introducing the concept of the healing of the relative wave function of two nucleons. Recent calculations of the effective interaction in nuclear matter are discussed which demonstrate that this concept of healing is not valid but instead particles scatter at all energies in the medium. It will be argued that the success of the shell model picture should instead be linked to the validity of the quasi-particle picture as used by Landau for liquid ^{3}He. Calculations of the spectral function for ^{16}O are discussed which demonstrate the essential difference between a finite nucleus and nuclear matter. The finiteness of a nucleus ensures that a relevant single-particle basis has shell structure, implying that there is an energy window in nuclei related to the difference between particle and hole energies in which this quasi-particle picture can be applied.

1. INTRODUCTION

The simplest version of the shell model considers nucleons bound in some attractive well moving independently from each other. This well can be thought of as being generated by the average interaction a nucleon experiences with the other

†This research was supported by the Condensed Matter Theory Program of the Division of Materials Research of the U.S. National Science Foundation under Grant No. DMR-8519077 (at Washington University) which also provided computer time for the calculations which were partly performed at the Pittsburgh Supercomputer Center. A travel grant from the U.S. Army Research Office is gratefully acknowledged. Additional support was provided in part by NATO under Grant No. RG 85/0684 and in part by CAYCIT Grant No. PB85-0072-C02-00 (Spain).

Condensed Matter Theories, Volume 5
Edited by V.C. Aguilera-Navarro
Plenum Press, New York, 1990

nucleons in the system. Since this average interaction is obviously attractive, nucleons can be assigned single-particle (sp) quantum numbers of a spherical well. When an average spin-orbit interaction is included, this nuclear shell model can account for all observed magic numbers as well as a number of other experimental data. This is essentially the picture as it has been published 40 years ago.[1,2] It has been very difficult to link this mean field picture of the nucleus and the strong repulsion present in the nucleon-nucleon (NN) interaction on a microscopic level. Historically, Brueckner[3] suggested to convert the bare interaction into a renormalized interaction (G matrix) which includes any number of interactions (ladder diagrams) between the particles in the medium, similar to the corresponding T matrix in free space. This G matrix is obtained by solving the Bethe-Goldstone equation.[4] This interaction has played an important role in the efforts to reconcile the strong interaction at short distance with the tranquil picture of nucleons moving in simple sp orbits. In ref. 5 the concept of healing of the relative wave function is introduced as an explanation for the validity of the shell model. This healing property results from the fact that the conventional solution of the Bethe-Goldstone equation does not allow scattering for energies below $2\varepsilon_F$ as a consequence of the Pauli principle and is therefore purely real.

One of the purposes of this contribution is to show that this analysis must be improved and extended. This new analysis will rely on explicit calculations which have been performed recently for nuclear matter.[6] In this work the short-range correlations of the Reid soft core interaction were carefully treated and the resulting sp properties were studied in detail. This study is also relevant for finite nuclei since short-range correlations should be similar in finite and infinite systems for the same bare interaction. It will be shown that the concept of healing as discussed in the literature has no strict validity since the particles in the medium do scatter at all energies. In sec. 2 of this contribution the results of these calculations which employ the Green's function method will be reviewed and its implications for the nuclear shell model indicated. Special attention will be paid to the nucleon spectral function and the relevance of the quasi-particle concept. In sec. 3 the modifications of these results due to the finiteness of the system will be discussed. Results for calculations of the spectral function for ^{16}O will illustrate the usefulness of the quasi-particle idea for finite nuclei and its relation to the success of the simple shell model.

2. SHORT-RANGE CORRELATIONS AND SINGLE-PARTICLE PROPERTIES

In ref. 6 the original idea of Brueckner how to treat the influence of short-range correlations, is embedded in a Self-Consistent Green's Function (SCGF) approach. The ladder equation is solved with inclusion of both particle-particle (pp) as well as hole-hole (hh) propagation. The connection between this interaction and the resulting sp properties is made by calculating the resulting self-energy. This self-energy can then be used to calculate the sp propagator or the particle and hole spectral functions in nuclear matter. In principle, this requires a self-consistent solution in which the full dressing of the sp propagator is included in the determination of the ladder correlations. At present this self-consistency has been established in ref. 6 for the quasi-particle energy which corresponds to the inclusion of only the real on-shell part of the self-energy in the self-consistency procedure.

$$\varepsilon(k) = \frac{k^2}{2m} + \mathrm{Re}\ \Sigma(k,\varepsilon(k)) \quad . \tag{1}$$

When this average self-consistency is achieved the complete energy and momentum dependence of the self-energy can be studied. This requires knowledge of the imaginary part of the self-energy above but also below the Fermi energy. It is this last contribution which has never been considered in the conventional treatment using the Bethe-Goldstone equation which only considers pp propagation in the ladder equation. Including hh propagation one naturally treats the coupling of sp degrees of freedom to 2h1p and more complicated states and therefore one obtains an imaginary contribution to the self-energy below ε_F. A Green's function procedure automatically treats hh propagation on the same footing as pp propagation. Recently, this hh propagation (to all orders) has also been shown[7] to be the crucial link in connecting the coupled ladder self-energy problem with the description of pairing, going even beyond the conventional BCS description.[8]

Including hh propagation in the ladder equation implies that the resulting interaction is complex not only for energies above $2\varepsilon_F$ but also below. This conflicts with the notion of healing which has been related to the G matrix being purely real for energies below $2\varepsilon_F$ in the case of the Bethe-Goldstone equation.[5] Obviously, collisions do take place in the correlated system at any energy. This statement must be regarded as general for a normal Fermi liquid. The validity of shell model concepts in finite nuclei can therefore not be linked to this healing property of the relative wave function.

Calculations have been performed for the central part of the $^3S_1 - {}^3D_1$ channel of the Reid soft core potential.[9] This avoids the strong pairing instability[10] in the deuteron channel and allows a careful study of the influence of short-range correlations on the sp self-energy.[6] The self-energy of a nucleon contains all the information necessary to obtain occupation probabilities, quasi-particle strength and broadening features which alternatively can be visualized in terms of hole and particle spectral functions. Results for the hole spectral function are given in Fig. 1 for $k = 0.828$, 1.258, 2.220, and 4.376 fm^{-1} at a density corresponding to $k_F = 1.4$ fm^{-1}. Results for the particle spectral function are given in Fig. 2 for the same momenta. Clearly one observes the increasing validity of the quasi-particle concept when k approaches k_F since the peak of the spectral function becomes sharper the closer the momentum is to k_F. This holds for momenta below k_F (Fig. 1) as well as for momenta above k_F (Fig. 2). One observes therefore in this actual calculation which uses a strongly repulsive NN interaction at short distances, that the notion of quasi-particles is valid in exactly the same sense as it has been used by Landau in his description of Fermi liquids.[11] In an infinite system this concept is strictly valid only for the Fermi momentum. For other momenta one observes therefore that the further the momentum value is removed from k_F the more the quasi-particle peak is broadened.

For momenta close to k_F most of the hole strength is under the quasi-particle peak when k is smaller than k_F. For $k = 1.258$ fm^{-1} 83% of the hole strength is under the peak whereas 3% is found at lower energies. The occupation of this momentum in the correlated ground state is obtained by integrating the hole spectral function up to ε_F and therefore corresponds to 86%. The remaining strength is moved beyond the Fermi energy and belongs to the particle spectral function shown in Fig. 2. The remaining 14% of the strength is smeared out over a very large energy domain which is intimately connected with the character of the interaction which couples sp degrees of freedom to very high-lying 2p1h etc. states. This removal of sp strength to high energy is observed for all momenta by the same amount (see Fig. 2) emphasizing the short-range nature of the effect. The jump in the momentum distribution at k_F is equal to the strength of the quasi-particle pole at k_F and a value of 0.83 is obtained here. The inclusion of tensor correlations is expected to decrease this value further by a few percent.[12]

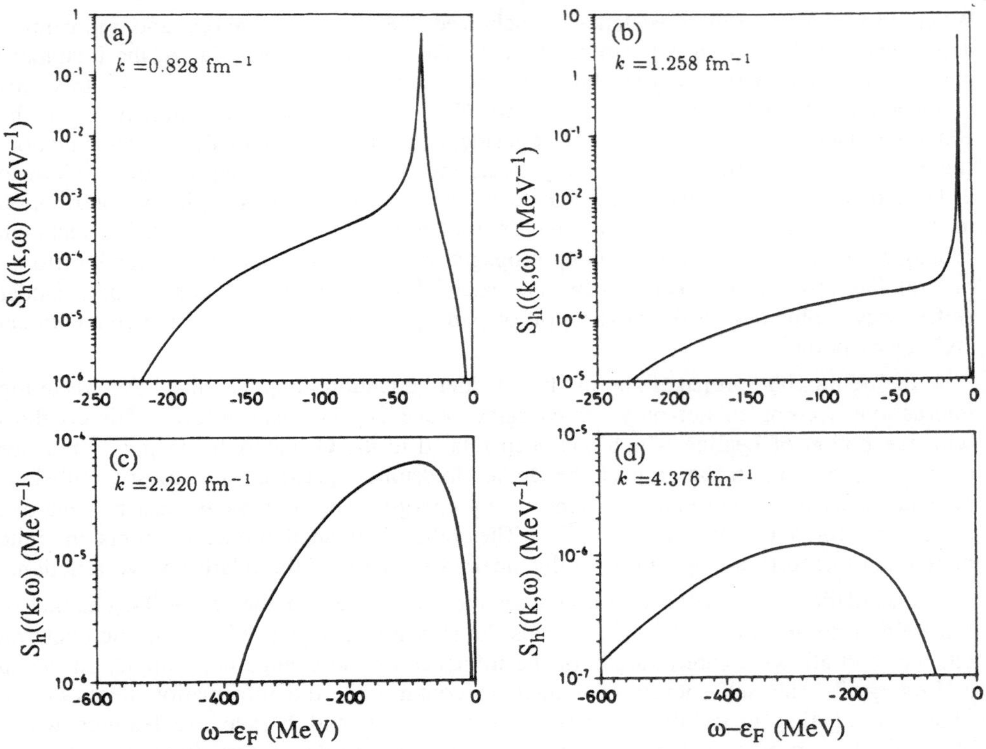

Fig. 1. Hole spectral functions at $k_F = 1.4\ fm^{-1}$ for k = 0.828 1,258, 2.220, and 4.376 fm^{-1}.

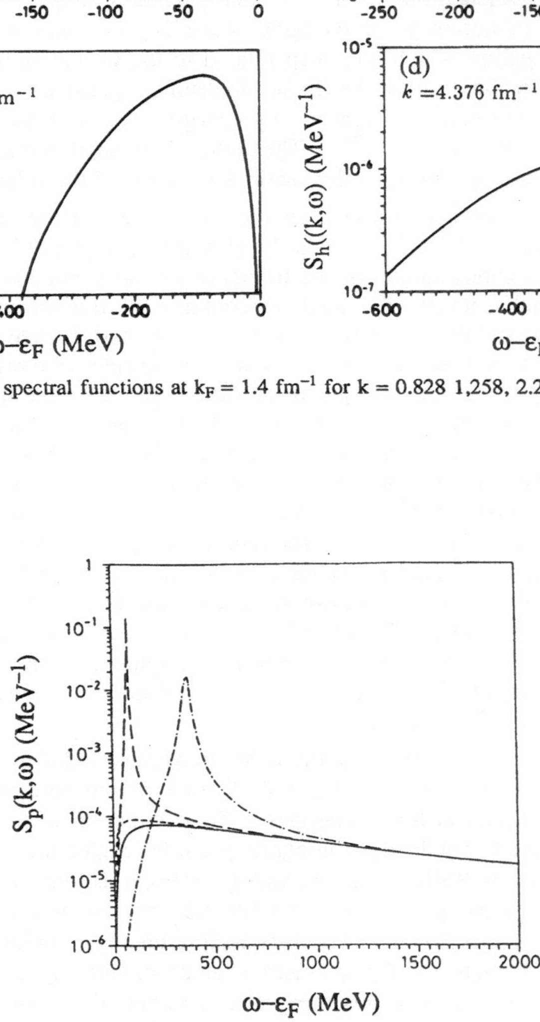

Fig. 2. Particle spectral functions at $k_F = 1.4\ fm^{-1}$ for k = 0.828 (solid), 1.258 (short dashed), 2.220 (long dashed), and 4.376 fm^{-1} (dot dashed).

These and other results are more fully discussed in ref. 6 together with the calculational details. One can conclude from these results that even though there is strong scattering in the medium, the concept of a quasi-particle picture is still extremely useful in exactly the same way as it has been used in Landau's Fermi liquid theory[11] for ^{3}He. For a normal Fermi liquid this validity of the quasi-particle concept is almost exclusively a phase space effect related to the presence of 2p1h and 2h1p states around the Fermi surface which can mix with sp states. The strength of the quasi-particle pole is then a good measure of the correlations, the closer to 1 the more the system can be described in terms of a mean field picture.

The role of short-range correlations should be similar in finite nuclei and infinite nuclear matter at the corresponding density. These results have therefore direct bearing on the situation in finite nuclei and depletion effects due to short-range correlations are inevitable.[13] From the present results one can also infer that the missing sp strength is not just "around the corner"; it is removed to very high energies as a consequence of the short-range correlations. This effect of the short-range correlations and its consequences have not been properly considered in Brueckner type calculations.

3. FINITE NUCLEI AND IMPLICATIONS FOR THE SHELL MODEL

Presently, the breakdown of the mean field picture is an experimental reality. Combined analysis of (e,e′p) reactions[14] together with absolute charge density distribution measurements[15] indicates that the occupation of the 3s1/2 proton shell in ^{208}Pb is 80 ± 10%. This should be compared to the theoretical results of refs. 13,16 and 17. Direct analysis of (e,e′p) cross sections with shell model momentum distributions suggests that the discrete low energy transitions carry about 50% of the hole strength.[18,19] This would imply that 30% of the hole strength is at even higher energies in the A−1 nucleus and cannot be detected. The analysis of ref. 16 for the neutron states in ^{208}Pb indicates that about 73% of the hole strength is found at low energy for such states with occupation numbers corresponding to 0.85. Assuming similar results for protons this indicates a 20% discrepancy between this theoretical result and the experimental analysis.

To address this question it is profitable to investigate the influence of low-lying excitations on the redistribution of sp strength directly in finite nuclei. One should keep in mind that the nuclear matter results discussed above indicate that somewhat more than 10% of the sp strength is at high energy. This result is due to a basic asymmetry between the influence of pp and hh contributions and should be taken into account in analyses like the one in ref. 16. The spectral functions in finite nuclei can be obtained directly by solving the Dyson equation in a suitable basis for the relevant nucleus. Results are reported here for ^{16}O. First a Hartree-Fock (HF) basis is generated by solving the HF equations using the G matrix interaction G^{II}_{NM} which is a good local and static approximation of a ^{16}O G matrix.[20] As in any HF calculation a mean field is generated in which one can define fully occupied and completely empty sp states for this nucleus. The G matrix that is used here does not include any shell structure information since it is calculated originally for nuclear matter with a standard gap in the sp spectrum.[21]

To calculate the influence of low energy excitations on the distribution of sp strength the second order self-energy contribution calculated with this interaction is used to solve the Dyson equation.

$$g(\alpha,\beta;\omega) = g^{(0)}(\alpha,\beta;\omega) + \sum_{\gamma,\delta} g^{(0)}(\alpha,\gamma;\omega)\Sigma^{(2)}(\gamma,\delta;\omega)g(\delta,\beta;\omega) \quad . \tag{2}$$

The sp quantum numbers α,β, etc. refer to the HF basis. The sums in Eq. (2) are restricted to HF states within 50 MeV of the Fermi energy of the HF calculation. This ensures that indeed the influence of low-lying 2p1h and 2h1p states on the redistribution of sp strength in a finite system is studied. Contributions from higher shells will eventually also lead to some form of double counting since these states have already been considered in the construction of the residual interaction. The presence of an explicit energy dependence in this second order self-energy contribution is responsible for the fragmentation of sp strength both above the Fermi energy by coupling to 2p1h states as well as below the Fermi energy by coupling to 2h1p states. This can be clearly seen by inspection of the explicit second order self-energy contribution

$$\Sigma^{(2)}(\gamma,\delta;\omega) = \frac{1}{2}\sum_{\varepsilon\mu\nu} \langle\gamma\mu|G^{\text{II}}_{\text{NM}}|\varepsilon\nu\rangle\langle\varepsilon\nu|G^{\text{II}}_{\text{NM}}|\delta\mu\rangle$$
$$\times\left[\frac{\theta(\varepsilon-F)\theta(\nu-F)\theta(F-\mu)}{\omega-(\varepsilon_\varepsilon+\varepsilon_\nu-\varepsilon_\mu)+i\eta}+\frac{\theta(F-\varepsilon)\theta(F-\nu)\theta(\mu-F)}{\omega-(\varepsilon_\varepsilon+\varepsilon_\nu-\varepsilon_\mu)-i\eta}\right] \quad . \tag{3}$$

The solution to Eq. (2) has the form

$$g(\alpha,\beta;\omega) = \sum_n \frac{\langle\Psi_0^A|a_\alpha|\Psi_n^{A+1}\rangle\langle\Psi_n^{A+1}|a_\beta^\dagger|\Psi_0^A\rangle}{\omega-(E_n^{A+1}-E_0^A)+i\eta} + \sum_m \frac{\langle\Psi_0^A|a_\beta^\dagger|\Psi_m^{A-1}\rangle\langle\Psi_m^{A-1}|a_\alpha|\Psi_0^A\rangle}{\omega-(E_0^A-E_m^{A-1})-i\eta} . \tag{4}$$

The poles in this equation give the energies of the states in the A ± 1 systems which can couple to the ground state of the A-body system through the creation and annihilation operators a and $a^\dagger$, respectively. The residues at the poles for the diagonal sp propagator correspond to the spectroscopic factors for the corresponding transition from the ground state in the A-body system to the state n(m) in the A+1(A-1)-body system. The energies and residues in Eq. (4) are obtained by solving an energy dependent eigenvalue problem which can be derived from Eq. (2).[22]

In Fig. 3 results for the sp strength distribution are shown for the relevant $s_{1/2}$ states in ^{16}O. The results are smeared by distributing the strength of each peak by 0.5 MeV. In the HF picture, the lowest $s_{1/2}$ state has full occupation, *i.e.*, a spectroscopic factor of 1.0 to one state in the A−1 nucleus. This result is also typical for Brueckner-Hartree-Fock calculations in which the energy dependence of the G matrix is treated explicitly. When no hh contributions in the ladder equation are included the resulting self-energy is real below the Fermi energy and therefore leads to HF like hole states. The next $s_{1/2}$ state is completely empty and has a spectroscopic factor of 1.0 to one state in the A+1 system. Including dynamic correlations in this second order self-energy calculation still leads to a concentration of the $2s_{1/2}$ strength in one peak as shown in Fig. 3. In contrast the $1s_{1/2}$ strength (represented by all the other peaks in the figure) is completely fragmented with most of its strength below the Fermi energy leading in this case to an occupation of 96%. The remaining 4% is located at higher energy shown by the smaller peaks to the right of the $2s_{1/2}$ peak.

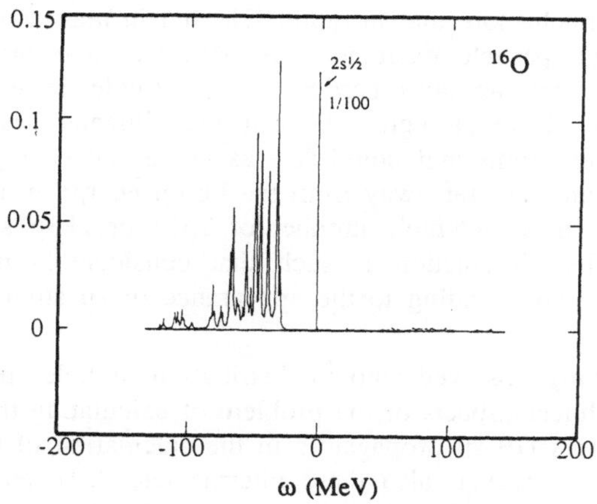

Fig. 3. Distribution of s1/2 strength in ^{16}O as a function of energy. The 2s1/2 strength is located mainly in one peak reduced here by a factor of 1/100. The rest of the 2s1/2 strength is not shown. All other peaks correspond to strongly fragmented 1s1/2 strength. Note that a small portion of 1s1/2 strength is also found above the Fermi energy.

This calculation therefore has the same typical features that one encounters in nuclear matter. For sp states which are close to the Fermi energy the dynamic coupling in the self-energy results in one strong peak which carries less than but of the order of 100% of the strength. In a finite nucleus like ^{16}O this is particularly obvious since there are no 2p1h or 2h1p states in the immediate vicinity of the $2s_{1/2}$ energy and the strength remains mainly concentrated in one state since more complicated states are too far away in energy. This feature emphasizes the important role of the finiteness of the system. It leads automatically to a finite ph energy gap which results in a larger window of validity of the quasi-particle approximation in finite nuclei. For the $1s_{1/2}$ state one observes a strong fragmentation since its energy (in HF) is located in an energy interval where (in this case) many 2h1p are located and it becomes relatively easy to mix with such more complicated states. The same result is observed in infinite systems for momenta far from the Fermi momentum. Experimental results on finite nuclei by means of the (e,e′p) reaction show a smooth $1s_{1/2}$ strength distribution for proton knock-out of ^{16}O.[23] Recent (p,2p) and (p,pn) experiments on ^{16}O show the same feature.[24] Clearly the knock-out of an $1s_{1/2}$ particle is possible for a broad range of energies.

The calculation gives similar results as discussed above for other orbitals. Particles in shells closer to the Fermi energy like the $1p_{3/2}$ and the $1p_{1/2}$ are found in large fractions in the low energy states of the A = 15 system but do not carry 100% of the strength. On the other side of the Fermi energy most of the $1d_{5/2}$, $2s_{1/2}$ (see Fig. 3) and $1d_{3/2}$ strength is found in the low energy states of the A = 17 system. Unoccupied states which in HF are far removed from the Fermi energy show a similar fragmentation as the $1s_{1/2}$ state on the other side of the Fermi energy.

These results are in sharp contrast with the mean field picture of nuclei. In an atom *e.g.*, where the Hartree-Fock approximation works very well, all hole states have a well defined energy and carry 100% of the strength. The situation in a nucleus therefore resembles much more that in liquid ^{3}He and in nuclear matter as discussed in sec. 2. Nuclear single-particle excitations as observed near the Fermi energy must therefore be reinterpreted as quasi-particle or quasi-hole excitations which carry a considerable amount of sp strength. Present experimental results put the nucleus somewhere in between atoms and liquid ^{3}He as far as the strength of correlations are concerned. States which are far away from the Fermi energy in mean field are located at energies at which a considerable number of 2p1h or 2h1p states are found. The strength of the nuclear interaction is such that considerable mixing between these degrees of freedom occurs leading to the appearance of sp strength in a broad energy range.

The experimentally observed smooth distribution of hole strength[18,19,23,24] further emphasizes the non-linear aspects of the problem of calculating the sp propagator. The present calculation uses HF sp propagators in the calculation of the second order self-energy contribution. However, also these intermediate 2h1p and 2p1h states are not good eigenstates (see Eq. (3)) and are broadened themselves. This effect can be incorporated by using self-consistent sp propagators in the calculation of the self-energy in the same non-linear fashion as the HF problem is solved. This self-consistent formulation is necessary to explain the observed experimental results of smooth hole strength distributions. This observation suggests that it is not very realistic to obtain information on the widths of such strength distributions from nuclear matter calculations. The role of the ph gap in a finite system in the fragmentation of sp strength should be explicitly treated.

The results obtained here for ^{16}O do not yet include the effect of short-range and tensor correlations on the sp propagator but instead emphasize the role of the finiteness of the system with its shell structure and surface. These calculations show that up to 5% of strength (using second order self-energy) can be found in tiny fragments below the Fermi energy as a result of the coupling of sp states to low-lying 2h1p states. Taking the global effect of short-range correlations (as discussed in sec. 2) into account and adding another 5% depletion effect due to tensor correlations,[12,17] one expects about 20% of the sp strength to be located above ε_F for normally occupied states. Still, most of the remaining strength for these valence hole states should be found in discrete peaks and not much more than 10% at lower energies (higher excitation energy in the A−1 system) according to the present theoretical results. This is in reasonable agreement with the results of ref. 16, but disagrees with the present analysis of the (e,e′p) experiment[18,19] which assigns about 50% of the hole strength to discrete states at low energy and from a combined analysis together with charge density distributions[15] leads to 80% occupation for the $3s_{1/2}$ state in ^{208}Pb.

From the results discussed here and in sec. 2 one can conclude that the success of the conventional shell model is related to the success of the quasi-particle picture for nuclear systems and the finiteness of the nucleus which extends the window of the validity of this picture. In a Fermi liquid this picture pertains strictly to the Fermi momentum and energy. Since particle-hole excitations start already at zero excitation energy, there is immediately this broadening effect when one moves away from the Fermi surface. In contrast to this, a finite particle-hole excitation energy exists in nuclei since these systems are confined in space which leads to discrete single-particle energies. For this reason there is a broader validity of the quasi-particle picture in nuclei and the shell model makes sense when it is not pushed beyond its window of validity. Finally, one should also remember the difficulty to explain the experimental charge density distribution of nuclei[25] when a mean field picture is employed. From partially occupied $s_{1/2}$ states in a nucleus one obtains an automatic reduction of the central charge density which can qualitatively explain the experimental data.

The conclusion one can draw from the above discussion is that a marriage of the idea of Brueckner to treat short-range correlations and the notion of quasi-particles employed by Landau both developed already in the fifties, leads to a fascinating picture of the nucleus as a correlated many-body system and also provides a basis for a theoretical foundation of the nuclear shell model in the presence of strong interactions. Present experimental and theoretical evidence suggests that nuclei are considerably more correlated than atoms and the mean field picture for nuclei is not valid.

REFERENCES

1. O. Haxel, J. H. D. Jensen, and H. E. Suess, Phys. Rev. **75** (1949) 1766.
2. M. Goeppert-Mayer, Phys. Rev. **75** (1949) 1969.
3. K. A. Brueckner, C. A. Levinson and H. M. Mahmoud, Phys. Rev. **95** (1954) 217.
4. H. A. Bethe and J. Goldstone, Proc. Roy. Soc. **A238** (1957) 1531.
5. L. C. Gomes, J. D. Walecka and V. F. Weisskopf, Ann. Phys. **3** (1958) 241.
6. A. Ramos, A. Polls, and W. H. Dickhoff, Nucl. Phys. **A** in press; in Condensed Matter Theories, vol. 3, ed. J. S. Arponen, R. F. Bishop and M. Manninen (Plenum, New York, 1987) p. 319.
7. W. H. Dickhoff, Phys. Lett. **210B** (1988) 15.
8. J. Bardeen, L. N. Cooper and J. R. Schrieffer, Phys. Rev. **108** (1957) 1175.
9. R. V. Reid, Ann. Phys.**50** (1968) 411.
10. W. H. Dickhoff, *Condensed Matter Theories Vol.4*, ed. J. Keller (Plenum, New York, 1989) in press.
11. L. D. Landau, Soviet Phys.**8** (1958) 70.
12. S. Fantoni and V. R. Pandharipande, Nucl. Phys. **A427** (1984) 473.
13. V. R. Pandharipande, C. N. Papanicolas and J. Wambach, Phys. Rev. Lett. **53** (1984) 1133.
14. E. N. M. Quint *et al.*, Phys. Rev. Lett. **58** (1987) 1088.
15. J. M. Cavedon *et al.*, Phys. Rev. Lett. **49** (1982) 978.
16. C. Mahaux and R. Sartor, Nucl. Phys. **A493** (1989) 157.
17. O. Benhar, A. Fabrocini and S. Fantoni, preprint.
18. E. N. M. Quint, Ph.D. Thesis (University of Amsterdam, 1988).
19. L. Lapikás, preprint NIKHEF (1989).
20. P. Czerski, W. H. Dickhoff, A. Faessler and H. Müther, Nucl. Phys. **A427** (1984) 224.
21. W. H. Dickhoff, Nucl. Phys. **A399** (1983) 287.
22. P. P. Domitrovich and W. H. Dickhoff, to be published.
23. J. Mougey, Nucl. Phys.**A335** (1980) 35.
24. W. J. McDonald *et al.*, Nucl. Phys. **A456** (1986) 577.
25. B. Frois and C. Papanicolas, Ann. Rev. Nucl. Part. Sci. **37** (1987) 133.

DEVELOPMENTS IN MULTIREFERENCE COUPLED-CLUSTER

APPLICATIONS TO MOLECULAR SYSTEMS*

Uzi Kaldor

School of Chemistry
Tel Aviv University
69 978 Tel Aviv, Israel

INTRODUCTION

The exp(S) or coupled-cluster (CC) method[1-4] has been used increasingly in recent years for atomic and molecular calculations. Most of the single-reference applications have been surveyed recently by Bartlett.[5] Our contributions to the two previous volumes in the present series document the development of applying the multireference version to atomic and molecular systems. The first of these[6] includes a detailed presentation of the methodology and preliminary applications, and the second[7] emphasizes the problem of intruder states and incomplete model spaces (using the Be $2p^2$ ^1S resonance as example) and describes the first application of the method to molecular potential functions. Recent developments, extending the scope of the method, are reported below. These include the caculation of 35 vertical excitation energies of N_2, a molecule with a relatively dense excitation spectrum; accurate potential functions for the lowest nine states of Li_2, using incomplete model spaces; and the first application to a chemical reaction, the prototype deprotonation reaction $NH_4^+ \rightleftarrows NH_3 + H^+$.

Many versions of the multireference coupled-cluster method appear in the literature,[8-24] although most of them have rarely been applied. The basic approach is to define an effective Hamiltonian in a low-dimensional model (or P) space, with eigenvalues approximating some desirable eigenvalues of the physical Hamiltonian. The effect of the complementary Q space is taken into account while calculating the matrix elements of the effective Hamiltonian, using an appropriate truncation of the wave operator.

*Supported in part by the U.S.-Israel Binational Science Foundation.

Most derivations of the coupled-cluster equations follow the original open-shell many-body perturbation theory (MBPT) of Brandow[25] in the choice of the model space. The orbitals are classified as core, valence, or particles (unoccupied). The core orbitals are always occupied, and all possible distributions of the remaining electrons in the valence orbitals give rise to determinants included in the P space (there may also be valence holes, i.e. unoccupied core orbitals, but the situation is not fundamentally different). Such model spaces have been called "complete".[26] This recipe is appropriate when the open-shell orbitals are close in energy, which is not the case for most atomic and molecular excited states. It is often impossible to select valence orbitals so that no Q space determinants (with one or more non-valence orbitals) lie close to or even within the energy range spanned by the P space. Perhaps the simplest example is the 1s2s ^1S state of He. If the 1s and 2s orbitals are selected as valence, the P space will include the $1s^2$, 1s2s and $2s^2$ determinants, but exclude the various 1sns terms lying within its energy span. This situation leads to the so-called "intruder states",[27] which destroy the convergence of the expansion. A general, incomplete model space MBPT has been proposed by Hose and Kaldor[26] and used in extensive calculations.[28] A similar CC method has been described by Jeziorski and Monkhorst.[16] Significant theoretical progress has been made in recent years in understanding incomplete model spaces.[13,19,29-31] The number of applications has also increased considerably.[6,19,32-36] The main interest in CC applications with incomplete model spaces has been aimed at calculating one-electron excitation energies,[32,34] where a natural choice of P includes determinants having one hole and one particle with respect to the closed-shell ground state. This is a special case of "quasicomplete" model spaces.[13,30] Sinha et al.[37] have shown that the energy calculation for the particular case of a 1h-1p space is operationally equivalent to the complete-space procedure. The calculation of N_2 excitations described below is of this type. More general model spaces are often necessary, as shown in the study of the Be $2p^2$ ^1S resonance.[7,38] The cases of Li_2 potential functions and NH_4^+ deprotonation reaction reported here fall into this category.

METHOD

The basic method used in previous work[32] follows Lindgren's[12] choice of a normal-ordered wave operator,

$$\Omega = \{\exp(S)\} = 1 + S + \frac{1}{2}\{S^2\} + \ldots \quad . \tag{1}$$

S is the excitation operator describing *connected* single, double, . . . excitations,

$$S = S_1 + S_2 + \ldots = \sum_{ij}\{a_i^\dagger a_j\}s_j^i + \frac{1}{2}\sum_{ijk\ell}\{a_i^\dagger a_j^\dagger a_\ell a_k\}s_{k\ell}^{ij} + \ldots \quad , \tag{2}$$

where s_j^i , $s_{k\ell}^{ij}$, . . . , are excitation amplitudes, and the curly brackets denote normal order with respect to a reference (core) determinant. The summation is carried over connected terms only. The equations determining the excitation amplitudes for a complete model space may be derived from

$$[S,H_0] = \{QV\Omega - \Omega V_{eff}\}_{conn} \quad , \tag{3}$$

$$V_{eff} = V\Omega \quad , \tag{4}$$

where H_0 and V result from the partitioning of the Hamiltonian in the usual way,

$$H = H_0 + V \quad . \tag{5}$$

Mukherjee[29] has shown that a similar formulation may be derived for a general model space. The model space $P^{(m)}$ with m valence electrons is chosen on physical grounds, and may be incomplete. Model spaces $P^{(k)}$, $k<m$, are then constructed by deleting m-k orbitals in all possible ways from the $P^{(m)}$ determinants. An operator is designated k-open if it corresponds to a $P^{(k)} \rightarrow Q^{(k)}$ transition, where $Q^{(k)}$ is the complement of $P^{(k)}$; otherwise it is k-closed. The construction of the $P^{(k)}$ spaces causes all m-closed operators to be k-closed for all $k<m$; m-open operators may however be k-closed (in other words, an orbital change transforming every $P^{(k)}$ determinant to another $P^{(k)}$ function may take some $P^{(m)}$ determinant to a $Q^{(m)}$ term). The basic equations for the k-valence sector are then[28]

$$[S,H_0]^{(k)}_{m\text{-}op} = \{V\Omega - \Omega V_{eff}\}^{(k)}_{m\text{-}op,conn} \tag{6}$$

$$\{\Omega V_{eff}\}^{(k)}_{m\text{-}cl} = \{V\Omega\}^{(k)}_{m\text{-}cl} \quad . \tag{7}$$

Two differences between Eqs. (6)-(7) and (3)-(4) should be noted. The classification of the transitions at the k-valence level into P→P and P→Q has to be done according to their effect on m-valence states; and the equations for V_{eff} are implicit [Eq. (7)] rather than explicit [Eq. (4)]. The former requires some additional, not very difficult bookkeeping. The latter involves a few diagrams not encountered in complete model spaces, and the solution of a set of equations for V_{eff} matrix elements. As the new diagrams are relatively simple and the equation system is of low dimension, incomplete model spaces are not more difficult to handle than complete ones.

EXCITED STATES OF N_2

N_2 has a dense excitation spectrum even at relatively low energies, presenting a challenge both to theoretical evaluation and to interpretation of experimental data. Not all of its low-lying states are well characterized. Orbital excitations of both valence and Rydberg character can mix strongly. A reliable theoretical prediction therefore requires *(i)* a sophisticated method which includes high-order effects, and *(ii)* a flexible basis which allows accurate description of both valence and Rydberg excitations. The coupled-cluster method in the singles and doubles approximation,

Table 1. N_2 excitation energies (eV)

method:[a]	Exp[b]	CCSD	TDHF	SOPPA	SOPPA	MBPT	MRDCI	MRDCI
Basis size		52	124	66	36-58	36	54	40
Reference			46	45	44	47	48	48
$A^3\Sigma_u^+$	7.75	7.56	7.64		7.91	7.49	7.96	7.93
$B^3\Pi_g$	8.04	8.05	8.17		7.87	8.39	8.27	8.28
$W^3\Delta_u$	8.88	8.93	8.86	9.01	8.93	8.85	9.25	9.39
$a^1\Pi_g$	9.31	9.27	9.60		9.32	9.90		9.63
$B'^3\Sigma_u^-$	9.67	9.86	10.07	9.99	9.96	9.48		9.98
$a'^1\Sigma_u^-$	9.92	10.09	10.40	10.05	10.02	10.02		10.56
$w^1\Delta_u$	10.27	10.54	10.76	10.54	10.51	10.26	10.57	10.73
$C^3\Pi_u$	11.19	11.19	11.43	11.10	11.05		11.19	11.36
$E^3\Sigma_g^+$	12.0	11.75				11.63	11.59	11.44
$a''^1\Sigma_g^+$	12.2	12.20				12.08		11.78
$c^1\Pi_u$	12.90	12.84			12.11			
$c'^1\Sigma_u^+$	12.98	12.82			12.30			
$b^1\Pi_u$	13.24	13.61			13.68			13.77
$o^1\Pi_u$	13.63	13.71		13.83				13.23
$b'^1\Sigma_u^+$	14.25	14.31		14.30	14.35			14.74
$e'^1\Sigma_u^+$	14.48	14.65						14.66
Average error, eight lowest states		0.11	0.27	0.19	0.14	0.22	0.23	0.35

[a]TDHF: multireference time-dependent Hartree-Fock. SOPPA: second order polarization propagator. MBPT: third order many-body perturbation theory, using the [2/1] Pade approximant. MRDCI: multireference configuration interaction, with extrapolation to estimate full CI.

[b]Values fitted in ref. 44 to experimental spectroscopic constants. $^1\Pi_u$ and $^1\Sigma_u^+$ states from ref. 43.

$$S \simeq S_1 + S_2 \quad , \tag{8}$$

is applied to the molecule at an internuclear separation of 2.074 bohr, the equilibrium separation of the ground state. The 6-311G basis[39] is used, with one set of d orbitals (ζ=0.913, 3s combination omitted) and two sets of diffuse s and p orbitals with exponents 0.05 and 0.013. The final basis is thus 6s5p1d and includes a total of 52 functions. The Hartree-Fock orbital energies are shown in figure 1. The selection of valence holes and valence particles, out of which the excited states are generated, is determined by our aim of calculating a considerable number of excited states, and by the desirability of having

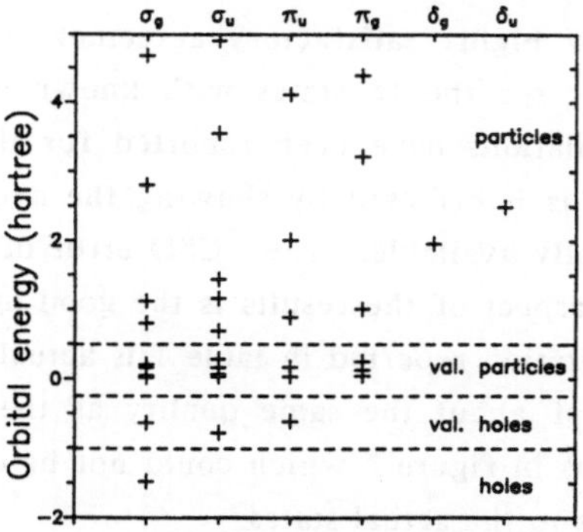

Figure 1. N_2 orbital energies

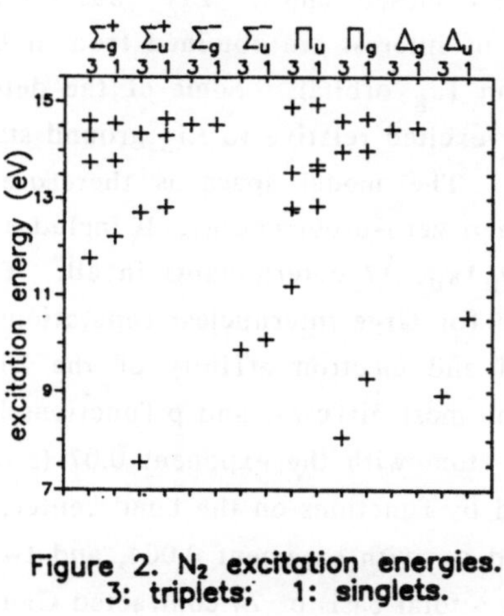

Figure 2. N_2 excitation energies.
3: triplets; 1: singlets.

the model space well separated from the Q space, to facilitate convergence. The selection shown in figure 1 gives a total of 96 states, but the higher ones are not meaningful. The 35 states below 15 eV are shown in figure 2, and the sixteen states for which we found experimental data are compared with experiment[40-43] and with other calculations[44-48] in table 1.

The data in table 1 show highly satisfactory agreement of the CCSD results with experiment, the average error for the 16 states with known experimental values being 0.13 eV. As very few calculations have been reported for the higher of these states, comparison with other methods is effected by showing the average error for the lowest eight states, which are generally available. The CCSD error of 0.11 eV is lower than all others. The most gratifying aspect of the results is the good agreement obtained for the higher states (the highest excitation reported in table 1 is actually the 33rd state found in our calculation), which are of about the same quality as the lower states. This may indicate that the 19 excitations in figure 2 which could not be compared with experiment also give good approximations to the actual states.

Ionization potentials are generated in the course of the calculations. The IP for $3\sigma_g$ is 15.44 eV, for $1\pi_u$ - 17.11 eV, and for $2\sigma_u$ - 18.66 eV. These may be compared with the corresponding values obtained from the potential curves of Lofthus and Krupenie,[42] which are 15.6, 17.2, and 18.7 eV. Very good agreement is given in this case as well.

Li_2 POTENTIAL FUNCTIONS

The potential functions of the nine lowest states were calculated. The reference state for the calculation must have closed shells. Li_2^{++} was selected because it dissociates correctly, and all the states of interest are obtained from it by adding two electrons in the $2\sigma_g$, $3\sigma_g$, $2\sigma_u$, $1\pi_u$, and $1\pi_g$ orbitals. Some of the determinants thus formed are very high in energy (doubly excited relative to Li_2 ground state), and a complete model space calculation diverges. The model space is therefore constructed by selecting determinants according to their zero-order energy. It includes functions with at least one $2\sigma_g$ electron, as well as $2\sigma_u 1\pi_u$, 17 determinants in all. The $2\sigma_u{}^2$ function must be included in the model space for large internuclear separations. The basis set developed for the ionization potential and electron affinity of the Li atom[32] was modified by replacing the exponent of the most diffuse s and p functions by 0.0093504, and adding a set of d functions on each atom with the exponent 0.07 (s component excluded). The basis was further augmented by functions on the bond center, taken from Schmidt-Mink et al.[49] These include s and p_σ with exponent 0.004, and two sets of p_π and $d_{\pi,\delta}$ with exponents 0.12 and 0.04, for a total basis of 74 contracted Gaussian-type orbitals.

The calculated potential functions are shown in figure 3, and table 2 presents a comparison of the CCSD molecular properties of $^7Li_2$ with experiment[50] and with previous calculations using the multi-configuration SCF[51] and effective core potential[49] methods. The latter method includes one adjustable parameter, selected to fit the atomic

Table 2. Molecular constants of $^7Li_2$.

	method[a]	R_e(A)	T_e(eV)	D_e(eV)	ω_e(cm^{-1})
$X^1\Sigma_g^+$	exp	2.673	--	1.056	351.4
	CCSD	2.67	--	1.061	351
	MCSCF	2.692	--	1.029	347.1
	ECP-CI	2.675	--	1.050	351.0
$a^3\Sigma_u^+$	CCSD	4.05	1.023	0.0380	75
	MCSCF	4.234		0.0362	61
	ECP-CI	4.182	1.010	0.0399	63.7
$b^3\Pi_u$	exp	2.591	1.394	1.510	345.7
	CCSD	2.58	1.375	1.533	349
	MCSCF			1.424	339
	ECP-CI	2.595	1.396	1.506	345.9
$A^1\Sigma_u^+$	exp	3.108	1.744	1.160	255.5
	CCSD	3.10	1.750	1.158	257
	MCSCF	3.13		1.153	254
	ECP-CI	3.108	1.742	1.160	256.1
$^3\Sigma_g^+$	CCSD	3.06	2.032	0.876	252
	MCSCF	3.096		0.845	245
	ECP-CI	3.067	2.025	0.877	252.2
$^1\Sigma_g^+$	CCSD	3.56	2.530	0.378	137
	ECP-CI	3.655	2.496	0.406	129
$B^1\Pi_u$	exp	2.936	2.534	0.370	270.7
	CCSD	2.94	2.557	0.350	269
	MCSCF	3.050		0.197	288
	ECP-CI	2.942	2.542	0.360	270.3
$^1\Pi_g$	CCSD	4.02	2.726	0.181	97
	MCSCF	4.161		0.161	83
	ECP-CI	4.073	2.728	0.174	91.8

[a] exp: experimental, reference 50.
CCSD: coupled cluster, single and double excitations (present work).
MCSCF: multiconfiguration self consistent field, ref. 51.
ECP: effective core potential, one adjustable parameter, reference 49.

ionization potential. The properties compared include the dissociation energy D_e, the equilibrium separation R_e, the adiabatic excitation energy T_e, and the vibrational constant ω_e. A cubic spline fit of the potential served to extract these constants, and ω_e was fitted from the lowest two or three vibrational levels. Excellent agreement is obtained for the four experimentally known states, with error limits of 0.01A for R_e, 0.02eV for D_e and T_e, and 3 cm^{-1} for ω_e. This is better than any previous *ab initio* results, and almost as good as the best effective core potential work.[49]

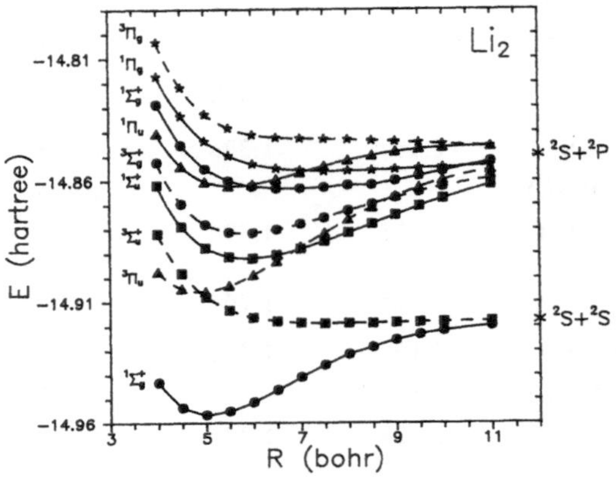

Figure 3. Li_2 potentials

DEPROTONATION OF NH_4^+: A TWO-SURFACE REACTION

The deprotonation of NH_4^+ (or $R_1R_2R_3NH^+$) is important in fields ranging from the decomposition of energetic compounds to proton transfer in biomolecules. From a methodological point of view, its interest lies in the existence of two crossing $^1A_1$ potential surfaces, dissociating respectively to $NH_3 + H^+$ or to $NH_3^+ + H$. The closed-shell Hartree-Fock function describes only the first process, which actually leads to dissociation products with higher energy.

The reaction path was calculated with the Huzinaga-Dunning[52] (9s5p)→[5s3p] set for N and (4s)→[3s] for H. Polarization functions (3d on N, 2p on H) were taken from Urban et al.[53] MRD-CI[54] and CCSD calculations were performed at several H--NH_3 separations. The angle θ between the C_3 axis and the NH bond in the NH_3 group was optimized at the MRD-CI level for the lowest state at each separation R. Figure 4 shows

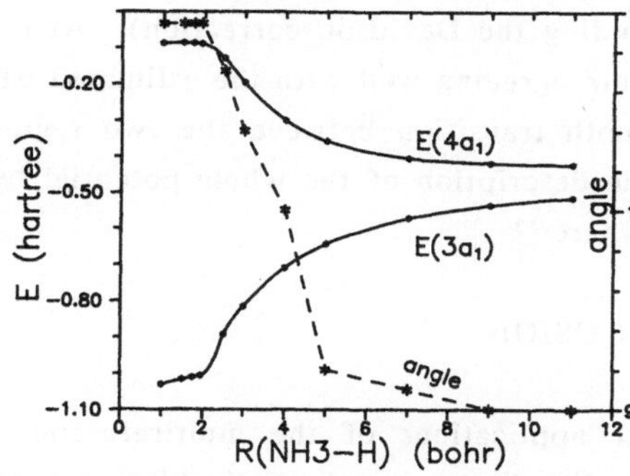

Figure 4. Hartree–Fock energies of NH_4^+ $3a_1$ and $4a_1$ orbitals, and angle of NH bond with C_3 axis

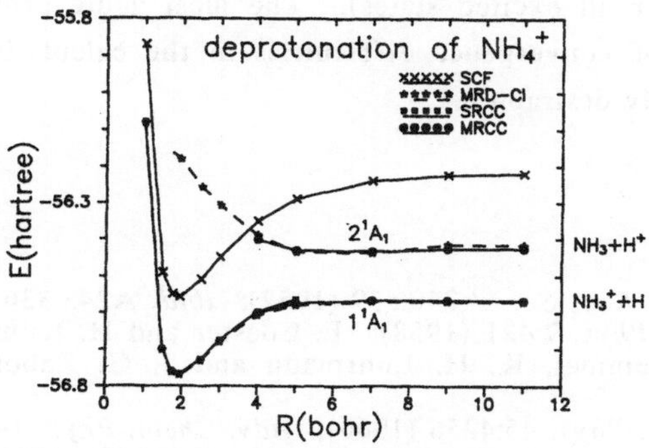

Figure 5. Potentials vs. $R(NH_3{-}H)$

this angle as a function of R. Also shown are the orbital energies of $3a_1$ and $4a_1$. These are the two valence orbitals, with the two surfaces dominated by the $1a_1^2 2a_1^2 1e^4 3a_1^2$ and $1a_1^2 2a_1^2 1e^4 3a_1 4a_1$ configurations. Obviously, the character of the lowest state changes rather abruptly near R=4 bohr, where the open-shell configuration becomes dominant. Figure 5 shows the SCF, MRD-CI and CCSD potentials. The SCF curve does not allow for the curve crossing, and dissociates incorrectly. The single-reference CCSD converges easily for R < 4 bohr, and gives energies within a few mhartree of the MRD-CI estimate of the basis set limit (including the Davidson correction). At larger R, a multireference CCSD is required, once again agreeing well with the estimated basis limit. The important point to notice is the smooth transition between the two regions at R = 4 bohr. This transition is essential to the description of the whole potential by the CC method. More details may be found elsewhere.[33]

SUMMARY AND CONCLUSION

Several new molecular applications of the multireference coupled-cluster method have been described, extending its scope of applicability. Systems treated include the relatively dense excitation spectrum of N_2, potential functions of the nine lowest states of Li_2, and the reaction profile for the deprotonation of NH_4^+, a two-surface reaction. Incomplete model spaces, both of the quasicomplete and of more general types, had to be employed. Very good results were obtained in all cases, which is highly encouraging for future applications. From the computational point of view, the coupled-cluster procedure is highly vectorizable, and most of it may be written in terms of scalar products. The computational resources required are therefore quite reasonable (thus, the post-Hartree-Fock stage of the N_2 excitation calculation took 15 minutes of Cyber 180/990 CPU time for all excited states). The most acute problem outstanding is the frequent occurrence of convergence difficulties in the calculations. A better iteration scheme would be highly desirable.

REFERENCES

1. J. Hubbard, *Proc. Roy. Soc.* A240:539 (1957); *ibid.* A243:336 (1958).
2. F. Coester, *Nucl. Phys.* 7:421 (1958); F. Coester and H. Kümmel, *Nucl. Phys.* 17:477 (1960); H. Kümmel, K. H. Lührmann and J. G. Zabolitzky, *Phys. Rept.* 36:1 (1978).
3. J. Cizek, *J. Chem. Phys.* 45:4256 (1966); *Adv. Chem. Phys.* 14:35 (1969).
4. J. Paldus, J. Cizek and I. Shavitt, *Phys. Rev. A* 5:50 (1972); J. Paldus, *J. Chem. Phys.* 67:303 (1977); B. G. Adams and J. Paldus, *Phys. Rev. A* 20:1 (1979).
5. R. J. Bartlett, *J. Phys. Chem.* 93:1697 (1989).
6. U. Kaldor, *in*: "Condensed Matter Theories", vol. 3, J. Arponen, R. F. Bishop, and M. Manninen, eds., Plenum, New York (1988).
7. U. Kaldor, *in*: "Condensed Matter Theories", vol. 4, J. Keller, ed., Plenum, New York (1989).
8. F. E. Harris, *Intern. J. Quantum Chem.* S11:403 (1977); H. J. Monkhorst, *Intern. J. Quantum Chem.* S11:421 (1977).
9. J. Paldus, J. Cizek, M. Saute and A. Laforgue, *Phys. Rev. A* 17:805 (1978); M. Saute, J. Paldus and J. Cizek, *Intern. J. Quantum Chem.* 15:463 (1979).

10. D. Mukherjee, R. K. Moitra and A. Mukhopadhyay, *Pramana* 4:247 (1975); *Mol. Phys.* 30:1861 (1975); A. Mukhopadhyay, R. K. Moitra and D. Mukherjee, *J. Phys. B* 12:1 (1979); D. Mukherjee and P. K. Mukherjee, *Chem. Phys.* 39:325 (1979); S. S. Adnan, S. Bhattacharyya and D. Mukherjee, *Mol. Phys.* 39:519 (1980); *Chem. Phys. Lett.* 85:204 (1981).
11. R. Offerman, W. Ey and H. Kümmel, *Nucl. Phys.* A273:349 (1976); R. Offerman, *Nucl. Phys.* A273:368 (1976); W. Ey, *Nucl. Phys.* A296:189 (1978).
12. I. Lindgren, *Intern. J. Quantum Chem.* S12:33 (1978); S. Salomonson, I. Lindgren and A. M. Martensson, *Phys. Scr.* 21:351 (1980); I. Lindgren and J. Morrison, "Atomic Many-Body Theory", Springer, Berlin, (1982).
13. I. Lindgren, *Phys. Scr.* 32:291, 32:611 (1985).
14. H. Nakatsuji, *Chem. Phys. Lett.* 59:362 (1978); *ibid.* 67:329 (1979); *Chem. Phys.* 75:425 (1983); *ibid.* 76:283 (1983); *J. Chem. Phys.* 80:3703 (1984).
15. H. Reitz and W. Kutzelnigg, *Chem. Phys. Lett.* 66:111 (1979); W. Kutzelnigg, *J. Chem. Phys.* 77:3081 (1981); *ibid.* 80:822 (1984).
16. B. Jeziorski and H. J. Monkhorst, *Phys. Rev. A* 24:1668 (1981); L. Z. Stolarczyk and H. J. Monkhorst, *Phys. Rev. A* 32:725, 32:743 (1985).
17. A. Banerjee and J. Simons, *Intern. J. Quantum Chem.* 19:207 (1981).
18. V. Kvasnicka, *Chem. Phys. Lett.* 79:89 (1981).
19. A. Haque and D. Mukherjee, *J. Chem. Phys.* 80:5058 (1984); *Pramana* 23:651 (1984).
20. P. Westhaus, *Int. J. Quantum Chem.* S7:463 (1973); P. Westhaus, E. G. Bradford, and D. Hall, *J. Chem. Phys.* 62:1607 (1975).
21. I. Shavitt and L. T. Redmon, *J. Chem. Phys.* 73:5711 (1980).
22. L. T. Redmon and R. J. Bartlett, *J. Chem. Phys.* 76:1938 (1972).
23. J. Arponen, *Ann. Phys. (NY)* 151:311 (1983).
24. K. Tanaka and H. Terashima, *Chem. Phys. Lett.* 106:558 (1984).
25. B. H. Brandow, *Rev. Mod. Phys.* 39:771 (1967).
26. G. Hose and U. Kaldor, *J. Phys. B* 12:3827 (1979).
27. T. H. Schucan and H. A. Weidenmuller, *Ann. Phys. (NY)* 73:108 (1972); *ibid.* 76:483 (1973).
28. G. Hose and U. Kaldor, *Phys. Scr.* 21:357 (1980); *Chem. Phys.* 63:165 (1981); *J. Phys. Chem.* 86:2133 (1982); *Phys. Rev. A* 30:2932 (1984); U. Kaldor, *J. Chem. Phys.* 81:2406 (1984).
29. D. Mukherjee, *Chem. Phys. Lett.* 125:207 (1986); *Intern. J. Quantum Chem.* S20:409 (1986).
30. I. Lindgren and D. Mukherjee, *Phys. Rep.* 151:93 (1987).
31. W. Kutzelnigg, D. Mukherjee, and S. Koch, *J. Chem. Phys.* 87:5902 (1987); D. Mukherjee, W. Kutzelnigg, and S. Koch, *J. Chem. Phys.* 87:5911 (1987).
32. A. Haque and U. Kaldor, *Chem. Phys. Lett.* 117:347 (1985); *Chem. Phys. Lett.* 120:261 (1985); *Intern. J. Quantum Chem.* 29:425 (1986); U. Kaldor and A. Haque, *Chem. Phys. Lett.* 128:45 (1986); U. Kaldor, *Intern. J. Quantum Chem.* S20:445 (1986); *J. Comput. Chem.* 8:448 (1987); *J. Chem. Phys.* 87:467 (1987); *J. Chem. Phys.* 87:4693 (1987).
33. U. Kaldor, S. Roszak, P. C. Hariharan, and J. J. Kaufman, *J. Chem. Phys.* 90:6395 (1989).
34. S. Pal, M. Rittby, R. J. Bartlett, D. Sinha, and D. Mukherjee, *Chem. Phys. Lett.* 137:273 (1987); *J. Chem. Phys.* 88:4357 (1988).
35. S. Ben-Shlomo and U. Kaldor, *J. Chem. Phys.* 89:956 (1988).
36. S. Koch and D. Mukherjee, *Chem. Phys. Lett.* 145:321 (1988).
37. D. Sinha, S. Mukhopadhyay, and D. Mukherjee, *Chem. Phys. Lett.* 129:369 (1986).
38. U. Kaldor, *Phys. Rev.* 38:6013 (1988).
39. R. Krishnan, J. S. Binkley, R. Seeger, and J. A. Pople, *J. Chem. Phys.* 72:650 (1980); M. J. Frisch, J. A. Pople, and J. S. Binkley, *J. Chem. Phys.* 80:3265 (1984).
40. K. P. Huber and G. Herzberg, "Constants of Diatomic Molecules", Van Nostrand, New York (1979).
41. F. R. Gilmore, *J. Quant. Spectr. Radiat. Transfer* 5:369 (1965).
42. A. Lofthus and P. H. Krupenie, *J. Phys. Chem. Ref. Data* 6:113 (1977).
43. D. Stahel, M. Leoni, and K. Dressler, *J. Chem. Phys.* 79:2541 (1983).
44. J. Oddershede, N. E. Grüner, and G. H. F. Diercksen, *Chem. Phys.* 97:303 (1985).
45. J. Oddershede, *Adv. Chem. Phys.* 69:201 (1987).
46. M. Jaszunski, A. Rizzo, and D. L. Yeager, *Chem. Phys.* in press.
47. U. Kaldor, *J. Chem. Phys.* 81:2406 (1984).
48. S. K. Shih, W. Butscher, R. J. Buenker, and S. D. Peyerimhoff, *Chem. Phys.* 29:241 (1978).

49. I. Schmidt-Mink, W. Müller, and W. Meyer, *Chem. Phys.* 92:263 (1985).
50. P. Kusch and M. M. Hessel, *J. Chem. Phys.* 67:586 (1977); M. M. Hessel and C. R. Vidal, *J. Chem. Phys* 70:4439 (1979); R. A. Bernheim, L. P. Gold, P. B. Kelly, T. Tipton, and D. K. Veirs, *J. Chem. Phys.* 76:57 (1982); J. Verges, R. Bacis, B. Barakat, P. Carrot, S. Churassy, and P. Crozet, *Chem. Phys. Lett.* 98:203 (1983).
51. M L. Olson and D. D. Konowalow, *Chem. Phys. Lett.* 39:281 (1976); *Chem. Phys.* 21:393 (1977); *Chem. Phys.* 22:29 (1977); D. D. Konowalow and M. L. Olson, *J. Chem. Phys.* 67:590 (1977); *J. Chem. Phys.* 71:450 (1979); D. D. Konowalow, M. E. Rosenkrantz, and D. S. Hochhauser, *J. Mol. Spectr.* 99:321 (1983); D. D. Konowalow and P. S. Julienne, *J. Chem. Phys.* 72:5817 (1980).
52. S. Huzinaga, *J. Chem. Phys.* 42:1293 (1965); T. H. Dunning Jr., *J. Chem. Phys.* 55:3958 (1971).
53. M. Urban, V. Kellö and P. Carsky, *Theor. Chim. Acta* 45:205 (1977).
54. R. J. Buenker, S. D. Peyerimhoff and W. Butscher, *Mol. Phys.* 35:771 (1978); R. J. Buenker, *in:* "Studies in Physical and Theoretical Chemistry" Vol. 21, R. Carbo, ed., Elsevier Scientific Publishing Co., Amsterdam (1982) pp. 17-34; R. J. Buenker and R. A. Phillips, *J. Mol. Struct. (Theochem)* 123:291 (1985).

ON THE BARGMANN SPACE APPROACH TO THE EXTENDED COUPLED CLUSTER METHOD FOR SIMPLE ANHARMONIC SYSTEMS

E. Aalto*, J. S. Arponen† and R. F. Bishop#

*Physics Computation Unit, University of Helsinki
Siltavuorenpenger 20 C, SF-00170 Helsinki, Finland

†Department of Theoretical Physics, University of Helsinki
Siltavuorenpenger 20 C, SF-00170 Helsinki, Finland

#Department of Mathematics, University of Manchester
Institute of Science and Technology, P. O. Box 88
Manchester M60 1QD, England

I. INTRODUCTION

Bosonic quantum field theories in a Hilbert space can be mapped into classical field theories of complex functions in a particular normed space, the Bargmann Hilbert space.[1] Since the theory of complex classical fields is well understood from complex analysis, the approach often affords an intuitively appealing alternative to the more usual canonical or path integral formulations. In the present paper we shall combine this approach with another well-known method of field theory, the coupled cluster method (CCM) of Coester and Kümmel.[2] The CCM focuses on the many-body correlations in a quantum system, and introduces a set of linked-cluster amplitudes, $\{S_n\}$, parametrizing the correlated many-body ground state.[3]

It has been pointed out[4] that the CCM equations for the ground state can be obtained from a dynamical variational principle, which can be used to extend the theory into a complete dynamical description of the system, including the CCM average-value functional for arbitrary operators as a central concept. It turned out that in this approach the set of variables, $\{\Omega_n\}$, dynamically conjugate to the amplitudes $\{S_n\}$, are not represented by linked diagrams in the ground state. In another formulation of CCM, the extended CCM (ECCM),[4,5] a new set of conjugate variables, $\{\sigma_n, \tilde{\sigma}_n\}$, was introduced for the parametrization of the quantum system and shown to be represented by linked diagrams only.

The importance of the linked-cluster properties are naturally related to the problem of size-extensivity and size-consistency (i.e. full separability) of the description

Condensed Matter Theories, Volume 5
Edited by V.C. Aguilera-Navarro
Plenum Press, New York, 1990

of the many-body system.[6,7] If we consider a local field theory or a many-body system with (sufficiently) short range interactions, the ECCM amplitudes $\{\sigma_n, \tilde{\sigma}_n\}$ in the ground state are quasilocal,[4,5] and can be regarded as multilocal coordinates in a classical symplectic complex phase space, the ECCM phase space. The original quantum problem thus becomes mapped into a classical Hamiltonian theory in an appropriate phase space.

Indeed, considering the locality properties of the basic conjugate variables, we may divide the conventional algebraic many-body methods into 3 main categories, in each of which the dynamical equations of motion are obtained from the generic canonical equations

$$i\frac{dx_n}{dt} = \frac{\partial \bar{H}[x,y]}{\partial y_n},$$
$$i\frac{dy_n}{dt} = -\frac{\partial \bar{H}[x,y]}{\partial x_n}.$$

The canonically conjugate operators $\{x, y\}$ have suitable configuration-space parametrizations in terms of c-number amplitudes $\{x_n, y_n\}$, respectively, and the parametrizations of the ket and bra ground states $|\psi\rangle$ and $\langle\tilde{\psi}|$ in these cases are

(I)	$\{x,y\} = \{F, \tilde{F}\};$	$\lvert\psi\rangle = F\lvert\phi\rangle;$	$\langle\tilde{\psi}\rvert = \langle\phi\rvert\tilde{F};$
(II)	$\{x,y\} = \{S, \Omega\};$	$\lvert\psi\rangle = e^{S}\lvert\phi\rangle;$	$\langle\tilde{\psi}\rvert = \langle\phi\rvert\Omega e^{-S};$
(III)	$\{x,y\} = \{\Sigma, \tilde{\Sigma}\};$	$\lvert\psi\rangle = e^{S}\lvert\phi\rangle;$	$\langle\tilde{\psi}\rvert = \langle\phi\rvert e^{S''}e^{-S},$

where $|\phi\rangle$ is some suitable reference state or cyclic vector. The cases are identified as (I) the configuration interaction (CI) method, (II) the normal CCM (NCCM), and (III) the ECCM. In the last case $\tilde{\Sigma} \equiv S''$ and $\Sigma|\phi\rangle = Qe^{S''}S|\phi\rangle$, where $Q \equiv I - |\phi\rangle\langle\phi|$. The classical phase space in each of the above cases is in principle equally complicated, but the full locality and separability feature in the ECCM should lead to an effective compactification of the phase space,[5a] allowing the physically important region to be described in terms of an effective mean field theory of reduced dimensionality.

To learn more about the basic mathematical aspects of the various coupled-cluster methods we undertake here a study of two very simple but nontrivial problems: the linear anharmonic oscillator (AO for short) with an x^4 anharmonicity, and an anharmonic spin system which is introduced as a finite-dimensional model to the anharmonic oscillator with an infinite-dimensional Hilbert space. The anharmonic oscillator is a system for which perturbation theory fails to converge, and the structure of the excitation spectrum changes drastically.[8-10] As an interacting field theory the AO is rather strange and singular because it is not a local field theory; rather, the interaction is maximally non-local. Therefore it is expected to pose an exceptionally hard test for a method such as the ECCM which — with its linked cluster properties — is tailored for systems with normal locality and separability properties.

II. ANHARMONIC MODEL SYSTEMS

II. 1. Anharmonic oscillator

In terms of the canonical creation and annihilation operators the Hamiltonian

$$H = \frac{1}{2}p^2 + \frac{1}{2}x^2 + \frac{\lambda}{4}x^4 \qquad (2.1)$$

of the anharmonic oscillator becomes

$$H = a^\dagger a + \frac{1}{2} + \frac{\lambda}{16}(a^\dagger + a)^4. \tag{2.2}$$

In numerical computation we usually start from the Bogoliubov-transformed optimized form[11,12)]

$$H = \omega\left[a^\dagger a + \frac{\lambda}{16\omega^3} : (a^\dagger + a)^4 : + \frac{3}{8} + \frac{1}{8\omega^2}\right], \tag{2.3}$$

in which ω is the positive root of equation $\omega^3 - \omega - \frac{3}{2}\lambda = 0$, and the colons denote normal ordering.

In the Bargmann representation we replace $a \rightarrow \frac{d}{dz}$, $a^\dagger \rightarrow z$, and the ket state, such as

$$|\psi\rangle = F(a^\dagger)|0\rangle, \tag{2.4}$$

where $|0\rangle$ is the vacuum state defined by $a|0\rangle = 0$, will be represented by functions $F(z)$ of the complex variable.[13)] In the SUB N truncation of the CI method the exact infinite-order holomorphic function $F(z) = \sum_{n=0}^{\infty} F_n z^n$ is approximated by a polynomial of order N. Similarly, in CCM the SUB N approximation to the exact function $S(z) = \sum_{n=1}^{\infty} S_n z^n$ defined through $F(z) = \exp S(z)$, is a polynomial of order N. We observe that the scalar product in the Bargmann representation can be given in any of the following forms,[13)]

$$\begin{aligned} \langle 0|g(a)f(a^\dagger)|0\rangle &= g(\frac{d}{dz})f(z)\Big|_{z=0} \\ &= \frac{1}{\pi}\int d^2z e^{-|z|^2} g^*(z)f(z) \\ &= \sum_n n! g_n^* f_n. \end{aligned} \tag{2.5}$$

Using the Hamiltonian $H(a^\dagger, a) \rightarrow H(z, \frac{d}{dz})$ we may write down the energy functional in each of the three methods:

$$\bar{H}(CI) = \tilde{F}(\frac{d}{dz})H(z, \frac{d}{dz})F(z)\Big|_{z=0} \Big/ \tilde{F}(\frac{d}{dz})F(z)\Big|_{z=0}; \tag{2.6}$$

$$\bar{H}(CCM) = \Omega(\frac{d}{dz})e^{-S(z)}H(z, \frac{d}{dz})e^{S(z)}\Big|_{z=0}; \tag{2.7}$$

$$\bar{H}(ECCM) = e^{S''(\frac{d}{dz})}e^{-S(z)}H(z, \frac{d}{dz})e^{S(z)}\Big|_{z=0}. \tag{2.8}$$

The exact values of the amplitudes $(F_n, \tilde{F}_n, S_n, \Omega_n, S''_n)$ are found in each case from the condition that $\bar{H}$ is required to be stationary against variations of the respective free variables.

After a lengthy but straightforward calculation the ground-state energy functional in the NCCM and ECCM cases is obtained in the form

$$\bar{H} = \sum_{n=0}^{M} n!\Omega_n \bar{H}_n[S], \tag{2.9}$$

where the coefficients are (here $g = \frac{\lambda}{16\omega^3}$),

$$
\begin{aligned}
\bar{H}_n[S]/\omega = \delta_{n,0}(\frac{3}{8} + \frac{1}{8\omega^2}) + nS_n \\
&+ g\{\delta_{n,4} + 4(n-2)S_{n-2} + 6n(n-1)S_n \\
&+ 4(n+2)(n+1)nS_{n+2} + (n+4)(n+3)(n+2)(n+1)S_{n+4} \\
&+ 6\sum_m m(n-m)S_m S_{n-m} \\
&+ 12\sum_m m(m-1)(n+2-m)S_m S_{n+2-m} \\
&+ \sum_m m(m-1)(n+4-m)(m+3n+1)S_m S_{n+4-m} \\
&+ 4\sum_{mk} mk(n+2-m-k)S_m S_k S_{n+2-m-k} \\
&+ 6\sum_{mk} mk(n+4-m-k)(n+3-m-k)S_m S_k S_{n+4-m-k} \\
&+ \sum_{mkl} mkl(n+4-m-k-l)S_m S_k S_l S_{n+4-m-k-l}\}.
\end{aligned}
\tag{2.10}
$$

Equations (2.9)-(2.10) give the energy functional $\bar{H}$ explicitely in terms of the NCCM canonical variables $\{S_n, \Omega_n\}$. However, in the ECCM SUB N approximation we do not express $\bar{H}$ explicitely in terms of the canonical variables $\{\sigma_n, \tilde{\sigma}_n\}$, but instead use the variables $\{S_n, S_n''\}$ for convenience. This corresponds to a one-to-one change of variables within the limits of computational accuracy. The ECCM canonical variables $\{\sigma_n\}$ are given by

$$\sigma_n = \sum_m \frac{m!}{n!} S_m \Omega_{m-n}, \tag{2.11}$$

where the upper limit of summation is N in a SUB N approximation and infinity in the exact definition.

Above, in equation (2.9) the upper limit of summation is $M = N$ in the NCCM SUB N, and $M = 4N - 4$ in the ECCM SUB N approximation. In the latter case the amplitudes Ω_n of $\Omega(z) \equiv \exp S''(z)$ must be given in terms of the coefficients $\{S_n'' | n = 1, \ldots, N\}$. To calculate Ω_n as functions of $\{S_m''\}$ and vice versa, we use the recursion formulae

$$n\Omega_n = \sum_{m=1}^{min(N,n)} mS_m''\Omega_{n-m}, \qquad n = 1, \ldots, M. \tag{2.12}$$

In practice, for the symmetric AO only the even coefficients are nonzero.

It should be stressed that the CI SUB N approximation simply corresponds to the Sturm-Liouville expansion of the eigenstates of the anharmonic oscillator; i.e. to the diagonalization of H in the subspace spanned by the $N + 1$ lowest eigenstates of the linear harmonic oscillator. Therefore, the sequence of CI SUB N results are expected to smoothly converge to the exact limit corresponding to the AO.

II. 2. Anharmonic spin model

The Hilbert space of the AO is infinite-dimensional. Therefore the summations in many of the exact untruncated mathematical expressions are infinite, and the question of convergence of such summations naturally arises (see section V). We have attempted to gain more understanding on these problems by constructing a finite-dimensional analog to the AO using spin algebra.

We consider the $(2J+1)$-dimensional spin space corresponding to spin quantum number J, and define the operators

$$\begin{aligned} j_0 &= J + J_z, \\ j_\pm &= \frac{1}{\sqrt{2J}} J_\pm . \end{aligned} \tag{2.13}$$

In the limit of large J, the commutation relations of j_0, j_+ and j_- are the same as those of $a^\dagger a$, $a^\dagger$ and a up to order $\mathcal{O}\left(\frac{\langle a^\dagger a\rangle}{J}\right)$. In the limit $J \to \infty$ the Hamiltonian

$$H_J = j_0 + \frac{1}{2} + \frac{\lambda}{16}(j_+ + j_-)^4 \tag{2.14}$$

should therefore go smoothly into the Hamiltonian (2.2) of the AO, in the sense that any energy eigenvalue E_n with fixed n converges to the corresponding AO energy.

The CI, NCCM and ECCM amplitudes can be defined in the same way as in the AO case. In particular, the expressions for the amplitudes σ_n of $\Sigma = \sum_n \sigma_n j_+^n$ are now finite sums,

$$\begin{aligned} \sigma_n &\equiv \frac{1}{K_n}\langle 0|j_-^n \Omega S|0\rangle \\ &= \sum_{m=n}^{2J} \frac{K_m}{K_n} S_m \Omega_{m-n}, \end{aligned} \tag{2.15}$$

where

$$K_n = \frac{n!(2J)!}{(2J-n)!(2J)^n} \quad \underset{J\to\infty}{\longrightarrow} \quad n! \tag{2.16}$$

III. NUMERICAL RESULTS

The NCCM equations for the anharmonic oscillator have been previously solved numerically,[11,12,14] but no ECCM solutions have been published so far. In order to compare the convergence and accuracy, we solved numerically the AO using all three methods. The ECCM and NCCM equations were solved by Newton's iteration, and the CI ground state by inverse iteration. We used VAX FORTRAN with quadruple precision for the ground state, and double precision NAG routines for the excitation energies (F02AFF for NCCM and ECCM, and F02AAF for CI). In each case the excited-state energies were calculated by using the dynamical matrix obtained from the second-order derivatives of the energy functional, as explained in an accompanying paper in this volume.[15]

In Fig. 1 we show the results for the first six eigenvalues corresponding to even-parity eigenstates at $\lambda = 0.05$. The accuracy is indicated by $-\lg|1 - \frac{E}{E_{exact}}|$ plotted as function of the truncation index N. Thus the ordinate represents directly the

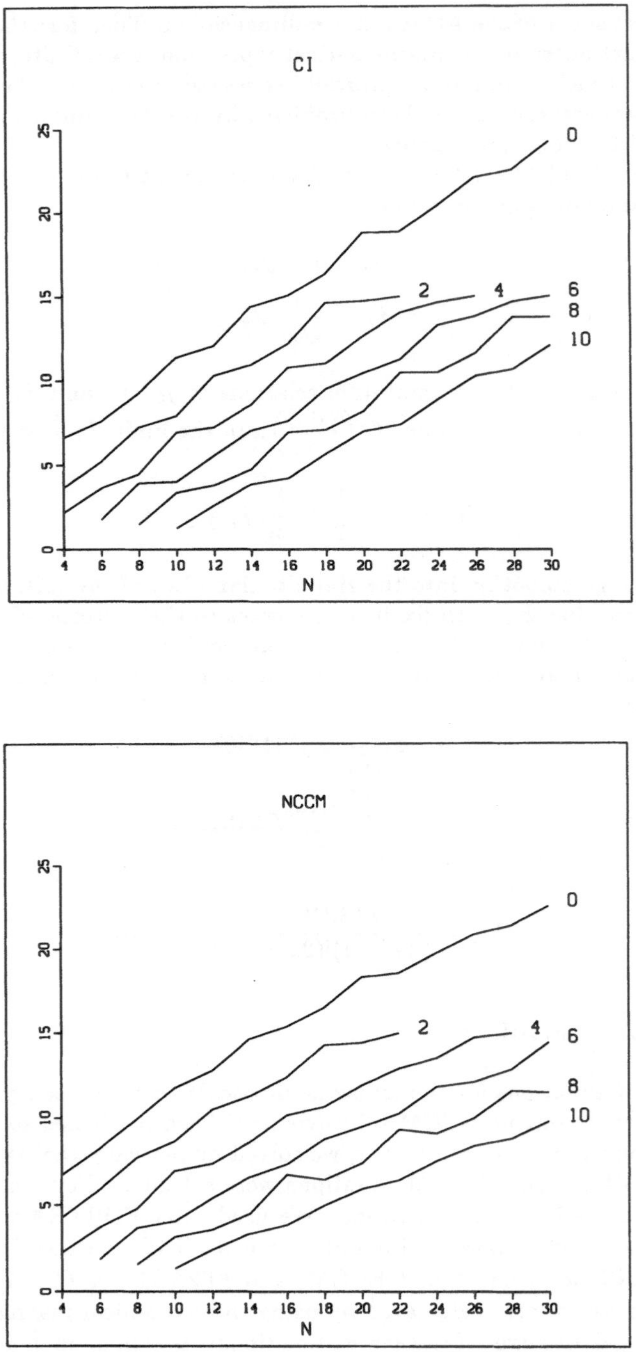

Fig. 1. Accuracy of the energies of the six lowest even eigenstates of the anharmonic oscillator with $\lambda = 0.05$ as functions of truncation index N. The excited-state results are given up to the point where full double-precision accuracy is reached.

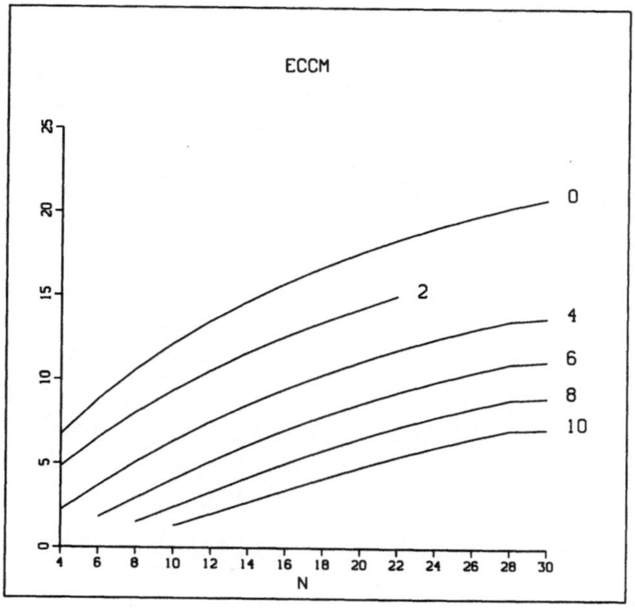

Fig. 1. Continued.

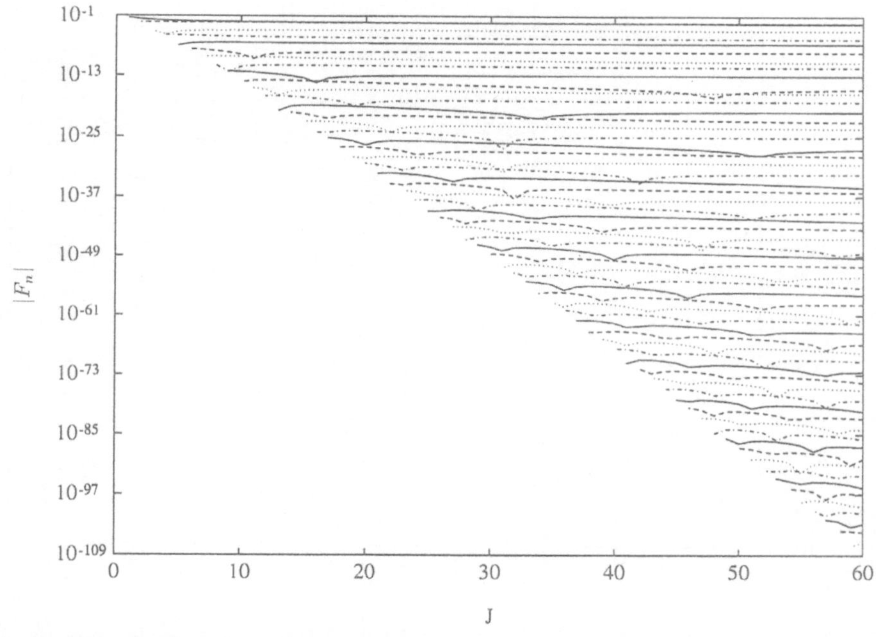

Fig. 2. Amplitudes for the anharmonic spin model as functions of J at $\lambda = 1$. Only even-indexed amplitudes up to $n \leq N = 2J$ exist.

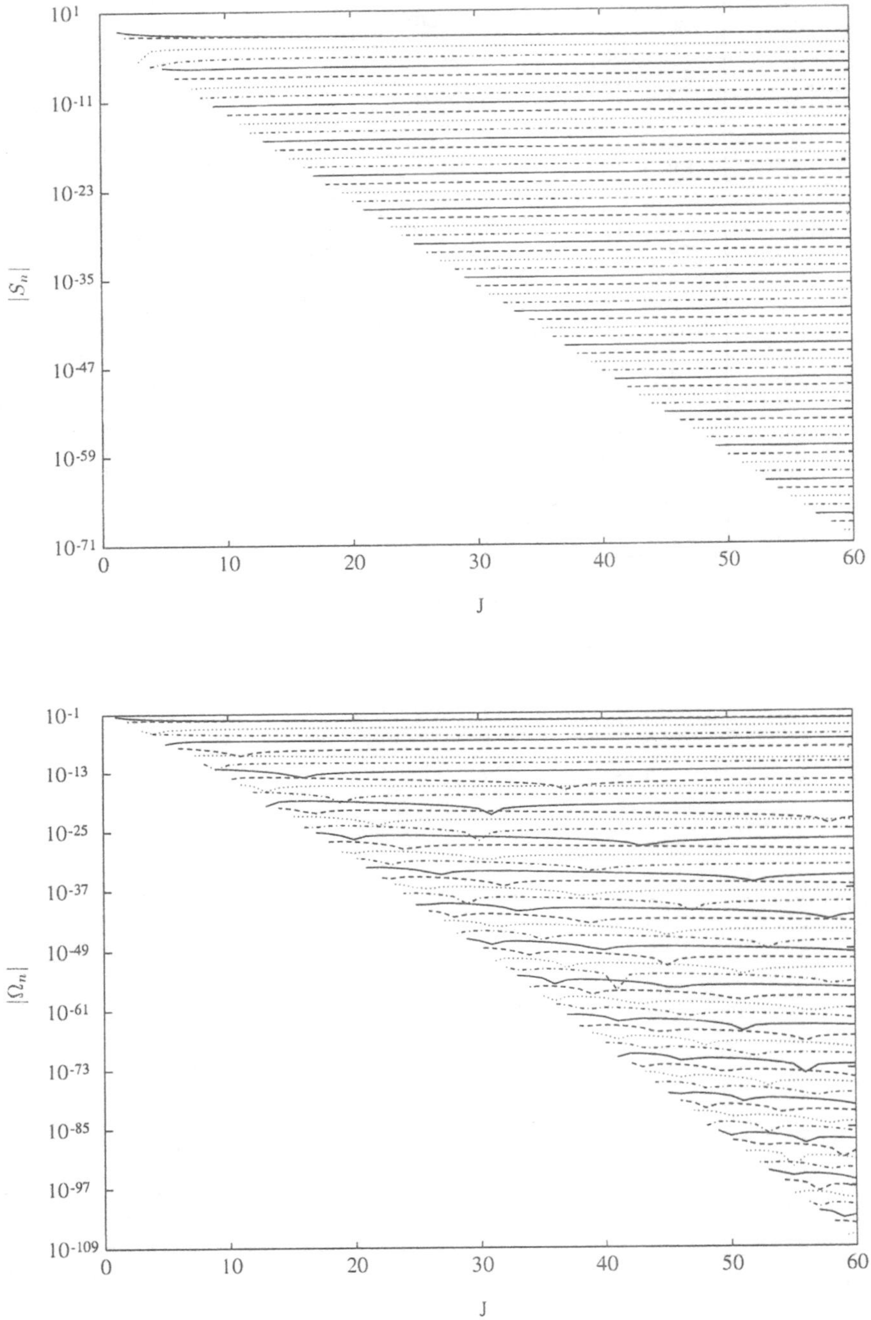

Fig. 2. Continued.

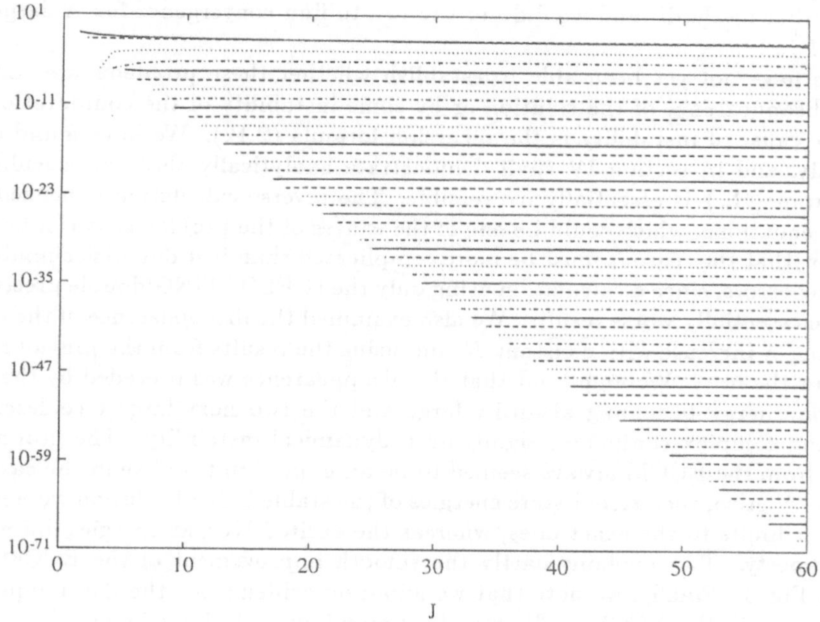

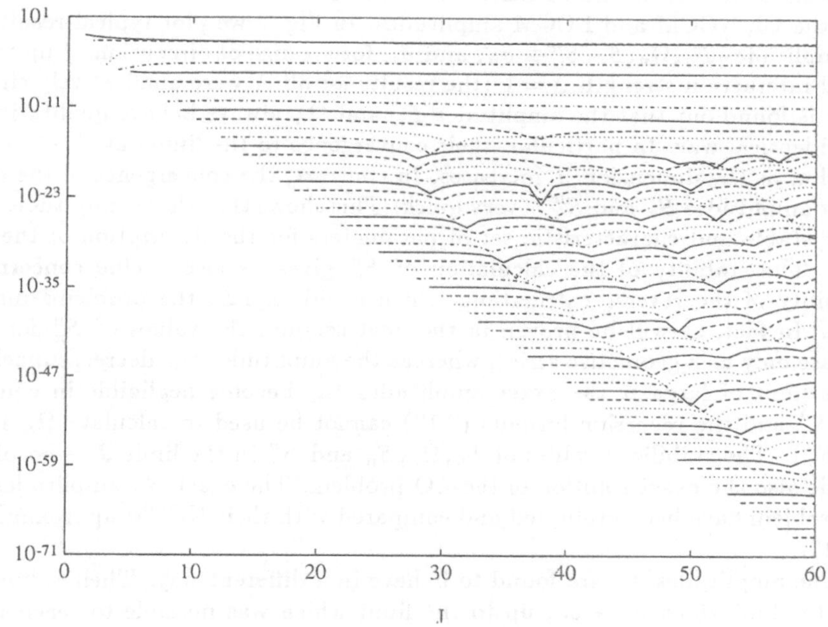

Fig. 2. Continued.

number of correct significant digits. It can be seen that for a small λ ECCM is competitive in accuracy with the other methods, and the increase of accuracy with increasing N is remarkably smooth. For larger λ ECCM gives results clearly inferior to the other methods, and we did not succeed to find convergence for N larger than 12 or 14.

We have not yet been able to establish whether these problems are caused by a real disappearence of the solution or by some instability of the computation. One possible cause of instability is the recursion formula (2.12). We have found out numerically, and in section IV shall demonstrate analytically that the calculation of $\{S_n''\}$ from $\{\Omega_n\}$ is essentially irreversible. The reverse calculation is needed in the ECCM equations, which could be one of the causes of the problems. It can be stated, however, that the reasons must be more complicated than just due to accumulation of rounding errors, since a calculation using only the G_FLOATING double precision did not give essentially worse results. We also examined the disappearence of the solution by letting λ increase with constant N and using the results from the previous step as input for the next. We found out that the disappearence was preceded by the largest exitation energy becoming absurdly large and the two next largest coalescing and becoming complex conjugates signifying a dynamical instability. The ground state energy from the ECCM always seemed to be an upper limit, unlike in the case of the NCCM.[11] Also, the excited-state energies of the stable ECCM solution were found to be upper limits to the exact ones, whereas the excited NCCM energies did not have this property. This explains partly the smooth improvement of the ECCM results seen in Fig. 1. Finally, we note that we found no evidence for the non-uniqueness of solutions, as in the NCCM[11,14]; we either found one solution or none.

The anharmonic spin system is solvable by ordinary matrix methods, like the ones used for the AO in the CI case, and it therefore allows us to find the exact values of all the CI, NCCM and ECCM amplitudes. In Fig. 2 we plot typical results for all the amplitudes $F_n, \Omega_n, S_n, S_n'' \equiv \tilde{\sigma}_n$, and σ_n for a range of dimensions J up to 60, for a fixed interaction strength $\lambda = 1$ (the results for all λ were qualitatively similar).

It is found out that the amplitudes F_n and Ω_n always behave qualitatively in a fairly identical way. In particular, their convergence to the limits at $J \to \infty$ is often very slow as the dimension J increases. In contrast, the convergence of the coupled-cluster amplitudes S_n and S_n'' is very quick. This shows that these amplitudes provide a very stable and characteristic set of parameters for the description of the ground state. The stability of the calculation of S_n'' gives us also a clue concerning the instability of the reverse calculation, which could explain the problems mentioned above. In fact, as will be proven in the next section, the values of S_n'' for large n decrease only as a geometric series, whereas the amplitudes Ω_n decrease much faster. In the limit of large n the exact amplitudes Ω_n become negligible in comparison with S_n'' and the recursion formula (2.12) cannot be used to calculate Ω_n from the set $\{S_n''\}$. The stabilized values of F_n, Ω_n, S_n and S_n'' in the limit $J \to \infty$ obviously provide also the exact solution of the AO problem. The exact S_n amplitudes for the AO problem have been evaluated and compared with their NCCM approximations in Ref. 12.

The amplitudes σ_n are found to behave in a different way. Their values do not seem to stabilize as $J \to \infty$, up to the limit which was possible to reach with our numerical accuracy. Also, it was found that the higher terms in the sum (2.15) yield significant contributions to σ_n at high J. Thus it seems that the exact σ_n-amplitudes of the AO case cannot be calculated from this particular finite-dimensional model in its infinite-dimensional limit.

IV. ANALYTIC CONSIDERATIONS

The holomorphic state function $F(z)$ is known, on general grounds, to be an entire function of order ≤ 2. Specifically, in our case it has the precise expression[8,13)]

$$F(z) = \pi^{-\frac{1}{4}} e^{-\frac{1}{2}z^2} \int_{-\infty}^{\infty} dx e^{\sqrt{2}zx - \frac{1}{2}x^2} \psi(x), \tag{4.1}$$

where $\psi(x)$ is the suitably normalized (i.e. $\langle 0|\psi\rangle = 1$) coordinate-space wave function. It is proven elsewhere[13)] that the function $F(z)$ corresponding to the ground state of the AO must have infinitely many distinct zeroes $\{z_m\}$ which, asymptotically far from the origin, are located at the points $\pm i y_m$ where

$$y_m \underset{m\to\infty}{\longrightarrow} \left(\frac{\lambda}{2}\right)^{\frac{1}{6}} \left(\frac{3\pi m}{2}\right)^{\frac{2}{3}}. \tag{4.2}$$

The function $e^{\frac{1}{2}z^2} F(z)$ is an entire function of order $\frac{3}{2}$ and, on using its Hadamard decomposition, it is straightforward to show that the exact CCM amplitude $S(z)$ of the AO must be of the form

$$S(z) = -\frac{1}{2}z^2 + \sum_m \left[\ln(1 - \frac{z}{z_m}) + \frac{z}{z_m}\right]. \tag{4.3}$$

It is now evident that the state $S(a^\dagger)|0\rangle$ is not normalizable, and therefore the CCM operator $S(a^\dagger)$ cannot be qualitatively considered as a 'small' correlation operator. Also, the coefficients S_n for high n are determined by the zero closest to the origin. The sequence nS_n must, indeed, be asymptotically a geometric series such that $\lim_{n\to\infty} \sqrt[n]{n|S_n|} = \rho^{-1}$, where ρ is the distance of the nearest zero.

The function $\Omega(z)$ can be given in terms of the wave function $\psi(x)$ as the integral[13)]

$$\Omega(z) = N^{-2} e^{-\frac{z^2}{4}} \int_{-\infty}^{\infty} dx e^{\frac{xz}{\sqrt{2}}} \psi^*(x)\psi(x - \frac{z}{\sqrt{2}}), \tag{4.4}$$

where $N^2 \equiv \int dx |\psi(x)|^2$. For the ground state of the AO this function is also an entire function.

We now calculate the coefficients σ_n of the ground-state amplitude $\Sigma(z) = \sum_n \sigma_n z^n$ from the definition (cf. section I and equation 2.11)

$$\sigma_n = \frac{1}{n!}\frac{d^n}{dz^n}\Omega(\frac{d}{dz})S(z)\Big|_{z=0} \qquad n \geq 1 \tag{4.5}$$

Using Eq. (4.3) for $S(z)$ one can proceed either by a Fourier integral or by a Borel summation method. Whichever method is chosen, the calculations are rather involved, and the details are given elsewhere.[13)] We first define an odd function $R(x)$ by

$$R(x) = \delta'(x) - \frac{1}{2}\mathrm{sgn}(x) \sum_m e^{-xz_m} \tag{4.6}$$

for $x \in \mathbb{R}$. Here $\delta(x)$ is the Dirac delta, and $R(x)$ is obviously a distribution. The infinite sum over the zeroes of $F(z)$ is formally divergent and must be properly interpreted. Indeed, the sum must first be partitioned into terms which are analytic in either the upper or the lower x-halfplanes. The boundary values of these analytic functions then enter in $R(x)$. Using this definition we find the NCCM and ECCM amplitudes in the form

$$S'(z) = \int_{-\infty}^{\infty} dx R(x) \sinh(xz), \tag{4.7}$$

$$\Sigma'(z) = \int_{-\infty}^{\infty} dx \rho(x) \sinh(xz), \tag{4.8}$$

where $\rho(x) = R(x)\Omega(x)$. The amplitudes S_n and σ_n are thus given by the moments of the distributions $R(x)$ and $\rho(x)$, respectively.

Using the above representations it is possible to formulate a generating function for the expectation values of arbitrary normal-ordered operators. We define

$$\begin{aligned} A(u,v) &\equiv \langle e^{ua^\dagger} e^{va} \rangle \\ &= \langle 0 | e^{S''(a)} e^{-S(a^\dagger)} e^{ua^\dagger} e^{va} e^{S(a^\dagger)} | 0 \rangle. \end{aligned} \tag{4.9}$$

Using this function the average value of e.g. the normal-ordered Hamiltonian $H(a^\dagger, a)$ is

$$\bar{H} = H\left(\frac{\partial}{\partial u}, \frac{\partial}{\partial v}\right) A(u,v)\Big|_{u=v=0}. \tag{4.10}$$

The results derived are

$$A(u,v) = \sum_{n=0}^{\infty} \frac{1}{n!} \prod_{i=1}^{n} \left[\int_{-\infty}^{\infty} dx_i R(x_i) \frac{e^{vx_i} - 1}{x_i} \right] \Omega(u + x_1 + \cdots + x_n) \tag{4.11}$$

for NCCM and

$$\begin{aligned} A(u,v) = \sum_{n=0}^{\infty} \frac{1}{n!} \prod_{i=1}^{n} &\left[\int_{-\infty}^{\infty} dx_i \rho(x_i) \frac{e^{vx_i} - 1}{x_i} \right] \\ &\times \exp[\tilde{\Sigma}(u + x_1 + \cdots + x_n) - \tilde{\Sigma}(x_1) - \cdots - \tilde{\Sigma}(x_n)] \end{aligned} \tag{4.12}$$

for ECCM. In each case the function gives expressions in closed form for the averages in terms of the pertinent free variables; the amplitudes S_n or σ_n appear through the moments of the functions $R(x)$ or $\rho(x)$, respectively. When formally expanded, these expressions represent the particular linking and connectivity properties of the corresponding NCCM or ECCM diagrams and their vertices. Precise parallels can clearly be drawn with the well known linked (L) or double-linked (DL) expansions for an average value of an arbitrary operator $\mathcal{O}$, given in the operator forms[3-5]

$$\begin{aligned} \bar{\mathcal{O}} &= \sum_{n=0}^{\infty} \frac{1}{n!} \langle 0 | \Omega \{ \mathcal{O} S^n \}_L | 0 \rangle \\ &= \sum_{n=0}^{\infty} \frac{1}{n!} \langle 0 | e^{\tilde{\Sigma}} \{ \mathcal{O} \Sigma^n \}_L | 0 \rangle_{DL} \end{aligned} \tag{4.13}$$

for the NCCM and ECCM, respectively.

V. DISCUSSION

It is evident that the SUB N truncations in the coupled cluster methods are potentially hazardous because the asymptotic analytic forms of the functions $e^{\pm S}$ or $e^{S''}$ become severely distorted from their exact behaviour. E.g., the states $e^S|0\rangle$ are not normalizable in the SUB N approximations with $N \geq 3$. Thus the SUB N approximations imply excursions into more general linear spaces outside the Hilbert space of normalizable functions. For this purpose the Bargmann representation in terms of functions of complex variable is very well suited. The exact function $F = e^S$, which in our case is an entire function of order 2, may nevertheless be very well represented by an approximate SUB N function $\exp S^{(N)}$, where $S^{(N)}$ is a polynomial of degree N, in a part of the complex plane around the origin, although far away the asymptotic behaviour is seriously wrong. For instance, the NCCM method focuses attention on the behaviour of the function $e^{-S}He^S$ just around the origin, requiring it to be constant up to the derivatives of order N at $z = 0$. This feature allows us to understand the practical success of the method in spite of the incorrect asymptotic behaviour. The normalizability problem of the NCCM SUB N approximation has been discussed also in Refs. 11-12. In the case of ECCM, the stationarity condition is slightly more complicated, and obviously more sensitive to the asymptotic behaviour, making the SUB N truncations even more risky.

The exact amplitudes S_n and S''_n for the AO problem can be easily computed either from the anharmonic spin analog or from the CI solution of the AO problem itself. However, it is doubtful whether they can be found from the SUB N solutions in the limit $N \to \infty$ (see also Ref. 12). This fact is rather significant and indicates that the sequence of SUB N truncations may not ultimately converge for either NCCM or ECCM. We interprete this to emphasize that 'particle number', or the number of correlated fields in a 'multilocal' amplitude, is not a good concept in a field theory with no particle number conservation. So far we have not tried to evaluate the exact ECCM amplitudes σ_n. They cannot be obtained (at least to our numerical accuracy) from the finite-dimensional spin analog. Nevertheless, they can be computed from the exact analytic expressions given above. This would obviously be a very challenging numerical enterprise.

On the basis of our numerical evaluations, the various coupled-cluster methods at low and intermediate truncation indices N up to SUB $6\ldots12$ provide a very good approximation for the energies of the ground state and the lowest excited states for all values of the coupling constant λ. In particular, for small values of the coupling constant the accuracy typically increases in the order CI$\to$NCCM$\to$ECCM, whereas the order is opposite for large coupling constants. However, for high SUB N ($N \to \infty$) the accuracy continues to improve indefinitely only in CI, whereas the NCCM stagnates to lower accuracy. In the ECCM the accuracy ceases to improve at an even lower truncation level (for small λ), or the solution is not even found for $N \gtrsim 14$ (for higher λ). Although the numerical difficulties in the limit $N \to \infty$ prevent us from very decisive conclusions, it seems obvious that the SUB N approximations are completely safe and convergent only in the CI case. At fixed λ, therefore, the convergence with respect to the limit $N \to \infty$ improves in the order ECCM$\to$NCCM$\to$CI.

Although the SUB N approximations are analytically somewhat questionable for the coupled-cluster methods, we have been able to show that the NCCM and ECCM methods perform rather satisfactorily even in the difficult case of the anharmonic oscillator, which has no attributes of locality, separability, or size-extensivity. We have not been able to demonstrate that the SUB N sequence of approximations ultimately converges for NCCM and ECCM; on the contrary, we were unable to find convergent solutions to ECCM for large N in the case of higher λ. Nevertheless, the ECCM results for all the eigenenergies E_n were, quite remarkably, found to be upper limits to the exact values. This is by no means an expected result, because the ECCM (as well as NCCM) is manifestly a non-Hermitian formulation of the many-body problem.

REFERENCES

1) V. Bargmann, Comm. Pure Appl. Math. 14: 180, 187 (1961); ibid. 20: 1 (1967); Rep. Math. Phys. 2: 221 (1971).
2) F. Coester, Nucl. Phys. 7: 421 (1958); F. Coester and H. G. Kümmel, Nucl. Phys. 17: 477 (1960).
3) For reviews see: R. F. Bishop and H. G. Kümmel, Phys. Today 40 (No. 3): 52 (1987); H. G. Kümmel in "Nucleon-Nucleon Interaction and Nuclear Many-Body Problems", S. S. Wu and T. T. S. Kuo (eds.), World Scientific, Singapore (1984), p. 46; H. G. Kümmel, K. H. Lührmann and J. G. Zabolitzky, Phys. Rep. 36C: 1 (1978). See also references 4,5.
4) J. Arponen, Ann. Phys. (NY) 151: 311 (1983).
5) J. S. Arponen, R. F. Bishop, and E. Pajanne, Phys. Rev. A36: 2519 (1987); ibid. 2539 (1987); J. Arponen, R. F. Bishop and E. Pajanne in "Condensed Matter Theories", Vol. 2, P. Vashishta, R. K. Kalia and R. F. Bishop (eds.), Plenum, New York (1987), p. 357.
6) H. Primas, in "Modern Quantum Chemistry", Vol. II, O. Sinanoglu (ed.), Academic, New York (1965), p. 45; R. J. Bartlett and G. D. Purvis, Phys. Scr. 21: 255 (1980).
7) I. Lindgren and D. Mukherjee, Phys. Rep. 151: 93 (1987); R. Chowdhuri, D. Mukherjee and M. D. Prasad, in "Aspects of Many-Body Effects in Molecules and Extended Systems", D. Mukherjee (ed.), Springer, Berlin (1989), p. 3.
8) C. M. Bender and T. T. S. Wu, Phys. Rev. 184: 1231 (1969); Phys. Rev. D7: 1620 (1973); Phys. Rev. Letters 27: 461 (1971).
9) B. Simon, Ann. Phys. (NY) 58: 76 (1970).
10) F. T. Hioe, D. MacMillen and E. W. Montroll, Phys. Rep. 43: 305 (1978); R. Balian, G. Parisi and A. Voros, Phys. Rev. Letters 41: 1141 (1978); S. K. Bose and D. N. Tripathi, Fortschr. Phys. 31: 131 (1983).
11) U. B. Kaulfuss and M. Altenbokum, Phys. Rev. D33: 3658 (1986).
12) H. G. Kümmel in "Condensed Matter Theories", Vol. 3, J. S. Arponen, R. F. Bishop and M. Manninen (eds.), Plenum, New York (1988), p. 21.
13) J. S. Arponen and R. F. Bishop, to be published.
14) R. F. Bishop and M. F. Flynn, Phys. Rev. A38: 2211 (1988); R. F. Bishop, M. C. Boscá and M. F. Flynn, Phys. Lett. A132: 440 (1988).
15) R. F. Bishop, N. I. Robinson, and J. Arponen, this volume.

QUANTUM MANY-BODY SYSTEMS: ORTHOGONAL COORDINATES

V. C. Aguilera-Navarro

Instituto de Fisica Teorica, UNESP
Rua Pamplona, 145
01405 Sao Paulo, SP - Brazil

1. INTRODUCTION

The algebraic properties of the Jacobi coordinates (JC) turned these coordinates very convenient in dealing with many-body systems. Originally introduced in Astronomy, the JC have also found forum in atomic and subatomic systems[1]. In these systems, they proved to be useful for the evaluation of both matrix elements (n-body interactions) and form factors. Besides, JC exhibit intrinsic transformation properties that make them particularly interesting in quantum problems where Pauli principle plays a central role. Kramer and Moshinsky proved[2] that the JC in a n-dimensional space carry the orthogonal irreducible representation {n-1,1} of the symmetric group S(n).

From the JC, a new set of coordinates was constructed by Aguilera-Navarro et al.[3] whose geometrical properties allied to the standard Racah algebra make them useful to treat many-body problems. When dealing with harmonic-oscillator states, for example, an iterative algorithm arises which can be readly introduced in computational procedures.

In the following, we present a brief self-contained introduction to the transformation defined in Ref. 3. A simple application to the alpha particle is also discussed.

Condensed Matter Theories, Volume 5
Edited by V.C. Aguilera-Navarro
Plenum Press, New York, 1990

2. THE VECTORS Y_s

For a system of n identical particles (the formalism can also be extended to systems of particles of different masses), the JC are defined as

$$X_s = [s(s+1)]^{-1/2}\left[\sum_{t=1}^{s} x_t - s x_{s+1}\right], \qquad 1 \le s \le n-1 \tag{2.1a}$$

$$X_n = n^{-1/2}\sum_{t=1}^{n} x_t, \tag{2.1b}$$

where $x_1, x_2, \ldots, x_n$ are the laboratory coordinates of the n particles.

In a compact form, the orthogonal transformation (2.1) can be written as

$$\mathbf{X} = \mathbf{M}(n)\mathbf{x}. \tag{2.2}$$

Consider now the vectors Y_s, s = 1,2,...,n, defined by

$$Y_s = \sum_{t=1}^{n-1} R_{st}(n)\, X_t, \qquad 1 \le s \le n-1 \tag{2.3a}$$

$$Y_n = X_n, \tag{2.3b}$$

where

$$\mathbf{R}(n) \equiv \mathbf{M}^{-1}(n-1). \tag{2.4}$$

Notice that $\mathbf{M}(n-1)$ is **not** a submatrix of $\mathbf{M}(n)$ but the one associated to a system of n-1 particles. $\mathbf{M}(n)$ is a $n \times n$ matrix while $\mathbf{R}(n)$ is $(n-1) \times (n-1)$. Both $\mathbf{M}$ and $\mathbf{R}$ are orthogonal matrices.

It can be proved[3] that the vectors Y_s, s = 1,2,...,n-1, as defined by (2.3a) carry the fundamental representation of the symmetric group S(n-1) acting on the <u>original</u> coordinates x_s. This result is particularly helpful because it turns easy to find the effect of any permutation $p \in S(n-1)$ -- we need to perform permutations in the symmetrization process.

Notice that Y_n is invariant under the whole S(n).

The transformation (2.3) has a nice and useful geometrical interpretation. The matrix R of (2.3a) can be decomposed as a product of successive transformations as

$$R = \sum_{s=2}^{n-1} R_s(\beta_s) \tag{2.5}$$

where R_s denotes a rotation by an angle β_s on the plane (s-1,s) defined by vectors u_{s-1} and u_s of a (n-1)-dimensional space. The rotation angle is given by

$$\cos\beta_s = (1/s)^{1/2} , \qquad \sin\beta_s = [(s-1)/s]^{1/2} \tag{2.6}$$

The decomposition (2.5) of R is extremely useful when dealing with n-body functions. In the particular case of n-body harmonic-oscillator functions it allows us to take advantage of the generalized Moshinsky coefficients[4] as will be shown in the next section.

3. MANY-BODY SYSTEMS. ALPHA PARTICLE AS A PARADIGM

As a simple application to illustrate the use of the vectors Y_s, we consider a variational analysis of the alpha particle using four-body harmonic-oscillator states. The example will also display an algorithm that can be straightforwardly extended to larger systems.

We will borrow the following notation from Ref. 3:

$w_i \equiv (n_i \ell_i)$ — denotes the pair of oscillator quantum numbers in any coordinate labelled i;

$(w_i' w_j' \ell | w_i w_j \ell)_\beta$ — denotes the generalized Moshinsky coefficient associated to the angle β;

$| \rangle$ angular kets denote harmonic oscillator in the JC;

$|)$ round kets denote harmonic oscillator in the Y coordinates;

$|\,\}$ curled kets denote harmonic oscillator in intermediate coordinates.

We start from easily built four-body harmonic-oscillator states in the Y_s coordinates. Through successive transformations we express them in terms of JC.

A four-body harmonic-oscillator state with defined angular momentum λ is given by

$$|w_1 w_2(\lambda_1); w_3; \lambda\mu). \tag{3.1}$$

This state is translationally invariant because so are the vectors Y_s. The notation in (3.1) indicates that we are coupling ℓ_1 with ℓ_2 to give λ_1, and then with ℓ_3 to get λ.

The rotation by the angle β_2, defined in (2.6), will affect only Y_1 and Y_2. The result is expressed in terms of generalized Moshinsky coefficients as

$$|w_1 w_2(\lambda_1); w_3; \lambda\mu) = \sum_{w_1' w_2'} |w_1' w_2'(\lambda_1); w_3; \lambda\mu\}\{w_1' w_2' \lambda_1 | w_1 w_2 \lambda_1)_{\beta_2} \tag{3.2}$$

Before rotating by β_3 in the plane (2,3), a recoupling is necessary. This can be done in a standard way by using Racah algebra[5], namely,

$$|w_1' w_2'(\lambda_1); w_3; \lambda\mu\} = \sum_{\lambda_1'} [\lambda_1][\lambda_1'] W(\ell_1' \ell_2' \lambda \ell_3; \lambda_1 \lambda_1') \, |w_1'; w_2' w_3(\lambda_1'); \lambda\mu\} \tag{3.3}$$

where

$$[\lambda] \equiv (2\lambda + 1)^{1/2}. \tag{3.4}$$

The rotation by β_3 can now be performed giving rise to

$$|w_1'; w_2' w_3(\lambda_1'); \lambda\mu\} = \sum_{w_2'' w_3'} |w_1'; w_2'' w_3'(\lambda_1'); \lambda\mu\rangle \, \langle w_2'' w_3' \lambda_1' | w_2' w_3 \lambda_1'\}_{\beta_3} \tag{3.5}$$

The states (3.5) are now expressed in terms of harmonic-oscillators states in the Jacobi coordinates.

We are now ready for evaluation of matrix elements of operators such as $V(x_1 - x_2)$. If we are also interested in form factors, a further recoupling of angular momenta is needed, since the form factors depend on the last relative JC in (2.1a). We get

$$|w'_1; w''_2 w'_3(\lambda'_1); \lambda\mu\rangle = \sum_{\lambda'_2} [\lambda'_1][\lambda'_2] W(\ell'_1 \ell''_2 \lambda \ell'_3; \lambda'_2 \lambda'_1) \, |w'_1 w''_2(\lambda'_2); w'_3; \lambda\mu\rangle \quad (3.6)$$

In order to apply the scheme to the alpha particle, we consider a very simple N-N interaction, namely,

$$V_{st} = \frac{1}{2}\left[{}^3V(r_{st}) + {}^1V(r_{st})\right] \quad (3.7)$$

where ${}^1V$ and ${}^3V$ are the singlet and triplet N-N interaction as given by Eikemeier and Hackenbroich[6]. The interaction is a sum of one repulsive and two attractive Gaussian functions for each spin multiplet. With such potential, the variational analysis can be entirely carried out in the configuration space. In such space, the four-nucleon states will be taken to be completely symmetric and with zero total angular momentum.

The states (3.1) are then written as

$$|n_1\ell_1, n_2\ell_2(\ell_3); n_3\ell_3; 00) \equiv |n_1\ell_1, n_2\ell_2, n_3\ell_3) \quad (3.8)$$

On the other hand, since the alpha-particle ground state has positive parity, for completely symmetric states, ℓ_1, ℓ_2 and ℓ_3 must be even[1]. So, a 4-quantum calculation requires a 5-dimensional space spanned by the harmonic-oscillator states

$$|00,00,00)_S, \; |10,00,00)_S, \; |20,00,00)_S, \; |10,10,00)_S, \; |02,02,00)_S \quad (3.9)$$

where the index S stands for symmetrized.

Consider, for example, the member $|10,00,00)$ of the component $|10,00,00)_S$. The rotation by β_2 produces

$$|10,00,00) = \sum_{w'_1 w'_2} |w'_1 w'_2(0); 00; 00\}\{w'_1 w'_2 0\,|\,10,00,0)_{\beta_2} \quad (3.10)$$

It can be shown that in this sum $\ell_1' = \ell_2' = 0$ are the only possibilities for ℓ_1' and ℓ_2'. Recoupling inside the ket we get

$$|w_1'w_2'(0);00;00\} = \sum_{\lambda_1'} [\lambda_1']W(0000;0\lambda_1')\,|w_1';w_2'00(\lambda_1');00\}$$

$$= |w_1';w_2'00(0);00\} \qquad (3.11)$$

Finally, the rotation by β_3 produces

$$|w_1';w_2'00(0);00\} = \sum_{w_2''w_3'} |w_1';w_2''w_3'(0);00\rangle\,\langle w_2''w_3'0|w_2'000\}_{\beta_3} \qquad (3.12)$$

Again, $\ell_2'' = \ell_3' = 0$.

Collecting all the results above we remain with

$$|10,00,00\rangle = \sum_{\substack{w_1'w_2'\\ w_2''w_3'}} |w_1';w_2''w_3'(0);00\rangle\,\langle w_2''w_3'0|w_2'000\}_{\beta_3}\,\{w_1'w_2'0|10,00,0\rangle_{\beta_2} \qquad (3.13)$$

with $\ell_1' = \ell_2' = \ell_2'' = \ell_3' = 0$. The ket in the RHS of (3.13) is ready for evaluation of matrix elements of operators like that in (3.7), since the coordinate X_1 is singled out. The other trial function components appearing in (3.9) are treated in the very same way.

The diagonalization of the Hamiltonian in the 5-dimensional subspace provided about 60% of the experimental ground-state energy of the alpha particle. This poor result can be mainly credited to the almost naive interaction considered and to the small subspace in which the variational calculation was carried out.

If we are also interested in the form factor, we just recouple the angular momenta in the ket in the RHS of (3.13) to obtain

$$|w_1';w_2''w_3'(0);00\rangle = \sum_{\lambda_2'} [\lambda_2']W(0000;\lambda_2'0)\,|w_1'w_2''(\lambda_2');w_3';00\rangle$$

$$= |w_1'w_2''(0);w_3';00\rangle \qquad (3.14)$$

Now, the last relative JC is singled out, what is convenient for form factors evaluation.

It is important to notice is that the previous algorithm can be easily extended to n-harmonic-oscillator basis. In the four-body case we used 2 generalized Moshinsky brackets and 2 Racah coefficients. In the general n-body case, we will need n-2 generalized Moshinsky brackets and 2(n-3) Racah coefficients. If R_s and W stand for a rotation and a recoupling of angular momenta, respectively, then the set of transformations to be performed from left to right, in the n-harmonic-oscillator states, can be symbolically indicated by

$$R_2 W R_3 W W R_4 W \ldots W R_{n-1} W \qquad (3.15)$$

which has the structure $R(WRW)^{n-3}$.

It is not too hard to implement the algorithm (3.15) in a computer program.

REFERENCES

1. M. Moshinsky, "*The harmonic oscillator in modern physics: from atoms to quarks*", (Gordon & Breach, New York, 1969).
2. P. Kramer and M. Moshinsky, "*Group theory of harmonic oscillator and nuclear structure*", in Group theory and its applications, E. M. Loebl (Ed.) (Academic Press, New York, 1968).
3. M. C. K. Aguilera-Navarro and V. C. Aguilera-Navarro, Kinam 4:25 (1982).
4. A. Gal, Ann. Phys. (NY) 49:341 (1968);
   L. Trlifaj, Phys. Rev. C 5:1534 (1972).
5. D. M. Brink and G. R. Satchler, "*Angular Momentum*", 2nd.ed. (Clarendom Press, London, 1968).
6. H. Eikemeier and H. H. Hackenbroich, Zeits. Phys. 195:412 (1966).

DISSIPATIVE EVOLUTIONS IN QUANTUM MECHANICS

J. Aliaga, G. Crespo and A.N. Proto

Laboratorio de Física
Comisión Nacional de Investigaciones Espaciales

I INTRODUCTION

The application of Information-Theory techniques to problems of Hamiltonian dynamics have provided one with interesting new insights.[1-6] The search of dissipative temporal evolutions, starting from a microscopic Hamiltonian remains an open field. Precisely, the aim of the present effort is to analyze different information-theoretic (IT) approaches to the quantum-mechanical description of dissipative temporal evolutions. We will study essentially two different cases: the possibility of obtaining dissipative evolutions for linear quantal Hamiltonians and for non-linear ones.

II DISSIPATIVE EVOLUTIONS FOR LINEAR HAMILTONIANS

The statistical description of temporal evolutions, within the IT point of view, can be summarized as follows: we start from the knowledge of the expectation values of, say, M operators $\hat{O}_j$. As it is usual in statistical mechanics, the mean values of operators are defined throughout evaluation of the trace of the product between these operators and the statistical operator $\hat{\rho}$, i.e. $< \hat{O}_j > = \mathrm{Tr} (\hat{\rho} \hat{O}_j)$. The operator $\hat{O}_0$ is taken equal to the identity in order to assure the normalization of the density matrix. From all possible statistical operators fulfilling these constraints, the IT formalism chooses the one which maximizes the entropy of the system, defined as

$$S = - \mathrm{Tr} (\hat{\rho} \ln \hat{\rho}) \tag{2.1}$$

Therefore, the density matrix reads

$$\hat{\rho} = \exp (- \lambda_0 \hat{I} - \sum_{j=1}^{M} \lambda_j \hat{O}_j) , \tag{2.2}$$

where λ_j are the Lagrange multipliers, determined so as to satisfy the constrains imposed by the mean values.

The temporal evolution of the density matrix is given by the Liouville equation

$$i \hbar \frac{\partial \hat{\rho}}{\partial t} = [\hat{H} , \hat{\rho}] . \tag{2.3}$$

Condensed Matter Theories, Volume 5
Edited by V.C. Aguilera-Navarro
Plenum Press, New York, 1990

Generally, the operators $\hat{O}_j$ do not commute with each other. So, in order to fulfill Eqs. (2.2) and (2.3) and the constancy of the entropy, we must modify the original set of operators $\{ \hat{O}_0, \ldots , \hat{O}_M \}$. It can be easily demonstrated that the set of operators must be extended in order to satisfy

$$[\hat{H} , \hat{O}_i] = i \hbar \sum_{j=0}^{N} \hat{O}_j G_{ij} , \tag{2.4}$$

Thus, Eq. (2.4) defines a new set of "relevant" operators $\{ \hat{O}_0, \ldots , \hat{O}_N \}$. Thus, the relevant operators are those that close a partial Lie algebra under commutation with the Hamiltonian $\hat{H}$. The G_{ij} are the elements of a NxN matrix $\underline{G}$. Moreover, this closure condition leads to the fact that the time-dependent Schrödinger equation can be replaced by a set of coupled equations for the λ_i's

$$\frac{d \lambda_i}{d t} = \sum_{l=0}^{N} G_{il} \lambda_l . \tag{2.5}$$

The temporal evolution of the expectation values of the operators can be obtained having recourse to Ehrenfest's theorem. For the purpose of this section we can assume that the relevant operators do not depend explicitly upon the time. Therefore, we find

$$\frac{d < \hat{O}_i >}{d t} = \sum_{l=0}^{N} < \hat{O}_l > G_{li} , \tag{2.6}$$

which provides us with a set of coupled differential equations that completely determine the temporal evolution of the expectation values $< \hat{O}_j >$, provided one knows the corresponding initial values $< \hat{O}_j >_0$. With all these elements at hand, we can try to determine in which cases we will obtain a dissipative evolution.

Dissipation is, in general a consequence of interactions between a given subsystem and the "rest of the universe", often referred to as the "reservoir". The formalism we are presenting consists in redefining what we are to understand both by the "system" and by the "rest of the universe", assuming that both are parts of a "super system", described by a Hamiltonian. The observables introduced by Eq. (2.4) represents both "our system" and a portion of the rest of the universe interrelated with the system by interactions. Thus, the closure of the semi-algebra allows to decouple this "super system" from the "real" external universe. Into this context, the energy and the entropy of the supersystem are time-dependent constant of motion.[4] Instead, the energy and the entropy of "our system" vary with time. Dissipation is the result of looking at a part of the supersystem, defined as the system under study.

As it was said before, Eq. (2.6) determines the temporal behavior of the mean values for a given Hamiltonian. The situation becomes particularly simple when $\hat{H}$ is independent of time, for which the coefficients G_{ij} of Eq. (2.6) are time independent. The $< \hat{O}_i >_t$ are then the solutions of a system of linear differential equations with constant coefficients. Consequently,

$$< \hat{O}_j >_t = \sum_{i=1}^{k} \exp^{R_i t} \sum_{m=0}^{\gamma} a_{im}^{(j)} t^m , \tag{2.7}$$

where k is the number of different roots R_i of the corresponding secular equations, the $a_{im}^{(j)}$ are constants to be determined by the initial conditions and $\gamma+1$ the multiplicity of the R_i. Thus, it can be clearly seen that the eigenvalues of the $\underline{G}$-matrix (R_i) determine the character of the solution at least in the case of time independent Hamiltonians. So, a necessary condition in order to obtain dissipative temporal evolutions is the existence of real eigenvalues of the $\underline{G}$ matrix.

One should here stress the importance of the initial conditions so as to give a correct description of the problem at hand. It can be seen that the algebra constructed under commutation with the Hamiltonian determines not only the dynamical aspects of the problem, but also the possible expectation values of the relevant operators. If we define the correlation operators as:

$$\hat{K}_{ij} = \frac{1}{2} [\hat{O}_i , \hat{O}_j]_+ - < \hat{O}_i > < \hat{O}_j > \tag{2.8}$$

we obtain:

$$\frac{d < \hat{K}_{ij} >}{d t} = - \sum_{l=0}^{q} (g_{li} < \hat{K}_{lj} > + g_{lj} < \hat{K}_{li} >) \tag{2.9}$$

From elementary statistical mechanics it is known that the mean values of operators are correlated by equations

$$K_{ii} \equiv (\Delta < \hat{O}_i >)^2 = < \hat{O}_i{}^2 > - < \hat{O}_i >^2 \geq 0 \tag{2.10}$$

and

$$K_{ii} K_{jj} \geq (\frac{1}{2} < [\hat{O}_i , \hat{O}_j]_+ > - < \hat{O}_i >< \hat{O}_j >)^2 \equiv K_{ij}{}^2. \tag{2.11}$$

The restrictions imposed to the possible initial mean values by Eqs. (2.10-11) are generally enhanced by the existence of Casimir operators of the Lie group, which is another interrelation between the initial conditions. Thus, they would play a crucial role in the dynamical behavior as they can not be selected in a completely independent way without violating the normalization condition or the probabilistic character of the density matrix.

So, two are the special features of our approach[6]: a) to assume the existence of a set of operators { $\hat{O}_i$, i=1,...,N }, relevant to the physical problem at hand, that closes a partial Lie-algebra under commutation with the Hamiltonian $\hat{H}$ (the G_{ij} elements of the matrix $\underline{G}$ determines the dissipative or conservative behavior of the temporal evolution of the operator's mean values (see Eq. (2.4))); and b) the existence of a coherent set of initial conditions for all those relevant operators. So, it can be shown that dissipative evolutions can be attained only for non-compact Lie groups. In order to stress the importance of both conditions, in the next paragraph we discuss an example which is characterized by real eigenvalues of the $\underline{G}$ matrix, satisfying only condition a) for dissipative temporal evolutions, but not condition b) on the initial mean values.

Let us consider a generalization of the well known Bateman's Hamiltonian[7-9]. The total Hamiltonian of the system is

$$\hat{H} = \hat{H}_a + \hat{H}_b + v_1 \hat{T}_{ab} + v_2 \hat{S}_{ab} , \tag{2.12}$$

where $\hat{H}_a = \varepsilon_a (\hat{a}^+ \hat{a} + 1/2)$ and $\hat{H}_b = \varepsilon_b (\hat{b}^+ \hat{b} + 1/2)$ are harmonic oscillator Hamiltonians representing the unperturbed systems, and $\hat{T}_{ab} = \hat{a}^+ \hat{b}^+ + \hat{b} \hat{a}$, $\hat{S}_{ab} = i(\hat{a}^+ \hat{b}^+ + \hat{b} \hat{a})$ take into account the interactions. It can be shown that the set of relevant operators { $\hat{I}$, $\hat{H}_a$, $\hat{H}_b$, $\hat{T}_{ab}$, $\hat{S}_{ab}$ } fulfill Eq. (2.4). Therefore, we evaluate the different roots of the secular equation corresponding to the $\underline{G}$ matrix which are the coefficients accompanying the exponential functions that characterize the appropriate temporal evolution of the system. We obtain that the eigenvalues of the $\underline{G}$ matrix are $\lambda^2 = [4(v_1{}^2 + v_2{}^2) - \varepsilon^2] / \hbar^2$, where $\varepsilon = \varepsilon_a + \varepsilon_b$. So, Two quite different regimes ensue according to the value of the coupling constant. For $\lambda \in \mathbb{C}$, the time-evolution of $<\hat{T}_a>_t$, $<\hat{T}_b>_t$, $<\hat{T}_{ab}>_t$, $<\hat{S}_{ab}>_t$ adopt a stationary character (oscillating functions). It is more interesting for us to study the alternative case, namely $\lambda \in \mathbb{R}$, in which we obtain real exponential functions for the evolution of the mean values. As usual, we call time decaying solutions "dissipative ones". The term dissipation should then be regarded within the context of works like the one by Dekker[7]. Notice that the total energy of the system is given by $<\hat{H}>$, a constant. The subsystems are represented by $<\hat{H}_a>$ and $<\hat{H}_b>$, that do evolve with time. Our work differs from previous literature in the fact that, in addition to $\hat{H}_a$ and $\hat{H}_b$, we are including two other operators in order to close a partial Lie algebra. We choose $\hat{H}_a$ as representing that portion of our total system in which an observer is located. The rest is to be considered as a heath bath. The closure above referred to guarantees that our total system is, both dynamically ($< \hat{H} >$ = constant) and thermodynamically (S = constant), a closed system.

For the purpose of avoiding divergences for $t \rightarrow \infty$ we must impose the initial condition that the coefficient related with the real positive exponential function vanishes. It is easy to show that it is impossible to achieve this condition. This means that for the particular case of the generalized quantal Bateman's Hamiltonian, even though the Hamiltonian dynamics allows one to obtain a dissipative behavior, this can not be attained due to the fact that initial conditions needed to cancel the divergent component can not be written down. This result is not restricted to the Bateman's Hamiltonian as Eq. (2.12) is the more general time-independent linear Hamiltonian which can be constructed for bilinear products of creation and annihilation operators. As it was said before, dissipative evolutions can be attained only for non-compact Lie groups. There exits only one group that can be constructed via bilinears products of creation and annihilation operators satisfying this property: it is the Simplectic Group, Sp_2, which has bosonic statistics and changes the number of particles. This is just the group considered in this work and so we can assure that dissipative evolutions can not be obtained with time independent linear Hamiltonians generated via any combination of bilinear products of creation and annihilation operators.

III NON LINEAR HAMILTONIANS AND QUANTAL DISSIPATION

An alternative approach to quantal dissipation is based on the introduction of a frictional potential which depends upon selected expectation values making the Schrödinger equation a non-linear one. In this case the most general hamiltonian is of the form[5,10-13]

$$H = \frac{-\hbar^2}{2\,m}\,\frac{\partial^2}{\partial\,x^2} + V(x) + \gamma\,W_g \tag{3.1}$$

$\hat{W}_g$ being the frictional potential, and γ an appropriate constant. Here the idea is to solve the set of equations $< \hat{H} > \cong < \hat{p}^2/2m > + < \hat{V}(x) >$, $d<\hat{p}>/dt + < \partial\hat{V}/\partial x > + \gamma < \partial\hat{W}_g/\partial x > = 0$ and $d< \hat{H} >/dt \cong - (\gamma/m) < \hat{p}^2 >$, where the $\cong$ symbol indicates that an additional term associated with $(\Delta\hat{p})^2$ (the momentum indeterminacy) can be added[10].

The first of these last equations is the counterpart of the classical relationship E = T + V which implies

$$< \hat{W}_g > = 0. \tag{3.2}$$

In similar fashion the second one can be related to $dp/dt + \partial V/\partial x + \gamma p = 0$, leading to $< \partial \hat{W}_g / \partial x > = < \hat{p} >$. Finally, the third one is associated to the classical expression $dE/dt = -(\gamma/m)\, p^2$.

As was pointed out by Hasse[10] and Dekker[7], both types of hamiltonians present ambiguities and unphysical features, although the expectation values of observables they reproduce are in correspondence with classical quantities.

A serious difficulty associated to non-linear[10] hamiltonians is their incompatibility with the superposition principle. This fact raises the question of how to evaluate expectation values, as they provide the only link with classical behaviour. It is not quite clear how to obtain them from the knowledge of the concomitant wave-packet. However, the vectorial space of the $\{ \hat{O}_j \}$ operators generated by Eq. (2.5), works as the dual space of the usual Hilbert space of wave functions and the $\underset{\sim}{g}$ matrix contains the dynamics so that the temporal evolution of the $< \hat{O}_j >(t)$ can be easily obtained through I.T.

This non linear hamiltonian is constructed so as to mimic a one-body approach (mean-field technique) to the damping problem in quantum mechanics. The general form of $\hat{W}_g$ is

$$\hat{W}_g = (\hat{x} - < \hat{x} >) \; [\; c \, \hat{p} + (1-c) < \hat{p} >] \; - \; i \hbar c / 2 \tag{3.3}$$

where c is a real constant. Several frictional potentials reported in recent literature can be encountered as special cases of Eq. (3.3). The value of c determines the particular potential one is concerned with. The form of Eq. (3.3) is due to Sussmann and is based on the identification of the classical momentum with the quantum operator $p = i \hbar \partial / \partial x$. Sussmann studied the case c = 1 , while Albrecht[11] did the same with the null c case. Hasse[10], instead, has studied the $c = \pm (1/2)$ situations.

The closure relation (2.5) with the Hamiltonian given by Eq. (3.1) can be fulfilled by the set of operators $\{ \hat{1} , \hat{x} , \hat{p} \}$, for either the choice of $\hat{V}(\hat{x})$ corresponding to a free particle, a particle subject to a constant force, or a harmonic potential. Other choices are possible in what respects to the operators involved, and shall be tackled later on. Notice that the temporal evolutions of $< \hat{x} >$ and $< \hat{p} >$ are independent of the choice of the constant c. The corresponding equations are linear ones, notwithstanding the fact that the hamiltonian is of the non linear sort. Moreover, $< \hat{x} >$ and $< \hat{p} >$ evolve like their classical counterparts. In the harmonic oscillator case we observe a frequency shift. The corresponding expression reads $\Omega = (w_o^2 - \gamma^2/4)^{1/2}$ independent of c, which, in fact, corresponds to the shifted frequency observed for the damped harmonic oscillator in classical mechanics. The pertinent value reported in the literature is, instead, $\Omega = (w_o^2 - c^2 \gamma^2)^{1/2}$ (Refs. 7 and 10). Notice that we obtain $d< \hat{x} >/dt = < \hat{p} >/m$, so that the identification of $\hat{p}$ with the usual kinetic momentum becomes obvious. The entropy is a constant of the motion as a result of having closed a semialgebra with $\hat{H}$.

Now, in order to better study the case of the damped harmonic oscillator, we define the dimensionless variables $\hat{Q} = (m \, w_o/\hbar)^{1/2} \, \hat{x}$ and $\hat{P} = (1/\hbar \, m \, w_o)^{1/2} \, \hat{p}$, where m is the oscillator mass, w_o its frequency and

$\hat{x}$, $\hat{p}$ the usual canonical variables. In terms of the new variables, the hamiltonian (3.1) (considering $\hat{V}(\hat{x}) = 1/2\, m\, w_o\, \hat{x}^2$ and $\hat{W}_g$ of the general form) reads:

$$\hat{H} / \hbar w_o = \frac{1}{2}\hat{P}^2 + \frac{1}{2}\hat{Q}^2 + c\,\Gamma\,\hat{Q}\,\hat{P} + (1-c)\,\Gamma < \hat{P} > \hat{Q} - \tag{3.4}$$
$$- c\,\Gamma < \hat{Q} > \hat{P} - (1-c)\,\Gamma < \hat{Q} > < \hat{P} > - \frac{i}{2} c\,\Gamma$$

with $\Gamma = \gamma / w_o$.

The set of observables $\{ -\hat{\mathbf{1}}, \hat{Q}, \hat{P}, \hat{Q}^2, \hat{P}^2, \hat{L} \}$, with $\hat{L} = (\hat{Q}\,\hat{P} + \hat{P}\,\hat{Q})/2$, closes a semialgebra with $\hat{H}$. As a result, we found that, while $< \hat{Q} >$ and $< \hat{P} >$ evolve linearly (as was previous indicated), the equations for $< \hat{Q}^2 >$, $< \hat{P}^2 >$ and $< \hat{L} >$ are coupled, non-linear. and they explicitly depend upon c.

On evaluating the time evolution of the quantities $(\Delta\hat{Q})^2$, $(\Delta\hat{P})^2$ and $\Delta\hat{Q}.\Delta\hat{P}$ we find that the indeterminacy principle is not violated, opposite to what is commented on the case of Kanai's approach[7,10]. However, we find a different behaviour for the dispersions of $\hat{Q}$ and $\hat{P}$ than that of Ref. 10. Moreover for $t \rightarrow \infty$ residual kinetic ($\sim < \hat{P}^2 >$) and potential ($\sim < \hat{Q}^2 >$) energies are found, that are equal, respectively, to the corresponding fluctuations $(\Delta\hat{P})^2$ and $(\Delta\hat{Q})^2$ notwithstanding the fact that, in that limit, both $< \hat{Q} >$ and $< \hat{P} >$ vanish as in the classical case. These residual energies may be associated with the temperature of a thermal bath into which the energy lost by the damped oscillator flows.

As for the temporal evolution of the expectation value of the hamiltonian we have

$$< \hat{H} / \hbar\, w_o > = \frac{1}{2} < \hat{P}^2 > + \frac{1}{2} < \hat{Q}^2 > + c\,\Gamma\,(<\hat{L}> - <\hat{Q}> <\hat{P}>) \tag{3.5}$$

As asserted by Dekker[7] one finds $< \hat{W}_g > \neq 0$, save for the cases $c = 0$ or $<\hat{L}> = <\hat{Q}> <\hat{P}>$. This should be contrasted with the original idea that led to the introduction of $\hat{W}_g$ (see Hasse, Ref. 10, and Eq. (3.2)) namely, that $< \hat{H} >$ represents the energy of the system, which in turn entails $< \hat{W}_g > \cong 0$.

Figs. 1, 2 and 3 exhibit the evolution of $< \hat{H} / \hbar\, w_o >$ for the three special values of c specified above. These energies were calculated applying Picard's method of integration. Both Eq. (3.5) and the value obtained with the approximation $< \hat{W}_g > \cong 0$ are illustrated. As discussed above, a residual energy appears in the limit $t \rightarrow \infty$. From the tentative identification

$$< \hat{H} >_{t=\infty} = \hbar\, w_o\, (\, (1/2)\, K\, T\,) \tag{3.6}$$

we could find the temperature of the bath into which energy is dissipated (k is Boltzmann's constant). Of course, Eq. (3.6) should be considered just a conjecture. The time-evolution of correlation coefficients according to Eqs. (2.8, 2.9) coincide with the system of equations tackled with a different prescription in Ref. 7.

Finally, one finds that, within the present context, the counter-part of the classical expression $dE/dt = -(\gamma/m)\, p^2$ is given by

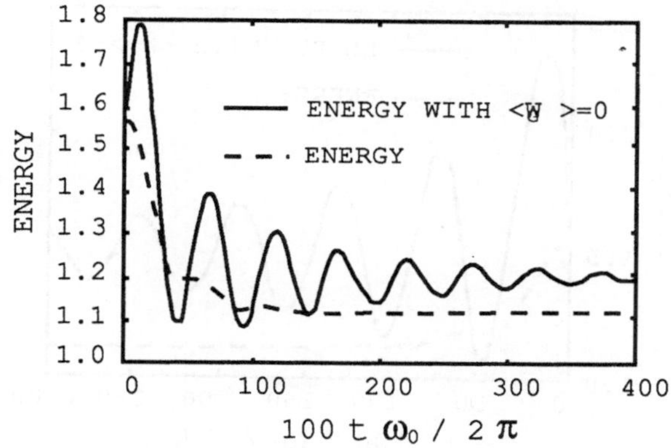

Fig. 1 Plot of $< \hat{H} / \hbar w_o >$ vs t, for c=1/2. The solid line corresponds to the approximation to $< \hat{H} / \hbar w_o >$ in which $< \hat{W}_g >=0$, and the dashed line corresponds to the expression of $< \hat{H} / \hbar w_o >$ (see Eq. (3.5)) without any approximation. Asympotic values for the solid line: 1.19; and for the dashed line 1.12.

$$dE/dt = -2 \pi \Gamma P^2 \tag{3.7}$$

where E is identified with $< \hat{H} / \hbar w_o >$. The explicit evolution is governed by

$$\frac{d}{dt} < \hat{H} / \hbar w_o > = 2\pi c \Gamma \{ (\Delta\hat{Q})^2 - (\Delta\hat{P})^2 \} - 2\pi \Gamma < \hat{P} >^2 \tag{3.8}$$

where (2.6) has been employed. If one identifies P^2 with $< \hat{P} >^2$, (3.8) will not yield (3.7), save for the case c = 0.

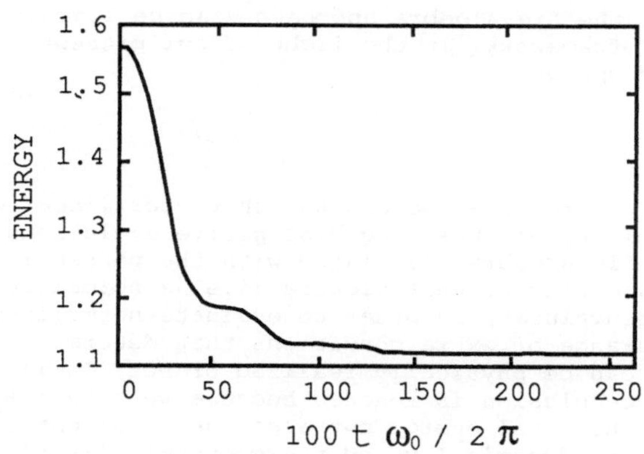

Fig. 2 Plot of $< \hat{H} / \hbar w_o >$ vs t, for c=0. In this case the result corresponding to the approximation to $< \hat{H} / \hbar w_o >$ in which $< \hat{W}_g >=0$, and the one corresponding to the expression of $< \hat{H} / \hbar w_o >$ (see Eq. (3.5)) without any approximation, coincide. Asympotic values: 1.12.

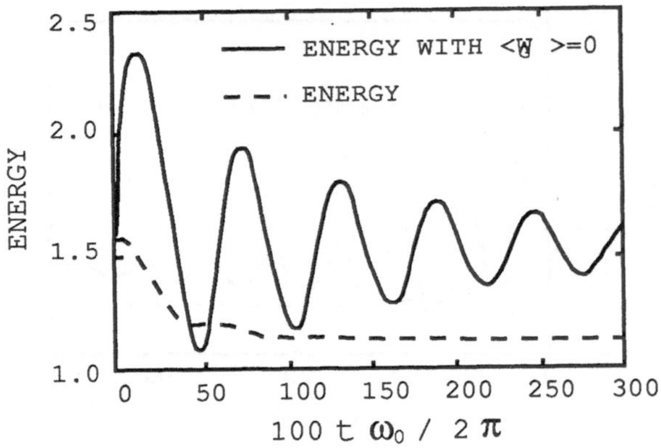

Fig. 3 Plot of $< \hat{H} / \hbar w_o >$ vs t, for c=1. The solid line corresponds to the approximation to $< \hat{H} / \hbar w_o >$ in which $< \hat{W}_g >=0$, and the dashed line corresponds to the expression of $< \hat{H} / \hbar w_o >$ (see Eq. (3.5)) without any approximation. Asympotic values for the solid line: 1.49; and for the dashed line 1.12.

An alternative fashion for computing the evolution of $< \hat{H} / \hbar w_o >$ is by a straightforward use of the Ehrenfest's theorem

$$\frac{d}{dt} < \hat{H} / \hbar w_o > \; = \; <[\hat{H} / \hbar w_o, \; \hat{H} / \hbar w_o]> + (1/\hbar w_o) < \partial\hat{H}/\partial t > \qquad (3.9)$$

Evaluating the second term of the r.h.s. with (3.4) one finds

$$\frac{d}{dt} < \hat{H} / \hbar w_o > \; = \; - 2\pi \;\; \Gamma < \hat{P} >^2 \qquad (3.10)$$

which agrees with (3.7). The question that comes up is whether the Ehrenfest's theorem or the Lie-algebra approach[2] can be legitimately used within this (non-linear) framework. In the light of our present results, we would prefer the second option.

IV CONCLUSIONS

Summarizing our results, we can say that, for linear quantal Hamiltonians, the possibility of obtaining dissipative evolutions is crucially dependent upon the Lie algebra associated with the physical problem but also the addition of a metric to this algebra (i.e. a normalized density matrix with positive eigenvalues), in order to evaluate normalized mean values, implies the appearance of extra conditions that determine whether such temporal evolutions can be physically realized or not. It is necessary to stress that this conclusion is reached because we have insisted upon conserving not only the total energy but also the total entropy, so that the total system is both dynamically and thermodynamically closed. This "double closure" distinguishes the present approach from previous ones[7-9].

On the other hand, recourse to a non-linear Hamiltonian constitutes another way to achieve irreversible leakage of energy to its surroundings. However, for this kind of Hamiltonians there exists a strong incompatibility with the superposition principle. When the Maximum Entropy Principle (M.E.P.) is used, most of the ambiguities emerging when dealing with these Hamiltonians disappear.We find different behaviors for the shifted frequency of the damped harmonic oscillator and the time evolution of the quantal dispersions from the ones previously reported in literature[10].As it was previously mentioned, also in this case, and as a result of the M.E.P., the information theoretic entropy S is conserved. Finally, on studying the behavior of the mean value of the energy $<\hat{H}/\hbar w_o>$, and not assuming that $<\hat{W}_g> = 0$, we find that subtle implications concerning the exact moment in which the commutator $< [\hat{H}, \hat{H}] >$ is calculated, apart from a tentative identification of the residual energy of the system with the thermal fluctations which in turn can be connected with the temperature of the bath.

J. A. thanks the Argentine National Research Council (CONICET) for its support. A. N. P. and G. C. acknowledge support from the Comisión de Investigaciones Científicas de la Provincia de Buenos Aires (CIC). A. N. P. also acknowledges support from Universidad Tecnológica Nacional (UTN).

REFERENCES

1. D. Otero, A. Plastino, A. Proto, G. Zannoli, Phys.Rev. **A26** (1982) 1209.
2. E. Duering, D. Otero, A. Plastino and A.N. Proto, Phys. Rev. **A32** (1985) 2455, 3661.
3. D. Otero, A. Plastino, A.N. Proto and S. Mizhari, Phys. Rev. **A33** (1986) 3446.
4. J. Aliaga, D. Otero, A. Plastino and A.N. Proto, Phys. Rev. **A35** (1987) 2304; Phys. Rev. **A37** (1988) 918.
5. G. Crespo, D. Otero, A.Plastino, and A. N. Proto,Phys. Rev. **A**39 (1989) 2133.
6. A.N. Proto, J. Aliaga, D.R. Napoli, D. Otero and A. Plastino, Phys. Rev. **A39** (1989) 4223.
7. H. Dekker, Physics Reports **80**, Nº 1 (1981).
8. H. Bateman, Phys. Rev. **38** (1931) 815.
9. H. Feshbach and Y. Tikochinsky, Trans. N.Y. Acad. Sci. **38** (1977) 44.
10. R. W. Hasse, J. Math. Phys. **16** (1975) 2005, J. Phys. **11A** (1978) 1245.
11. K. Albrecht, Phys. Lett. **56B** (1975) 127.
12. K. K. Kan, J. J. Griffin, Phys. Lett. **50B** (1974) 241.
13. M. D. Kostin, J. Stat. Phys. 12 (1974) 145, M. D. Kostin, J. Chem. Phys. **57** (1973) 3589.
14. C. Cohen-Tannoudji, B. Diu, F. Laloe, Quantum Mechanics, Vol I (John Wiley & Sons and Hermann, 1977).

EXTENDED COUPLED CLUSTER TECHNIQUES FOR EXCITED STATES: APPLICATIONS TO QUASISPIN MODELS

R.F. Bishop,* N.I. Robinson* and J. Arponen†

*Department of Mathematics
University of Manchester Institute of Science and Technology
P.O. Box 88, Manchester M60 1QD, England

†Department of Theoretical Physics, University of Helsinki
Siltavuorenpenger 20C, SF-00170 Helsinki, Finland

1. INTRODUCTION

We have described both in an earlier volume in this series[1] and elsewhere,[2] how the so-called extended coupled cluster method (ECCM)[3] can be used to give an extremely general microscopic description of quantum-mechanical many-body and field-theoretical systems. The formalism is in principle exact, capable of being systematically implemented at various levels of approximation, and essentially applicable to all systems governed by some underlying Schrödinger dynamics. The reader is referred to the earlier article[1] for a full discussion of its relationships with both the earlier (normal) coupled cluster method (NCCM) or exp(S) approach of Coester and Kümmel,[4] and the more primitive configuration-interaction (CI) method.[5] Special attention was focussed there on the interpretation of each method within the context of time-independent perturbation theory, in terms of suitably defined generalized tree-diagram structures. Particular applications of the ECCM which have been explored in previous volumes of this series, have provided fully microscopic effective gauge-field descriptions both of a charged impurity in a polarizable medium (applicable, for example, to the important experimental tool of positron annihilation in metals),[6] and of the zero-temperature quantum fluid dynamics of a strongly-interacting condensed Bose fluid.[7]

We have shown elsewhere[1,2] how the ECCM provides a biorthogonal formulation of the bra and ket states of an N-body system in terms of a set of quasi-local and c-number ("classical") generalized mean fields or generalized order parameters. A key feature of the method is that *all* of these amplitudes, which between them completely describe the full system in terms of its various n-body subsystems (n = 1,2,...,N), are linked-cluster quantities. Since they all exactly obey the cluster property, the formalism is, in principle, capable of describing such phenomena as spontaneous symmetry breaking and phase transitions. We have further demonstrated how the ECCM enables an arbitrary quantum many-body problem to be mapped exactly onto a well-defined classical Hamiltonian mechanics for these same many-body, classical, multi-local, configuration-space amplitudes, which fully parametrize the original system.

Given that the original description of the ECCM[1-3] as reviewed above, was

Condensed Matter Theories, Volume 5
Edited by V.C. Aguilera-Navarro
Plenum Press, New York, 1990

based on a very general time-dependent formulation, it is possible to discuss the linear response of the system in the usual way by considering small oscillations around the (ground-state) equilibrium. In this way, we have shown elsewhere[8] how the collective eigenmodes may be found by thus linearizing the equations of motion. It has furthermore been demonstrated how this formally exact ECCM description may be viewed as an exact generalization of the well-known random phase approximation (RPA) of Bohm and Pines.[9] It is, in this sense that the ECCM cluster amplitudes referred to above, may be precisely interpreted as a set of generalized multi-local mean fields acting in the many-body configuration space. We have also shown[8] how the stationary excited energy-eigenstates may be described within the ECCM by analogy with their parametrization in the NCCM due to Emrich.[10] Finally, in the same paper it has also been shown that there is an exact one-to-one correspondence between the excited states so obtained as the solution to a set of linear ECCM eigenvalue equations, and the collective eigenmodes of the effective Hamiltonian obtained from expanding, up to second-order terms, the average-value functional for the energy in powers of the deviations from their equilibrium values of the ECCM cluster amplitudes.

One of our primary aims here is to introduce a new variational functional for the excitation energies. This new formulation not only retains all of the positive attributes of the previous versions of the ECCM, but it also provides both a different description of the excited states and one which allows for further flexibility in the various approximate implementations of the method in practice.

The basic ingredients of the stationary (time-independent) version of the (ground-state) ECCM are first briefly summarized in Sec.2. After its extension to excited states is similarly reviewed in Sec.3, the new variational functional for the excited states and their excitation energies is introduced in Sec.4. As a concrete application of the method, it is tested in Sec.5 with considerable success on the SU(2) quasispin model of Lipkin, Meshkov and Glick,[11] which models the spherical to deformed shape transition of atomic nuclei at high spins, via the spontaneous breakdown of parity at high values of the coupling parameter which characterizes the model. It is explicitly shown how *both* phases are very well described by the formalism in a *single* calculational scheme. Finally, the results are discussed in Sec.6.

2. ELEMENTS OF THE ECCM GROUND-STATE PARAMETRIZATION

The static ground-state ECCM has at its heart a double similarity transformation, $A \longrightarrow \hat{A}$, for an arbitrary operator A in the Hilbert space,

$$\hat{A} \equiv e^{S''} e^{-S} A e^{S} e^{-S''} , \tag{1}$$

which is induced by the following basic parametrizations of the exact ground-state ket and bra energy eigenvectors,

$$|\Psi_0\rangle \equiv e^{S}|\Phi\rangle = e^{S} e^{-S''}|\Phi\rangle , \tag{2a}$$

$$\langle\Psi'_0| \equiv \langle\Phi|e^{S''} e^{-S}, \tag{2b}$$

in terms of cluster operators S and S″ and some suitable (normalized) model state $|\Phi\rangle$. This model state is often, but not necessarily, chosen to be the appropriate non-interacting ground state in the limit where the interparticle forces are adiabatically switched off. Equations (2a,b) clearly imply the definite normalization, $\langle\Psi'_0|\Psi_0\rangle = 1$. The exact eigenstates of Eq.(2) are assumed to be the non-degenerate ket and bra ground states of the many-body Hamiltonian H,

$$H|\Psi_0\rangle = E_0|\Psi_0\rangle \quad , \quad \langle\Psi'_0|H = E_0\langle\Psi'_0| \ . \tag{3}$$

The prime in the bra-state notation introduced above is intended to remind us that although $\langle\Psi'_0| = \text{const.} \times (|\Psi_0\rangle)^\dagger$ as usual for a hermitian Hamiltonian, the parametrizations of Eq.(2) are clearly not manifestly hermitian adjoints of each other. Indeed, in subsequent truncations of the (otherwise exact) theory, this hermiticity cannot be guaranteed *a priori*. The ground-state Schrödinger equations (3) are thus exactly mapped to the forms,

$$\hat{H}|\Phi\rangle = E_0|\Phi\rangle \quad , \quad \langle\Phi|\hat{H} = E_0\langle\Phi| \ , \tag{4}$$

where the doubly similarity-transformed operator $\hat{H}$ is again not manifestly hermitian, since the transformation of Eq.(1) is not unitary.

Turning to the two cluster operators defined in Eqs.(2a,b), S (S'') is defined to be constructed solely from creation (destruction) pieces with respect to the state $|\Phi\rangle$, so that

$$\langle\Phi|S = 0 = S''|\Phi\rangle \ . \tag{5}$$

We note in passing that the second equation in Eq.(2a) follows from the first defining relation by making use of Eq.(5). This latter equation also implies the intermediate normalization for the ket eigenstate, $\langle\Phi|\Psi_0\rangle = \langle\Phi|\Phi\rangle = 1$. Implicit in Eq.(5) and the statement which precedes it is that the otherwise rather free choice of model state $|\Phi\rangle$ is made so as to fulfill the only requirement made of it, namely that the algebra of all possible operators in the many-body Hilbert space is spanned by the two Abelian subalgebras of creation and destruction operators defined with respect to $|\Phi\rangle$. In other words, one assumes that these two subalgebras and the state $|\Phi\rangle$ are *cyclic*, in the sense that an arbitrary ket (bra) state in the Hilbert space may be constructed from some suitable linear combination of states obtained from $|\Phi\rangle$ ($\langle\Phi|$) by pre-multiplication (post-multiplication) with elements of the set of creation (destruction) operators.

It is thus possible to define complete orthonormal sets of creation operators $\{C_i^\dagger\}$ and their hermitian adjoint destruction counterparts $\{C_i\}$, where the many-body configuration space index i represents a suitable *set* of single-particle labels. Exámples have been given elsewhere,[2] and their choice for the LMG model used as illustration here, is made explicit in Sec.5. With respect to some suitable (generalized) Kronecker symbol $\delta(i,j)$ in the set-indices i and j, we have the orthonormality relation,

$$\langle\Phi|C_iC_j^\dagger|\Phi\rangle = \delta(i,j) \ , \tag{6}$$

and the completeness relation for the resolution of the identity operator I,

$$I = \sum_i C_i^\dagger|\Phi\rangle\langle\Phi|C_i \equiv |\Phi\rangle\langle\Phi| + \sum_i{}' C_i^\dagger|\Phi\rangle\langle\Phi|C_i \ . \tag{7}$$

The prime on the sum in Eq.(7) is used henceforth to exclude from the sum the term $i = 0$, where, by definition, $C_0^\dagger \equiv I$.

The ECCM cluster operators S and S'', defined by Eqs.(2a,b), now have the exact representations,

$$S = \sum_i{}' s_iC_i^\dagger \quad , \quad S'' = \sum_i{}' s''_iC_i \ . \tag{8}$$

An alternative pair of cluster operators to $\{S,S''\}$ is denoted by $\{\Sigma,\tilde{\Sigma}\}$, and defined as follows,

$$\Sigma = \sum_i{}' \sigma_i C_i^\dagger \quad , \quad \tilde{\Sigma} = \sum_i{}' \tilde{\sigma}_i C_i \; ;$$
$$\sigma_i \equiv \langle\Phi|C_i e^{S''} S|\Phi\rangle \quad , \quad \tilde{\sigma}_i \equiv S''_i \; . \tag{9}$$

One of the most important attributes of the ground-state ECCM is that *both* sets of c-number amplitudes $\{S_i,S''_i\}$ and $\{\sigma_i,\tilde{\sigma}_i\}$, either one of which completely characterizes the ground-state many-body problem, are comprised of wholly-linked cluster amplitudes. Each amplitude thereby obeys the cluster property, and has a definite linked diagrammatic representation. The formalism is hence also manifestly size-extensive.[12]

The ground-state expectation value of an arbitrary operator A,

$$\langle A\rangle \equiv \bar{A} \equiv \langle\Psi'_0|A|\Psi_0\rangle = \langle\Phi|\hat{A}|\Phi\rangle \; , \tag{10}$$

may thus be regarded as a functional of either complete set of amplitudes $\{S_i,S''_i\}$ or $\{\sigma_i,\tilde{\sigma}_i\}$. In both cases, it has a well-defined linked structure.[2,3] In particular, the ground-state energy functional, $\langle H\rangle = E_0$, may be so expressed in terms of fully connected diagrams, thereby making explicit the Goldstone linked cluster theorem.[13] We have shown[1,2] that the ground-state Schrödinger equations (3) are fully equivalent to the requirement that $\langle H\rangle$ be stationary with respect to small variations in each of the cluster amplitudes $\{\sigma_i,\tilde{\sigma}_i\}$, say,

$$\delta\langle H\rangle/\delta\sigma_i = 0 = \delta\langle H\rangle/\delta\tilde{\sigma}_i \; , \tag{11}$$

which equations may formally be used to solve for the cluster amplitudes. Finally, it has been shown in Ref.[2] how such general matrix elements involving $\hat{A}$ as $\langle\Phi|C_i\hat{A}C_j^\dagger|\Phi\rangle$, which are needed for the further algebraic development of the formalism, may be expressed in terms of the first- and second-order functional derivatives of the expectation value $\langle A\rangle$ with respect to the cluster amplitudes. This relation takes the especially simple form for the Hamiltonian operator at the stationary ground-state equilibrium,

$$\langle\Phi|C_i\hat{H}C_j^\dagger|\Phi\rangle = E_0\delta(i,j) + \delta^2\langle H\rangle/\delta\tilde{\sigma}_i\delta\sigma_j + \sum_k{}' \sigma_{i+k}\delta^2\langle H\rangle/\delta\sigma_k\delta\sigma_j \; , \tag{12}$$

for $i \neq 0 \neq j$, in terms of second-order derivatives only, in view of Eq.(11). The cluster amplitude in Eq.(12) with the compound set-index $(i+k)$ is defined as,

$$\sigma_{i+k} \equiv \langle\Phi|C_iC_k e^{S''} S|\Phi\rangle \; . \tag{13}$$

3. ECCM DESCRIPTIONS OF THE EXCITED STATES

The corresponding excited-state Schrödinger equations,

$$H|\Psi_\lambda\rangle = (E_0+\varepsilon_\lambda)|\Psi_\lambda\rangle \quad , \quad \langle\Psi'_\lambda|H = (E_0+\varepsilon_\lambda)\langle\Psi'_\lambda| \; , \tag{14}$$

may also be written in terms of the similarity-transformed Hamiltonian as,

$$\hat{H}|\chi_\lambda\rangle = (E_0+\varepsilon_\lambda)|\chi_\lambda\rangle \; , \; \langle\chi'_\lambda|\hat{H} = (E_0+\varepsilon_\lambda)\langle\chi'_\lambda| \; . \tag{15}$$

A comparison of Eqs.(14) and (15) thus implies the connections,

$$|\Psi_\lambda\rangle = e^{S}e^{-S''}|\chi_\lambda\rangle \quad , \quad \langle\tilde{\Psi}_\lambda'| = \langle\chi_\lambda'|e^{S''}e^{-S} \ . \tag{16}$$

Motivated by the comparable NCCM parametrization of Emrich[10] for the excited ket states, we introduce the convenient parametrizations of the states $|\chi_\lambda\rangle$ and $\langle\chi_\lambda'|$,

$$|\chi_\lambda\rangle = X^\lambda|\Phi\rangle \quad , \quad \langle\chi_\lambda'| = \langle\Phi|Y^\lambda \ , \tag{17}$$

in terms of excitation and de-excitation operators, X^λ and Y^λ respectively,

$$X^\lambda = \sum_i{}' X_i^\lambda C_i^\dagger \quad , \quad Y^\lambda = \sum_i{}' Y_i^\lambda C_i \ . \tag{18}$$

By combining Eqs.(4), (15) and (17), the excited-state Schrödinger equations may be written in the explicit forms,

$$[\hat{H},X^\lambda]|\Phi\rangle = \varepsilon_\lambda X^\lambda|\Phi\rangle \ , \quad \langle\Phi|[Y^\lambda,\hat{H}] = \varepsilon_\lambda\langle\Phi|Y^\lambda \ , \tag{19}$$

in terms of the excitation energies ε_λ directly. Projection of the former and latter of Eqs.(19) with the states $\langle\Phi|C_i$ and $C_i^\dagger|\Phi\rangle$ respectively, then gives the ECCM equations for the c-number amplitudes $\{X_i^\lambda\}$ and $\{Y_i^\lambda\}$ which completely characterize the excited states, as coupled sets of linear eigenvalue equations. By making use of the completeness relation (7) and Eq.(12), one may also readily rewrite these eigenvalue equations in the form,

$$\sum_j{}' \left(\frac{\delta^2\langle H\rangle}{\delta\tilde{\sigma}_i\delta\sigma_j} + \sum_k{}' \sigma_{i+k}\frac{\delta^2\langle H\rangle}{\delta\sigma_k\delta\sigma_j}\right) X_j^\lambda = \varepsilon_\lambda X_i^\lambda \ , \tag{20a}$$

$$\sum_j{}' Y_j^\lambda \left(\frac{\delta^2\langle H\rangle}{\delta\tilde{\sigma}_j\delta\sigma_i} + \sum_k{}' \sigma_{j+k}\frac{\delta^2\langle H\rangle}{\delta\sigma_k\delta\sigma_i}\right) = \varepsilon_\lambda Y_i^\lambda \ . \tag{20b}$$

All of the input ground-state quantities inside the parentheses in Eqs.(20a,b) are evaluated at the stationary equilibrium found from Eq.(11).

We have also considered[8] the dynamics of small oscillations of the system around this stationary point, in terms of an effective second-order Hamiltonian $\bar{H}^{(2)}$, which is bilinear in the deviations from their equilibrium values of the cluster amplitudes $\{\sigma_i,\tilde{\sigma}_i\}$, and which is obtained by linearizing the equations of motion. It is self-evidently found by expanding $\bar{H}$ around the stationary equilibrium defined by Eq.(11), up to second order. It has been shown[8] how the eigenfrequencies of $\bar{H}^{(2)}$ are identical to the excitation energies ε_λ found above. We have also explicitly demonstrated how this latter procedure produces an exact generalization of RPA, and hence how the ground-state amplitudes $\{\sigma_i,\tilde{\sigma}_i\}$ may thus be viewed as a set of (multi-local) generalized mean many-body fields in the configuration space where they are now labelled by the (sets of single-particle) indices $\{i\}$.

4. AN EXCITATION-ENERGY FUNCTIONAL

Up to this point our parametrizations of both the ground and excited many-body states have been formally exact. In practice one must, however, approximate. Perhaps the most natural such approximation scheme is the so-called SUB(n) scheme for the ground-state formalism and the SUB(m,n) scheme for excited states. In the SUB(n) scheme, the configuration-space indices {i} in either the set $\{S_i, S''_i\}$ or $\{\sigma_i, \tilde{\sigma}_i\}$ of cluster amplitudes, are restricted to at most n single-particle labels. Similarly in the SUB(m,n) scheme, the ground-state amplitudes are truncated at $n^{\underline{th}}$-order clusters as above, and at the same time the excitation amplitudes $\{X_i^\lambda, Y_i^\lambda\}$ are similarly truncated at $m^{\underline{th}}$ order. In both cases, the higher amplitudes are set identically to zero and the remaining truncated sets of equations are then solved, ideally without further approximation. We note that although the ground-state formalism is based on the variational principle expressed in Eq.(11), the resulting approximate values for E_0 in the SUB(n) scheme are *not*, in general, guaranteed to be upper bounds, since our whole formalism is a biorthognal one, which is not manifestly hermitian at an arbitrary level of truncation.

We also note that although the NCCM, which is based upon a single similarity transformation in terms of the single cluster operator S, has the distinctly advantageous attribute that at any level of truncation the ground-state equations are of finite order in the cluster amplitudes, the double similarity transformation of the ECCM spoils this property. It transpires that in order both to keep the practical applications of the ECCM as compact as possible, and to circumvent this problem as far as practicable, the ground-state parametrization is best considered in terms of the original cluster amplitudes $\{S_i, S''_i\}$. This is in spite of the fact that the set $\{\sigma_i, \tilde{\sigma}_i\}$ forms a much more symmetric and canonical set of coordinates for the theoretical discussions and formal applications of the method. We thus consider the ground-state energy functional ⟨H⟩ from Eq.(10) in the form $E_0 = E_0[S_i, S''_i]$, where

$$E_0[S_i, S''_i] = \langle\Phi|e^{S''}e^{-S}He^{S}|\Phi\rangle \ . \tag{21}$$

Stationarity of this ground-state energy functional with respect to each of the independent amplitudes $\{S_i, S''_i\}$ leads to the set of equations which determine these amplitudes, and which are fully equivalent to the Schrödinger equations (3),

$$\langle\Phi|e^{S''}e^{-S}[H,C_i^\dagger]e^{S}|\Phi\rangle = 0 \ , \quad i \neq 0 \ , \tag{22a}$$

$$\langle\Phi|C_i e^{S''}e^{-S}He^{S}|\Phi\rangle = 0 \ , \quad i \neq 0 \ . \tag{22b}$$

We now attempt to construct a comparable functional for the excitation energies. With this aim in mind we introduce a new excitation operator S^λ, defined to be the creation part of the operator product $\exp(-S'')X^\lambda$,

$$S^\lambda|\Phi\rangle \equiv e^{-S''}X^\lambda|\Phi\rangle \quad , \quad S^\lambda = \sum_i{}' \ S_i^\lambda C_i^\dagger \ . \tag{23}$$

It is clear from a comparison of Eqs.(9) and (23) that the relationship between the excitation amplitudes S_i^λ and X_i^λ is thus the precise counterpart for the excited states of that between the amplitudes S_i and σ_i for the ground state. Since the operators S and S^λ commute, we may once again combine the Schrödinger equations (3) and (14) for the ground and excited ket states respectively, to obtain the relation

$$e^{-S}[H,S^\lambda]e^{S}|\Phi\rangle = \varepsilon_\lambda S^\lambda|\Phi\rangle \ . \tag{24}$$

By taking the inner product of Eq.(24) with the state $\langle\Phi|Y^\lambda\exp(-S'')$, we readily obtain the functional,

$$\varepsilon_\lambda[S_i^\lambda, Y_i^\lambda] = \frac{\langle\Phi|Y^\lambda e^{S''} e^{-S}[H, S^\lambda]e^S|\Phi\rangle}{\langle\Phi|Y^\lambda e^{S''} S^\lambda|\Phi\rangle} , \tag{25}$$

for the excitation energy.

If the excitation-energy functional ε_λ of Eq.(25) is now required to be stationary with respect to each of the independent amplitudes $\{S_i^\lambda, Y_i^\lambda\}$, we readily arrive at the generalized eigenvalue equations,

$$\langle\Phi|Y^\lambda e^{S''} e^{-S}[H,C_i^\dagger]e^S|\Phi\rangle - \varepsilon_\lambda\langle\Phi|Y^\lambda e^{S''} C_i^\dagger|\Phi\rangle = 0 , \tag{26a}$$

$$\langle\Phi|C_i e^{S''} e^{-S}[H,S^\lambda]e^S|\Phi\rangle - \varepsilon_\lambda\langle\Phi|C_i e^{S''} S^\lambda|\Phi\rangle = 0 , \tag{26b}$$

for all $i \neq 0$, which now determine these excited-state amplitudes, and which may be compared with their ground-state counterparts in Eqs.(22a,b). By making use of both the exact ground-state equations (2) and (3), and the completeness relation of Eq.(7), it is not difficult to show that the stationarity conditions of Eqs.(26a,b) are completely equivalent to the original Schrödinger equations for the excited bra and ket states. Again, this is just as expected, since the set of amplitudes $\{S_i^\lambda, Y_i^\lambda\}$ now gives a complete specification of the excited states in terms of the corresponding ground states. We note that the overall normalization of the excited states has never been specified. This fact is reflected in the homogeneity of the (linear) eigenvalue equations (26a,b) for the excitation amplitudes, which are thereby determined only up to an overall multiplicative constant. Finally, we remark that the replacement $\exp(S'') \longrightarrow I$ in Eqs.(22b) and (26b) leads to two sets of equations which are themselves trivially obtained from Eqs.(4) and (24) respectively. Furthermore, these resulting equations are simply the corresponding NCCM counterparts, due respectively to Coester and Kümmel,[4] and to Emrich.[10]

The SUB(m,n) approximation scheme that we have outlined above is now performed in practice by solving the excited-state equations (26a,b) for the cases $i = 1,2,\ldots,m$ only, and with all higher amplitudes $\{S_i^\lambda, Y_i^\lambda;\ i > m\}$ set to zero. The ground-state amplitudes $\{S_i, S_i''\}$ which are needed as input to these equations are similarly taken from the SUB(n) approximation to Eqs.(22a,b). We note that the (truncated) ground-state equations (22a,b) are nonlinear in the amplitudes $\{S_i, S_i''\}$, and the possibility of multiple solutions cannot be discounted. Furthermore, the excited-state equations (26a) and (26b) form decoupled sets of equations for the amplitudes $\{Y_i^\lambda\}$ and $\{S_i^\lambda\}$ respectively. The eigenvalue spectrum ε_λ obtained from either set of equations is the same. Clearly, a SUB(m,n) approximation will yield estimates for m excitation energies.

5. APPLICATION TO THE LMG MODEL

In the LMG model,[11] N identical fermions are distributed between two single-particle energy levels, each of which is N-fold degenerate, and which are separated by an energy gap of unit magnitude. The Hamiltonian is given in terms of the ususal canonical fermion creation and destruction operators, $a_{p,m}^\dagger$ and $a_{p,m}$ respectively, as

$$H = \tfrac{1}{2} \sum_{p,m} m a^\dagger_{p,m} a_{p,m} + \tfrac{1}{2} V \sum_{p,p',m} a^\dagger_{p,m} a^\dagger_{p',m} a_{p',-m} a_{p,-m} , \tag{27}$$

where the index $m = \pm 1$ labels the two levels and the quantum number $p = 1,2,\ldots,N$ labels single-particle states within each level. As is by now very well-known, the model is readily mapped into its spin-algebraic equivalent,

$$H = J_z + \frac{g}{2N} (J_+^2 + J_-^2) , \qquad g \equiv NV , \tag{28}$$

where the quasispin operators J_+, J_- and J_z, defined as

$$\begin{aligned} J_+ &\equiv \sum_p a^\dagger_{p,+1} a_{p,-1} \equiv J_x + iJ_y , \\ J_- &\equiv J_+^\dagger = J_x - iJ_y , \\ J_z &\equiv \tfrac{1}{2} \sum_{p,m} m a^\dagger_{p,m} a_{p,m} , \end{aligned} \tag{29}$$

are readily checked to obey the usual SU(2) algebra,

$$[J_+,J_-] = 2J_z , \quad [J_z,J_\pm] = \pm J_\pm . \tag{30}$$

The Hamiltonian of Eq.(28) is easily seen to commute with the total quasispin operator $J^2 \equiv J_x^2 + J_y^2 + J_z^2$ which has eigenvalues $j(j+1)$ as usual, the parity operator $\Pi \equiv \exp[i\pi(J_z+j)]$, and each of the N operators $n_p \equiv \sum_m a^\dagger_{p,m} a_{p,m}$. In particular, the eigenstates of H may thus be classified into multiplets according to the quantum number j. In the limit as $g \longrightarrow 0$, the unperturbed ground state $|\Phi\rangle$ is clearly the member of the multiplet with the maximal value of j, namely $\tfrac{1}{2}N$, which is also an eigenstate of J_z with minimal eigenvalue, $-\tfrac{1}{2}N$. This state satisfies the relation $J_-|\Phi\rangle = 0$, and in terms of the original formulation of Eq.(27) it corresponds to the state in which all N fermions occupy the lower level.

The ket Hilbert space of the N-body problem is thus clearly spanned by the (N+1) unnormalized vectors $\{J_+^k|\Phi\rangle \; ; \; k = 0,1,\ldots,N\}$, and the bra space is similarly spanned by their hermitian adjoints. In terms of this basis, in which $|\Phi\rangle$ is the cyclic vector, the ground-state cluster operators S and S″ now have the representations

$$S = N \sum_{k=1}^{N} (iN)^{-k} S_k J_+^k \quad ; \quad S'' = N \sum_{k=1}^{N} (-iN)^{-k} S''_k J_-^k . \tag{31}$$

The excitation operators S^λ and Y^λ have the corresponding representations,

$$S^\lambda = N \sum_{k=1}^{N} (iN)^{-k} S^\lambda_k J_+^k \quad ; \quad Y^\lambda = N \sum_{k=1}^{N} (-iN)^{-k} Y''_k J_-^k . \tag{32}$$

The factors of N in Eqs.(31) and (32) have been chosen so that each term is a thermodynamically extensive operator in the large-N limit when the c-number amplitudes $\{S_k,S''_k\}$ and $\{S^\lambda_k,Y^\lambda_k\}$ are of order unity. The imaginary factors of i similarly permit us to use real values for these amplitudes.

The above choice of $|\Phi\rangle$ as the cyclic ket vector, and the corresponding basis $\{J_+^k|\Phi\rangle\}$ built on it, is not however the only possible choice. While it is true that the choice of model state or cyclic vector is irrelevant for an *exact* calculation (i.e., a SUB(N) calculation for the ground state), the best choice of the model state for an approximate calculation at the SUB(n) level, with $n < N$, is not *a priori* obvious. In particular, one knows that the LMG model undergoes a phase transition in the vicinity of some critical coupling strength $g \longrightarrow g_c$.[14] Thus, while in the limit $N \longrightarrow \infty$, the exact ground state $|\Psi_0\rangle$ remains qualitatively close to the perturbative ground state $|\Phi\rangle$ for $g < 1$, and in particular has $\langle J_y\rangle = 0$, in the region $g > 1$ the state $|\Psi_0\rangle$ becomes, in the same large-N limit, deformed and doubly-degenerate, with order parameter $\langle J_y\rangle/j \longrightarrow \pm(g^2-1)^{\frac{1}{2}}/g$. This transition is perhaps most simply studied at the lowest SUB(1) level of approximation, which is identical to Hartree-Fock approximation. In a time-dependent HF calculation it is easy to show that as $g \longrightarrow g_c^{HF} = N/(N-1)$, the frequency of the collective excitation approaches zero, thereby signalling the phase (or shape[15]) transition. For $g > g_c^{HF}$, two degenerate HF solutions exist. Neither of them any longer has parity as a good quantum number. Of course, for finite values of N these two broken-symmetry solutions can communicate via quantum tunnelling through the intervening potential barrier of finite height, and hence the transition is not sharp. Although the *exact* ground state has even parity, the first excited state (with odd parity) becomes exponentially close to it for increasing N. Indeed for large N and large g, the lowest several states form into such closely-spaced parity doublets.

While it is thus highly suggestive on physical grounds that either one of these two degenerate broken-symmetry or deformed HF states might form a better model state or cyclic vector in the deformed region $g > g_c$ than the previous symmetric solution $|\Phi\rangle$, it is interesting to enquire whether the coupled cluster formalism can itself suggest the most appropriate starting wavefunction $|\Phi'\rangle$. A particularly appealing argument has been discussed by Kümmel.[15] He suggests that an appropriate optimal choice for $|\Phi'\rangle$ is that state which maximizes the overlap M with the exact ground state,

$$M \equiv \frac{|\langle\Phi'|\Psi_0\rangle|^2}{\langle\Phi'|\Phi'\rangle\langle\Psi_0|\Psi_0\rangle} \quad . \tag{33}$$

The suggestion is implemented via the Thouless theorem,[16] which states that any Slater determinant $|\Phi'\rangle$ in the neighbourhood of $|\Phi\rangle$, and hence not orthogonal to it, may be written in the form $|\Phi'\rangle = \text{const.} \times \exp(T)|\Phi\rangle$, where T is a one-body operator (i.e., one which excites 1p-1h pairs out of $|\Phi\rangle$). In the case of the LMG model, the appropriate choice is clearly $T = i\gamma J_+$, for some constant γ. By making use of the SU(2) algebra we may then show that

$$|\Phi'\rangle = \exp(i\alpha J_x)|\Phi\rangle \ , \quad \gamma \equiv \tan(\tfrac{1}{2}\alpha) \ . \tag{34}$$

We may thus define a new set of (unitarily transformed) quasispin operators K_+, K_- and K_z, where

$$K_i \equiv \exp(i\alpha J_x)\, J_i\, \exp(-i\alpha J_x) \ ; \quad i = x,y,z \ , \tag{35}$$

such that $K_-|\Phi'\rangle = 0$.

The state $|\Phi'\rangle$ of Eq.(34) may then be used as an alternative cyclic vector for any value of the angle α, and the operators $\{S,S''\}$ and $\{S^\lambda,Y^\lambda\}$ may be comparably expanded as in Eqs.(31) and (32) but with the replacements $|\Phi\rangle \longrightarrow |\Phi'\rangle$ and $J_\pm \longrightarrow K_\pm$. In particular, the HF choice is obtained by minimizing the expectation value of the Hamiltonian in the model state,

$$\langle\Phi'|H|\Phi'\rangle = -\tfrac{1}{2}N\left[\cos\alpha + \tfrac{1}{2}g\left(\frac{N-1}{N}\right)\sin^2\alpha\right] . \tag{36}$$

It is trivial to show that for $g < g_c^{HF} = N/(N-1)$, the minimal value of Eq.(36) is with $\alpha = 0$, i.e. $|\Phi'\rangle \longrightarrow |\Phi\rangle$, the symmetric state. Conversely, for $g > g_c^{HF}$, the two values $\alpha = \pm\cos^{-1}[g^{-1}N/(N-1)]$ give the two degenerate minima. Alternatively, a full maximum-overlap calculation would choose the value of α by maximizing M from Eq.(33). We show examples below of numerical results obtained with both the symmetric and deformed HF solutions for the cyclic vector.

The implementation of the ECCM for the LMG model is now straightforward. The parametrizations of Eqs.(31) and (32), or the corresponding expansions in terms of $|\Phi'\rangle$ and $K_\pm$, are inserted into Eqs.(21) and (25) for the ground-state and excitation energy functionals $E_0[S_i,S''_i]$ and $\varepsilon_\lambda[S_i^\lambda,Y_i^\lambda]$ respectively. The resulting expressions are then made stationary with respect to each of their respective parameters, as already described above. We omit all details of the remaining somewhat laborious algebraic manipulations, although the interested reader may be referred to similar calculations by one of us which have been described more fully elsewhere.[17] We prefer instead to dwell here on the results obtained at the various SUB(m,n) levels of approximation and with both symmetric (or normal) and symmetry-breaking (or deformed) model states as cyclic vectors. Some analytical results can be given in the limit $N \to \infty$. For example, if the ground-state results from a SUB(2) approximation based on the symmetric model state $|\Phi\rangle$ for *all* values of g, are inserted into a SUB(1,2) excited-state calculation, the (first) excitation energy is found to be given as,

$$\varepsilon_1 \xrightarrow[N \to \infty]{} \begin{cases} (1-g^2)^{\frac{1}{2}} & ; \quad g < 1 \\ [2(g^2-1)]^{\frac{1}{2}} & ; \quad g > 1 . \end{cases} \tag{37}$$

This result is identical to that obtained in time-dependent HF theory, and it is also exact in this large-N limit.

Typical numerical results based on the symmetric model state $|\Phi\rangle$ are shown for the N = 14 system in Table 1 and Fig.1. The ground-state ECCM provides excellent results in both phases of the LMG model so long as the critical value g_c is not approached too closely. At a given SUB(n) level, an approximate critical value is determined by the point beyond which a new branch of solutions opens up in which the odd-indexed parameters $\{S_i,S''_i\}$ become non-zero, to yield a parity-violating state. Even near this critical value, the method gives results which are well-behaved and still reasonably accurate. Figure 1 illustrates this behaviour and also demonstrates the lack of a sharp transition. Results for higher values of N are similar. We show in Fig.2 results for a higher value of g (well into the deformed region), based on the same symmetric model state. One sees that whereas the ECCM now readily obtains estimates for the average values of the very nearly degenerate low-lying parity doublets, the level-splitting can only be seen at all in a SUB(m,n) approximation with values of the truncation index m close to the maximal value N. The other striking feature of our results in the deformed

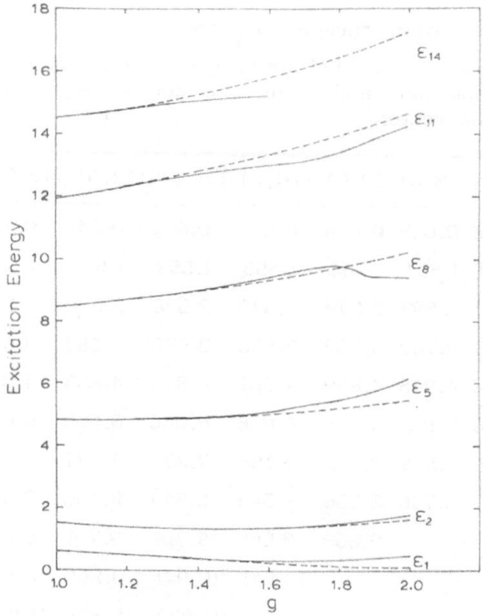

Fig.1. Selected ECCM excitation energies ε_n (full lines) in a SUB(14,6) approximation based on the symmetric model state as cyclic vector, for the N = 14 system in the critical region $1 \leq g \leq 2$. The dashed lines show the corresponding exact results.

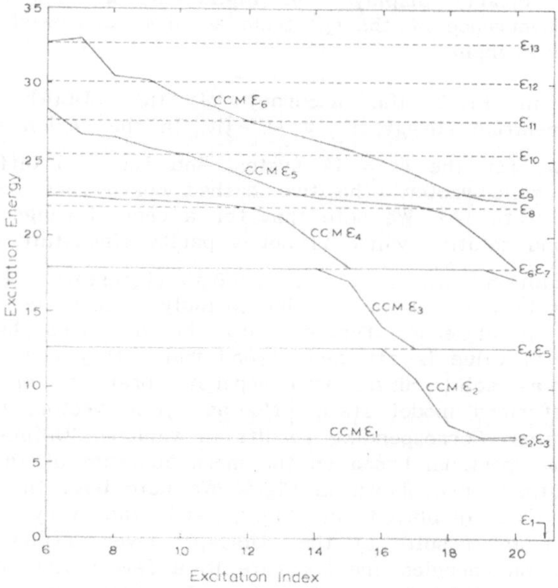

Fig.2. Convergence properties of the six lowest ECCM excitation energies ε_n (full lines) for the N = 30, g = 5 system, as a function of the truncation index m, in a SUB(m,6) approximation based on the symmetric model state as cyclic vector. The horizontal dashed lines show the corresponding exact results, with the four lowest parity doublets unable to be resolved.

Table 1. The excitation energies ε_n for the N = 14 system with g = 1.0. Columns labelled (m,n) give the ECCM results in SUB(m,n) approximation, and using the normal (symmetric) model state as cyclic vector.

	(6,6)	(7,6)	(8,6)	(9,6)	(10,6)	(11,6)	(12,6)	(13,6)	(14,6)	Exact
ε_1	0.649	0.648	0.648	0.648	0.648	0.648	0.648	0.648	0.648	0.648
ε_2	1.557	1.557	1.554	1.554	1.553	1.553	1.553	1.553	1.553	1.553
ε_3	2.619	2.587	2.587	2.579	2.579	2.578	2.578	2.578	2.578	2.578
ε_4	3.759	3.759	3.702	3.702	3.688	3.688	3.687	3.687	3.687	3.686
ε_5	5.169	4.948	4.948	4.869	4.869	4.853	4.853	4.852	4.852	4.852
ε_6	6.398	6.398	6.153	6.153	6.066	6.066	6.053	6.053	6.053	6.052
ε_7		7.599	7.599	7.355	7.355	7.277	7.277	7.271	7.271	7.270
ε_8			8.756	8.756	8.543	8.543	8.490	8.490	8.489	8.488
ε_9				9.868	9.868	9.714	9.714	9.689	9.689	9.688
ε_{10}					10.941	10.941	10.860	10.860	10.854	10.853
ε_{11}						11.982	11.982	11.961	11.961	11.962
ε_{12}							12.984	12.984	12.984	12.987
ε_{13}								13.889	13.889	13.892
ε_{14}									14.539	14.540

phase, which is clearly displayed in Fig.2, is the extremely non-uniform nature of the convergence of the spectrum as m is increased at fixed index n for the ground-state input.

We display in Fig.3 the accuracy of the SUB(4) results for the ground-state correlation energy, $E_c \equiv -E_0-\frac{1}{2}N$, in the region around and above the critical point, for the N = 14 system and for two different choices of model state as cyclic vector. The two dashed curves are both based on the symmetric model state $|\Phi\rangle$. We note that for a certain range of values of $g > g_c \approx 1.6$, a second solution which is not a parity eigenstate coexists with the normal-phase solution which is a parity eigenstate. Presumably these normal-phase solutions become increasingly unstable against small perturbations as g increases further into the deformed-phase region, until some upper critical value is attained beyond which they cease to exist. Figure 3 also displays as solid curves two separate branches of SUB(4) solutions based on the deformed model state $|\Phi'\rangle$ as cyclic vector, with the angle α taking the HF value. Corresponding results in various SUB(m,4) approximations for the excitation spectrum based on the more accurate of these two deformed ground-state solutions, are shown in Fig.4. We note both the similarity of the convergence to that displayed in Fig.2, and the very good quantitative agreement with exact results at the SUB(14,4) level. Typically, these results for all 14 excitation energies are accurate to a few tenths of 1%. The results change very little if the HF value for the deformation angle α is replaced by that from a full maximum-overlap calculation.

6. DISCUSSION

It is clear from the numerical results that the ECCM is easily capable of

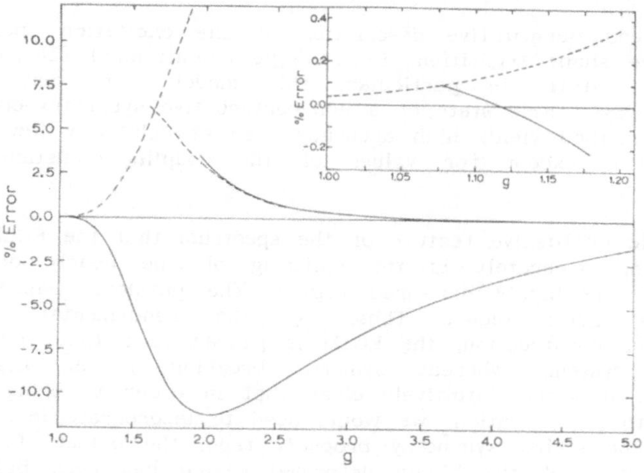

Fig.3. The error, $(1-E_c/E_c^{exact})$, in the SUB(4) ECCM correlation energy, $E_c \equiv -E_0 - \frac{1}{2}N$, for the N = 14 system, as a function of coupling strength g. The two dashed (solid) curves are based on the symmetric (deformed HF) model state as cyclic vector.

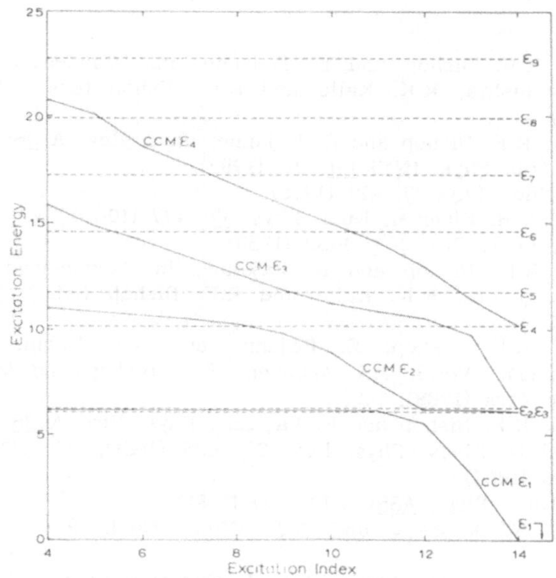

Fig.4. Convergence properties of the four lowest ECCM excitation energies ε_n (full lines) for the N = 14, g = 5 system, as a function of the truncation index m in a SUB(m,4) approximation based on the deformed HF model state as cyclic vector. The ground-state SUB(4) input corresponds to the upper solid curve in Fig.3. The horizontal dashed lines show the corresponding exact results.

giving a good quantitative description of the excitation spectrum, on both sides of the shape transition, in a single calculational scheme based upon a fixed model state. In particular, this model state may either be the perturbative symmetric state or a non-perturbative symmetry-breaking deformed state. The method yields high accuracy even at relatively low SUB(m,n) levels of truncation, except for values of the coupling constant close to the critical value.

The sole qualitative feature of the spectrum that the ECCM has difficulty in describing accurately is the splitting of the nearly degenerate parity doublets in the highly deformed region. The physical reasons for this are nevertheless quite clear. Thus, by the fundamental nature of its linked-cluster construction, the ECCM is geared to a *local* description of the many-body system, whereas symmetry-breaking is an essentially *global* phenomenon. It seems intuitively clear that in order to bring these disparate features into a unification, we would need to incorporate into the ECCM some aspect of the global symmetry property from the outset. In particular, our best description of the highly deformed region has been built on a *single* cyclic vector, which is either one of the two degenerate deformed model states $|\Phi'\rangle$ for some choice of the angle $\pm\alpha$. It would be of considerable interest to develop a generalized version which incorporates *both* on an equal footing. Such a formulation would have even wider applicability.

ACKNOWLEDGEMENT

One of us (RFB) gratefully acknowledges support for this work in the form of a research grant from the Science and Engineering Research Council of Great Britain.

REFERENCES

1. J. Arponen, R.F. Bishop and E. Pajanne, in: "Condensed Matter Theories," Vol.2, P. Vashishta, R.K. Kalia and R.F. Bishop (eds.), Plenum, New York (1987), p.357.
2. J.S. Arponen, R.F. Bishop and E. Pajanne, Phys. Rev. A 36: 2519 (1987).
3. J. Arponen, Ann. Phys. (NY) 151: 311 (1983).
4. F. Coester, Nucl. Phys. 7: 421 (1958);
F. Coester and H. Kümmel, Nucl. Phys. 17: 477 (1960).
5. R.K. Nesbet, Phys. Rev. 109: 1632 (1958).
6. J. Arponen, R.F. Bishop and E. Pajanne, in: "Condensed Matter Theories," Vol.2, P. Vashishta, R.K. Kalia and R.F. Bishop (eds.), Plenum, New York (1987), p.373.
7. J. Arponen, R.F. Bishop, E. Pajanne and N.I. Robinson, in: "Condensed Matter Theories," Vol.3, J.S. Arponen, R.F. Bishop and M. Manninen (eds.), Plenum, New York (1988), p.51.
8. J.S. Arponen, R.F. Bishop and E. Pajanne, Phys. Rev. A 36: 2539 (1987).
9. D. Bohm and D. Pines, Phys. Rev. 82: 625 (1951); 92: 609 (1953); D. Pines, *ibid.* 92: 626 (1953).
10. K. Emrich, Nucl. Phys. A351: 379, 397 (1981).
11. H.J. Lipkin, N. Meshkov and A.J. Glick, Nucl. Phys. 62: 188, 199, 211 (1965).
12. R.J. Bartlett and G.D. Purvis, Int. J. Quantum Chem. 14: 561 (1978); Phys. Scripta 21: 251 (1980).
13. J. Goldstone, Proc. Roy. Soc. (London) A239: 267 (1957).
14. D. Agassi, H.J. Lipkin and N. Meshkov, Nucl. Phys. 86: 321 (1966).
15. H.G. Kümmel, Nucl. Phys. A317: 199 (1979).
16. D.J. Thouless, "The Quantum Mechanics of Many Body Systems," Academic, New York (1961); Nucl. Phys. 21: 225 (1960).
17. J. Arponen and J. Rantakivi, Nucl. Phys. A407: 141 (1983).

TEMPORAL EVOLUTION OF FLUCTUATIONS

N. Canosa*, A. Plastino* and R. Rossignoli**

Physics Department, National University of La Plata
(1900) La Plata, Argentina

INTRODUCTION

The Time Dependent Hartree-Fock (TDHF) approximation[1,2] constitutes the basic tool for dealing with the evolution of an uncorrelated many-body wave function. The approach has been extensively applied to a wide variety of many-fermion problems, yielding in general a reliable description of single particle (s.p.) expectation values. However, in situations where fluctuations become relevant, as for instance in the prediction of spreading widths of s.p. operators, TDHF fails to provide an accurate picture, due to the inherent neglect of correlations. The addition of involved collisional terms becomes thus necessary[3-5].

CORRECTED MEAN FIELD APPROACH (CMFA)

In ref. 6 a systematic and tractable procedure for improving TDHF predictions was developed, based on a suitable approximate closure of the semialgebra formed by the Hamiltonian $H = H_o + V$ with the observables of interest. Starting with a set of one-body observables $O_i^{(1)}$, the ensuing scheme can be cast as (we set $\hbar = 1$)

$$-i d\langle O_i^{(j)}\rangle/dt = \langle[H,O_i^{(j)}]\rangle, \quad j = 1,\ldots,m-1 \tag{1a}$$

$$-i d\langle O_i^{(m)}\rangle/dt = \langle[H_o,O_i^{(m)}]\rangle + \langle[V,O_i^{(m)}]\rangle_{hf} \tag{1b}$$

*Supported by Consejo Nacional de Investigaciones Científicas y Técnicas (CONICET) of Argentina.
**Supported by Comisión de Investigaciones Científicas de la Provincia de Buenos Aires (CIC).

Condensed Matter Theories, Volume 5
Edited by V.C. Aguilera-Navarro
Plenum Press, New York, 1990

where $O_i^{(j)}$, $j \geq 2$, denote the j-body operators arising in the commutators with H in the j-1 step. The subindex hf indicates an uncorrelated evaluation (i.e., by means of Wick's theorem). H_0 denotes an unperturbed s.p. Hamiltonian (it can be for instance a static HF Hamiltonian) and V the corresponding residual interaction. Thus, the semialgebra with H is exactly closed up to m-1 body operators, while in the last step, it is closed just with H_0. System (1) is then complete if all one-body operators entering the uncorrelated evaluation are included in the original set.

For m=1, the scheme becomes equivalent to TDHF. For m>1, we attain thus an improved description, which yields exact (m-j-1)th order temporal derivatives at t=0 for operators $O_i^{(j)}$. In TDHF, only the initial first order time derivatives of the one-body observables $O_i^{(1)}$ are exactly evaluated, due to the violation of Ehrenfest theorem for higher order observables.

If the non-linear evaluation in the r.h.s. of (1b) is omitted, we attain a linear perturbative scheme[7], in which the mth power of V is discarded in the time evolution of $O_i^{(1)}$. In this case, it is necessary to go up to m+1 order in (1) to obtain the correct mth order time derivatives of observables $O_i^{(1)}$ at t=0.

THE EVOLUTION OF FLUCTUATIONS

In view of what has been said above, it becomes clear that the temporal evolution of the fluctuation of a one-body observable cannot be correctly described, even for short times, using TDHF. The exact equation of motion for the fluctuation F of an operator O_i, $F = \langle O_i^2 \rangle - \langle O_i \rangle^2$, can be cast as

$$-i dF^{ex}/dt = \langle [H, O_i^2 - 2\langle O_i \rangle O_i] \rangle = 2C\{[H, O_i], O_i\} \tag{2}$$

with $C\{O_i, O_j\} = \frac{1}{2}\langle O_i O_j + O_j O_i \rangle - \langle O_i \rangle \langle O_j \rangle$ (quantum covariance), whereas in TDHF,

$$-i dF^{hf}/dt = \langle [h, O_i^2 - 2\langle O_i \rangle O_i] \rangle_{hf} \tag{3}$$

since $-i d\langle O_i^2 \rangle_{hf}/dt = \langle [h, O_i^2] \rangle_{hf}$, where $h = \sum_i (\partial \langle H \rangle_{hf} / \partial \langle O_i^{(1)} \rangle) O_i^{(1)}$ is the s.p. mean field effective Hamiltonian (with the sum running over all one-body observables appearing in $\langle H \rangle_{hf}$). Thus, even at t=0 we attain a non-vanishing difference between both evolutions, given by

$$-i d(F^{ex} - F^{hf})/dt|_{t=0} = \langle [V_{res}, O_i^2] \rangle_{hf} \tag{4}$$

where $V_{res} = H - h$ is the residual interaction. In fact, the l.h.s. of (4) is identical with the initial rate of increase of the correlation $\langle O_i^2 \rangle_c = \langle O_i^2 \rangle - \langle O_i^2 \rangle_{hf}$.

On the other hand, the CMFA yields exact initial temporal derivatives of fluctuations already for m=2, providing at least the correct initial trend. Perturbative treatments require m=3. Higher quality predictions of s.p. fluctuations can be obtained in CMFA for m=3, in which case the second temporal derivative of F coincides with the exact value at t=0.

As a specific example, we have calculated the temporal evolution of the fluctuation of the operator J_x, under the action of the Hamiltonian $H = \varepsilon J_z - V(J_x^2 - J_y^2)$, within the framework of the Lipkin model[8], where

$$J_z = \tfrac{1}{2}\sum_{p,\nu}^{n} \nu c^+_{p\nu} c_{p\nu} \quad , \quad J_+ = \sum_{p}^{n} c^+_{p+} c_{p-} = J_-^+ \quad , \tag{5}$$

with $p = 1,\ldots,N$, $\nu=\pm1$ (N is the number of particles). The s.p. density matrix $\langle c^+_{q\nu} c_{p\nu'}\rangle = \delta_{pq} x_\nu x^*_{\nu'}$, $\sum_\nu |x_\nu|^2 = 1$, provides the allowed initial conditions to solve the system (1).

It can be easily seen that in this case the initial difference (4) is given by

$$\langle [V_{res}, J_x^2]\rangle_{hf} = 4v\langle J_x\rangle\langle J_y\rangle\langle J_z\rangle/[N(N-1)] \tag{6}$$

(where $v = V(N-1)$), which is of the same order of magnitude of the fluctuation ($O(N)$), for fixed v (in (6) we have neglected terms of order 1). The evolution of the fluctuation of J_x in TDHF is given by

$$-i dF^{hf}/dt = -2\varepsilon\langle J_x\rangle\langle J_y\rangle/N + 4v\langle J_x\rangle\langle J_y\rangle\langle J_z\rangle/N^2, \tag{7}$$

so that a situation in which the sign of the exact initial derivative differs from that given by (7) may occur. A typical situation is illustrated in Fig. 1, for real initial values of x_ν. In this case, (6) and (7) vanish at t=0, but nevertheless TDHF fails to provide the correct initial trend. It also predicts a wrong amplitude, and the extrema are out of phase with the exact ones.

Results obtained with CMFA and the perturbative closure (for m = 2 and m = 3 respectively) are also shown, and are of similar quality, yielding both an acceptable agreement with exact results, at least for short times. Time is given in units of $\hbar/\varepsilon$, which for ε = 500 Kev yields 1.3×10^{-21}s, larger than the nucleon transversal time ($\cong 10^{-22}$s).

In conclusion, we have shown that TDHF predictions for fluctuations are in general inadequate, even for short times. However, a corrected mean field approach which does not violate Ehrenfest theorem for two-body operators, provides a reliable picture, being at the same time sufficiently tractable and simple. This fact allows possible applications to more complex and realistic systems.

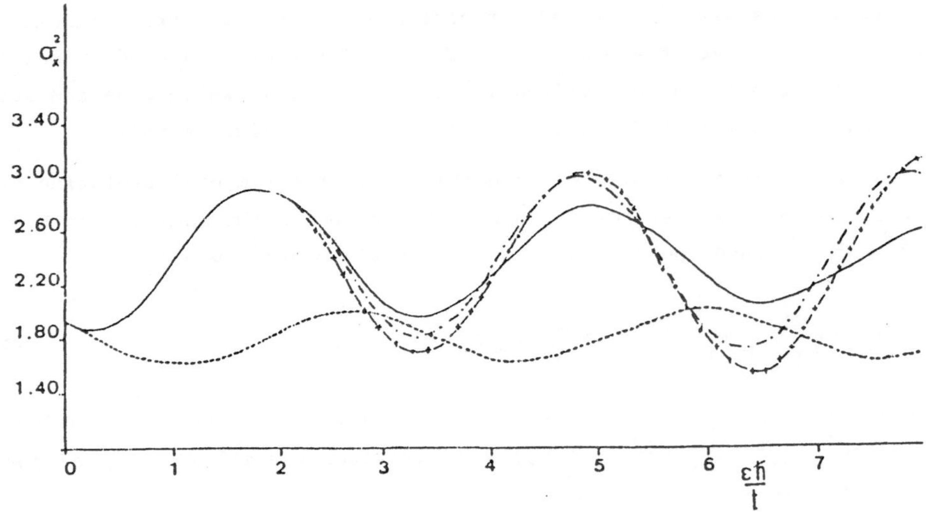

Fig. 1. Temporal evolution of the fluctuation of the operator J_x for N=8 and v/ε = -0.30. The initial conditions correspond to x_μ real, with x_-^2 = 0.034. Exact results,———; TDHF, ...; CMFA,._._._.; perturbative treatment,-+-+-+-.

References

1. P. Ring and P. Schuck, "The Nuclear Many-Body Problem", Springer Verlag, New York (1980).
2. J. Negele, Rev. Mod. Phys. 54:913 (1982).
3. H. Reinhardt, R. Balian, and Y. Alhassid, Nucl. Phys. A422:349 (1984); R. Balian, H. Reinhardt and Y. Alhassid, Phys. Rep. 131 1&2 (1986).
4. S. Ayik and M. Dworzecka, Phys. Rev. Lett. 54:534 (1985); Nucl. Phys. A440:424 (1985).
5. H. Reinhardt, P. Reinhard, and K. Goeke, Phys. Lett. 151B:177 (1985).
6. N. Canosa, A. Lõpez, A. Plastino, and R. Rossignoli, Phys. Rev. C37:320 (1988).
7. N. Canosa, A. Lõpez, D. Otero, A. Plastino, and A. Proto, Z. Phys. A326 :195 (1987).
8. H. Lipkin, N. Meshkov, and A. Glick, Nucl. Phys. 62:188 (1965).

SQUEEZED STATES REPRESENTATION; AN $\hbar$ -EXPANSION OF STATISTICAL MECHANICS

Salomon S. Mizrahi

Departamento de Fisica, Universidade Federal de São Carlos
Via Washington Luiz Km 235,13560,São Carlos,SP,Brazil

Diógenes Galetti

Instituto de Física Teórica, Universidade Estadual Paulista
Rua Pamplona 145,01405,São Paulo,SP,Brazil

ABSTRACT

An alternative semiclassical approach (to the Wigner-Kirkwood Method) that makes use of the Squeezed States Phase Space Representation is derived. As an illustration, some statistical mechanics formulae are written as an $\hbar$-expansion.

INTRODUCTION

Semiclassical expansion methods of quantum mechanics have greatly helped understanding the mechanisms by which one can separate from original quantum expressions a classical part plus quantum corrections. In fact the approaches by Weyl-Wigner[1] and Kirkwood[2] have led to two somewhat different trends (although with similar results) in treating this kind of problem. If on the one hand one treats the mapping of operators and quantum dynamical equations directly by an integral transformation[1] and separates a classical contribution by selecting the proper term in a power series in $\hbar$, on the other hand one initially writes the quantum partition function factored as the classical partition part times a power series in $\hbar$ [2], whose coefficients can be calculated for specific choices of the single particle potential of the many-body system. In practice the analytical calculation of the high order terms of the series is very laborious and can only be calculated for very few potentials.

The Bloch equation for the canonical ensemble statistical operators together with the Laplace transform techniques applied in the Wigner-Kirkwood (WK) partition function expansion permit a derivation of a density matrix for a quantum system also written in power series in $\hbar$ [3]. This scheme has been sucessful in describing, for instance, many-body properties using single particle potentials, specifically it permits the separation of the spatial density in a semiclassical part (Thomas-Fermi) plus quantum correction terms (extended Thomas-Fermi)[4].

In this work we want to show how one can use an alternative scheme[5,6] to put the Bloch equation in a phase space representation where a new series arises for a semiclassical description of the density of a many-body system. To this end we begin with a brief discussion of the squeezed states.

Condensed Matter Theories, Volume 5
Edited by V.C. Aguilera-Navarro
Plenum Press, New York, 1990

II. THE SQUEEZED STATES

The squeezed states[7] are defined by a successive application of two transformations on the vacuum state: the squeezing and the coherent state generators

$$S(y) = \exp\left[\frac{y}{2}(a^{+2} - a^2)\right] \tag{1}$$

and

$$D(\alpha) = \exp(\alpha a^{+} - \alpha^{*} a) \tag{2}$$

respectively; then

$$|\alpha, y) = D(\alpha) S(y) |0\rangle \tag{3}$$

is the squeezed state and y is the real valued squeezing parameter. When $y = 0$, one is led back to the usual coherent states of the Weyl-Heisenberg group[8]

$$|\alpha\rangle = D(\alpha)|0\rangle . \tag{4}$$

The operators a and a^{+} are the usual bosonic operators and $S(y)$ generates on these operators a canonical transformation of the Bogoliubov-Valatin type

$$S(y)\begin{pmatrix} a \\ a^{+} \end{pmatrix} S^{-1}(y) = \cosh\begin{pmatrix} a \\ a^{+} \end{pmatrix} - \sinh\begin{pmatrix} a^{+} \\ a \end{pmatrix} . \tag{5}$$

Since the operators a and a^{+} are related to the position and momentum operators through

$$a = \frac{1}{(2\hbar)^{1/2}}\left[(m\omega)^{1/2}\hat{q} + \frac{i}{(m\omega)^{1/2}}\hat{p}\right] \tag{6}$$

$$a^{+} = (a)^{+} , \tag{7}$$

where m and ω are respectively the mass and angular frequency associated to a harmonic oscillator, the effect of the $S(y) \cdot S^{-1}(y)$ transformation on $\hat{q}$ and $\hat{p}$ is to scale - on these operators

$$S(y)\begin{pmatrix} \hat{q} \\ \hat{p} \end{pmatrix} S^{-1}(y) = \begin{pmatrix} e^{-y}\,\hat{q} \\ e^{y}\,\hat{p} \end{pmatrix} . \tag{8}$$

The scaling factor e^{y} is responsible for the squeezed character of the phase space elementary cell and it is related to the parameter a_0, introduced in an ad hoc manner in Refs. (5,6) by $a_0 = (m\omega)^{1/2} e^{-y}$ as shown in ref. (9). a_0 is a squeezing parameter (with dimensions $M^{1/2} T^{-\frac{1}{2}}$) that characterizes the Wave-Packet Phase Space Representation (WPPSR), where the mapped operators and the dynamical equations depend on a_0 in a crucial way. Therefore the WPPSR is completely equivalent to a Squeezed States Phase Space Representation[9] (SSPSR) where deformation of the fundamental cell is allowed although its area is constant.

III. THE BLOCH DENSITY AND ITS SEMICLASSICAL APPROXIMATION IN THE SQUEEZED STATES PHASE SPACE REPRESENTATION

In this way the map of the Bloch equation,

$$\frac{\partial}{\partial\beta}\hat{\rho} = -\frac{1}{2}\{\hat{H},\hat{\rho}\}, \tag{9}$$

where $\hat{\rho}$ is the canonical ensemble statistical operator and $\beta=(kT)^{-1}$, using this scheme, leads to

$$\frac{\partial}{\partial\beta}P(\vec{q},\vec{p},\beta) = -\frac{1}{2}\mathcal{H}(\vec{q},\vec{p})(\overleftrightarrow{\Gamma}+\overleftrightarrow{\Gamma}^*)\,P(\vec{q},\vec{p},\beta) \tag{10}$$

$$P(\vec{q},\vec{p},\beta) = \langle\vec{p}\vec{q};a_0^2|\hat{\rho}|\vec{p}\vec{q};a_0^2\rangle$$

and

$$\mathcal{H}(\vec{q},\vec{p},\beta) = \langle\vec{p}\vec{q};a_0^2|\hat{H}|\vec{p}\vec{q};a_0^2\rangle$$

are the covariant mapped equivalent of $\hat{\rho}$ and $\hat{H}$ in the SSPSR, respectively. On the other hand, the operator

$$\overleftrightarrow{\Gamma} = 2\,e^{\frac{\hbar}{2}\overleftrightarrow{\Omega}}\,e^{i\frac{\hbar}{2}\overleftrightarrow{\Lambda}} \tag{11}$$

depends on the two-side operators

$$\overleftrightarrow{\Omega} = \frac{1}{a_0^2}\overleftarrow{\nabla}_q\cdot\overrightarrow{\nabla}_q + a_0^2\overleftarrow{\nabla}_p\cdot\overrightarrow{\nabla}_p \tag{12}$$

$$\overleftrightarrow{\Lambda} = \overleftarrow{\nabla}_p\cdot\overrightarrow{\nabla}_q - \overleftarrow{\nabla}_q\cdot\overrightarrow{\nabla}_p \tag{13}$$

It is worthwhile to observe that the operator given by (11) presents, in addition to the well known sharp-states Weyl-Wigner contribution $\exp(i\hbar\overleftrightarrow{\Lambda}/2)$, a new series of derivatives, (12), that take into account the characteristics of the SSPSR. This contribution also gives rise to modifications in the final expression (10) when compared with those coming from an usual Weyl-Wigner mapping[10].

We now propose the solution of eq. (10) to be of the form

$$P(\vec{q},\vec{p},\beta) = P_0(\vec{q},\vec{p},\beta)\sum_{n=0}^{\infty}\left(\frac{\alpha\hbar}{2}\right)^n \chi_n(\vec{q},\vec{p},\beta), \tag{14}$$

where $P_0(\vec{q},\vec{p},\beta)$ <u>is not</u> the classical function but instead

$$P_0(\vec{q},\vec{p},\beta) = e^{-\beta\mathcal{H}(\vec{q},\vec{p},\beta)} \tag{15}$$

and α is a parameter taken equal to one at the end of the calculations. It is important to observe that this ansatz allows us to put quantum fluctuation contributions due to the Gaussian smearing already in this zeroth-order term because $\mathcal{H}(\vec{q},\vec{p})$ is the mapped quantum Hamiltonian[5].

Retaining terms up to second order in α, the first terms of expression (14) are

$$\chi_0 = 1 \tag{16.a}$$

$$\chi_1(\vec{q},\vec{p},\beta) = \frac{\beta^2}{2}\left(\mathcal{H}\overleftrightarrow{\Omega}\mathcal{H}\right) \tag{16.b}$$

and

$$\chi_2(\vec{q},\vec{p},\beta) = \frac{1}{2}\Big\{\frac{\beta}{2}\left(\mathcal{H}\overleftrightarrow{\Omega}^2\mathcal{H}\right) - \frac{\beta^3}{3}\Big(\Big[\frac{1}{2a_0^4}(\nabla_q\mathcal{H}).\nabla_q\Big]^2\mathcal{H} +$$

$$+\left[(\nabla_p\mathcal{H}).\nabla_p\right]\left[(\nabla_q\mathcal{H}).\nabla_q\right]\mathcal{H} + \frac{a_0^4}{2}\left[(\nabla_p\mathcal{H}).\nabla_p\right]^2\mathcal{H} + \left(\mathcal{H}\overleftrightarrow{\Omega}\left(\mathcal{H}\overleftrightarrow{\Omega}\mathcal{H}\right)\right)\Big)\Big\} +$$

$$+\frac{1}{2}\Big\{-\frac{\beta^2}{2}\left(\mathcal{H}\overleftrightarrow{\Lambda}^2\mathcal{H}\right) + \frac{\beta^3}{3}\Big(\Big(\frac{\partial}{\partial q_i}\mathcal{H}\Big)\Big[(\nabla_q\mathcal{H}).\nabla_p\frac{\partial}{\partial p_i}\mathcal{H} - (\nabla_p\mathcal{H}).\nabla_q\frac{\partial}{\partial p_i}\mathcal{H}\Big] +$$

$$-\Big(\frac{\partial}{\partial p_i}\mathcal{H}\Big)\Big[(\nabla_q\mathcal{H}).\nabla_p\frac{\partial}{\partial q_i}\mathcal{H} - (\nabla_p\mathcal{H}).\nabla_q\frac{\partial}{\partial q_i}\mathcal{H}\Big]\Big)\Big\} + \frac{\beta^4}{8}\left(\mathcal{H}\overleftrightarrow{\Omega}\mathcal{H}\right)^2.$$

(16.c)

Particularizing the form of the Hamiltonian to

$$H(\hat{q},\hat{p}) = \frac{\hat{p}^2}{2m} + V(\hat{q}). \tag{17}$$

then its covariant mapped form is given by

$$\mathcal{H}(\vec{q},\vec{p}) = \frac{1}{2m}\left(p^2 + \frac{\hbar}{2}a_0^2\right) + V\left(q + \frac{\hbar}{2a_0^2}\frac{\partial}{\partial q}\right).1$$

$$= H(\vec{q},\vec{p}) + \mathcal{V}_Q(\vec{q}), \tag{18}$$

where $H(\vec{q},\vec{p})$ is the classical Hamiltonian and the extra potential $\mathcal{V}_Q(\vec{q})$ embodies the quantum corrections since $\lim_{\hbar\to 0}\mathcal{V}_Q(\vec{q})\to 0$. Eqs. (16.b) and (16.c) write now

$$\chi_1(\vec{q},\vec{p},\beta) = \frac{\beta^2}{2}\left[\frac{1}{a_0^2}|\nabla U|^2 + a_0^2\frac{p^2}{m^2}\right] \tag{19.a}$$

$$\chi_2(\vec{q},\vec{p},\beta) = \frac{1}{2}\left\{-\beta^2\frac{\nabla^2 U}{m} + \frac{\beta^3}{3}\left[\frac{1}{m}|\nabla U|^2 + \frac{1}{m^2}(\vec{p}\cdot\nabla)^2 U\right]\right\} +$$

$$+ \frac{1}{2}\left\{\frac{\beta^2}{2}\left[\frac{1}{a_0^4}(\nabla\nabla U):(\nabla\nabla U) + \frac{3a_0^4}{m^2}\right] +\right.$$

$$\left. - \frac{\beta^3}{3}\left[\frac{1}{2a_0^4}(\nabla U)\cdot\nabla|\nabla U|^2 + a_0^4\frac{p^2}{m^3}\right] + \frac{\beta^4}{4}\left[\frac{1}{a_0^2}|\nabla U|^2 + \frac{a_0^2}{m^2}p^2\right]^2\right\}, \tag{19.b}$$

where

$$U(\vec{q}) = V(\vec{q}) + V_Q(\vec{q}) \tag{20}$$

and

$$(\nabla\nabla U):(\nabla\nabla U) = \sum_{ij}\frac{\partial^2 U}{\partial q_i\partial q_j}\cdot\frac{\partial^2 U}{\partial q_i\partial q_j}. \tag{21}$$

Now we can compare Eqs. (19) with the ones obtained by the WK method[11].

$$\chi_1^{WK}(\vec{q},\vec{p}) = \frac{-i\beta^2}{2m}\,\vec{p}\cdot\nabla V(\vec{q}) \tag{22.a}$$

$$\chi_2^{WK}(\vec{q},\vec{p}) = \frac{\beta^2}{8m}\nabla^2 U + \frac{\beta^3}{24}\left[\frac{1}{m}(\nabla V)^2 + \frac{(\vec{p}\cdot\nabla)^2}{m^2}V\right] \tag{22.b}$$

In the WK method, χ_1^{WK} is odd in the momentum $\vec{p}$, so when integrated over this variable in order to obtain the first order correction to the local Bloch density, it gives no contribution

$$\rho_{WK}^{(1)}(\vec{q},\beta) = \hbar\int d^3p\; e^{-\beta H(\vec{q},\vec{p})}\,\chi_1^{WK}(\vec{q},\vec{p}) = 0\,. \tag{23}$$

On the other hand, in the SSPSR mapping, $\chi_1(\vec{q},\vec{p},\beta)$ is even in $\vec{p}$, so its contribution to the local Bloch density is

$$\rho^{(1)}(\vec{q},\beta) = \frac{\hbar}{2}\int d^3p \; e^{-\beta \mathcal{H}(\vec{q},\vec{p})} \chi_1(\vec{q},\vec{p})$$

$$= \left(\frac{m}{2\pi\hbar^2\beta}\right)^{3/2} \frac{\hbar\beta}{4}\left[\frac{\beta}{a_0^2}(\nabla U)^2 + \frac{3a_0^2}{m}\right] e^{-\beta U(\vec{q})}$$

(24)

For the second order correction (third term in the WK series) χ_2^{WK} coincides with the terms contained in the first rounded bracket in eq. (19.b) (aside from a $\frac{1}{4}$ factor coming from the different $\hbar$ -expansions in the two methods); the additional terms in that equation are contributions due to the squeezed character of the phase space (each term has an a_0 dependence). Therefore the second order corrections to the Bloch density for the two approaches are:

$$\rho^{(2)}_{WK}(\vec{q},\beta) = \hbar^2\int d^3p \; e^{-\beta H(\vec{q},\vec{p})} \chi_2^{WK}(\vec{q},\vec{p},\beta)$$

$$= \left(\frac{m}{2\pi\hbar^2\beta}\right)^{3/2} e^{-\beta V(\vec{q})} \left(\frac{\beta\hbar}{2}\right)^2\left[-\frac{1}{3}\frac{\nabla^2 V}{m} + \frac{(\nabla V)^2}{6m}\beta\right]$$

(25)

and

$$\rho^{(2)}(\vec{q},\beta) = \left(\frac{\hbar}{2}\right)^2\int d^3p \; e^{-\beta \mathcal{H}(\vec{q},\vec{p})} \chi_2(\vec{q},\vec{p},\beta)$$

$$= \left(\frac{m}{2\pi\hbar^2\beta}\right)^{3/2} e^{-\beta U}\left(\frac{\beta\hbar}{2}\right)^2\left\{\left[-\frac{1}{3}\frac{\nabla^2 U}{m} + \frac{(\nabla U)^2}{6m}\beta\right] + \right.$$

$$+\left[\frac{1}{4a_0^4}(\nabla\nabla U):(\nabla\nabla U) + \frac{9}{8}\frac{a_0^4}{m^2} - \frac{\beta}{4}\left(\frac{1}{a_0^4}(\nabla U)\cdot\nabla(\nabla U)^2 + \right.\right.$$

$$\left.\left.\left. -\frac{3}{m}(\nabla U)^2\right) + \frac{1}{8a_0^4}\beta^2(\nabla U)^4\right]\right\}$$

(26)

As an illustrative example we consider now the three-dimensional harmonic oscillator (HO) potential $V(\vec{q})=\frac{m}{2}\omega^2\vec{q}^2$ which gives for the local Bloch density in the SSPSR

$$\rho(\vec{q},\beta)=\frac{1}{(2\pi\hbar)^3}\int d^3p\; P_0(\vec{q},\vec{p},\beta)\left[1+\frac{\hbar}{2}\chi_1(\vec{q},\vec{p},\beta)+\frac{\hbar^2}{4}\chi_2(\vec{q},\vec{p},\beta)+\Theta(\hbar^3)\right.$$

$$=\left(\frac{m}{2\pi\hbar^2\beta}\right)^{3/2} e^{-\beta\left(\frac{m\omega^2}{2}q^2+\frac{3}{2}\hbar\omega\right)}\left\{1+\frac{\hbar\omega\beta}{4}\left(3+\beta m\omega^2 q^2\right)+\right.$$

$$\left.+\left(\frac{\hbar\omega\beta}{4}\right)^2\left[\frac{7}{2}+\frac{5}{3}\left(\beta m\omega^2 q^2\right)+\frac{1}{2}\left(\beta m\omega^2 q^2\right)^2\right]+\Theta(\hbar^3)\right\}. \qquad (27)$$

In the calculation of eq. (27) we have particularized the squeezing parameter a_0 to the value $(m\omega)^{\frac{1}{2}}$ since this minimizes the mapped Hamiltonian $\mathcal{H}(\vec{q},\vec{p})$, giving the zero-point energy of the HO. The main differences with the WK density are: the zero-point energy term in the exponential, that originates from the ansatz (14) and (15), and the odd contributions in the expansions. The integration of $\rho(\vec{q},\beta)$ over the coordinate variables leads to the partition function

$$Z(\beta)=\frac{e^{-\frac{3}{2}\beta\hbar\omega}}{(\beta\hbar\omega)^3}\left[1+\frac{3}{2}\beta\hbar\omega+(\beta\hbar\omega)^2+\Theta(\hbar^3)\right] \qquad (28)$$

which coincides with the usual series when we expand the exponential and retain the terms up to second order in $\hbar$, namely

$$Z^{WK}(\beta)=\frac{1}{(\beta\hbar\omega)^3}\left[1-\frac{1}{8}(\beta\hbar\omega)^2+\Theta(\hbar^3)\right]. \qquad (29)$$

Therefore, our ansatz is equivalent to performing a resummation of an infinite number of terms in the WK series, leading to the factor $\exp(-\frac{3\beta\hbar\omega}{2})$. In the present example the zero-point energy obtained by considering the squeezing parameter a_0 as a variational parameter turned out to be the exact one because the sharp-x representation of the coherent state wave packet is Gaussian. For any other potential well the actual zero-point energy is going to be a lower bound to the quantum energy correction calculated by the present method.

Finally, for a given Fermi energy we can write the spatial density for the system of N particles in the following series form (Mellin's transform)[3]

$$\rho_{\epsilon_F}(\vec{q})=\frac{2}{2\pi i}\int_{C-i\infty}^{C+i\infty} d\beta\;\frac{e^{\beta\epsilon_F}}{\beta}\;\rho(\vec{q},\beta)$$

$$=2\left(\frac{m}{2\pi\hbar^2}\right)^{3/2}\frac{1}{\pi^{1/2}}\left\{4\,\frac{\left(\epsilon_F-\frac{3\hbar\omega}{2}-\frac{m\omega^2q^2}{2}\right)^{3/2}}{3}+\frac{3}{4}\hbar\omega\,2\left(\epsilon_F-\frac{3\hbar\omega}{2}-\frac{m\omega^2q^2}{2}\right)^{1/2}+\right.$$

$$+\left[\frac{\hbar\omega}{4}m\omega^2q^2+\frac{7}{32}(\hbar\omega)^2\right]\left(\epsilon_F-\frac{3\hbar\omega}{2}-\frac{m}{2}\omega^2q^2\right)^{-\frac{1}{2}}-\frac{5}{48}(\hbar\omega)^2\frac{m\omega^2q^2}{2}\left(\epsilon_F-\frac{3\hbar\omega}{2}-\frac{m}{2}\omega^2q^2\right)^{-3/2}$$

$$+\frac{3}{32}(\hbar\omega)^2\frac{(m\omega^2q^2)^2}{4}\left(\epsilon_F-\frac{3\hbar\omega}{2}-\frac{m}{2}\omega^2q^2\right)^{-5/2}\Big\}. \qquad (30)$$

This expression exhibits interesting features; it shows explicitly the effect of the zero-point quantum fluctuation in shifting the zero of the potential energy as is indeed necessary because in the present framework q and p are mapped values within the SSPSR and it contains additional contributions for each order of $\hbar$ in the series which have no counterparts in the other approaches, although the singularities at the turning points remain.

IV- FURTHER USEFUL RESULTS

In this section we present some results, for any dimension d=1,2,3, of typical use in statistical mechanics obtained with the present method up to second order in parameter α. The results are independent of the choice of the single particle potential, U, which is left as an arbitrary function of the coordinates.

i) The Bloch density is written as

$$\rho^{(d)}(\vec{q},\beta)=\left(\frac{m}{2\pi\hbar^2\beta}\right)^{d/2}e^{-\beta U}\Big\{1+\frac{1}{2}\left(\frac{\hbar}{2}\right)\beta^2\left[\frac{|\nabla U|^2}{a_o^2}+\frac{a_o^2 d}{m\beta}\right]+$$

$$+\left(\frac{\hbar}{2}\right)^2\beta^2\Big[\frac{1}{4a_o^4}\nabla\nabla U:\nabla\nabla U+\frac{d^2a_o^4}{8m^2}-\frac{1}{4a_o^4}\beta\,\nabla U.\nabla|\nabla U|^2 +$$

$$+\frac{d}{4}\beta\frac{|\nabla U|^2}{m}+\frac{1}{8a_o^4}\beta^2|\nabla U|^4+\left(-\frac{\nabla^2U}{3m}+\frac{\beta}{6}\frac{|\nabla U|^2}{m}\right)\Big]\Big\}. \qquad (31)$$

In this expression the potential U depends on $\hbar$ too, since it is the mapped covariant form of the potential $V(\hat{q})$.

ii) The spectral density is defined as

$$g^{(d)}(\vec{q},\epsilon)=f\,\mathcal{L}^{-1}_{\beta\to\epsilon}\left[\rho^{(d)}(\vec{q},\beta)\right]$$
$$=\sum_n|\phi_{\epsilon_n}(\vec{q})|^2\,\delta(\epsilon-\epsilon_n), \qquad (32)$$

where $\phi_{\epsilon_n}(\vec{q})$ are the Hamiltonian eigenfunctions, f is the spin degeneracy and $\mathcal{L}^{-1}_{\beta\to\epsilon}[\cdot]$ is the Mellin transform. For the Bloch density given by (31) the spectral density reads

$$g^{(d)}(\vec{q},\epsilon)=f\left(\frac{m}{2\pi\hbar^2}\right)^{d/2}\Big\{1+\frac{\hbar}{4}\left[\frac{|\nabla U|^2}{a_o^2}\frac{d^2}{d\epsilon^2}+\frac{a_o^2d}{m\beta}\frac{d}{d\epsilon}\right]+$$

$$+\left(\frac{\hbar}{2}\right)^2\Big[\left(-\frac{\nabla^2U}{3m}\frac{d^2}{d\epsilon^2}+\frac{|\nabla U|^2}{6m}\frac{d^3}{d\epsilon^3}\right)+\frac{1}{4a_o^4}\nabla\nabla U:\nabla\nabla U\frac{d^2}{d\epsilon^2} +$$

$$+ \frac{a_0^4 d^2}{8 m^2} \frac{d^2}{d\epsilon^2} - \frac{1}{4a_0^4} \nabla U . \nabla(\nabla U)^2 \frac{d^3}{d\epsilon^3} + \frac{d}{4m} |\nabla U|^2 \frac{d^3}{d\epsilon^3} +$$

$$+ \frac{1}{8a_0^4} |\nabla U|^4 \frac{d^4}{d\epsilon^4} \Big\} (\epsilon - U)^{\frac{d-2}{2}} \Theta(\epsilon - U) . \qquad (33)$$

It is worth to mention that for the two-dimensional case the derivatives in the brackets apply only to the theta function, therefore the spectral density is written as combinations of delta functions and derivatives, exception to the first term (Thomas-Fermi contribution).

iii) Level density

$$g^{(d)}(\epsilon) = Tr\left[\delta(\epsilon - \hat{H})\right] = \int d^{(d)}q \; g^{(d)}(\vec{q}, \epsilon) \qquad (34)$$

iv) The partition function may be computed by either of the two following expressions

$$Z^{(d)}(\beta) = \int d^{(d)}q \; \rho^{(d)}(\vec{q}, \beta) \qquad (35.a)$$

$$Z^{(d)}(\beta) = \mathcal{L}_{\epsilon \to \beta}\left[g^{(d)}(\epsilon)\right] , \qquad (35.b)$$

where $\mathcal{L}_{\epsilon \to \beta}[\cdot]$ is the Laplace transform.

v) The spatial density is obtained from

$$\rho^{(d)}(\vec{q}, \epsilon_F) = f \, \mathcal{L}^{-1}_{\beta \to \epsilon_F}\left[\frac{\rho^{(d)}(\vec{q}, \beta)}{\beta}\right] . \qquad (36)$$

vi) The number of particles and Fermi energy are related by

$$N^{(d)} = \int_0^{\epsilon_F} d\epsilon \; g^{(d)}(\epsilon) = \int d^{(d)}q \int_0^{\epsilon_F} d\epsilon \; g^{(d)}(\vec{q}, \epsilon) . \qquad (37)$$

vii) The total energy for the N particle system is

$$E^{(d)} = \int_0^{\epsilon_F} d\epsilon \; \epsilon \; g^{(d)}(\epsilon) \qquad (38)$$

viii) Connection with the statistics of level occupation

Define the operator

$$G(\hat{H}) = f \int d\epsilon \; \delta(\epsilon - \hat{H}) \, n(\epsilon ; \beta) , \qquad (39)$$

where the mean occupation number is

$$n(\epsilon ; \beta) = \frac{1}{e^{\beta(\epsilon - \mu)} + a} , \qquad (40)$$

and μ is the chemical potential. For $a=+1, 0, -1$ we have the Fermi-Dirac, Boltzmann and Bose-Einstein statistics respectively.

Computing the trace of (39) and using the Mellin transform representation for the delta, we get

$$\begin{aligned} \mathrm{Tr}\, G(\hat{H}) &= \int d\epsilon\, n(\epsilon;\beta) \int d\vec{q}\, \frac{1}{2\pi i} \oint d\beta'\, e^{\beta'\epsilon} \rho(\vec{q};\beta') \\ &= \int d\epsilon\, n(\epsilon;\beta) \int d\vec{q}\, \mathcal{L}^{-1}_{\beta' \to \epsilon} \left[\rho(\vec{q},\beta')\right] \\ &= \int d\vec{q} \int_0^\infty d\epsilon\; n(\epsilon;\beta)\, g(\vec{q},\epsilon) \end{aligned} \tag{41}$$

Now we identify the spatial density

$$\begin{aligned} \rho_a(\vec{q},\beta) &= \int_0^\infty d\epsilon\; n(\epsilon;\beta)\, g(\vec{q},\epsilon) \\ &= \sum_\epsilon n(\epsilon;\beta) \left|\phi_\epsilon(\vec{q})\right|^2, \end{aligned} \tag{42}$$

which embodies the statistics of level occupation.

V. SUMMARY AND CONCLUSIONS

We have presented a novel technique in mapping the Bloch equation in phase space using the squeezed states representation. This scheme differs from the usual Wigner-Kirkwood method by taking into account the zero-point energy of the potential and, thus, all orders in $\hbar$ appear in the series expansion of the Bloch density.

We applied this mapping to the harmonic oscillator Hamiltonian in three dimensions and pointed out the differences, on the results, with the Wigner-Kirkwood method. Furthermore we have presented an $\hbar$-expansion for several statistical relevant expressions for dimensions d=1,2,3, including the introduction of the occupation level statistics in the spatial density.

Finally it is worth to stress that the present scheme permits the obtention of semiclassical expansions which exhibit different behavior as compared with those obtained with the usual Wigner-Kirkwood method.

REFERENCES

1. M.Hillery et.al., Phys.Rep. C106:121(1984)
2. J.G.Kirkwood, Phys.Rev. 44:31(1933)
3. P.Ring and P.Schuck, Chapter 13 and references therein, in: "The Nuclear Many-Body Problem", Springer Verlag, Berlin(1980)
4. R.Kubo, Jour.Phys.Soc. Japan 19:2127(1964), R.Kubo, Some topics on the orbital magnetism, in: "Lectures in Theoretical Physics", Vol.VIII-A:239, Ed. W.E.Brittin, Boulder Colorado (1966)
5. S.S.Mizrahi, Physica 127A:241(1984)
6. S.S.Mizrahi, Physica 135A:237(1986)
7. M.Martin Nieto, What are squeezed states really like?, in: "Proc. of NATO Advanced Study Institute: Frontiers of Nonequilibrium Statistical Mechanics" (1984)
8. A.Perelomov, "Generalized Coherent States and their Applications", Springer Verlag", Berlin(1986)
9. S.S.Mizrahi and D.Galetti Physica A153:567(1988)
10. S.R. de Groot and L.G.Suttorp, Chapter VI and Appendix, in: "Foundations of Electrodynamics", North Holland, Amsterdam(1972)
11. L.Landau and E.Lifchitz, "Physique Statistique", Ed. MIR, Moscou(1984)

MAXIMUM ENTROPY PRINCIPLE AND QUANTUM MECHANICS

Araceli N. Proto

Laboratorio de Física
Comisión Nacional de Investigaciones Espaciales

I. INTRODUCTION

The concept of entropy is one of the most important ones in physics. It plays a central role both, in thermodynamics and in statistical mechanics, and has been subject of an enormous amount of work (See for example the review articles cited in ref. 1-3). Entropy may also play a *dynamical* role although the amount of work in this respect is scarce, at least in comparison with the broadly treated thermodynamical aspects. In this last sense, a very interesting alternative is to be found in the pioneer work of Jaynes[4] concerning to the Maximum Entropy Principle (MEP) and Information Theory (IT). Further extensions to quantum-mechanical systems have been made[5-15], that explicitly exploit the dynamical relevance of the entropy, S.

II THE MAXIMUN ENTROPY PRINCIPLE. BASIC CONCEPTS

A brief summary concerning IT and the least biased probability assignment criterium, usually referred to as the Maximum Entropy Principle (MEP) is presented here. Given the expectation values, O_j, of operators $\hat{O}_j$, the statistical operator $\hat{\rho}(t)$ is defined by:

$$\hat{\rho}(t) = \exp\,(-\,\lambda_0 - \sum_{j=1}^{M} \lambda_j(t)\,\hat{O}_j)\ . \tag{2.1}$$

The λ_i's, M + 1 of them, are Lagrange multipliers which will be determined to fulfill the set of constraints

$$<\hat{O}_j> = O_j = \mathrm{Tr}(\,\hat{\rho}\,\hat{O}_j\,),\ j = 1,2,\ldots,M \tag{2.2}$$

and the normalization condition

$$\mathrm{Tr}(\,\hat{\rho}\,\hat{I}\,) = \mathrm{Tr}(\,\hat{\rho}\,\hat{O}_0\,) = \mathrm{Tr}\,\hat{\rho} = 1, \tag{2.3}$$

where $\hat{I}$ is the unity operator. The time evolution of the statistical operator is given by

Condensed Matter Theories, Volume 5
Edited by V.C. Aguilera-Navarro
Plenum Press, New York, 1990

$$i \hbar \frac{d\hat{\rho}}{dt} = [\hat{H}, \hat{\rho}], \tag{2.4}$$

the entropy, defined by

$$S(\hat{\rho}) = -k \, \mathrm{Tr} (\hat{\rho} \ln \hat{\rho}) \tag{2.5}$$

is maximum, and a constant of the motion[5-15]. One should endeavor to find those (relevant) operators entering Eq.(2.1) so as to satisfy Eq.(2.4), in order to guarantee that S is a constant of the motion. Using Eqs.(2.1) and (2.4) it is easy to verify that the relevant operators are those that close a partial Lie algebra under commutation with the Hamiltonian $\hat{H}$,

$$[\hat{H}, \hat{O}_j] = \sum_{i=0}^{M} \hat{O}_i g_{ji} . \tag{2.6}$$

In this way, defining a Lie algebra of (M+1) elements, whose structure constants are given by the matrix $\underline{G}$, which can be time dependent if $\hat{H}$ is. For our present purposes we shall consider the case M = q (M < q was discussed in ref.9). Equation (2.6) constitutes the central requirement to be fulfilled by the operators entering in the density matrix. One of the principal objections made to Jaynes' formulation of statistical mechanics is referred to the ambiguity of the concept of "relevant information". Equation (2.6) properly establishes which are the "relevant operators". It is worth to mention that the ambiguity is removed when a quantum mechanical formulation of MEP is given. As can be seen from refs.6-15 *the entire dynamics of the problem under consideration is embedded in the value of the structure factors G_{li},* Equation (2.6) not only guarantees the conservation of S but also that the time evolution of the relevant observables is totally decoupled from the rest of the universe[8].

The Liouville equation (2.4) can be substituted by a set of coupled equations for the Lagrange multipliers λ_i, as follows:

$$\frac{d\lambda_i}{dt} = \sum_{j=0}^{M} g_{ij} \lambda_j, \ i = 1,...,M \tag{2.7}$$

Solving this set of equations and using the Ehrenfest theorem:

$$\frac{d< \hat{O}_i >}{dt} = \sum_{j=0}^{M} < \hat{O}_j > g_{ji} , \tag{2.8}$$

the entropy becomes

$$S (\hat{\rho}(t)) = \sum_{j=0}^{M} \lambda_j(t) < \hat{O}_j > . \tag{2.9}$$

Since the entropy is a constant of the motion, Eq. (2.9) provides the constraint in the time evolution of the $\hat{O}_j$'s.

The mean value of the operators and the Lagrange multipliers are related by

$$- \frac{\partial \lambda_0}{\partial \lambda_i} = < \hat{O}_i > \tag{2.10}$$

Following Duering et al[7] we can write

$$< \hat{O}_j >_t = \sum_i < \hat{O}_i >_{t_0} F_{ij}(t,t_0), \quad (2.11)$$

or, in vector notation

$$< \hat{O} >_t = < \hat{O} >_{t_0} \underline{F}(t,t_0) \ . \quad (2.12)$$

$\underline{F}(t,t0)$ is a square matrix defined by

$$- \frac{\partial F}{\partial t} = \underline{F} . \underline{G} \ . \quad (2.13)$$

In the Heisemberg representation we can write

$$\hat{O}_t = \hat{O}_{t_0} \ \underline{F}(t,t_0) \ . \quad (2.14)$$

Initial conditions (IC) play a crucial role in the dynamical behavior of the system. As it was shown in A.N.Proto et al[12] even when the G matrix determines the character of the solution (dissipative or not, at least in the case of time independent hamiltonians), *the initial conditions impose additional restrictions not contained explicitly in the dynamics of the problem* [6,9,12].

For the Lagrange multipliers we have

$$\lambda^t = \underline{F}^{-1}(t,t_0) \ \lambda^{t_0} \quad (2.15)$$

The subscript or superscript t, t_0 in Eqs. (2.11) - (2.15) indicates whether we are working with a covariant or a contravariant vector respectively. We shall obtain this information by examining the transformation of properties of the spaces, and their invariants, which the matrices $\underline{F}$ or $\underline{G}$ define. In order to define vectorial Reimann spaces with $< \hat{O} >_t$ (or $\hat{O}_t$) and λ^t as elements, we need first to specify the form of the scalar products, which are invariants, as

$$< \hat{O} >_t < \hat{O} >^t = < \hat{O} >_0 \ \ < \hat{O} >^0 \quad (2.16)$$

or

$$\hat{O}_t \ \hat{O}^t = \hat{O}_0 \ \ \hat{O}^0 \quad (2.17)$$

and

$$\lambda_t \ \lambda^t = \lambda_0 \ \lambda^0 \quad (2.18)$$

where the index zero refers to the time $t = t_0$ before the transformation is applied. Given a covariant vector $\hat{O}_t$ we need to find the metric tensor of the space, $\underline{e}$ (such that $\underline{e} \ \underline{e}^{-1} = \underline{1}$ for which

$$< \hat{O} >^t = \underline{e} < \hat{O} >_t \ . \quad (2.19)$$

In this way we shall be able to construct the invariants of the motion of the system[12-15]. Using the invariance of the scalar product we get

$$\underline{F} \ \underline{e} \ \underline{\tilde{F}} = \underline{e} \quad (2.20)$$

where $\tilde{\underline{F}}$ indicates transposed matrix. And

$$\tilde{\underline{F}}^{-1}\ \underline{e}'\ \underline{F}^{-1} = \underline{e}' \tag{2.21}$$

where

$$\lambda_t = \lambda^t\ \underline{e}' \tag{2.22}$$

Using Eqs. (2.14) and (2.15), Eqs. (2.20) and (2.21) can be replaced by:

$$\underline{G}\ \underline{e} = -\ \underline{e}\ \tilde{\underline{G}} \tag{2.23}$$

and

$$\tilde{\underline{G}}\ \underline{e}' = -\ \underline{e}'\ \tilde{\underline{G}} \tag{2.24}$$

which are easier to handle. Then, through the use of the metric matrices (which are not uniquely defined) we obtain the invariants (Eqs. (2.16) - (2.18)) as[13]

$$\hat{O}_t\ \hat{O}^t = \hat{O}_t\ \underline{e}\ \hat{O}_t \tag{2.25}$$

and

$$\lambda_t\ \lambda^t = \lambda^t\ \underline{e}'\ \lambda^t \tag{2.26}$$

The metric tensor $\underline{e}$ plays an important role. From Eq (2.10) we can write

$$\frac{\partial^2\ \lambda_0}{\partial\ \lambda_m\ \lambda_n} = -\ \frac{\partial < \hat{O}_n >}{\partial\ \lambda_m} = -\ \frac{\partial < \hat{O}_m >}{\partial\ \lambda_n} = K_{nm} = K_{mn} \tag{2.27}$$

with

$$K_{mn} = \frac{1}{2} < [\hat{O}_m, \hat{O}_n]_+ > - <\hat{O}_m > < \hat{O}_n > \tag{2.28}$$

The K-space is the direct, Kroenecker product of the $< \hat{O}_m >_t$, having a covariant character. In ref.7, we have demonstrated that it is possible to define the divergence and the rotor of $< \hat{O} >_t$ with respect to λ^t, if we consider the later as a "coordinate", remembering that we are working with an N+1 dimensional "space". Then

$$\nabla_\lambda < \hat{O} > = -\sum_m K_{nn} \tag{2.29}$$

and

$$\nabla_\lambda \times < \hat{O} > = 0 \tag{2.30}$$

Moreover,

$$\Delta\lambda_0 = -\sum_m K_{nn} \tag{2.31}$$

So, we have defined the gradient and the rotor of the $< \hat{O} >$ vector and we show that the Massieu-Planck function, λ_0, is the potential function of a vectorial field of operators where the sources are the quantal dispersions.

III.QUANTUM THERMODYNAMICS

As it is well-known, thermodynamics is based on the existence of "state variables". In order to reobtain thermodynamic equations from quantum mechanical considerations, the first step is to recognize Eq. (2.5) as the entropy of the system, and *to include the hamiltonian as a relevant operator.* Obviously, the inclusion of $\hat{H}$ as such an operator does not alter the dynamics of the system, as $\hat{H}$ commutes with itself. Once this step has been made, as in classical thermodynamics it is possible to define the system's temperature using Eq. (2.5) (k=1)

$$\beta = 1 / T = \frac{\partial S}{\partial < \hat{H} >} \Bigg|_{\{< \hat{O}_i >\}} , \quad i = 2,\ldots,N \tag{3.1}$$

and it is possible then to rewrite

$$S / \beta = \lambda_0 / \beta + \sum_{j=2}^{N} \lambda'_j < \hat{O}_j > + < \hat{H} > \tag{3.2}$$

where $\lambda'_j = \lambda_j/\beta$. Our second step is that of associating mean values with extensive variables, and the λ'_j with the intensive ones.

Canonical process

In reference 14, J. Aliaga et al called "canonical" such process for which the mean values and its associated Lagrange multipliers evolved as Eqs.(2.12, 2.15) prescribes. In this case S is a constant of the motion For this process it is possible to consider two cases: a) all the operators entering in Eqs.(2.6) commute with H or b) operators which do not commute with H are taken into account. Let us began with case b) as it is the more general situation. For this case $\hat{\rho}(t)$ is an off-equilibrium density matrix (OEDM) within the MEP context. The entropy and the specific heat read

$$S = \lambda_0 + \beta \left(\sum_{j=2}^{N} \lambda'_j < \hat{O}_j > + < \hat{H} > \right) \tag{3.3}$$

$$C\ (\beta,\ \lambda_2, \ldots,\ \lambda_N) = -\ \beta^2 \sum_{j=2}^{N} \lambda'_j\ K_{1j} = -\ \beta \sum_{j=2}^{N} \lambda_j\ K_{1j}\ , \tag{3.4}$$

where the K_{ij} are defined by Eq.(2.28). The specific heat is always a constant of motion, even when an OEDM is considered. This can be easily demonstrated using the results obtained in ref.13. Further details can be found in ref.14.

Thermodynamical process

If some constraints on $\{\lambda'_2,..,\lambda'_N\}$ or $\{<\hat{O}_2>,..,<\hat{O}_N>\}$ becomes operative and both, the temperature and the mean energy are allowed to change, the process is called "thermodynamical".[14] For them, entropy can vary. The usual thermodynamical process take place at "constant volume" (or "constant pressure") and are able to modify only the temperature (energy) and the pressure (or volume). Into the present quantum context of MEP, the analog situation is given when only one extensive (intensive) variable is allowed to vary, and the remainder ones rest unchanged. Then, we obtain

$$\left.\frac{\partial \lambda_0}{\partial \beta}\right|_F = - < \hat{H} > - \sum_{i=2}^{N} \left.\frac{\partial \lambda_i}{\partial \beta}\right|_F < \hat{O}_i > , \tag{3.5}$$

$$\left.\frac{\partial S}{\partial \beta}\right|_F = \left.\frac{\partial \lambda_o}{\partial \beta}\right|_F + <\hat{H}> + \beta \left.\frac{\partial <\hat{H}>}{\partial \beta}\right|_F + \sum_{i=2}^{N} \left.\frac{\partial \lambda_i}{\partial \beta}\right|_F <\hat{O}_i> + \sum_{i=2}^{N} \lambda_i \left.\frac{\partial <\hat{O}_i>}{\partial \beta}\right|_F . \tag{3.6}$$

where the symbol F means that the variation respect to β should be evaluated keeping constant extensive or intensive variables. The specific heat at "constant-extensive-variables", an extension of C_V is given by

$$C_{\{<\hat{O}_i>\}} = - \beta \left.\frac{\partial S}{\partial \beta}\right|_{\{<\hat{O}_i>\}} = - \beta^2 \left.\frac{\partial <\hat{H}>}{\partial \beta}\right|_{\{<\hat{O}_i>\}}$$

$$= \beta^2 \left(\tilde{K}_{11} + \sum_{r=2}^{N} \left.\frac{\partial \lambda_r}{\partial \beta}\right|_{\{<\hat{O}_i>\}} \tilde{K}_{r1} \right), \tag{3.7}$$

with

$$\tilde{K}_{ji} = \frac{1}{2} \operatorname{Tr} (\hat{\rho} [\bar{\hat{O}}_j, \hat{O}_i]_+) - < \hat{O}_j > < \hat{O}_i > . \tag{3.8}$$

where $\bar{\hat{O}}_j$ is the Kubo transform, which appears as a consequence of the non-commutativity of the relevant operators. The analog to C_p results

$$C_{\{\lambda_i'\}} = - \beta \left.\frac{\partial S}{\partial \beta}\right|_{\{\lambda_i'\}} = - \beta \left(\beta \left.\frac{\partial <\hat{H}>}{\partial \beta}\right|_{\{\lambda_i'\}} + \sum_{r=2}^{N} \lambda_r \left.\frac{\partial <\hat{O}_r>}{\partial \beta}\right|_{\{\lambda_i'\}} \right)$$

$$= \beta^2 \sum_{k;r=1}^{N} \lambda_k' \lambda_r' K_{rk} \quad , \lambda_1' = 1 . \tag{3.9}$$

In both process, canonical or thermodynamical, specific heats described by Eqs (3.7), (3.9) remain constants of motion for time independent hamiltonians. It is important to note that all the usual thermodynamical quantities (S, < $\hat{H}$ >, C), are time independent *even if the state is described by an OEDM*. Notice that the closure relation procedure guarantees that the system is, quantically, a closed one characterized by S= const, <$\hat{H}$>= const. However, as many of the relevant operators are not constant of motion their mean values evolve with time. The formalism is consistent enough so as to produce time independent specific heats. The above results have been applied to the Zeeman effect, and to hamiltonians expresed in terms of unbounded operators, like (<$\hat{q}$>, <$\hat{p}$>). These examples can be found in Ref.14.

IV DENSITY MATRIX APPROACH TO COHERENT AND SQUEEZED STATES

Now it will be described a simple general method to find squeezing[21], based on a density matrix formalism, and exemplified it with the harmonic oscillator. Three are our main results: a) squeezing is a property of the Hamiltonian dynamics; b) they are also present for non-zero temperatures (i.e. mixed density matrix); c) their appearance is related to the knowl edge included in the density matrix. As it is well known the squeezed states fulfill

$$(\Delta \hat{O}_i)^2 < \frac{1}{2} \left| [\hat{O}_i , \hat{O}_j]_- \right| \quad \text{or} \quad (\Delta \hat{O}_j)^2 < \frac{1}{2} \left| [\hat{O}_i , \hat{O}_j]_- \right| , \tag{4.1}$$

if dimensionless operators are considered. We are going to evaluate the dispersions of $\hat{q}$ and $\hat{p}$ for the particular case of the harmonic oscillator. The more general algebra is $\{\hat{1}, \hat{q}, \hat{p}, \hat{q}^2, \hat{p}^2, \hat{l}\}$, where $\hat{l} = \frac{1}{2}\{\hat{q}\hat{p} + \hat{p}\hat{q}\}$. So, following eq. (2.1), the density matrix becomes

$$\hat{\rho} = \exp(-\lambda_0 \hat{1} - \lambda_q \hat{q} - \lambda_p \hat{p} - \lambda_{q^2} \hat{q}^2 - \lambda_{p^2} \hat{p}^2 - \lambda_l \hat{l} - \beta \hat{H}) \tag{4.2}$$

It is important to note that we have included the Hamiltonian as a relevant operator in the density matrix. As we are considering a set of noncommuting relevant operators, we need to diagonalize the logarithm of the density matrix $\hat{\rho}$ (generalizing a method developed in Refs. 16, 17), before applying Eq. (2.2). We define a new creation operator

$$\hat{b}^+ = |\cosh r|\, e^{i\varphi}\, \hat{a}^+ + |\sinh r|\, e^{-i\theta}\, \hat{a} + |\gamma|\, e^{-i\psi} \tag{4.3}$$

with

$$[\hat{b} , \hat{b}^+]_- = 1 . \tag{4.4}$$

It is easy to prove that the density matrix can be written in terms of the new operators $\hat{b}^+$ and $\hat{b}$ as

$$\begin{aligned}\hat{\rho} &= \exp(-\lambda_0 \hat{1} - \lambda_q \hat{q} - \lambda_p \hat{p} - \lambda_{q^2} \hat{q}^2 - \lambda_{p^2} \hat{p}^2 - \lambda_l \hat{l} - \beta \hat{H}) \\ &= \exp\left(-\lambda_0 \hat{1} - \beta \hbar \omega \left(\hat{b}^+ \hat{b} + \frac{1}{2}\right) + \beta \hbar \omega |\gamma|^2\right) ,\end{aligned} \tag{4.5}$$

and λ_0 can be easily calculated. Equations (2.2,2.7-8) allow us to evaluate the product of the dispersions of $\hat{q}$ and $\hat{p}$

$$\Delta\hat{q}\ \ \Delta\hat{p} = \frac{\hbar}{2} \left(1 + \frac{\lambda'^2_{l_0}}{\omega^2}\right)^{1/2} \left(1 + \frac{\exp(-\beta \hbar \omega / 2)}{\sinh(\beta \hbar \omega / 2)}\right) \tag{4.7}$$

where

$$\lambda'^2_{l_0} = \hbar \omega \left(\frac{\lambda'_{q_0^2}}{\gamma_1^2} + \frac{\lambda'_{p_0^2}}{\gamma_2^2}\right) + 4 \lambda'_{q_0^2} \lambda'_{p_0^2} \geq 0 \tag{4.8}$$

and

$$\lambda_{q^2} \gamma_1^2 + \lambda_{p^2} \gamma_2^2 \geq 0 . \tag{4.9}$$

It can be shown that $\lambda_{q^2} \gamma_1^2 + \lambda_{p^2} \gamma_2^2$ is an invariant of motion (see Ref. [13]). Besides, we can easily evaluate

$$<\hat{H}> = \frac{\hbar\omega}{2} \left\{ \frac{1}{2}\left[\left(\frac{<\hat{q}>}{\gamma_1}\right)^2 + \left(\frac{<\hat{p}>}{\gamma_2}\right)^2\right] + \left(1 + \frac{\lambda_{q^2}\gamma_1^2 + \lambda_{p^2}\gamma_2^2}{\beta \hbar \omega}\right)\left(1 + \frac{\exp(-\beta \hbar \omega / 2)}{\sinh(\beta \hbar \omega / 2)}\right)\right\} \tag{4.10}$$

and

$$S = -\,\mathrm{Tr}\,(\hat{\rho}\ \ln \hat{\rho})$$

$$= -\ln\left[2\sinh(\beta\hbar\omega/2)\right] + \frac{\beta\hbar\omega}{2}\left(1+\frac{\exp(-\beta\hbar\omega/2)}{\sinh(\beta\hbar\omega/2)}\right) \qquad (4.11)$$

where $\lambda'_i = \lambda_i / \beta$ are the intensive variables in the formalism[14]. The information about the initial mean values $<\hat{q}^2>_0$, $<\hat{p}^2>_0$ and $<\hat{1}>_0$, is introduced into the density matrix via the Lagrange multipliers $\lambda'_{q_0^2}$, $\lambda'_{p_0^2}$ and λ'_{1_0}[9,14]. To analyze the possibility of squeezing, we write down the uncertainty relations (using eqs. (2.8,2.10))

$$\Delta^2\hat{q} = \gamma_1^2 \quad \left(1+\frac{4\,\gamma_2^2}{\hbar\,\omega}\,\lambda'_{p^2}\right)\quad\left(1+\frac{\exp(-\beta\hbar\omega/2)}{\sinh(\beta\hbar\omega/2)}\right) \qquad (4.12)$$

$$\Delta^2\hat{p} = \gamma_2^2 \quad \left(1+\frac{4\,\gamma_1^2}{\hbar\,\omega}\,\lambda'_{q^2}\right)\quad\left(1+\frac{\exp(-\beta\hbar\omega/2)}{\sinh(\beta\hbar\omega/2)}\right)\ . \qquad (4.13)$$

As it can be seen, Equations (4.7, 4.10-13) are temperature dependent because we introduce the Hamiltonian as a relevant operator. First, we shall analyze the T = 0 case. We can write, using the variables r, φ and θ

$$\frac{4\,\gamma_1^2}{\hbar\,\omega}\lambda'_{q^2}(t) = 2\,[(\sinh r)^2 + |\sinh r|\,|\cosh r|\,\cos(2\omega t-\theta-\varphi)] \qquad (4.14a)$$

$$\frac{4\,\gamma_2^2}{\hbar\,\omega}\lambda'_{p^2}(t) = 2\,[(\sinh r)^2 - |\sinh r|\,|\cosh r|\,\cos(2\omega t-\theta-\varphi)] \qquad (4.14b)$$

$$\lambda'_1(t) = -\,2\,\omega\,|\sinh r|\,|\cosh r|\,\sin(2\omega t-\theta-\varphi)\ . \qquad (4.14c)$$

As it is always valid that

$$(\sinh r)^2 \le |\sinh r|\,|\cosh r|\ , \qquad (4.15)$$

we obtain squeezing for *all possible initial conditions*, if T = 0. Notice also that the uncertainty product given by Eq. (4.7) becomes minimum *twice* each period. This can be seen in Fig. 1. As it was said before, r is an invariant of the motion and this determines that the evolution develops through lines with slope -1 (e.g. c in Fig. 1), limited by the hyperbola of $\lambda'_1 = 0$. We can also see that there is a zone with initial squeezing on $\hat{q}$ (a in Fig. 1) and another with initial squeezing on $\hat{p}$ (b in Fig. 1). The squeezing factor **s**, introduced in ref [18], corresponds, in our theory, to

$$\mathbf{s} = |\cosh r| + |\sinh r|,\ \mathbf{s}>1;\quad \mathbf{s} = |\cosh r| - |\sinh r|,\ \mathbf{s}<1. \qquad (4.16)$$

Finally, we remark that at T = 0, for all possible initial conditions we have squeezed states characterized by the condition

$$(|\cosh r|\ e^{i\varphi}\,\hat{a}^+ + |\sinh r|\ e^{-i\theta}\,\hat{a}\ )\ |0>_b = |\gamma|\ e^{-i\psi}\ |0>_b\ . \qquad (4.17)$$

Thus, the ground state of the transformed harmonic oscillator $\hat{b}^+\,\hat{b}$ is the most general squeezed state, being the eigenstate of the operator $(|\cosh r|\ e^{i\varphi}\,\hat{a}^+ + |\sinh r|\ e^{-i\theta}\,\hat{a})$.

We shall now study the *thermodynamical case*, where $T \neq 0$. From eqs. (4.7,4.12-13) it follows that the dispersions are those obtained from the T = 0 case, multiplied by the *thermodynamical* factor "$\coth(\beta\hbar\omega/2)$".

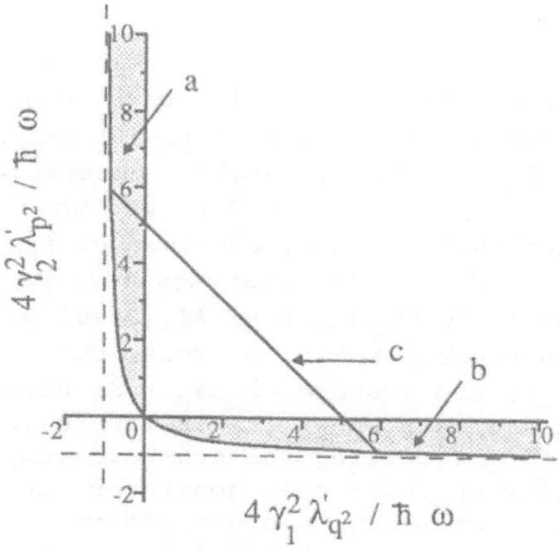

Fig. 1. Space of independent Lagrange multipliers: λ'_{q^2} and λ'_{p^2}. The gray zone, named "a" ("b"), shows the region where squeezing in $\hat{q}$ ($\hat{p}$) is obtained. The evolution of the system is shown by line "c". It is bounded by the hyperbola given by eq. (14) (i.e. the uncertainty principle).

Thus, there will be no more minimum uncertainty states (see Eq. (4.7)), and the uncertainty of $\hat{q}$ and $\hat{p}$ will be always greater than

$$\Delta^2\hat{q}_{min} = \gamma_1^2\ [|\cosh r| - |\sinh r|]^2 \coth(\beta\ \hbar\ \omega\ /\ 2) \qquad (4.18a)$$

$$\Delta^2\hat{p}_{min} = \gamma_2^2\ [|\cosh r| - |\sinh r|]^2 \coth(\beta\ \hbar\ \omega\ /\ 2) \quad . \qquad (4.18b)$$

If these dispersions are smaller than one, we shall have squeezing *although we are considering neither pure states nor minimum uncertainty ones*. This is possible if we have

$$[|\cosh r| - |\sinh r|]^2 < \tanh(\beta\ \hbar\ \omega\ /\ 2), \qquad (4.19)$$

or

$$\frac{4\left(\gamma_2^2\ \lambda'_{q^2} + \gamma_1^2\ \lambda'_{p^2}\right)}{\hbar\ \omega} = \frac{4\ c}{\hbar\ \omega} > \left(2\sinh\left(-\frac{1}{2}\ \ln\left(\tanh(\beta\hbar\omega/2)\right)\right)\right)^2, \qquad (4.20)$$

meaning that for *any initial* condition above a critical line (e.g. c in Fig. 1) of slope -1, which will be β dependent, there will be squeezing. Then, the domain of initial conditions which give squeezing are restricted by Eq. (4.20). Usually, the squeezed states are compared with coherent states. Let us consider the case in which the knowledge about the system at $t = 0$ is restrained to $<\hat{q}>_0$ and $<\hat{p}>_0$. Then, the expectation values $<\hat{q}^2>_0$, $<\hat{p}^2>_0$ and $<\hat{1}>_0$ should be obtained using Eq. (2.2). So, we can consider that the density matrix is $\hat{\rho} = \exp(-\lambda_0\hat{1}-\lambda_q\hat{q}-\lambda_p\hat{p})$, i.e. $\lambda'_{q^2}(0)=0$, $\lambda'_{p^2}(0)=0$ and $\lambda'_1(0)=0$, and we obtain a pure state with minimum uncertainty and classical temporal evolution which correspond to the coherent state. This can be easily proved using Eq. (4.3). At $T = 0$ the new oscillator $\hat{b}^+\ \hat{b}$ condense to its ground state, $|0>_b$, for which ($r = 0$ in Eq. (4.3))

$$\hat{b} \mid 0 >_b = (e^{i\varphi} \hat{a} + |\gamma| e^{-i\Psi}) \mid 0 >_b = 0 . \quad (4.21)$$

So, this state is an eigenstate of $\hat{a}$ and this is one of the definitions of a coherent state. We can use Eqs. (4.7, 4.12-13) and $\lambda'^2_q = \lambda'^2_p = \lambda'_1 = 0$ to conclude that for $T \neq 0$ there are no minimum uncertainty states. If we know, the mean values $<\hat{q}^2>_0$, $<\hat{p}^2>_0$ and $<\hat{1}>_0$, and when these mean values are different from the ones that would be obtained by applying Eq. (2.2), with $\hat{\rho} = \exp(-\lambda_0\hat{1}-\lambda_q\hat{q} -\lambda_p\hat{p})$, this implies that new information is to be included in $\hat{\rho}$ by letting $\lambda'_{q^2}(0)\neq 0$ or $\lambda'_{q^2}(0)\neq 0$ or $\lambda'_1(0)\neq 0$. So, we obtain a completely different dynamical behavior. Thus, our density matrix (Eq. (4.2)) plays the role of the squeeze operator $\hat{S}$, which is the only way to obtain squeezing that has been shown in literature [16, 18-20]. This method not only allows to define squeezing for non-zero temperature cases, but also relates the density operator that generates squeezing to the relevant operators of the system. Recently, a great effort was devoted in order to obtain squeezed vacuum states for the SU(1,1) group [13]. It is important to notice that this group is isomorphic to the group formed by $\{\hat{q}^2, \hat{p}^2, \hat{1}\}$.

A.N. P. acknowledges support from the Comisión de Investigaciones Científicas de la Provincia de Buenos Aires (CIC).

REFERENCES

1. H.D. Zeh, Foundations of Quantum Mechanics (academic Press New York, 1971) p263.
2. Mehra and E.C.G. Sudarshan, Nuovo Cim. **11B** (1972) 215.
3. A.Wehrl, Rev. Mod. Phys.**50** (1978) 221.
4. E.T. Jaynes, Phys. Rev. **106** (1957) 620; **108** 171 (1957).
5. Y. Alhassid and R.D. Levine, J. Chem. Phys. **67** 4321 (1977).
6. D. Otero, A. Plastino, A.N. Proto, G. Zannoli, Phys.Rev. **A26** 1209 (1982)
7. E. Duering, D. Otero, A. Plastino and A.N. Proto, Phys. Rev. **A32** 2455,(1985)
8. D. Otero, A. Plastino, A.N. Proto and S. Mizrahi, Phys. Rev. **A33** 3446 (1986).
9. J. Aliaga, M. Negri, D. Otero, A. Plastino and A.N. Proto, Phys. Rev. **A36** (1987) 3427.
10. J. Aliaga, D. Otero, A. Plastino and A.N. Proto, Phys. Rev. **A35** 2304 (1987)
11. J. Aliaga, D.Otero, A.Plastino and A.N.Proto **A36** 3427 (1987).
12. A.N. Proto, J.Aliaga, D. R.Napoli, D.Otero and A. Plastino, Phys. Rev. **A 39** (1989) 4212.
13. E. Duering, D. Otero, A. Plastino, and A.N. Proto. Phys. Rev. **A35** (1987) 2314.
14. J. Aliaga, D. Otero, A. Plastino and A.N. Proto, Phys. Rev. **A37** (1988) 918.
15. J. Aliaga, H. Cerdeira, A.N. Proto and D. Otero, Phys. Rev. **B** (1989) to be published.
16. H. P. Yuen, Phys. Rev. **A 13** (1976) 2226; C. M. Caves, Phys. Rev. **D 23** (1981) 1693; D.F. Walls, Nature **306** (1983) 141.
17. W.H. Louisell, Radiation and Noise in Quantum Electronics (R. Kriege, New York, 1977).
18. M.M. Nieto and L.M. Simmons Jr., Phys. Rev. **D 20** (1979) 1321.
19. C.C. Gerry and E.R. Vrscay, Phys. Rev. **A 37** (1988) 4265; P. Tombesi and A. Mecozzi, Phys. Rev. **A 37** (1988) 4778.
20. A.M. Perelomov, Jour. Math. Phys. **26** (1972) 222.
21. J. Aliaga and A.N. Proto submitted to Phys. Lett.
22. J. Aliaga, G. Crespo and A.N. Proto submitted to Phys. Rev. A.
23. J. Aliaga, G. Crespo and A.N. Proto Statphys 17 Satellite Conference, Iguazu Falls, 7-10 Aug. 1989.

BAYM-KADANOFF THEORY MADE EVEN PLANAR

Roger Alan Smith

Center for Theoretical Physics
Physics Department
Texas A&M University
College Station, TX 77843

INTRODUCTION

In the last workshop in this series, I presented some results obtained by expressing the Baym-Kadanoff algorithm[1-2] in the language of the parquet theory[3-14]. That paper[15] showed (i) that the parquet theory is not a conserving theory in the sense of the Baym-Kadanoff theory, (ii) applied the Baym-Kadanoff theory in this form to some simple self-energy diagrams, and (iii) observed that none of the resulting vertex functions was antisymmetric. This contribution presents some new results obtained using the same formalism. The main result is a proof that any set of diagrams which is both conserving and antisymmetric under exchange of the outgoing legs must include all parquet diagrams obtained using the bare interaction as well as an infinite number of irreducible diagrams.

NOTATION

The notation has been changed slightly since the last workshop, so a definition of terms is in order. The diagrammatic constructs of interest are the two-body Green's function and the single-particle self-energy. Diagrams for the two-body Green's function are to be written in terms of one-body Green's functions which use the self-energy; hence explicit inclusion of self-energy insertions in the diagrams for the two-body Green's function should be avoided. The lowest-order contribution to the two-body vertex is taken to be the bare interaction and its exchange; the terms in the two-body Green's function which are just the direct and exchange products of two one-body Green's functions are not included in this category. A direct diagram is one in which the line of propagators entering one side at the bottom ultimately can be traced through to the same side as it goes out the top,

Condensed Matter Theories, Volume 5
Edited by V.C. Aguilera-Navarro
Plenum Press, New York, 1990

while for an exchange diagram this line leaves on the opposite side. If α is a diagram, then $\overline{\alpha}$ is the diagram obtained by crossing the outgoing legs and $\widetilde{\alpha}$ is the diagram obtained by crossing the ingoing legs and the outgoing legs. Greek letters are used to represent typical diagrams, while Roman letters are used to denote collections of diagrams. Typically X will represent a collection of direct diagrams. If a collection includes both direct and exchange diagrams, it is denoted by $\check{X}$; the direct diagrams included are X and $\hat{X}$ is the set of direct diagrams obtained by applying a further exchange to the exchange diagrams in $\check{X}$.

PARQUET

The parquet theory sums two-body-reducible diagrams for the two-body vertex Γ which can be made starting with some set of irreducible diagrams I. To lowest order, I is just the bare interaction V, and in this paper the parquet Γ is understood to mean the one which is generated from just this one irreducible diagram. There are five ways to join two two-body diagrams, these are labelled Ⓢ, Ⓤ, ©, Ⓣ and Ⓓ in fig. 1.

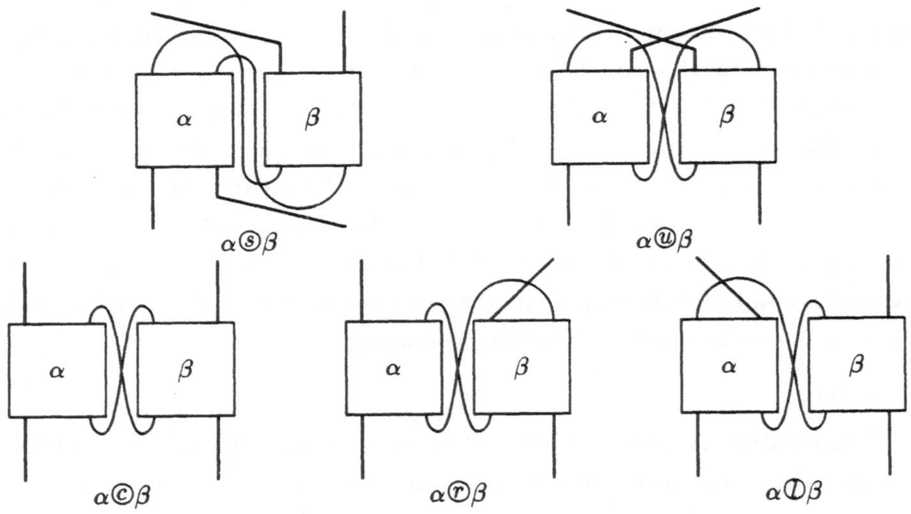

Fig. 1. The basic parquet operations.

Operationally, these represent ladder sums of two kinds, ring sums, and vertex corrections. For present purposes, it is convenient to write the parquet equations in two different but equivalent forms. If we let a capital letter denote the sum of direct diagrams composed of direct diagrams built using the corresponding operation, then a complete set of reducible diagrams $\Gamma = \hat{\Gamma}$ which can be built from an antisymmetric set of irreducible diagrams $I = \hat{I}$ can be generated by the asymmetric-looking equations[8]

$$
\begin{aligned}
\Gamma &= V+S+U+C+L+R\\
S &= \Gamma \textcircled{s} (V+U+C+L+R)\\
U &= \Gamma \textcircled{u} (V+S+C+L+R)\\
C &= \Gamma \textcircled{c} (V+S+U)\\
L &= \Gamma \textcircled{l} (V+S+U)\\
R &= \Gamma \textcircled{r} (V+S+C+L+R).
\end{aligned}
\tag{1}
$$

The apparent asymmetry in these equations arises from the fact that some diagrams generated by the $\textcircled{c}$, $\textcircled{r}$, and $\textcircled{l}$ operations could be generated in several different ways. The equations, as written, guarantee that each such diagram will be generated but once. A more symmetric set of equations, which is equivalent, is written in the form [12,14,16]

$$
\begin{aligned}
\Gamma &= I+S+U+T\\
S &= (\Gamma - S)\textcircled{s}\Gamma\\
U &= (\Gamma - U)\textcircled{u}\Gamma\\
T &= B+\Gamma\textcircled{l}(I+S+U+B)+(I+S+U+B)\textcircled{r}\Gamma+\Gamma\textcircled{l}(I+S+U+B)\textcircled{r}\Gamma\\
B &= (I+S+U)\textcircled{b}(I+S+U+B),
\end{aligned}
\tag{2}
$$

where

$$\alpha \textcircled{b} \beta = \alpha \textcircled{c} \beta + \alpha \textcircled{c} \Gamma \textcircled{l} \beta. \tag{3}$$

Each form has its utility; eq. 2 is more symmetric, but it also has two operations explicitly on the right-hand side.

A convenient way to compute the single-particle self-energy Σ is to take some approximation for the exact vertex $\check{\Gamma}$, add an interaction V using the operation $\textcircled{s}$, and then close off the two right-hand lines with a one-body propagator. If $\check{\Gamma}$ is antisymmetric, the self-energy can be computed from the direct part of Γ by

$$\Sigma = d(\Gamma \textcircled{s} V) + e(\Gamma \textcircled{s} V), \tag{4}$$

where the direct and exchange lines contributions are formed by applying the two operations illustrated in fig. 2.

If $\check{\Gamma}$ is not antisymmetric, then

$$\Sigma = d(\check{\Gamma} \textcircled{s} V). \tag{5}$$

The exact theory and the parquet theory both have an antisymmetric $\check{\Gamma}$, so that $\Gamma = \hat{\Gamma}$ and eq. 4 is appropriate. However, the Baym-Kadanoff algorithm generates an approximate vertex which is not necessarily antisymmetric.

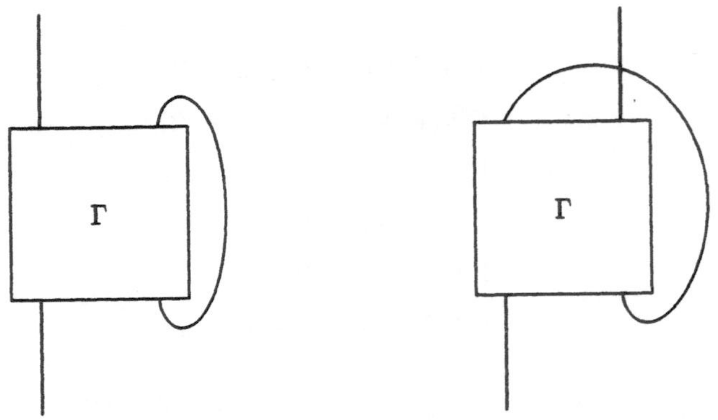

$d(\Gamma)$ $e(\Gamma)$

Fig. 2. Forming the self-energy Σ.

BAYM-KADANOFF THEORY

The Baym-Kadanoff (BK) algorithm takes an approximation to the self-energy Σ which is taken to be constructed diagrammatically from one-body Green's functions which use the same Σ. The source of the approximate Σ need not be specified; it could be computed from some approximate vertex using eqs. 4 or 5, but it need not be. The conserving vertex is obtained as the functional derivative of the self-energy with respect to a weak external two-point non-local perturbation. The one-body Green's function is itself a functional of this interaction, and the functional derivative of the one-body Green's function can be expressed back in terms of the functional derivative of the self-energy with respect to this small non-local perturbation. We note that any self-energy contribution can always be expressed in the form of fig. 2, where the Γ may be irreducible. The algorithm, in the parquet notation, proceeds in two stages.

First, one generates a two-body kernel

$$\check{\Xi} = \Sigma,' \tag{6}$$

where the prime denotes the functional derivative with respect to the Green's function. Careful inspection shows that $\check{\Xi}$ can have both direct and exchange parts, but they cannot be of all reducibility types. Specifically, only reducibility types I, S, U, $\hat{I}$, $\hat{S}$, $\hat{C}$, $\hat{L}$, $\hat{R}$ and $\hat{A}$ can come out of this operation. To show this, we first write

$$\begin{aligned} d'(\alpha \textcircled{s} \beta) &= d(\alpha' \textcircled{s} \beta) + d(\alpha \textcircled{s} \beta') + \alpha \textcircled{s} \beta + \alpha \textcircled{u} \beta + \overline{\alpha \textcircled{c} \tilde{\beta}} \\ e'(\alpha \textcircled{s} \beta) &= e(\alpha' \textcircled{s} \beta) + e(\alpha \textcircled{s} \beta') + \overline{\alpha \textcircled{s} \beta} + \overline{\alpha \textcircled{l} \beta} + \overline{\alpha \textcircled{r} \tilde{\beta}} \end{aligned} \tag{7}$$

The immediate functional derivatives are all of the types indicated. For computing $\check{\Xi}$, $\beta = V$, so that $\beta' = 0$. The only remaining terms are the $d(\alpha' \textcircled{s} \beta)$ or $e(\alpha' \textcircled{s} \beta)$. If α is irreducible, then these terms are as well. If α is reducible, then the functional derivative can be transformed using

$$\begin{aligned} d((\alpha \textcircled{x} \beta)' \textcircled{s} \gamma) &= d((\alpha' \textcircled{x} \beta) \textcircled{s} \gamma) + d((\alpha \textcircled{x}' \beta) \textcircled{s} \gamma) + d((\alpha \textcircled{x} \beta') \textcircled{s} \gamma) \\ e((\alpha \textcircled{x} \beta)' \textcircled{s} \gamma) &= e((\alpha' \textcircled{x} \beta) \textcircled{s} \gamma) + e((\alpha \textcircled{x}' \beta) \textcircled{s} \gamma) + e((\alpha \textcircled{x} \beta') \textcircled{s} \gamma). \end{aligned} \tag{8}$$

The last term on the right-hand sides of eq. 8 is always irreducible. The middle term can be expressed in reducible form using table 1. All of these table entries are in one of the stated categories.

Table 1: Diagrams from the middle term of the right-hand sides of eq. 8.

$\textcircled{x}$	$d((\alpha\textcircled{x}'\beta)\textcircled{s}\gamma)$		$d((\alpha\textcircled{x}'\beta)\textcircled{s}\gamma)$	
$\textcircled{s}$	$\overline{\alpha\textcircled{c}(\tilde{\beta}\textcircled{s}\tilde{\gamma})}$	$\alpha\textcircled{u}(\beta\textcircled{s}\gamma)$	$\overline{\alpha\textcircled{t}(\tilde{\beta}\textcircled{s}\tilde{\gamma})}$	$\overline{\alpha\textcircled{I}(\beta\textcircled{s}\gamma)}$
$\textcircled{u}$	$\overline{\alpha\textcircled{c}(\tilde{\gamma}\textcircled{u}\tilde{\beta})}$	$\alpha\textcircled{s}(\beta\textcircled{u}\gamma)$	$\overline{\alpha\textcircled{t}(\tilde{\gamma}\textcircled{t}\tilde{\beta})}$	$\overline{\alpha\textcircled{s}(\beta\textcircled{I}\gamma)}$
$\textcircled{c}$	$\alpha\textcircled{u}(\gamma\textcircled{c}\tilde{\beta})$	$\alpha\textcircled{s}(\gamma\textcircled{c}\tilde{\beta})$	$\alpha\textcircled{s}(\gamma\textcircled{I}\tilde{\beta})$	$\alpha\textcircled{u}(\gamma\textcircled{I}\tilde{\beta})$
$\textcircled{t}$	$\alpha\textcircled{s}(\gamma\textcircled{t}\tilde{\beta})$	$\alpha\textcircled{u}(\gamma\textcircled{t}\tilde{\beta})$	$\overline{\alpha\textcircled{I}(\gamma\textcircled{u}\tilde{\beta})}$	$\overline{\alpha\textcircled{s}(\gamma\textcircled{u}\tilde{\beta})}$
$\textcircled{I}$	$\overline{\alpha\textcircled{s}(\beta\textcircled{c}\tilde{\gamma})}$	$\overline{\alpha\textcircled{t}(\gamma\textcircled{c}\tilde{\beta})}$	$\overline{\alpha\textcircled{t}(\gamma\textcircled{I}\tilde{\beta})}$	$\overline{\alpha\textcircled{s}(\beta\textcircled{t}\tilde{\gamma})}$

The final case is the first term on the right-hand sides of eq. 8. These can be rearranged using the diagrams in fig. 3 which correspond to the right-hand sides of the identities

$$
\begin{aligned}
d((\alpha\textcircled{s}\beta)\textcircled{s}\gamma) &= d(\alpha\textcircled{s}(\beta\textcircled{s}\gamma)) \\
d((\alpha\textcircled{u}\beta)\textcircled{s}\gamma) &= d(\alpha\textcircled{s}(\beta\textcircled{u}\gamma)) \\
d((\alpha\textcircled{c}\beta)\textcircled{s}\gamma) &= d(\alpha\textcircled{s}(\gamma\textcircled{c}\tilde{\beta})) \\
d((\alpha\textcircled{t}\beta)\textcircled{s}\gamma) &= d(\alpha\textcircled{s}(\gamma\textcircled{t}\tilde{\beta})) \\
d((\alpha\textcircled{I}\beta)\textcircled{s}\gamma) &= e(\alpha\textcircled{s}(\beta\textcircled{c}\tilde{\gamma})) \\
e((\alpha\textcircled{s}\beta)\textcircled{s}\gamma) &= e(\alpha\textcircled{s}(\beta\textcircled{s}\gamma)) \\
e((\alpha\textcircled{u}\beta)\textcircled{s}\gamma) &= e(\alpha\textcircled{s}(\beta\textcircled{I}\gamma)) \\
e((\alpha\textcircled{c}\beta)\textcircled{s}\gamma) &= d(\alpha\textcircled{s}(\gamma\textcircled{I}\tilde{\beta})) \\
e((\alpha\textcircled{t}\beta)\textcircled{s}\gamma) &= e(\alpha\textcircled{s}(\gamma\textcircled{u}\tilde{\beta})) \\
e((\alpha\textcircled{I}\beta)\textcircled{s}\gamma) &= e(\alpha\textcircled{s}(\beta\textcircled{t}\tilde{\gamma}))
\end{aligned}
\tag{9}
$$

These diagrammatic identies are equally valid if α is replaced by α', and hence the first terms on the right-hand sides of eq. 8 can be transformed into the form of the left-hand sides of eq. 8. This process cannot go on indefinitely, since each time eq. 9 is used to make a transformation, the diagram α is of lower order in perturbation theory.

The full BK conserving vertex is generated from $\check{\Xi}$ by solving the equation

$$\check{X} = \check{\Xi} + \check{\Xi}\textcircled{c}\check{X} + \check{\Xi}\textcircled{I}\check{X} + \check{\Xi}\textcircled{t}\check{X}. \tag{10}$$

Sorting out the direct and exchange parts and doing some diagrammatic rearrangement, we cast the resulting components of the BK vertex in the form of eq. 2,

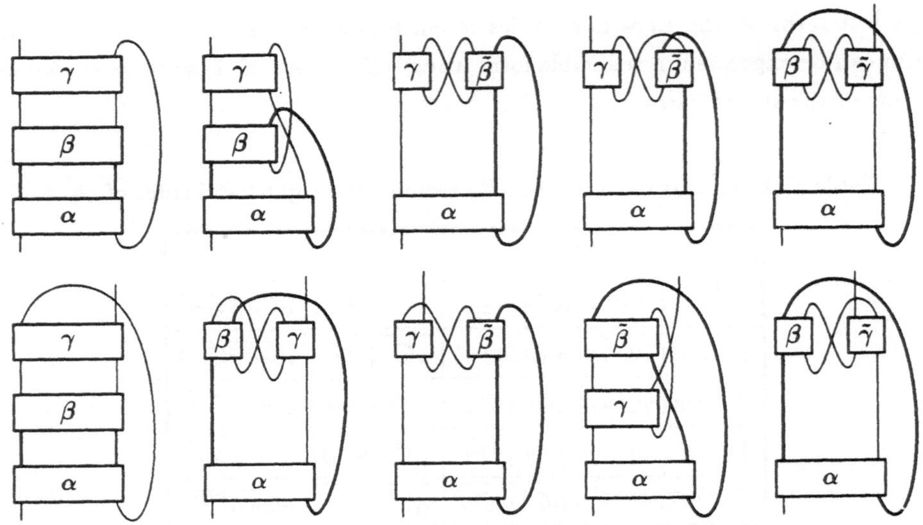

Fig. 3 Some diagrammatic rearrangements

obtaining exchange components

$$
\begin{aligned}
\hat{\Gamma}_X &= \hat{I}_X + \hat{S}_X + \hat{U}_X + \hat{T} \\
\hat{I}_X &= \hat{I}_\Xi \\
\hat{S}_X &= \hat{S}_\Xi \\
\hat{U}_X &= (\hat{\Gamma}_X - \hat{U}_X) \textcircled{u} \hat{\Gamma}_X \\
\hat{T}_X &= \hat{T}_\Xi
\end{aligned}
\tag{11}
$$

and direct components

$$
\begin{aligned}
\Gamma_X &= I_X + S_X + U_X + T_X \\
I_X &= I_\Xi \\
S_X &= S_\Xi \\
U_X &= U_\Xi \\
T_X &= B_X + \hat{\Gamma} \textcircled{t} (I_X + S_X + U_X + B_X) + (I_X + S_X + U_X + B_X) \textcircled{t} \hat{\Gamma} \\
&\quad + \hat{\Gamma} \textcircled{t} (I_X + S_X + U_X + B_X) \textcircled{t} \hat{\Gamma} \\
B_X &= (I_X + S_X + U_X) \textcircled{b} (I_X + S_X + U_X + B_X)
\end{aligned}
\tag{12}
$$

The $\textcircled{b}$ operation in eq. 5 requires the Lindhard bubble to be dressed by $\hat{\Gamma}$.

These expressions for the conserving vertex basically can be easily understood: the self-consistency implicit in the BK algorithm is a t-channel operation as seen in eq. 10. Thus, the integral equations build the t-channel chains and vertex corrections, while the exchange channel is the u-channel. The s-channel diagrams can only come from the explicit functional differentiation with respect to the one-body

Green's function, since they can't be generated by a direct or exchange t-channel operation.

As we noted before, the vertex generated by the BK algorithm need not be antisymmetric; indeed, eq. 11 and eq. 12 show what the diagrammatic content actually is.

CONSERVING AND ANTISYMMETRIC IS MORE THAN PARQUET

In this section, we prove that any approximation to the self-energy which is used to generate a vertex which is both antisymmetric and conserving and which includes the Hartree term must include a class of diagrams which includes all parquet diagrams generated from the bare interaction and a large class of irreducible diagrams. The key to this argument is that antisymmetry of the conserving vertex implies no need to distinguish between Γ and $\hat{\Gamma}$. This provides a significant constraint, since eq. 11 and eq. 12 then can be used to identify certain functional derivative results with the results of integral equations in the ⓢ and ⓣ channels.

At the Hartree-Fock level (Fock is included if Hartree is), I will include V. Now assume that all of the parquet terms are generated up to order n in perturbation theory by some set of self-energy diagrams. (A minimum set of self-energy diagrams can always be found by using the algorithm of the following section). The exchange channels (eq. 11) then generate U correctly to order $n+1$ using only the V in I, since all of the inputs are correct to order n and the first-order term is correct. The direct channels (eq. 12) generate T correctly to order $n+1$ for the same reason. Finally, we note that the functional differentiation of any diagram which produces αⓤβ also produces αⓢβ, so that since U is correct to order $n+1$, so must be S. But if all of the diagrams are correct to order $n+1$, then they are correct to all orders. Hence all parquet diagrams must be generated.

As we argued previously, parquet is not conserving; a set of self-energy diagrams which generates parquet must also generate an infinite set of irreducible diagrams. As a consequence, parquet theory is not a sufficiently large class of diagrams to be both conserving and antisymmetric. We speculate that the only set of diagrams which satisfies both criteria is the complete set of diagrams.

PARQUET DIAGRAMS GENERATE SELF-ENERGY DIAGRAMS

Here we present the algorithm which determines, for any reducible diagram μ, the parquet diagram whose direct or exchange contribution to the self-energy gives μ on functional differentiation with respect to the Green's function. It is convenient to use the notation of eq. 1 in this section. The first step is to determine whether the diagram is a member of the acceptable types $\{S, U, \hat{S}, \hat{C}, \hat{L}, \hat{R}\}$.

If so, then $\mu = \alpha x \lambda$, and the starting point is obtained by looking in tables 2-3.

In tables 2-3, the $\rightarrow$ entries indicate that the direct diagram can be rearranged, and one should look up the transformed form.

Table 2. The starting point for direct diagrams.

λ	$\alpha\textcircled{s}\lambda$	$\alpha\textcircled{u}\lambda$
V	$d(\alpha\textcircled{s}V)$	$d(\alpha\textcircled{s}V)$
$\beta\textcircled{s}\gamma$	$\rightarrow(\alpha\textcircled{s}\beta)\textcircled{s}\gamma$	$d((\alpha\textcircled{s}\beta)\textcircled{s}\gamma)$
$\beta\textcircled{u}\gamma$	$d((\alpha\textcircled{u}\beta)\textcircled{s}\gamma)$	$\rightarrow(\alpha\textcircled{u}\beta)\textcircled{u}\gamma$
$\beta\textcircled{c}\gamma$	$d((\alpha\textcircled{c}\tilde{\gamma})\textcircled{s}\beta)$	$d((\alpha\textcircled{c}\tilde{\gamma})\textcircled{s}\beta)$
$\beta\textcircled{t}\gamma$	$d((\alpha\textcircled{t}\tilde{\gamma})\textcircled{s}\beta)$	$d((\alpha\textcircled{t}\tilde{\gamma})\textcircled{s}\beta)$
$\beta\textcircled{i}\gamma$	$e((\alpha\textcircled{c}\tilde{\gamma})\textcircled{s}\beta)$	$e((\alpha\textcircled{c}\tilde{\gamma})\textcircled{s}\beta)$

Table 3. The starting point for exchange diagrams.

λ	$\overline{(\alpha\textcircled{s}\lambda)}$	$\overline{(\alpha\textcircled{c}\lambda)}$	$\overline{(\alpha\textcircled{i}\lambda)}$	$\overline{(\alpha\textcircled{t}\lambda)}$
V	$e(\alpha\textcircled{s}V)$	$d(\alpha\textcircled{s}V)$	$e(\alpha\textcircled{s}V)$	$e(\alpha\textcircled{s}V)$
$\beta\textcircled{s}\gamma$	$\rightarrow\overline{((\alpha\textcircled{s}\beta)\textcircled{s}\gamma)}$	$d((\alpha\textcircled{s}\tilde{\beta})\textcircled{s}\tilde{\gamma})$	$e((\alpha\textcircled{s}\beta)\textcircled{s}\gamma)$	$e((\alpha\textcircled{s}\tilde{\beta})\textcircled{s}\tilde{\gamma})$
$\beta\textcircled{u}\gamma$	$e((\alpha\textcircled{t}\tilde{\gamma})\textcircled{s}\beta)$	$d((\alpha\textcircled{u}\tilde{\gamma})\textcircled{s}\tilde{\beta})$	$e((\alpha\textcircled{t}\tilde{\gamma})\textcircled{s}\beta)$	$\rightarrow\overline{((\alpha\textcircled{t}\beta)\textcircled{t}\gamma)}$
$\beta\textcircled{c}\gamma$	$d((\alpha\textcircled{i}\beta)\textcircled{s}\tilde{\gamma})$	$\rightarrow\overline{((\alpha\textcircled{c}\beta)\textcircled{c}\gamma)}$	$\rightarrow\overline{((\alpha\textcircled{i}\beta)\textcircled{c}\gamma)}$	$d((\alpha\textcircled{i}\tilde{\gamma})\textcircled{s}\beta)$
$\beta\textcircled{t}\gamma$	$e((\alpha\textcircled{i}\beta)\textcircled{s}\tilde{\gamma})$	$\rightarrow\overline{((\alpha\textcircled{c}\beta)\textcircled{t}\gamma)}$	$\rightarrow\overline{((\alpha\textcircled{i}\beta)\textcircled{t}\gamma)}$	$e((\alpha\textcircled{u}\tilde{\gamma})\textcircled{s}\tilde{\beta})$
$\beta\textcircled{i}\gamma$	$e((\alpha\textcircled{u}\beta)\textcircled{s}\gamma)$	$\rightarrow\overline{((\alpha\textcircled{t}\beta)\textcircled{c}\gamma)}$	$\rightarrow\overline{((\alpha\textcircled{u}\beta)\textcircled{i}\gamma)}$	$e((\alpha\textcircled{i}\tilde{\gamma})\textcircled{s}\beta)$

The second kind of entry is indicated by the first line: if λ is V, then we are really done, since then α is the parquet diagram whose functional derivative gives the desired diagram.

The final case is a little more complicated: the table entry is used as a starting point to carry out a reduction using table 4. That is, a direct or exchange diagram from table 2 or 3 is used to look up a new diagram from table 4. If this diagram has the form $d(\nu\textcircled{s}V)$ or $e(\nu\textcircled{s}V)$, the procedure terminates with ν as the parquet diagram whose functional derivative gives the desired diagram. If not, the table entry is used to generate a new table entry. For parquet diagrams, this procedure eventually must terminate.

DISCUSSION

We are at present at work determining a good form for the three-body integral equations which sum all irreducible diagrams generated by the BK algorithm using parquet as input. The present form of the equations is not particularly enlightening, and will not be presented here.

The main conclusion to be drawn is that two properties of the exact vertex which one would desparately like to maintain, antisymmetry and the conserving

Table 4. Iterate using this table until the right-hand term is V.

λ	$d(\lambda) \rightarrow$	$e(\lambda) \rightarrow$
$\alpha \textcircled{S} (\beta \textcircled{S} \gamma)$	$d((\alpha \textcircled{S} \beta) \textcircled{S} \gamma)$	$e((\alpha \textcircled{S} \beta) \textcircled{S} \gamma)$
$\alpha \textcircled{S} (\beta \textcircled{U} \gamma)$	$d((\alpha \textcircled{U} \beta) \textcircled{S} \gamma)$	$e((\alpha \textcircled{T} \tilde{\gamma}) \textcircled{S} \beta)$
$\alpha \textcircled{S} (\beta \textcircled{C} \gamma)$	$d((\alpha \textcircled{C} \tilde{\gamma}) \textcircled{S} \beta)$	$d((\alpha \textcircled{D} \beta) \textcircled{S} \tilde{\gamma})$
$\alpha \textcircled{S} (\beta \textcircled{T} \gamma)$	$d((\alpha \textcircled{T} \tilde{\gamma}) \textcircled{S} \beta)$	$e((\alpha \textcircled{D} \beta) \textcircled{S} \tilde{\gamma})$
$\alpha \textcircled{S} (\beta \textcircled{D} \gamma)$	$e((\alpha \textcircled{C} \tilde{\gamma}) \textcircled{S} \beta)$	$e((\alpha \textcircled{U} \beta) \textcircled{S} \gamma)$

property, are incompatible at any level less than parquet. This has been pursued for the purpose of finding out a satisfactory approximation scheme for fermion parquet theory. It appears that to satisfy these properties where they are needed, it will be necessary to use different approximations in different contexts.

ACKNOWLEDGEMENTS

The work done here was supported in part by the NSF under grant PHY-8806265. I am grateful to the U. S. Army Research Office for a grant supporting travel to this workshop.

REFERENCES

1. G. Baym and L. P. Kadanoff, Phys. Rev. **124**, 287-299 (1961).
2. G. Baym, Phys. Rev. **127**, 1391-1401 (1962).
3. A. D. Jackson, A. Lande and R. A. Smith, Phys. Report **86**, 55-111 (1982).
4. R. A. Smith, Proceedings of the Sixth Pan-American Workshop on Condensed Matter Theories, ed. J. M. C. Chen, J. W. Clark, P. Suntharok-Priesmeyer, (Department of Physics, Washington Univ., St. Louis), 153-154 (1982).
5. A. Lande and R. A. Smith, Phys. Lett. **131B**, 253-256 (1983).
6. A. D. Jackson, A. Lande, R. W. Guitink and R. A. Smith, Phys. Rev. **B31**, 403-415 (1985).
7. A. D. Jackson, A. Lande and R. A. Smith, Phys. Rev. Lett. **54**, 1469-1471 (1985).
8. R. A. Smith, Proceedings of the IX International Workshop on Condensed Matter Theories, San Francisco, Aug. 1985, ed. F. B. Malik, (Plenum Press, New York, 1986), 9-18.
9. E. Krotscheck, R. A. Smith, and A. D. Jackson, Phys. Rev. **A33**, 3535-3536 (1986).
10. R. A. Smith and A. D. Jackson, Nucl. Phys. **A476**, 448-470 (1988).
11. A. D. Jackson and R. A. Smith, Phys. Rev. **A36**, 2517-2518 (1987).
12. R. A. Smith and A. Lande, Proceedings of the XI International Workshop on Condensed Matter Theories, Oulu, July 1987, ed. J. Arponen, R. F. Bishop and M. Manninen (Plenum Press, New York, 1988), 1-9.

13. R. A. Smith and A. D. Jackson, Proceedings of the V International Conference on Recent Progress in Many-Body Theories, Oulu, August 1987, ed. A. Kallio, J. Arponen and R. F. Bishop (Plenum Press, New York, 1987), 327-333.

14. A. Lande and R. A. Smith, Proceedings of the V International Conference on Recent Progress in Many-Body Theories, Oulu, August 1987, ed. A. Kallio, J. Arponen and R. F. Bishop (Plenum Press, New York, 1987), 335-342.

15. R. A. Smith, Proceedings of the XII International Workshop on Condensed Matter Theories, Taxco, August 1988, ed. J. Keller (Plenum Press, New York, 1989).

16. A. Lande and R. A. Smith, Phys. Rev, to be submitted.

CONTRIBUTORS AND PARTICIPANTS

SUBJECT INDEX